4th International Conference on Governance and Accountability (ICGA 2016)

SHS Web of Conferences Volume 36 (2017)

Melaka, Malaysia
25 - 27 September 2016

Editors:

Mohd Yassir Jaafar
A'leshah Abdullah Sani
Anuar Muhammad

ISBN: 978-1-5108-4517-6

Printed from e-media with permission by:

Curran Associates, Inc.
57 Morehouse Lane
Red Hook, NY 12571

Some format issues inherent in the e-media version may also appear in this print version.

Printed by Curran Associates, Inc. (2017)

For additional information, please contact EDP Sciences – Web of Conferences at the address below.

EDP Sciences – Web of Conferences
17, Avenue du Hoggar
Parc d'Activité de Courtabœuf
BP 112
F-91944 Les Ulis Cedex A
France

Phone: +33 (0) 1 69 18 75 75
Fax: +33 (0) 1 69 28 84 91

contact-edps@webofconferences.org

Additional copies of this publication are available from:

Curran Associates, Inc.
57 Morehouse Lane
Red Hook, NY 12571 USA
Phone: 845-758-0400
Fax: 845-758-2633
Email: curran@proceedings.com
Web: www.proceedings.com

TABLE OF CONTENTS

APPENDIX 1:

LIST OF SCIENTIFIC EDITORS FOR 2016 4TH ICGA ARTICLES:

1. Dato' Prof Dr Mustafa Mohamed Zain – Faculty of Accountancy, Universiti Teknologi MARA, Puncak Alam, Selangor, Malaysia
2. Prof Dr Kiymet Calyurt – Trakya University, Erdine, Turkey
3. Prof Dr Rohana Othman – Accounting Research Institute, Universiti Teknologi MARA, Shah Alam, Selangor, Malaysia
4. Prof Dr Nafsiah Mohamed – Accounting Research Institute, Universiti Teknologi MARA, Shah Alam, Selangor, Malaysia
5. Prof Dr Ruhaya Atan - Faculty of Accountancy, Universiti Teknologi MARA, Puncak Alam Selangor, Malaysia
6. Prof Dr Azlan Amran – Graduate School of Business, Universiti Sains Malaysia, Penang, Malaysia
7. Assoc Prof Dr Hadrian Djajadikerta - Edith Cowan University, Australia
8. Assoc Prof Dr Hasan Fauzi – Sebelas Maret University, Solo, Indonesia
9. Assoc Prof Dr Roshayani Arshad – Accounting Research Institute, Universiti Teknologi MARA, Shah Alam, Selangor, Malaysia
10. Assoc Prof Dr Jamaliah Said – Accounting Research Institute, Universiti Teknologi MARA, Shah Alam, Selangor, Malaysia
11. Assoc Prof Dr Faizah Darus - Accounting Research Institute, Universiti Teknologi MARA, Shah Alam, Selangor, Malaysia
12. Assoc Prof Dr Haslinda Yusoff - Faculty of Accountancy, Universiti Teknologi MARA, Puncak Alam, Selangor, Malaysia
13. Assoc Prof Dr Zuraidah Sanusi – Accounting Research Institute, Universiti Teknologi MARA, Shah Alam, Selangor,Malaysia
14. Assoc Prof Dr Intan Salwani Mohamed – Accounting Research Institute, Universiti Teknologi MARA, Shah Alam, Selangor, Malaysia
15. Assoc Prof Dr Nawal Kasim - Accounting Research Institute, Universiti Teknologi MARA, Shah Alam, Selangor, Malaysia
16. Assoc Prof Dr Corina Joseph - Faculty of Accountancy, Universiti Teknologi MARA, Sarawak, Malaysia
17. Assoc Prof Dr Aliza Ramli - Faculty of Accountancy, Universiti Teknologi MARA, Puncak Alam, Selangor, Malaysia
18. Assoc Prof Dr Roshima Said - Faculty of Accountancy, Universiti Teknologi MARA, Kedah, Malaysia
19. Assoc Prof Dr Hanim Norza Baba- Faculty of Accountancy, Universiti Teknologi MARA, Melaka, Malaysia
20. Assoc Prof Dr Loo Ern Chen - Faculty of Accountancy, Universiti Teknologi MARA, Melaka, Malaysia
21. Dr Fadzlina Mohd Fahmi - Faculty of Accountancy, Universiti Teknologi MARA, Puncak Alam, Selangor, Malaysia
22. Dr Ruslaina Yusoff - Faculty of Accountancy, Universiti Teknologi MARA, Kelantan, Malaysia
23. Dr Soheil Kazemian – Accounting Research Institute, Universiti Teknologi MARA, Shah Alam, Selangor, Malaysia
24. Dr Mehran Nejati - Graduate School of Business, Universiti Sains Malaysia, Penang, Malaysia
25. Dr Rani Diana Othman- Faculty of Accountancy, Universiti Teknologi MARA, Melaka, Malaysia
26. Dr Hamidah Bani- Faculty of Accountancy, Universiti Teknologi MARA, Melaka, Malaysia
27. Dr Hanafiah Hasin- Faculty of Accountancy, Universiti Teknologi MARA, Melaka, Malaysia
28. Dr Sharifah Norhafiza Syed Ibrahim- Faculty of Accountancy, Universiti Teknologi MARA, Melaka, Malaysia
29. Dr Saiyidi Mat Roni - Faculty of Accountancy, Universiti Teknologi MARA, Melaka, Malaysia
30. Cik Rozaiha Abdul Majid - Faculty of Accountancy, University Teknology MARA, Melaka, Malaysia

LIST OF PROCEEDING EDITORS FOR 2016 4TH ICGA PUBLICATION IN SHS WOC:

1. Mohd Yassir Jaaffar - Faculty of Accountancy, Universiti Teknologi MARA, Melaka, Malaysia
2. Cik A'ieshah Abdullah Sani - Faculty of Accountancy, Universiti Teknologi MARA, Melaka, Malaysia
3. Anuar Muhammad - Faculty of Accountancy, Universiti Teknologi MARA, Melaka, Malaysia

edp sciences

SHS Web of Conferences

Statement of Peer review

In submitting conference proceedings to SHS Web of Conferences, I certify to the Publisher that I adhere to the **Policy on Publishing Integrity** of the journal in order to safeguard good scientific practice in publishing.

1. All articles have been subjected to peer review administered by the proceedings editors.
2. Reviews have been conducted by expert referees, who have been requested to provide unbiased and constructive comments aimed, whenever possible, at improving the work.
3. Proceedings editors have taken all reasonable steps to ensure the quality of the materials they publish and their decision to accept or reject a paper for publication has been based only on the merits of the work and the relevance to the journal.

Title, date and place of the conference

THE 2016 4th INTERNATIONAL CONFERENCE ON GOVERNANCE AND ACCOUNTABILITY (2016 ICGA), HOLIDAY INN HOTEL MELAKA (25 - 27 SEPTEMBER 2016)

Proceedings editor(s):

MOHD YASSIR JAAFFAR
A'IESHAH ABDULLAH SANI
ANUAR MUHAMMAD

Date and editor's signature

MOHD YASSIR JAAFFAR 20/6/2017
A'IESHAH ABD. SANI 20/6/2017
ANUAR MUHAM 20/6/2017

TAKAFUL OPERATORS' CORPORATE SOCIAL PERFORMANCE (CSP): AN INDUSTRY PERSPECTIVE

Amirul Afif Muhamat, Mohamad Nizam Jaafar, Sharifah Faigah Syed Alwi
Faculty of Business Management, Universiti Teknologi MARA

ABSTRACT

Takaful operators which are part of Islamic financial institutions (IFIs) derive their fundamental principles from *shariah*. These religious based institutions are expected to fulfill the two important roles in their business operations: commercially profitable and socially responsible. Nevertheless, their societal role is rarely measured and discussed. Therefore, this study appraised the societal role of *takaful* operators by assessing the components which have been proposed under the corporate social performance (CSP) theme for IFIs. This study has arranged structured interview sessions with the Chief Investment Officers and Heads of Investment of each of the eleven *takaful* operators in Malaysia. The Delphi-style technique was adopted when developing the interview questions. The questions were developed in the form of a five-point Likert scale, addressing specific issues on CSP of *takaful* operators. In addition, information on *takaful* operators' CSR activities, zakat and tax payment were gathered from the companies' websites and annual report of *takaful* operators. The study concludes that *takaful* operators in Malaysia have achieved their societal role through two channels: CSP programmes financed from companies' profits and fulfillment of CSP as a result of business-community agenda. This study covers every *takaful* operator in Malaysia and the results reflect industry opinion.

Keywords: *Takaful* operator, *Takaful*, corporate social performance, Islamic insurance, Islamic financial institution

INTRODUCTION

Takaful embraced the concepts of mutual assistance and brotherhood by bearing each other risk in the event of calamity (Gönülal 2012; Siddiqui & Hunt-Ahmed 2013). Over the years, the scheme which was started as community based initiative is now seen as a lucrative business sector, as part of the Islamic finance industry. The *takaful* sector is expected to achieve USD 20 billion in term of gross contribution by 2017 (The World Takaful Report 2014). In Malaysia, the sector is forecasted to achieve nearly USD 5 billion in 2016 (Malaysian Takaful Association 2014).

Takaful operators which are part of Islamic financial institutions (IFIs), derive their fundamental principles from *shariah*. These religious based institutions are expected to fulfil the two important roles in their business operations by being commercially profitable and socially responsible.

This research argues that for contribution towards the society, such performance is categorized as corporate social performance (CSP) of *takaful* operators which is received limited attention

Corresponding author's email: amirulafif@salam.uitm.edu.my

compared to financial performance of *takaful* operators. Therefore, this study appraised the societal role of *takaful* operators by assessing their CSP components as suggested in previous literature for the IFIs.

The structure of this paper after this introduction section is review on the philosophy and principles of *takaful*. Next is overview on *takaful* operators in Malaysia, followed by discussion on *takaful* operators' CSP, research method, findings and conclusion.

Philosophy and Principles of *Takaful*

Takaful early practices were dated back to a system known as *aqilah* which provides compensation by paying money to the victim's family (Alhabshi & Razak 2011). This scheme was initiated by pooling money (*diyyah*) purposely to compensate for the unintentional injury or death caused by member of a particular tribe to other tribes. Interestingly, this has been practiced even before the advent of Islam. Therefore, instead of 'compensation' was decided on the battlefield which was normally practiced during that time, *takaful* scheme in form of *aqilah* model has managed to some extent to avoid the instigation of tribal war – as a peaceful option for them (Engku Ali, Odierno & Ismail 2008; Swartz & Coetzer 2010).

With the spread of Islam and acceptance of the faith by the Arabs, some of the early practices were retained, since they were compatible with Islam. *Takaful* was then developed during the Prophet Muhammad (pbuh) and his companions period, and such practices were continued by various Islamic dynasties from Umayyad, Abbasid, and lastly the Ottomans (Pamuk 2004; Wilson 2008; Çizakça 2011; Kuran 2012).

Tabarru' (donation) is the underpinning element that differentiates *takaful* from conventional insurance. This is because the former business operates on a risk sharing basis, whereas the latter provides risk transfer (Wahab, Lewis & Hassan 2007; Bakar 2011). *Tabarru'* will relinquish *takaful* contributors' claims on the fund as it is considered a gratuitous contract* and not a sale contract. The *tabarru'* fund also known as premium, contributed by the *takaful* contributors and segregated into two main accounts: Participant Special Account (PSA), or the risk fund, and Participant Account (PA), or the investment fund.

The basic principles in *takaful* policy are equivalent to the conventional insurance, albeit with some philosophical differences as explained earlier. A potential *takaful* contributor (policyholder) is expected to exhibit the features of basic insurance (*takaful*) principles in order to be eligible to sign a *takaful* policy or contract. Hassan (258: 2014) describes six basic principles which must exist before a person takes up a *takaful* policy: utmost good faith; insurable interest; indemnity; proximate cause; beneficiary issue; and underwriting. In spite of philosophical differences as explained earlier, the basic principles in *takaful* policy are equivalent to the conventional insurance.

*Engku Ali, Odierno and Ismail (2008) clarify that *tabarru'* has caused the *takaful* transaction to be permissible according to *Shariah*, since it changes the basis of insurance contract from an exchange contract which is bilateral in nature to a charitable contract which is unilateral in nature. Being a charitable contract, the *takaful* contributors are not aiming for profit from uncertainty, rather they intend to pool sufficient funds to be used for assisting other members in case of misfortune.

Corresponding author's email: amirulafif@salam.uitm.edu.my

Takaful Operators in Malaysia

First *takaful* operator in Malaysia was established in 1984, and since then, the industry has gone through rapid development by having 12 *takaful* operators operating side by side the conventional insurance companies. The numbers then dropped to 11 *takaful* operators when merger and acquisition (M&A) process took place between AIA and ING operations in Malaysia (ING Website 2012). *Takaful* operators are not only competing amongst themselves for the same market, but also neck-to-neck with conventional insurance companies in a dual financial system landscape. This situation caused stiff competition amongst the market players; however it promotes better services and products for the customers. There are five basic *takaful* models in the world, which are *ta'awuni* (cooperation), *mudharabah* (profit sharing), *wakalah* (agency), *waqf* (endowment) and hybrid. Each model has its own features and limitations. In Malaysia, the *wakalah* model is adopted by every *takaful* operator which have been customized to meet specific needs of *takaful* operators' products (Htay & Zaharin 2012). The *wakalah* model compared to other models is capable to realize the commercial objective of *takaful* operators, less controversial from *shariah* point of view and it can be combined with other *takaful* model (hybrid model) to make it flexible for other purpose like investment (Wahab, Lewis & Hassan 2007; Mohd Kassim 2012).

Takaful operators' corporate social performance (CSP)

Much has been debated on the periphery of corporate social performance (CSP) in the Western literature and such positive vibes transmitted to the Islamic business. Accordingly, *takaful* companies which derived their philosophical foundation from Islam should behave parallel with the religion's teaching (Asutay 2008; Dar 2012). Islam views the use of profit for society's needs is preferable compared to individual consumption (Siddiqi 1981; Metwally 1997). It even has higher priority than the state (Cantori & Lowrie 1992). In this particular context, firms have to be aware that their achievements are not measured alone in term of financial performance; rather the greatest achievement is when the firms can use their profits for the betterment of the society. Raghib (1995) suggests that profit is closely associated with business performance; but it should not be the only factor that firms consider since there are other responsibilities which firms need to reflect. This consistent with Ibn Khaldun comments that "...economic endeavour must generate income to market actors and enable and motivate them to serve their community," cited in Ali, Al-Aali and Al-Owaihan (2012).

Most of the literature discuss CSP within the periphery of corporate social responsibility (CSR) practiced by the companies. This assumption is partially accurate as Frederick (1994) emphasises that CSP is beyond that; and CSR is part and parcel of the CSP broad agenda. He suggested three steps which need to be fulfilled before the company can be regarded as achieving its CSP: CSR_1; CSR_2; and CSR_3. Sairally (2013) describes CSR_3 as the companies' CSP which eventually achieved after formulating the strategies and plans in CSR_1, and implementing the ideas which described as CSR_2. Carroll (1979) concisely describes CSR_1 as the companies tasks to realize their duties: legal, ethical, economic and discretionary responsibilities. There are four components under CSR_1 which are legal, ethical, economic and discretionary responsibilities as suggested by Carroll (1979). Companies have to plan for strategies and programs under each theme before they can implement them in the next level, CSR_2. This specific components help to

Corresponding author's email: amirulafif@salam.uitm.edu.my

bring focus on the wide concept of CSR which Frederick (1994) describes as the implementable and measurable tasks.

The economic responsibilities deemed as the basic or first layer requirement of companies as business entities which required to provide goods and services needed by the society and expected to be profitable at the end of the process. Every other business roles are assigned to this function. The next function is legal responsibilities regarded as the framework which the society expected the companies to be operated by adhering to the rules and regulations imposed on them. After that is ethical responsibilities which described as the extension of the early two responsibilities (economic and legal) which are actually expected to be exercised ethically; however the difference under this precept is the society expects the companies to be ethically responsible beyond the prescribed ethics in the early two responsibilities. Lastly, the discretionary responsibilities, which are ethical actions that above the ethical responsibilities. This is exercised by the companies voluntarily, which even the society does not expect the companies to fulfil this. Carroll (1979: 500) gives instances on companies' contributions on this aspect such as day care for the working mothers and training for the unemployed people.

In term of CSR_2 which Frederick (1994) describes it as corporate social responsiveness, dubbed as the managerial responds towards the precepts laid down in CSR_1 and how the companies respond to the society demands over this issue. Thus, management implementation of CSR_2 will produce CSR_3 or CSP, the outcomes or results from the implementation of activities in CSR_1 and CSR_2 which give opportunity for measuring the success of overall firms' CSR based on the empirical available data.

The benefits of CSP are not only received by the targeted groups in the CSR programmes, moreover, it brings benefits to the companies which have proper implementation of CSR programmes in form positive financial performance for firms that invest in CSR activities as suggested by Bushee (2001). This relationship somewhat virtuous cycle, as described by Dean (1998) due to the situation either financial performance influenced by CSR activities or CSR activities influenced by firms' good financial performance. Nelling and Webb (2009) concur with relationship between CSR and financial performance, but they caution such relationship becomes weaker with the use of fixed effects Granger test. Recent study on CSP by Gregory et. al. (2016) suggest that CSP influences positively the firm's value, earnings persistence and acts as competitive edge. Their study were based on the 44 industries in the United Kingdom. Nevertheless, the unanswered question is how much should be spent or contributes for CSR programmes in order for the company to be seen as achieving their CSP is remained puzzled (Dean 1998). Sairally (2013) posits the suggested CSP components to be used when assessing the CSP of IFIs, as laid down by Carroll (1979): legal, ethical, economic and discretionary responsibilities.

Therefore, by exercising these four responsibilities, it signifies the fulfilment of CSR_1 and CSR_2 conditions. Thus, from these four required responsibilities, it can be postulated that *takaful* operators seemed have acted as responsible corporate citizen which utilize various sources and consistently looking for avenues in order to contribute towards the society. If this really materializes in practise – *takaful* operators have conformed to the IFIs' societal objective.

Corresponding author's email: amirulafif@salam.uitm.edu.my

DOI: 10.1051/shsconf/20173600001

RESEARCH METHOD

This study has arranged structured interview sessions with the Chief Investment Officers and Heads of Investment of each of the eleven *takaful* operators in Malaysia. The Delphi-style technique was adopted when developing the interview questions – by gathering a panel of experts' opinions on the questions' draft. This is to ensure that the questions and scope of the study are relevant and capable of providing meaningful information for this research. The questions were developed in the form of a five-point Likert scale, addressing specific issues on CSP of *takaful* operators. In addition, information on *takaful* operators' CSR activities, zakat and tax payment were gathered from the companies' websites and annual report of *takaful* operators.

FINDINGS[†]

Findings gathered from the structured interviews indicate that *takaful* operators measure their corporate social performance in two ways: *zakat* payment and CSR activities. Interestingly, the interviewees did not mention or consider the way their businesses being managed as part and parcel of their CSP. This is partly accurate since CSP for *takaful* operators is actually beyond *zakat* and charitable works. This situation substantiated Lu and Castka (2009) finding when they describe the confusion amongst the corporate sector in Malaysia when defining their companies' CSP scope. In contrast, Lu and Castka (2009) also did not delineate clearly in their study what can be constituted as CSR especially in the Malaysian context.

In the other end of the spectrum, the *raison d'etre* of the company can categorized as one of the items for the company's CSP. This is particularly important for institutions like *takaful* operators which regarded as religiously and commercially driven business entities. Therefore, they have to simultaneously serve these two needs without isolation to any of the objectives (Asutay 2008; Dar 2012). Thus, the presence of *takaful* operators in Malaysia for offering *Shariah* compliant insurance products and services are part of the companies' CSP – providing services required by the society – to satisfy commercial and religious needs. Previous literatures suggest that the CSP of IFIs can be summarized in two categories: business activities with society related agenda[‡] and actions purposely taken to fulfil the CSR activities. These two broad categories were explored in this study by having questions on how investment activities and the business model of *takaful* operator possibly help to achieve the societal objectives, as exhibited in the interview's five Likert-scale questions. As discussed earlier, Carroll (1979) breaks down the two categories into four components of CSP which a corporation expected to realize: economic; legal; ethical; and discretionary responsibilities. From the open-ended interview questions on CSR, the interviewees (as clarified earlier) mentioned that their *takaful* operators involved in various CSR activities for the community and every year they contribute for *zakat* payment. The CSR activities financed either from *zakat*'s fund or from companies' dedicated budget for such programs. In addition for *zakat*, the payment either been made to the state Islamic council or they distribute them directly to the eight groups (*asnaf*) which qualified to receive *zakat* as mentioned in Holy Quran, with advised from their *takaful* operator's *Shariah* advisory committee.

† *(Dummy takaful operators' names are used to maintain the confidential aspect of the information)*
‡ This can be seen from *raison d'etre* of takaful operators to provide *Shariah* compliant insurance products.

Corresponding author's email: amirulafif@salam.uitm.edu.my

The delicate part is to measure how much is actually being spent or allocated for social programs yearly in order to assess the *takaful* operators' CSP. Most of the interviewees hardly able to brief (even rough estimation) agreeable percentages; except that citing *zakat*'s amount (2.5% from the prescribed *zakat* formula[§]) as their bases. Likewise, *zakat* is only imposed to the shareholders' fund since the paid-up capital provided by them, if there is surplus (Abdullah et al. 2012). A simple comparison between *takaful* operators and conventional insurers will signify that in term of contribution to the nation in forms of taxes – *takaful* operator contributes more due to *zakat* payment (pseudo religious tax) and corporate tax compared to conventional insurers, albeit in term of amount the latter might contribute more. Likewise, it signifies that *takaful* operators have fulfilled the first CSR component: legal responsibility.

In another perspective, one may argue that *zakat* is a kind of compulsory religious tax in Islam and if there is any CSR tax impose on the conventional insurer; then the same situation will exist. Nonetheless, it is important to evoke the sacred philosophy and principles of Islamic financial institution in order to understand why such religious tax is imposed on *takaful* operators. Table 6(a) displays the yearly *zakat* contributions, tax payments and gross profits for five years, from 2010-2014. Prior to 2012, *takaful* operators prepared their financial statements according to the funds managed, which are: shareholders; family; and general funds. Accordingly, *zakat* and tax payment have contributed as low as MYR 100k to MYR 70 million to the country respectively; without coalescing specific CSR projects organized by the firms as displayed in Table 6(b). Tax is essential for every government in the world in order to run a country effectively for the citizen. Freedman (2003) emphasises that under CSR requirement, companies are expected to pay their tax to the government to show them good corporate citizen.

		General *Takaful* Fund (amount in MYR million)					**Family *Takaful* Fund (amount in MYR million)**					**Shareholder's Fund (amount in MYR million)**				
		2010	**2011**	**2012**	**2013**	**2014**	**2010**	**2011**	**2012**	**2013**	**2014**	**2010**	**2011**	**2012**	**2013**	**2014**
TO1 Berhad	*Zakat*											0.1	0.1	0.1		
	Tax		4	8				2	9			4	7	2		
	Gross Profit		9	20				6	13			-21	19	17		
TO2 Berhad	*Zakat*											0.2	0.2	0.2	0.3	
	Tax		0.7	3	1			1	2	1		-	3	3	3	
	Gross Profit		7	10	3			16	21	3		7	15	15	16	
TO3 Berha	*Zakat*													-	-	-

§ In general, zakat can be calculated by using growth or working capital models. Growth Model: Owner's equity + Long term liability – fixed assets – intangible assets +/- adjustments = 2.5% x amount left for zakat x % Muslim's share in the business. Capital working model: Current assets – current liability +/- adjustments = 2.5% x amount left for zakat x % Muslim's share in the business. If partner is non-Muslim, then in both models, the share must be excluded from zakat.

Corresponding author's email: amirulafif@salam.uitm.edu.my

DOI: 10.1051/shsconf/20173600001

d	Tax								0.8	1	-			-	0.3	0.3
	Gross Profit								-3	-17	-4			-11	-8	-9
TO4 Berhad	*Zakat*													-	-	
	Tax								0.01	0.06				-	-	
	Gross Profit								-1	-0.3				-8	-0.8	
TO5 Berhad	*Zakat*	-	-	-	-	-	-	-	-	-	-	0.6	0.3	0.09	0.05	0.3
	Tax	-	1	0.9	2	3	0.09	0.9	0.5	0.06	0.6	-	0.2	1	2	0.8
	Gross Profit	-0.07	-6	-2	7	9	30	-120	-13	1	0.6	0.8	-0.2	-0.01	-5	-3
TO6 Berhad	*Zakat*													0.3	0.3	
	Tax								1	2				8	11	
	Gross Profit			16	8				210	139				40	23	
TO7 Berhad	*Zakat*															
	Tax												3		0.8	
	Gross Profit												-15	-18	-3	
TO8 Berhad	*Zakat*											0.1	0.4	0.4	0.8	
	Tax	0.4	0.5				0.1	0.4	0.2			6	8	10	12	
	Gross Profit	2	2				102	139	102			20	28	32	41	
TO9 Berhad	*Zakat*											0.4	0.4	0.4	0.4	
	Tax	3	1	2	18		3	3	4	4		5	5	5	8	
	Gross Profit	14	6	13	25		200	241	256	63		15	14	13	19	
TO10 Berhad	*Zakat*											4	4	8	8	
	Tax		10					29				43	20	71	70	
	Gross Profit	91	66				202	182				106	67	205	267	
TO11 Berhad	*Zakat*											0.4	0.5	1	2	
	Tax											43	22	25	43	
	Gross Profit											93	94	127	178	

Corresponding author's email: amirulafif@salam.uitm.edu.my

 DOI: 10.1051/shsconf/20173600001

Table 6(a): *Takaful* Operators' Gross Profit, *Zakat* Contributions and Tax Payments

Move on to the next component which is ethical responsibility as suggested by Carroll (1979), it is appropriate to denote that *takaful* operators in Malaysia involved in various CSR activities voluntarily and their charity works were noticeable especially during Ramadhan (the fasting month) due to religious sentiment on that holy month. The number of CSR activities of *takaful* operators in Malaysia are displayed in Table 6(b) based on the information gathered from the companies' websites from 2010 – 2014. However, total money spent for the activities are not available except for TO11 Berhad – the company spent nearly MYR 1 million every year for CSR activities. For the rest of *takaful* operators, some only listing their CSR projects, but nearly half of them did not inform any CSR activities in their websites and annual reports.

Takaful Operators	Number of CSR Activities and Budget Spent				
	2010	2011	2012	2013	2014
TO1 Berhad	2	2	2	2	2
TO2 Berhad	n/a	n/a	n/a	n/a	4
TO3 Berhad	1	n/a	n/a	n/a	n/a
TO4 Berhad	n/a	n/a	n/a	n/a	n/a
TO5 Berhad	n/a	n/a	n/a	n/a	1
TO6 Berhad	n/a	n/a	n/a	n/a	n/a
TO7 Berhad	1	2	1	1	3
TO8 Berhad	1	2	4	2	n/a
TO9 Berhad	7	19	28	28	17
TO10 Berhad	n/a	n/a	n/a	n/a	n/a
TO11 Berhad	6 (MYR 2.5mil)	5 (MYR 1mil)	12 (MYR 1mil)	9 (MYR 1mil)	4

Table 6(b): *Takaful* Operators' CSR Activities & Budget Spent, accessed from companies' websites as at 7 Jan 2015

The next component is economic responsibility which previous studies have covered in-depth on this in form financial performance based on the efficiency management of cost, structural form and technological effect (Saad et al. 2006; Kader, Adams & Hardwick 2010; Ismail, Alhabshi & Bacha 2011; Saad & Idris 2011). In other words, each year *takaful* operators have performed this responsibility well by achieving business growths and profits for their principals: shareholders and *takaful* contributors, as shown in Table 6(a).

The last component is discretionary responsibility (Carroll 1979), which defined by Gitman and McDaniel (2008) as the endeavour took by the firms to fulfil the early three components (legal, economic and ethical responsibilities) and move to the highest level of philanthropy by ensuring the company's business model and its activities meet able to serve the society. This part may sound utopia or difficult to achieve. However, looking back at the early three responsibilities under CSR, *takaful* operators can be regarded have achieved and fulfilled them. Thus, for

Corresponding author's email: amirulafif@salam.uitm.edu.my

discretionary responsibility, arguably *takaful* operators have managed to achieve it. Even though philanthropically, the *waqf* (endowment) model looks preferable since it has the capacity to reach and deliver better social services, but the currently adopted *wakalah* version of *takaful* model also capable to fulfil its social obligation within its spectrum. It is noteworthy to mention that while *waqf* model established solely for charitable, in the other side, *wakalah* model shows how commercial features enjoined with social obligations able to serve both needs: profit and social welfare.

It is understandable that there are groups of people who critics and claims that such social performance is not sufficient for Islamic financial institution. The criticisms normally address to the Islamic banks. As for the *takaful* operators, although they share some similarities with the Islamic banks, however in term of operational wise they are different.
All things considered, it is convenient to remark that *takaful* operators have fulfilled their CSR_1 and CSR_2. However, the CSR_3 component is hardly measured empirically in this study as it requires longer time frame and more active involvement from *takaful* operators in order to track their CSR activities and deriving the positive outcomes resulted from such activities.
Nevertheless, it is appropriate to denote in general such activities have given benefits to the receivers based on the information and news retrieved from newspapers and companies' websites in relations to their CSR activities, also from the local Islamic councils that manage *zakat* fund.

CONCLUSION

Takaful operators have achieved their CSP within the limited features of *wakalah takaful* model. This aspect can be derived from the findings presented in Table 6(a) on *zakat* contributions and tax payment, and Table 6(b) on the number of CSR activities and budget spent for the programs. The figures displayed indicate that *takaful* operators' CSP is a "profit driven social programs." In other words, the profits derived from the business were used to conduct social programs in forms of *zakat* contributions, tax payment and CSR activities. Nevertheless, this only represents part of the broad social responsible aspect which the companies have done. Another important aspect which is rarely considered and mentioned under the *takaful* operators' social responsible theme is their *raison d'etre* or business agenda. It should be emphasised that *takaful* operators were established to continue the classical notion of *takaful*[**] operation; but in a different setting, as a result of the needs to embrace modern corporation concept. Nevertheless, the main purpose is remained similar which is to offer alternative to the Muslims to be able to utilize insurance services which comply with *Shariah* (Husin & Rahman 2013; Souiden & Jabeur 2015), and to eliminate financial exclusion due to religious constraint (Coolen-Maturi 2013). This noble objective is stipulated in the white paper of steering committee comprising 20 experts in the field which was formed by the Malaysian government to study the feasibility of Islamic finance in Malaysia in 1981 (Laldin 2008).

Therefore, from the discussion above, this study views that it has managed to assess *takaful* operators' corporate social performance (CSP) from four angles as suggested in previous literature: economic; legal; ethical; and discretionary responsibilities. Moreover, *takaful* operators have fulfilled expectations bestowed upon them as Islamic financial institutions (IFIs)

[**] *Takaful* was initially formed based on brotherhood and mutual assistance in helping each other in the event of calamity.

Corresponding author's email: amirulafif@salam.uitm.edu.my

 DOI: 10.1051/shsconf/20173600001

by delivering their CSP well and attaining profit every year – socially responsible and financially profitable.

REFERENCES

Abdullah, M, Amanbayev, Y, Omer, GS & Elobied, A 2012, *An overview of takaful in Malaysia*, INCEIF, Kuala Lumpur.

Al-Quran 5:2 *Surah Al-Maidah*.

Alhabshi, SO & Razak, SH 2011, 'Takaful insurance: concept, history and development challenges', in Ariff, M & Iqbal, M (eds), *The Foundations of Islamic Banking : Theory, Practice and Education*, Edward Elgar Publishing, Cheltenham.

Ali, A, Al-Aali, A & Al-Owaihan, A 2012, 'Islamic perspectives on profit maximization', *Journal of Business Ethics*, 2012/11/01, pp. 1-9.

Asutay, M 2008, 'Islamic banking and finance: social failure', *New Horizon: Global Perspective on Islamic Banking and Insurance*.

Bakar, MD 2011, 'Shariah principles governing takaful models', in Archer, S, Karim, RAA & Nienhaus, V (eds), *Takaful Islamic Insurance : Concepts and Regulatory Issues*, 1 edn, Wiley, Hoboken.

Bushee, BJ 2001, 'Do institutional investors prefer near-term earnings over long-run value?', *Contemporary Accounting Research,* vol. 18, no. 2, pp. 207-246.

Cantori, L & Lowrie, A 1992, 'Islam, democracy, the state and the west', *Middle East Policy,* vol. 1, pp. 49-50.

Carroll, AB 1979, 'A three-dimensional conceptual model of corporate performance', *Academy of Management Review,* vol. 4, no. 4, pp. 497-505.

Çizakça, M 2011, *Islamic capitalism and finance: origins, evolution and the future*, Edward Elgar Publishing,

Coolen-Maturi, T 2013, 'Islamic insurance (takaful): demand and supply in the UK', *International Journal of Islamic and Middle Eastern Finance and Management,* vol. 6, no. 2, pp. 87-104.

Dar, H 2012, 'Identity crisis: Islamic banking - in search of a lost soul', viewed 14 January 2015, <http://islam.ru/en/content/story/identity-crisis-islamic-banking-search-lost-soul>.

Corresponding author's email: amirulafif@salam.uitm.edu.my

Dean, KL 1998, 'The chicken and the egg revisited: Ties between corporate social performance and the financial bottom line', *Academy of Management Perspectives,* vol. 12, no. 2, pp. 99-100.

Engku Ali, ER, Odierno, HS & Ismail, A 2008, *Essential guide to takaful (Islamic Insurance)*, CERT Publications Sdn. Bhd, Kuala Lumpur.

Frederick, W 1994, 'From CSR1 to CSR2', *Business and Society,* vol. 33, no. 2, p. 150.

Freedman, J 2003, 'Tax and corporate responsibility', *Tax Journal,* vol. 695, no. 2, pp. 1-4.

Gitman, L & McDaniel, C 2008, *The future of business: the essentials*, Cengage Learning,

Gönülal, SO 2012, *Takaful and mutual insurance : alternative approaches to managing risks*, World Bank Publications, Washington, <http://UNISA.eblib.com.au/patron/FullRecord.aspx?p=1069840>.

Gregory, A, Whittaker, J & Yan, X 2016, 'Corporate social performance, competitive advantage, earnings persistence and firm value', *Journal of Business Finance & Accounting*, vol. 43, no. 1 & 2.

Hasan, Z 2014, *Islamic banking and finance: an integrative approach,* Oxford University Press, Kuala Lumpur.

Htay, SNN & Zaharin, HR 2012, 'Critical analysis on the choice of takaful (Islamic Insurance) operating models in Malaysia', *World,* vol. 2, no. 2, pp. 112-127.

Husin, MM & Rahman, AA 2013, 'What drives consumers to participate into family takaful schemes? A literature review', *Journal of Islamic Marketing,* vol. 4, no. 3, pp. 264-280.

ING Website 2012, *Newsroom*, AIA PublicTakaful, viewed 29 May, <http://www.ing.com/Newsroom/All-news/Press-releases/PR/ING-to-sell-insurance-businesses-in-Malaysia-to-AIA.htm>.

Ismail, N, Alhabshi, SO & Bacha, O 2011, *Cost Efficiency and Investment Performance: Mutual and Stock Form in Malaysian Insurance Industry*, Working paper, International Centre for Education in Islamic Finance (INCEIF), Malaysia.

Kader, HA, Adams, M & Hardwick, P 2010, 'The cost efficiency of takaful insurance companies*', *Geneva Papers on Risk & Insurance,* vol. 35, no. 1, pp. 161-181.

Kuran, T 2012, *The Long Divergence*, Princeton University Press, United States of America.

Laldin, MA 2008, 'Islamic financial system: the Malaysian experience and the way forward', *Humanomics,* vol. 24, no. 3, pp. 217-238.

Corresponding author's email: amirulafif@salam.uitm.edu.my

Lu, JY & Castka, P 2009, 'Corporate social responsibility in Malaysia – experts' views and perspectives', *Corporate Social Responsibility and Environmental Management,* vol. 16, no. 3, pp. 146-154.

Malaysian Takaful Association 2014, *Malaysian Takaful Dynamics*, Malaysian Takaful Association and Ernst & Young Kuala Lumpur.

Metwally, MM 1997, 'Economic consequences of applying Islamic principles in Muslim societies', *International Journal of Social Economics,* vol. 24, no. 7, pp. 941-957.

Mohd Kassim, ZA 2012, *The primary insurance models*, World Bank Publications, Washington.

Nelling, E & Webb, E 2009, 'Corporate social responsibility and financial performance: the "virtuous circle" revisited', *Review of Quantitative Finance and Accounting,* vol. 32, no. 2, pp. 197-209.

Pamuk, Ş 2004, 'Institutional change and the longevity of the Ottoman Empire, 1500–1800', *Journal of Interdisciplinary History,* vol. 35, no. 2, pp. 225-247.

Raghib, H 1995, 'Islamic values and beliefs and their impact on marketing policy: Comparative analysis study', *Management in Islam*, pp. 307-355.

Saad, NM, Majid, MSA, Yusof, RM, Duasa, J & Rahman, A 2006, 'Measuring Efficiency of Insurance and Takaful Companies in Malaysia Using Data Envelopment Analysis (DEA)', *Review of Islamic Economics,* vol. 10, no. 2, p. 5.

Saad, NM & Idris, NEH 2011, 'Efficiency of life insurance companies in Malaysia and Brunei: a comparative analysis'.

Sairally, BS 2013, 'Evaluating the corporate social performance of Islamic financial institutions: an empirical study', *International Journal of Islamic and Middle Eastern Finance and Management,* vol. 6, no. 3, pp. 238-260.

Siddiqi, MN 1981, *Muslim economic thinking: a survey of contemporary literature*, International Centre for Research in Islamic Economics, King Abdul Aziz University,

Siddiqui, F & Hunt-Ahmed, K 2013, 'Takaful', in Hunt-Ahmed, K (ed), *Contemporary Islamic Finance*, John Wiley & Sons, Inc., pp. 203-213.

Souiden, N & Jabeur, Y 2015, 'The impact of Islamic beliefs on consumers' attitudes and purchase intentions of life insurance', *International Journal of Bank Marketing,* vol. 33, no. 4, pp. 423-441.

Swartz, NP & Coetzer, P 2010, 'Takaful: An Islamic Insurance Instrument', *Journal of Development and Agricultural Economics,* vol. 2, no. 10, pp. 333-339.

Corresponding author's email: amirulafif@salam.uitm.edu.my

The World Takaful Report 2014, *Growth momentum continues*, Ernst & Young.

Wahab, ARA, Lewis, MK & Hassan, MK 2007, 'Islamic takaful: business models, shariah concerns, and proposed solutions', *Thunderbird International Business Review,* vol. 49, no. 3, pp. 371-396.

Wilson, R 2008, 'Islamic economics and finance', *World Economics,* vol. 9, no. 1, p. 177.

Corresponding author's email: amirulafif@salam.uitm.edu.my

THE DETERMINANTS OF VOLUNTARY RISK DISCLOSURES: THE CASE OF SHARIAH COMPLIANT COMPANIES IN MALAYSIA

Nurul Fida Rosli*, Nor Farizal Mohammed, Zuraidah Mohd Sanusi[1]
**Faculty of Accountancy Universiti Teknologi MARA, [1]Accounting Research Institute, Universiti Teknologi MARA*

ABSTRACT

Risk reporting is essential in notifying investors on how companies manage the risks they engage in. Even though companies nowadays release information through press releases, corporate websites, and other forms of communications, they still have to disclose more information in their annual reports. The objective of this study is to examine the determinants of voluntary risk disclosures of Shariah compliant companies (ShCCs) in Malaysia, focusing on the religiosity factor and the ownership structure. A quantitative analysis using secondary data was employed as a method to assess the annual reports of 116 Shariah compliant companies in Malaysia for the financial years of 2012 and 2013. The findings of this study reveal that the voluntary risk disclosures in ShCCs on average are not more than sixty per cent. This study indicates no relationship between the religiosity of board members and directors' ownership to the level of voluntary risk disclosure. More importantly, this study finds the relationship between the existence of government ownership and voluntary risk disclosure in ShCCs. The study adds value to the current body of knowledge in voluntary risk disclosures and suggests the importance of risk management information in the annual reports of ShCCs as Malaysia grows as a leading country in the Islamic economy.

Keywords: Voluntary Risk Disclosures; Corporate Governance; Shariah Compliant Companies; Risk Reporting; Malaysia

INTRODUCTION

The absence of risk management information in companies' annual reports have steered regulators as well as researchers (Ismail, 2013; Bursa Malaysia, 2012a; Cabedo & Tirado, 2004) to demand for information on risks from these companies. The ever increasing corporate scandals have stunned the confidence among investors and creditors in companies' financial reporting (Ismail & Rahman, 2013). Therefore, risk disclosure has become an important element in corporate communication (Beretta & Bozzolan, 2004), where companies are encouraged to provide more voluntary information regarding their risk management (Securities Commission, 2012). Indeed, risk reporting and disclosures have become a main concern of many international accounting standard-setters and regulators (Atan et al., 2010). Companies that provide comprehensive reporting of their risk management will enhance their investor's confidence, as voluntary disclosure is viewed as credible information (Healy & Palepu, 2001). In this case, voluntary disclosure may act as a mechanism to differentiate superior and significant information in

*Corresponding author's email: norfa783@salam.uitm.edu.my

explaining mandatory information (Healy & Palepu, 2001). Further, lack of comprehensive voluntary risk information will give impact on the business' future as well as investor's confidence (Cabedo & Tirado, 2004; Ismail, Arshad, & Othman, 2012). These disclosure practices regarding risks are expected more in Shariah Compliant Companies (ShCCs) as the principle of being transparent is in adherence to the Islamic value.

According to prior studies, a good governance practice enhances monitoring, thus provides an excellent economic communication, thereby enhancing voluntary disclosure in annual reports (Chen & Janggi, 2000; Cheng & Courtenay, 2006; Cheung, Connelly, Limpaphayom, & Zhou, 2006; Eng & Mak, 2003; Ho & Wong, 2001). Corporate governance mechanism in terms of board of directors, holds the responsibility in realising the excellence of corporate governance practices, in which boards should establish clear roles and responsibilities in discharging its fiduciary and leadership functions (Securities Commission, 2012). Corporate governance mechanism as a monitoring tool ensures that financial statements prepared can deliver information that could help investors make decisions, for example, by disclosing important information on risks. A review of prior literature on corporate governance in Malaysia suggests that corporate governance factors such as directors' ownership, government ownership and ethnicity, are worthy to be investigated further in relation to the disclosure of financial statements.

Voluntary corporate disclosure has been comprehensively explored in recent years due to the growing importance of voluntary risk disclosures in the non-financial information of the annual reports (Ahmed H & Mohd Ghazali, 2013; Amalina Wan Abdullah, Percy, & Stewart, 2013; Ousama & Fatima, 2010;). Nevertheless, the lack of studies on voluntary risk disclosure focusing on Shariah Compliant Companies in Malaysia is evident as most previous literatures focused more on Public Listed Companies and Islamic Financial Institutions (Ismail, 2013; Atan et al., 2010; R. F. Ismail et al., 2012; R. Ismail & Rahman, 2011; Rahman, Kighir, Oyefeso, & Salam, 2013). Hence, this study seeks to explore the impacts of social values of the companies' accomplishments, focusing on the determinants of voluntary risk disclosures in Shariah compliant companies. Therefore, the main objective of this study is to examine the determinants of voluntary risk disclosures of ShCCs in Malaysia. The sub-objectives of this study are; to examine the association of ethnicity of board with voluntary risk disclosure; to examine the association of boards' ownership with voluntary risk disclosure; and to examine the association of government ownership with voluntary risk disclosure. The following section provides the literature review and hypotheses development. Section 3 discusses the research design and methodology. Section 4 explains the findings, while section 5 concludes the study.

LITERATURE REVIEW AND HYPOTHESES DEVELOPMENT

The post 1997 Asian financial crisis has solved the liquidity problem resulting from surplus funds from the Islamic finance industries, which has led to the emergence of the Islamic Capital Market (ICM) (Bursa Malaysia, 2012b). The aim of setting up the ICM is to globalise Malaysia as a centre for ICM in the Islamic economy. The ICM represents an assertion of religious law in capital market transactions where the "market is free from prohibited activities and elements such as riba (usury), maisir (gambling) and gharar (ambiguity)" (Bursa Malaysia, 2012b). The pace of development in the Islamic capital market has created the formation of various international Islamic organisations to study and promote this alternative market and has brought up Malaysia as an international Islamic finance hub (Nazri et al., 2014). Today, the Islamic capital market runs parallel to the conventional capital market and provide investors an alternative investment philosophy that is rapidly gaining acceptance and entrance by Muslims and non-Muslims (Ousama & Fatima, 2010b).

Malaysia has put on a continuous effort in promoting the country as a regional Islamic capital market. There are needs for continuous changes and improvements in meeting the complex and different demands of various sectors of the economy (Nazri et al., 2014). One of the efforts is the classification of companies as Shariah Compliant Companies (ShCCs), aiming to help investors who are concerned about the halal status of their investments to make their decisions. Shariah compliant companies are companies that are regulated and monitored by the Shariah Advisory Council (SAC). These companies refrain from being involved in activities such as gambling and alcoholic business matters in order to comply with the Islamic Shariah principles. They are also more likely to adhere to Islamic values such as accountability and transparency, as Islam requires a total commitment of Shariah in all aspects of life. Thus, in order to support the government's efforts, ShCCs must provide an excellent form of communication through their mandatory reports and voluntary reports of financial results regarding risks and other vital information. Consequently, this study hopes to contribute towards the understanding of risk management in the annual reports of ShCCs, hence assisting in strengthening the Islamic economy of this country. The Financial Accounting Standard Board describes "voluntary disclosures" as "information primarily outside of the financial statements that are not explicitly required by accounting rules or standards" (MASB, 2009). As such, voluntary disclosures involve disclosures in excess of the requirements and include accounting and other information that managers of a company deem relevant to the needs of various stakeholder groups (Jensen, 1993). Furthermore, risk disclosure is an example of voluntary disclosure and has become an essential part of corporate disclosure where it will enhance the transparency level of the company as well as the investor's confidence towards them (Linsley & Shrives, 2005, 2006b). Similarly, Yazid and Muda (2006) found that companies with risk disclosure provided better assessments concerning the risks they faced. By disclosing the information, they are more likely to take the necessary actions to reduce their exposure to losses.

Organisational legitimacy is the congruence between the social values associated with or implied by their activities and the norms of acceptable behaviour in the larger social system in which they are a part of (Tilling, 2004). Voluntarily disclosing information is one way to reduce the legitimacy gap (R. M. Haniffa & Cooke, 2005; Tilling, 2004). The legitimacy theory is defined as ''a generalised perception or assumption that the actions of any entity are desirable, proper, or appropriate within some socially constructed system of norms, values, beliefs and definitions'' (Suchman, 1995). There are four broad legitimation strategies identified by prior studies that firms may use to secure organisational legitimacy, which are: informing stakeholders about intended improvements in performance; seeking to change stakeholder's perceptions of an event; distracting attention away from an issue; and changing external expectations about its performance (Lindblom, C.K., 1994). As a result, this study is expected to assist the literature in exploring the behaviour of voluntary risk disclosure through these companies' responses to the organisational legitimacy gap. ShCCs are expected to provide more information on voluntary risks in order to respond to the society on their responsiveness towards the changing environments of the economy.

This study investigates the variation of Muslim directors on board and examines whether directors' and government ownership in the companies affect their percentage of voluntary risk disclosure. Market to book ratio for each Shariah compliant company is used to control the effect of size differences. Figure 1.1 illustrates the framework of this research. This section provides the linkage between theoretical as well as empirical and research questions proposed in this study, from which the research hypotheses are then constructed

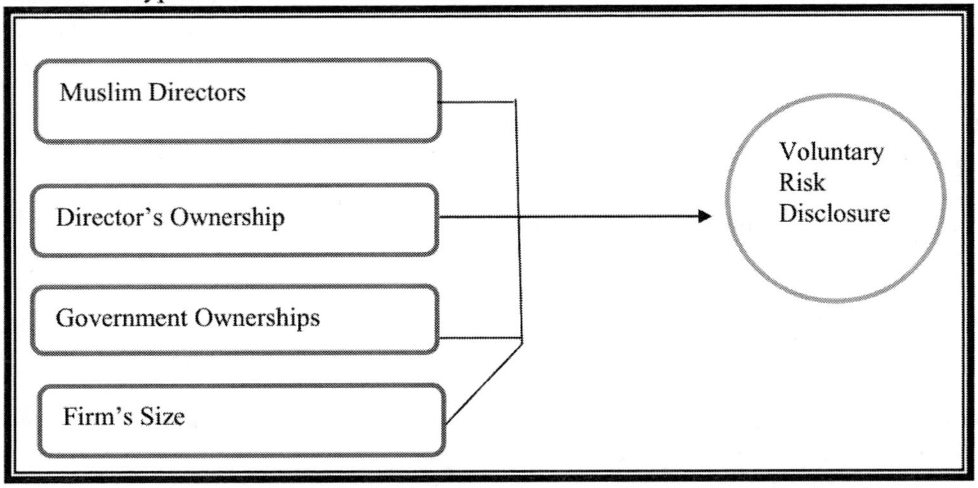

Figure 1.1: The framework of the research

The following hypotheses are built up based on the findings and arguments from prior studies. Shariah compliant companies are expected to be in line with Shariah rules and regulations set by the SAC. Thus, it is expected that these companies are operated by people who are knowledgeable about the Islamic values. However, in this multi-racial country, not all ShCCs are operated by Muslims. In the context of Malaysia, which is a multicultural country, it is very important to distinguish the attitudes and behaviours of the different races in society that influence the extent of voluntary disclosure (Haniffa & Cooke, 2000). The presence of religiosity as proxy for culture of decision makers within an organisation is important in some countries because the traditions of a nation are instilled in its people and might help explain why things are as they are (R. M. Haniffa & Cooke, 2005). The existence of clearly identifiable ethnic domination gives rise to differential monitoring in a company (Yatim, Kent, & Clarkson, 2006). Muslims are obliged to be transparent in showing their business matters, which is stated in the Holy Quran, Chapter Al- Baqarah, verse 282;

> *"O ye who believe! When ye deal with each other, in transactions involving future obligations in a fixed period of time, reduce them to writing. Let a scribe write down faithfully as between the parties…."*

Corporations are responsible for a wide spectrum of stakeholders and their aims should not strictly be on monetary values, but also on social responsibilities to the Ummah. Corporations should disclose information regarding its policies, especially on the risks they are facing. Haniffa & Cooke (2000) indicated cultural and corporate governance as determinants in voluntary disclosures. According to Yatim, Kent and Clarkson (2006), Bumiputera-dominated companies appear to have a better risk management system than non-Bumiputera dominated companies. Consequently, the presence of Islamic Malay directors is expected to give influence to the company's disclosure culture and practices due to their devotion to the Islamic values. In Malaysia, quite a number of public companies are family owned and managed, and it is more common among the Chinese (Ibrahim & Abdul Samad, 2011; Marimuthu, 2010). When family members own and manage the companies, they are less likely to accept their obligations to society (Haniffa & Cooke, 2000). They would prefer to have managers and directors from their own group, and this would limit the reporting practices to meet the bare minimum legal requirements. Thus, it may be expected that Chinese-managed companies would be less transparent or more secretive in their disclosure practices.

H1: Ceteris paribus, there is a positive relationship between voluntary risk disclosures and the existence of Muslim directors in boards of the company.

According to MCCG (2012), the board of directors is responsible in managing all risks involved in the business and ensuring an effective system to monitor and manage the risks. The responsibilities held by the board of directors include determining the company's level of risk tolerance and actively identify, assess and monitor key business risks to safeguard shareholders' investments and the company's assets, and ensure that their risk management system is viable and robust (Securities Commission, 2012). The board's duty is to serve the interest of shareholders as well as make the company known to the outside environment through disclosure (M. Akhtaruddin & Haron, 2010). The rationale for the existence of director ownership is to make the directors have the perspectives of real owners of the company in which they are expected to provide a strategic leadership that aligns with the company's interest (Chatterjee, 2009). The decisions on disclosure are in the boards' hands (M. Akhtaruddin & Haron, 2010), therefore, the existence of director ownership in the company will enable them to serve the shareholders better as strategic leaders and thus provide more information on voluntary disclosure. However, studies found managerial ownership to be negatively associated with the extent of voluntary disclosure in Malaysian and Singaporean listed companies (Eng & Mak, 2003; Mohd Ghazali, 2007) and the results are significant. There are many reasons to explain this negative relationship. According to Darmadi (2013) and Ghazali (2007), voluntary disclosure is the director's decision and they will disclose them voluntarily when the benefits outweigh the associated costs. Similarly, Jiang et al. (2009) indicated that directors provide minimum disclosures and less credible disclosures to outsiders to control and exploit their private benefits. Additionally, a highly concentrated ownership in a company results in less disclosure, especially on risks, as they discuss it in board meetings and not via annual reports (Konishi & Ali, 2007; Mokhtar & Mellett, 2013). Due to inconsistent arguments and findings, it is expected that director ownership has a relationship with voluntary disclosures.

H2: Ceteris paribus, there is a significant relationship between voluntary risk disclosures and director's ownership held in the company.

The government of Malaysia owns a significant portion of corporate ownership (Claessens, Djankov, & Lang, 2000; Mohd Ghazali, 2010). The other ownership structure that gives a significant influence to voluntary disclosure is government ownership. In the sense of risk disclosure, government owned companies may disclose less in annual reports in order to avoid bad perceptions from the public, as they are strongly politically associated and may want to guard their political connections (Jiang, Habib, & Hu, 2011; Mohd Ghazali, 2007). Conversely, Ghazali (2007) argued that government owned companies provide more voluntary disclosure because they are accountable to the public. This argument is supported by Mohd Ghazali and Weetman (2006), where companies that are held by the government are forced to provide more disclosure in their annual reports because the government is accountable to the public at large. Similarly, government owned companies' disclosure quality is higher than other companies because they are monitored by the government. Therefore, this study expects that ShCCs, in which the government owns a portion of their shares, will provide additional disclosure.

H3: Ceteris paribus, there is a positive relationship between voluntary risk disclosures and government ownership held in the company.

Previous studies found that the volume of voluntary disclosures are related to the size of the company (Ahmed Haji & Mohd Ghazali, 2013; Haniffa & Cooke, 2005; Ismail et al., 2012; Rahman et al., 2013). Normally, large companies tend to disclose more as they have a wider ownership base than smaller companies and they are more sensitive to political costs, giving them additional incentives for voluntary disclosures (Al-Janadi, Rahman, & Omar, 2013). Therefore, the voluntary risk disclosure is related to the company's size.

H4: There is a positive relationship between voluntary risk disclosures and the size of the company.

RESEARCH DESIGN AND METHODOLOGY

Sample and Data Collection

This study focuses on the disclosure made by Shariah Compliant Companies (ShCCs) that are listed in Bursa Malaysia which was obtained from it's website as well as from the Security Commission Malaysia's website. As of 31st June 2014, there are a total of 666 ShCCs that are listed under Islamic securities in Bursa Malaysia. Only the top one hundred and fifty Shariah Compliant Companies are used as an initial sample based on capitalisation ranking on 31st of December 2013. Companies with bigger market capitalisation are usually the top companies. The top companies are expected to disclose more information in their annual reports because they are more likely to be inspected by the regulators. However, financial industry companies such as banking, insurance, trust, closed-end funds and securities are excluded from the sample due to their nature of being governed by the additional regulation aside from Malaysian Financial Reporting Standards (Ismail & Rahman, 2013).

The table below shows the sample selection process. From the initial sample of 150 companies, three of them are financial companies, seventeen are delisted companies from the 2012 listing, and fourteen of the companies' data are not available. After excluding these companies, the final sample consists of 116 companies.

	No. of Companies
Initial sample	150
Less: companies in the finance industry	3
Less: delisted companies	17

Less : companies with unavailable data	14
Total usable data	116

Table 3.1: Data sampling process

The data used in this study were gathered from multiple sources of secondary data which were downloaded from the annual reports available on the Bursa Malaysia website. The information about the voluntary risk disclosure was extracted and collected from these annual reports, while the financial information of the companies, which were used as the control variable, were extracted from the Bloomberg database. Specifically, it was used to collect data on the control variable, which is size of the companies. All the collected data from the annual reports were hand collected. The annual reports for the period of 2012 to 2013 were collected, including the data of dependent and independent variables. Thus, this study observed the voluntary risk disclosures for the years of 2012 and 2013 because these two years were the latest and had complete annual reports of the selected sample company. Prior to these, total observations for this study were 232. This study incorporated both corporate governance and personal characteristics of directors into the voluntary disclosure model using company-specific characteristics as control variables, which is similar to prior studies of Haniffa and Cooke (2000) and Haniffa and Cooke (2005) as the determinants to the voluntary risk disclosures of ShCCs in Malaysia.

Variables

The following model was used primarily to test the hypotheses discussed and the variables are defined in Table 3.1:

$$VRD = \beta o + \beta 1MD + \beta 2DO + \beta 3GO + \beta 4SIZE + \varepsilon it$$

Types of Variables	Variables	Acronym	Measurement Scale
Dependent variable	Voluntary Risk Disclosure	VRD	Voluntary Risk Disclosure Index
Independent variable	Muslim Directors	MD	Percentage of Muslim directors on board (Muslim Directors / Total Directors) x 100
Independent variable	Directors Ownership	DO	Percentage of total directors' ownerships (Directors Shares / Total Shares) x 100
Independent variable	Government Ownership	GO	Percentage of government ownership
Control Variable	Firm Size	SIZE	Log (Total Assets)

Table 3.2: List of variables

FINDINGS AND DISCUSSION

Descriptive Statistics

	N	Minimum	Maximum	Mean	Median	Std. Deviation
VRD	232	20.83	62.5	35.722	33.3333	7.11068
MD	232	0	100	38.1724	30.0000	27.9205
DO	232	0	57.1	10.5664	1.9000	16.04113
GO	232	0	87	9.7819	5.3500	15.87198
SIZE	232	2.18	7.25	3.3221	3.2102	0.67968

Table 4.1 – Descriptive Statistics of all variables

The findings section begins with a discussion on the descriptive statistics for dependent, independent and control variables. Table 4.1 presents the descriptive statistics of the sampled companies. On average, the level of voluntary risk disclosure (VRD Score) is 35.72%, which implies that companies disclosed less than fifty per cent of the voluntary risk information to the users of the annual reports. This percentage is relatively low considering the importance of extra risk information for the users' decision-making. The highest percentage of disclosure for this category is 62.50% and the lowest is 20.83%. The range of Muslim directors (MD) are from 100% to 0%. The average of Muslim directors is 30% which indicates that not all ShCCs are governed by Muslims. The statistics on director ownership (DO) indicate that on average, only 1.9% is owned by their board. On the other hand, these companies are 5.35% owned by the government (GO). The firm size (SIZE), proxies by logged total assets, indicates that the sample firms were widely distributed from 7.25 to 2.18.

VOLUNTARY RISK DISCLOSURE

The table below outlines the companies that highly disclosed voluntary risk disclosure. Since the mean of voluntary risk disclosure is only 35.72% as referred in Table 4.1, these companies actually presented far more than the average compared to the other companies. However, these results are considered low by Ismail & Rahman (2011), considering the importance of risk management for users to make decisions. The highest percentage of VRD is 56.25% by Maxis Berhad whereas the lowest is 22.92% by White Horse Berhad.

Company	Mean Score Of VRD
Maxis Berhad	56.25
Tenaga Nasional Bhd	54.17
YTL Corporation Berhad	54.17
Axiata Group Berhad	47.92
Petronas Gas Berhad	47.92
Digi.Com Berhad	47.92
UMW Holdings Berhad	47.92

DOI: 10.1051/shsconf/20173600002

Gamuda Berhad	47.92
Dialog Group Berhad	47.92
Sime Darby Berhad	45.83
Telekom Malaysia Berhad	45.83
Petronas Dagangan Bhd	43.75

Table 4.2: List of Companies with the highest voluntary risk disclosed on average for 2012-2013

Variables	Beta	t	Sig.
MD	0.059	0.919	0.359
DO	0.045	0.746	0.456
GO	0.119	1.877	0.062
SIZE	0.469	8.058	0.000
R	.489		
R Squared	0.239		
Adjusted R Squared	0.226		
F-Statistic	17.87		
Significance	0.000		

Table 4.3: Regression analysis on Voluntary Risk Disclosure

The above table tabulates the summary of multiple regression results conducted. Based on the regression analysis, the adjusted R squared value for voluntary risk disclosure is 22.6%. This value indicates that 22.6% of variation in the dependent variable is caused by variation in independent variables. This regression model is also significant at $F (4,227) = 17.87$, $p < 0.01$. Based on the regression results, Muslim directors do not give any significant contributions toward voluntary risk disclosure at 0.350, where it is higher than 0.05 of significant value. In contrast with Haniffa and Cooke (2005), the existence of Muslim directors is not crucial towards practicing voluntary risk disclosure. According to Haniffa and Cooke (2005), a policy of positive discrimination exists in business opportunities as ethnicity affects corporate behaviour, including its disclosure practice. The mean of Muslims' existence in ShCCs is only 38.2%, which is respectively low considering that these companies are guided by the Shariah Advisory Council. The low mean score of MD indicates that most ShCCs are owned by non-Muslims. In the study of Haniffa and Cooke (2005), companies dominated by Malay directors will disclose more. This could explain the results of this study where most ShCCs dominated by non-Muslims or non-Malays did not provide additional disclosure of their risks. The first hypothesis is not supported because the P-value is 0.359. There is no significant relationship between voluntary risk disclosures and the existence of Muslim directors in Shariah Compliant Companies according to the regression result. According to the legitimacy theory, the community expects organisations to be responsive to the environment in which they operate (Tilling, 2004). In the current environment and economy where everything is uncertain, the public expects ShCCs to provide more information as they are known to be guided by the Shariah. ShCCs are widely accepted as companies that operate within the Islamic boundaries, thus, they are expected to be more transparent than normal type companies. In contrast, non-Muslim directors may act like Muslim directors in meeting the expectations of the stakeholders to provide supplementary information in their annual reports. Therefore, no difference is seen between Muslim and non-Muslim directors' domination in terms of providing the risk management report.

In addition, director's ownership does not provide any significant contribution towards the voluntary risk disclosure at 0.456, where it is higher than 0.05 of significant value. This result is similar to prior studies (Ahmed Haji & Mohd Ghazali, 2013b; Eng & Mak, 2003). Jensen and Meckling (1976) stated that monitoring directors who have ownership interest might not disclose more to create pressure on the management. According to prior studies, these directors may not disclose more information because they already know the inside information and might be interested to maximise their interest, hence conceal some of the material information (Ahmed, 2013; Chau & Gray, 2002; Eng & Mak, 2003; Leung & Horwitz, 2004). This result also supports Akhtaruddin et. al (2009), where it is actually outsider's ownership that significantly affects company's disclosure. Outside ownership exists to serve as monitors on managers' actions and reduces the likelihood of managers withholding information for their self-interest. Information disclosure is likely to be greater in firms where ownership is dispersed widely (Akhtaruddin, Hossain, & Yao, 2009). Thus, this study also rejects the hypothesis, as there is no relationship between voluntary risk disclosures and the existence of director's ownership in Shariah Compliant Companies.

Conversely, government ownership shows a significant contribution towards voluntary risk disclosure at 0.062 where it is lower than 0.10 of significant value. This result contradicts with the results of previous studies regarding the impact of government ownership in voluntary disclosure (Ahmed Haji & Mohd Ghazali, 2013b; Boesso & Kumar, 2007). Similarly, this study is also in contrast with the result of Haniffa and Cooke (2005), whereby Malay dominated firms owned by the government implemented a responsive legitimation strategy to change perceptions and divert the attention of its various stakeholders from the close relationship enjoyed with the government by increasing social responsibility disclosures. Conversely, most ShCCs are dominated by non-Muslims that are non-Bumiputera. However, they maintain their affiliations with the government using disclosures. In this case, regardless of variation in the board's religions, organisations seek to establish congruence between the social values associated with or implied by their activities and the norms of acceptable behaviour in the larger social system in which they are a part of (Tilling, 2004). This argument is also supported by Ahmed and Mohd Ghazali (2013), where companies that have governmental bodies owning significant portions of their shares provide more information in their annual reports since the government is accountable to the public at large. Thus, this study supports the hypothesis that there is a significant relationship between voluntary risk disclosures and the existence of government ownership in Shariah Compliant Companies.

According to the regression results, firm size has a significant contribution towards the voluntary risk disclosure at 0.000, where it is lower than 0.05 of significant value. This result is consistent with most previous literatures on voluntary disclosures, where it is indeed affected by the firm size (Ahmed Haji & Mohd Ghazali, 2013; Al-Janadi et al., 2013; Amran et al., 2009; Ismail & Rahman, 2013). As companies get bigger, the number of stakeholders also increases. With this increase, the burden of disclosure becomes heavier for the company since it has to cater to the needs of a bigger group of people. Larger companies are under greater public scrutiny and are more likely to face political costs. Consequently, they have the encouragement to give greater devotion to legitimisation management through higher disclosure (Liu & Taylor, 2008).

CONCLUSION

Islamic teachings have started to be accepted globally. This may be the reason for non-Muslim directors to act accordingly and adhere to the Islamic values in managing companies. Moreover, there are also managerial implications of the results. Managers who engage in good practices of information disclosure are recommended to continue doing so. It could also be a wakeup call to other companies that do not provide voluntary disclosure. Accordingly, the existence of government policy has encouraged them to voluntarily disclose their information. Several contributions of this research are identified. This study adds to the literature regarding voluntary risk disclosure in Malaysia, specifically in Shariah compliant companies. Voluntary risk disclosure is expected to enhance ShCCs' transparency towards a more comprehensive risk reporting. This study has found that companies with a high percentage of government ownership also provide more information on risk disclosure in their annual reports. It confirms prior studies that voluntary risk disclosure is also used as a reactive legitimation strategy aimed at diverting attention from questionable business practices, cronyism, nepotism and close affiliations with the government as well as a proactive legitimation strategy to ensure a continued influential voice on both governmental and institutional levels. This result contributes useful information to the regulatory bodies where they are expected to guide their firms toward the best practices of disclosures.

The findings of this research are subjected to several limitations. Firstly, this study only focused on disclosures in corporate annual reports although it is known that managements utilise other mass communication mechanisms too. Hence, future research may consider disclosures in other forms of media such as newspapers, the Internet, and in-house magazines. Secondly, this study only observed Shariah Compliant Companies in Malaysia. Future researchers might want to find the influence of these variables on voluntary risk reporting in other countries as well. Additionally, given the exceedingly complex nature of the business environment as well as time and geographical constraints, there are inherent limits in the ability of positive empirical research to capture all dimensions that influence voluntary risk disclosure policies and practices. Hence, further studies regarding this matter need to be more detailed in order to understand the issues better.

REFERENCES:

Ahmed Haji, A., & Mohd Ghazali, N. A. (2013a). The quality and determinants of voluntary disclosures in annual reports of Shari'ah compliant companies in Malaysia. *Humanomics*, *29*(1), 24–42. http://doi.org/10.1108/08288661311299303

Ahmed Haji, A., & Mohd Ghazali, N. A. (2013b). The quality and determinants of voluntary disclosures in annual reports of Shariah compliant companies in Malaysia. *Humanomics*, *29*(1), 24–42. http://doi.org/10.1108/08288661311299303

Ahmed, S. (2013). Impact of Corporate Governance on Disclosure Quality : Empirical Evidence from Listed Banks in Malaysia, International Journal of Economics and Finance, *7*(2), 242–279.

Akhtaruddin, M., & Haron, H. (2010). Board ownership, audit committees' effectiveness, and corporate voluntary disclosures. *Asian Review of Accounting*, *18*(3), 245–259.

Akhtaruddin, M., Hossain, M. A., Hossain, M., & Yao, L. (2009). Corporate Governance and Voluntary Disclosure in Corporate Annual Reports of Malaysian Listed Firms. *Journal of Applied Management Accounting Research*, *7*(1), 1–20.

Al-Janadi, Y., Rahman, R. A., & Omar, N. H. (2013). Corporate Governance Mechanisms and Voluntary Disclosure in Saudi Arabia. *Research Journal of Finance and Accounting*, *4*(4), 25–36.

Amran, A., Bin, A. M. R., & Hassan, C. H. M. (2009). Risk reporting: An exploratory study on risk management disclosure in Malaysian annual reports. *Managerial Auditing Journal*, *24*(1), 39–57. http://doi.org/10.1108/02686900910919893

Atan, R., S. Maruhun, E. N., Kadir W. A., W. H., & Jusoff, K. (2010). Annual risk reporting of listed companies in Malaysia, *Journal of Modern Accounting and Auditing*, *6*(8), 26–38.

Beretta, & Bozzolan. (2004). A Framework for the Analysis of Firm Risk Communication. *The International Journal of Accounting*, *39*(3), 265–288.

Boesso, G., & Kumar, K. (2007). Drivers of corporate voluntary disclosure: A framework and empirical evidence from Italy and the United States. *Accounting, Auditing & Accountability Journal*, *20*(2), 269–296. http://doi.org/10.1108/09513570710741028

Bursa Malaysia. (2012b). THE ISLAMIC CAPITAL MARKET. *Heterocycles*, *85*(7). http://doi.org/10.3987/Contents-12-85-7

Cabedo, J. D., & Tirado, J. M. (2004). The disclosure of risk in financial statements. *Accounting Forum*, *28*(2), 181–200. http://doi.org/10.1016/j.accfor.2003.10.002

Chau, G. K., & Gray, S. J. (2002). Ownership structure and corporate voluntary disclosure in Hong Kong and Singapore. *The International Journal of Accounting*, *37*(2), 247–265. http://doi.org/10.1016/S0020-7063(02)00153-X

Chen, J. P., & Janggi, B. L. (2000). The Association between independent non-executive directors, family control and financial disclosures. *Journal of Accounting and Public Policy*, *19*, 285–310.

Cheng, E. C. M., & Courtenay, S. M. (2006). Board composition, regulatory regime and voluntary disclosure. *The International Journal of Accounting*, *41*, 262–290.

Cheung, S. Y., Connelly, J., Limpaphayom, P., & Zhou, L. (2006). *Determinants of corporate disclosure and transparency: Evidence from Hong Kong and Thailand.*

Darmadi, S. (2013). Corporate governance disclosure in the annual report: An exploratory study on Indonesian Islamic banks. *Humanomics*, *29*(1), 4–23. http://doi.org/10.1108/08288661311299295

Dobler, M., Lajili, K., & Ze, D. (2011). Dobler, M., Lajili, K., & Ze, D. (2011). Attributes of Corporate Risk Disclosure: An International Investigation in the Manufacturing Sector. *Journal of International Accounting Research: Fall 2011*, Vol. 10, No. 2, pp. 1-22.

Eng, L. L., & Mak, Y. T. (2003). "Corporate governance and voluntary disclosure",. *Journal of Accounting and Public Policy, Vol. 22,* pp. 325–45.

Ferrero-ferrero, I. , Fernández-Izquierdo, M A & Muñoz-Torres, M J (2012). The impact of the board of directors characteristics on corporate performance and risk-taking before and during the global financial crisis, *Review of Managerial Science*, Volume 6, Issue 3, pp 207-226

Ghazali, N. a. M. (2007). Ownership structure and corporate social responsibility disclosure: some Malaysian evidence. *Corporate Governance*, *7*(3), 251–266. http://doi.org/10.1108/14720700710756535

Haniffa, & Cooke, T. E. (2005). The impact of culture and governance on corporate social reporting, *24*, 391–430. http://doi.org/10.1016/j.jaccpubpol.2005.06.001

Haniffa, R., & Cooke, T. (2000). CULTURE, CORPORATE GOVERNANCE AND DISCLOSURE IN MALAYSIAN CORPORATIONS, *2000*(November), 28–30.

Healy, P. M., & Palepu, K. G. (2001). Information asymmetry, corporate disclosure, and the capital markets: A review of the empirical disclosure literature. *Journal of Accounting and Economics*, *31*(1-3), 405–440. http://doi.org/10.1016/S0165-4101(01)00018-0

Ho, S. S. M., & Wong, K. S. (2001). A study of the relationship between corporate governance structures and the extent of voluntary disclosure. *Journal of International Accounting Auditing and Taxation*, *10*, 139–158.

Ibrahim, H., & Abdul Samad, F. M. (2011). Corporate Governance and Agency Costs. *Advances in Financial Economics*, (2006), 109–130.

Ismail, R. F., Arshad, R., & Othman, S. (2012). THE INFLUENCE OF VOLUNTARY RISK DISCLOSURE ON FIRMS' MARKET VALUE. *3rd INTERNATIONAL CONFERENCE ON BUSINESS AND ECONOMIC RESEARCH (3rd ICBER 2012) PROCEEDING*, (March 2012), 2907–2934.

Ismail, R., & Rahman, R. A. (2011). Institutional Investors and Board of Directors ' Monitoring Role on Risk Management Disclosure Level in Malaysia, 37–62.

Ismail, R., & Rahman, R. A. (2013). Risk Management Disclosure In Malaysian Islamic Financial Institutions : Pre- And Post-Financial Crisis. *Journal of Applied Business Research, 29.2* (Mar/Apr 2013): 419-431.

Jensen, M. C. (1993). The modern industrial revolution, exit, and the failure of internal control systems. *The Journal of Finance, 48*(3), 831–880.

Jensen, M. C., & Meckling, W. H. (1976). Theory of the firm: managerial behavior, agency costs and ownership structure. *Journal of Financial Economics*, *3*, 305–60.

Jiang, H., Habib, A., & Hu, B. (2011). Ownership concentration, voluntary disclosures and information asymmetry in New Zealand. *The British Accounting Review*, *43*(1), 39–53. http://doi.org/10.1016/j.bar.2010.10.005

Jiang, H., Habib, A., & Smallman, C. (2009). The effect of ownership concentration on CEO compensation-firm performance relationship in New Zealand. *Pacific Accounting Review*, *21*(2), 104–131. http://doi.org/10.1108/01140580911002053

Konishi, N., & Ali, M. . (2007). Risk reporting of Japanese companies and its association with corporate characteristics. *Nternational Journal of Accounting, Auditing and Performance Evaluation I*, *4*(3), 263–285.

Leung, S., & Horwitz, B. (2004). Director Ownership and Voluntary Segment Disclosure: Hong Kong Experience. *Journal of International Financial Management and Accounting*, *15*(3), 235–260.

Lindblom, C.K. (1994). *The implications of organizational legitimacy for corporate social performance and disclosure, Working Paper for the American Accounting Association Public Interest Section, USA*.

Linsley, P. M., & Shrives, P. J. (2005). Examining risk reporting in UK public companies. *The Journal of Risk Finance*, *6*(4), 292–305. http://doi.org/10.1108/15265940510613633

Linsley, P. M., & Shrives, P. J. (2006b). Risk reporting: A study of risk disclosures in the annual reports of UK companies. *The British Accounting Review*, *38*(4), 387–404. http://doi.org/10.1016/j.bar.2006.05.002

Liu, J., & Taylor, D. (2008). Legitimacy and corporate governance determinants of executives ' remuneration disclosures. *Corporate Governance: The International Journal of Business in Society*, *VoL. 8 NO.*, pp. 59–72. http://doi.org/10.1108/14720700810853400

Lu, C., Liao, G. M., & Yang, Y. (2007). Ownership Structure , Information Disclosure and Corporate Value : An Empirical Analysis of Taiwan Companies, 698–704.

Marimuthu, M. (2010). Bumiputera-Controlled Companies: Performance Evaluation Using A Non-Parametric Approach. *International Journal of Economics and Finance*, *2*(2), 178–185. http://doi.org/10.5539/ijef.v2n2p178

MASB. (2009). Financial Reporting from an Islamic Perspective.

Mokhtar, E. S., & Mellett, H. (2013). Competition, corporate governance, ownership structure and risk reporting. *Managerial Auditing Journal*, *28*(9), 838–865. http://doi.org/http://dx.doi.org/10.1108/MAJ-11-2012-0776

Nazri, M., Mohd, B., Tamkin, J., & Borhan, B. (2014). The Global Hub Process : Malaysia ' s Vision towards Becoming the Islamic Finance Global Hub Country, *5*(8), 206–223.

Ousama, a. a., & Fatima, A. H. (2010b). Voluntary disclosure by Shariah approved companies: an exploratory study. *Journal of Financial Reporting and Accounting*, *8*(1), 35–49. http://doi.org/10.1108/19852511011055943

Rahman, R. A., Kighir, A., Oyefeso, L. O., & Salam, O. A. (2013). Risk Management Disclosure Practices of Islamic Banks in the Mena Region : An Empirical Analysis, *15*(1), 152–160. http://doi.org/10.5829/idosi.mejsr.2013.15.1.2188

Rajab, B. & Handley-Schachler, M. (2009). Corporate risk disclosure by UK firms: trends and determinants,World Review of Entreprenuership, Management and Sustainable Development, 5(1). http://dx.doi.org/10.1504/WREMSD.2009.026801

Said, R. A., Zainuddin, Y. H., & Haron, H. (2009). The relationship between corporate social responsibility disclosure and corporate governance characteristics in Malaysian public listed companies. *Social Responsibility Journal*, *5*(2), 212–26.

Securities Commission. (2012). Malaysian Code on Corporate Governance 2012.

Subramaniam, N., & Mcmanus, L. (2009). Corporate governance , firm characteristics and risk management committee formation in Australian companies, 316–339. http://doi.org/10.1108/02686900910948170

Suchman, M. C. (1995). Managing legitimacy: strategic and institutional approaches. *Academy of Management Review*, *20*(3), 571–610.

Tilling, M. V. (2004). Refinements to Legitimacy Theory in Social and Environmental Accounting. *COMMERCE RESEARCH PAPER SERIES NO. 04-6*, (ISSN: 1441-3906).

Yatim, P., Kent, P., & Clarkson, P. (2006). Governance Structures , Ethnicity , and Audit Fees of Malaysian Listed Firms.

Yazid, A. S., & Muda, M. S. (2006). The Role of Foreign Exchange Risk Management in Malaysia. *The Irish Journal of Management*, *26*(2), 45 – 63.

CSR INFLECTIONS: AN OVERVIEW OF CSR PRACTICES ON FINANCIAL PERFORMANCE BY PUBLIC LISTED COMPANIES IN MALAYSIA

Norwazli Bt Abdul Wahab, Noryati Bt Ahmad, Haslinda Bt Yusoff
Faculty of Business Management, Faculty of Accountacy, Universiti Technology of MARA,

ABSTRACT

Of late, there has been growing demand for companies to intensify its corporate social responsibility (CSR). It is no longer one of the responsibilities of the companies but has become the responsibility of companies to give back to the society in which it has been operating and making huge profit. Unfortunately there appears to be a wide gap between the level of CSR awareness between companies from developed countries and those from the developing companies. Companies from developed countries have higher CSR level of awareness and faced no difficulty in disclosing their CSR activities relatively to the companies in the developing countries (including Malaysia). Thus, the aim of this paper is to examine the level of CSR disclosure of consumer products and trading and services industries from Malaysia. In addition, the study also would like to ascertain if the level of CSR disclosure influences the performance of the company in terms of its return on asset (ROA) and firm value (Tobin's Q). The study starts with the development of CSR disclosure index and using the index to determine if company's performance is related to its contribution to the society. Panel data is used to achieve the second objective. Findings of the study provide positive, negative and insignificant results between CSR activities on the company performance.

Keywords: Corporate social responsibility, corporate social responsibility disclosure (CSRD), corporate financial performance (CFP), content analysis

INTRODUCTION

There has been a rapid evolution on Corporate Social Responsibility (CSR) over the last few decades among companies. Today, the investors and stakeholders are paying attention not only to the bottom line but also company's contribution towards the society. As a result, the companies are expected to be more responsible in doing their business. In line with this development, more and more companies have taken CSR seriously and incorporate this element in corporate strategy. They realized that embedded CSR as part of their corporate strategy has alleviated their image and competitive advantage. These practices have become a significant aspect to compete locally and globally. Based on the literature review, as for developed countries, CSR practices are considered as a business strategy in order to improve the company's financial performance. According to Kay (1993), there are three reasons for CSR to be pertinent to business, (1) provides input to strategy as it provides an understanding about the different key elements in the business milieu, (2) a support for the value of chain and value creation activities and (3) a middle-of-the- road management task that to be supervised as well as any other business process. This is supported by Ruf, Muralidhar, Brown, Janney and Paul (2001) stating that "the dominant stakeholder group, shareholders, financially benefit

Corresponding author's email: nor_1143@yahoo.com

when, management meets the demands of multiple stakeholders. They conclude that by improving subsequent corporate social performance (CSP), it is positively related to sales growth for the current and subsequent years. Perhaps, it generates short term benefits and also create potential for the long term benefits. Conversely, Mahoney and Roberts (2007) argued that besides improving the financial performance of companies, CSR practices also attracts the positive response from institutional investors. However, although slowly gaining attention, CSR practices among companies in developing countries have been rather slow or lacking. Most of these companies viewed CSR as additional costs and found it not to be relevant to their business strategy.

To what extent have Malaysian companies been involved in CSR practice? In general, corporate social responsibility disclosure (CSRD) has been used as a proxy to CSR studies. It discloses information of what the companies have contributed and plan for future to the welfare of the society. Usually the information will be stated in a social responsibility report and publishes in company's websites or annual report of public listed companies (Zainuddin and Nasruddin, 2006; Mohamed Zain and Janggu, 2006). Additionally, their concern whether CSRD is positively or negatively influence the shareholders' insight has led companies to think not only their financial performance but also their environmental and societal performance. Therefore, to attract investors, companies begin to coordinate their CSR activities and disclose more information concerning CSR activities in annual report and stand-alone sustainability report. This triggers interest among researchers to conduct studies on the relationship between CSRD and company financial performance (CFP) (Mustaruddin, Norhayah and Rusnah, 2011; Nor Hawani, Mustaffa and Norhasnah, 2011; Shveta and Sandhu, 2010; Choi, Kwak and Choe, 2010; Mustaruddin, 2009). However, empirical results on the relationship between CSRD and company's financial performance have been inconclusive where some studies show positive results, others have negative relationship between the two variables and there are those that produced mixed relationship. The validity of the findings on the direction of the impact of CSR activities on future corporate performance is continuously being questioned (Margolis and Walsh, 2003; Orlitzky et al, 2003) due to various measures over a substantial time span for instance how different measures of socially responsible practices can generate better financial performance.

In Malaysia, starting on 31st December 2007, it is compulsory for public listed companies in Malaysia to disclose their CSR activities. The focus of the CSR practices are mainly on four areas namely; community, environment, marketplace and workplace with no order of priority since Bursa Malaysia acknowledged the diversity of those listed companies. Nevertheless these guidelines are flexible and require minimum disclosure in their report. There is no detail specific information to be disclosed and the amount of disclosure required by the companies but rather gives companies the flexibility to provide information relating to their CSR activities. Since involvement of CSR practices are still relatively new among listed companies in Bursa Malaysia, it is therefore interesting to investigate the extent of CSR disclosure that were made by these companies and which activities are mostly been disclosed. In addition, this study would like to determine the relationship between CSR disclosure and financial performance. The area of study is considered relevant since the topic has created an attention to several organizations and the Malaysian government particularly seeing that CSR is the national agenda (Othman, Darus and Arshad, 2011).

REVIEW OF PREVIOUS RELATED STUDIES

Corresponding author's email: nor_1143@yahoo.com

There motive behind corporate social responsibility (CSR) goes beyond the financial gains of company. CSR attempts to promote socially beneficial activities into the business that would ensure sustainability and survivorship of the company. Unfortunately very few companies are willing to engage in social responsibility activities and disclosure. For those who are championing the CSR practices do so because they realize the positive impacts on their products, services and other activities. Highlighted by Tsoutsoura (2004), being socially responsible should have bottom-line benefits in order to be sustainable. This include the increased in the productivity has resulted from the improved working conditions and labour practices. For instance, by implementing simple measures such as reducing wrapping material and planning optimal route for delivery trucks are not only could reduce a company's operating costs but also as a motivation for companies to increase their CSR activities as it helps to protect the environment. Furthermore, Jenkins (2005) emphasized that brand image and reputation is greater in companies that implement CSR practices than those who are not, or in socially irresponsible companies since the customers and suppliers are looking forward to deal with companies with good CSR practices (Fombrun and Shanley, 1990).

In addition, the business risk could be reduced as soon as the company able to attract and retain employees (Turban and Greening, 1997). Perhaps, the companies may increase the market share relative to the competitors by providing important information about how the products have been manufactured in a socially or environmentally responsible manner (Miller, 1994). As we know, consumers are willing to pay more for products and services from socially responsible companies (Auger, Devinney and Louveniere, 2003).

Nonetheless, level of awareness of CSR activities is increasing among companies particularly those of developed countries. These companies have become more sensitive to social, environmental and economic issue and stakeholder concerns. It becomes a challenge for the companies as they find ways to improve their performance and integrate social responsibility impacts and financial performance into day-to-day management decision making. However, there are arguments in relation to the relationship between company's corporate social responsibility and their financial performance.

The emerging interests among companies to incorporate CSR in their corporate strategy have attracted studies to examine either the extent of CSR disclosure provided by companies (Ibrahim, Zam Zuriyati, Jamal and Norlia (2013) and Dalilawati, Norhayah and Zakiah (2013)) or the effect of CSR on financial performance of the company relationship (Ioannou and Serafeim, 2012; Inoue and Lee, 2011 and Choi Kwak and Choe, 2010).

Ibrahim, Zam Zuriyati, Jamal and Norlia (2013) carried out a content analysis to examine the importance of corporate social responsibility disclosure (CSRD) of Malaysia consumer products and plantation industry for the year ended 2009. The authors concluded that. Although the number of companies disclosed their CSR activities was high for both industries but the level of disclosure is still low level category in disclosing their CSR activities. It also revealed a gap between the maximum and minimum number of sentences for CSR disclosure in the annual report.

Furthermore, Dalilawati, Norhayah and Zakiah (2013) study, compares the quantity and quality of CSR disclosure disclosed in annual reports of 300 top companies on the main board of Bursa Malaysia between Shariah and non-Shariah approved companies. The analysis is done through a content analysis reflecting the voluntary period (2005-2006) and mandatory period of CSR

Corresponding author's email: nor_1143@yahoo.com

disclosure (2007-2009). Findings from the analysis can be concluded that there is no significant difference in the overall quantity and quality of CSRD disclosed between Shariah and non-Shariah approved companies for the period of 2005-2006. However, through the introduction of the mandatory of CSR disclosures requirement by the Bursa Malaysia in 2007, there are a significance difference particularly on two categories of CSR that is for environment and community related disclosures.

However, the empirical studies have inconclusive, as some studies found a positive correlation; some other determined a negative while others found no correlation at all due to a number of relevant theories and the inconsistencies of research methods. Some researchers reported a positive, negative and neutral impact of CSR on financial performance. The inconsistency may be due to flawed empirical analysis. There are studies indicate a positive relationship (Ioannou and Serafeim, 2012; Inoue and Lee, 2011 and Choi Kwak and Choe, 2010), whilst some studies found a negative and mixed relationship. For instance, Mustaruddin, Norhayah Zulkifli & Rusnah (2011) found a positive and significant related to the CSR on financial performance but there is limited evidence of the relationship in the long term. Besides, McWilliams and Siegel (2000) did not find any relationship between CSRD and financial performance.

Furthermore, study conducted by Hackston and Milne (1996) on annual reports of 47 companies in New Zealand for the year 1992 is to find evidence on the relationship between CSR disclosure and financial performance. They counted number of sentences related to six types of CSR areas (Environment, human resources, products, energy, community and others) and tested correlated coefficients with profitability measures (ROA and ROE), company size (capitalization and total assets) and industries. It can be concluded that CSR disclosure is significantly associated with size and industry of the company, while there is no correlation with the profitability. Moreover, Everaert, Bouten, Van Liedekerke, De Moor and Christiaens (2009) did find positive and significant relationship between CSR disclosure and profitability and the size of the company on 117 listed Belgium companies for the year 2005.

21st century owing competitiveness to all national and multinational companies to put the element of CSR in their decision making and to deliver high corporate citizenship. No doubt, to involve in CSR require substantial amount of money and resources to be allocated hence, will lower the company's profitable margin. Therefore, top management of the company need to ensure the resources are being allocated and fully utilized in effective and efficient manner. Although there are lacking of commitment towards CSR practice in emerging countries, specifically in Malaysia, however, it is gradually increase due to emergences of Non-Governmental Organization (NGO) and consumer interest group. In addition, the importance of CSR has triggered. The empirical analysis of the relationship between CSR and CFP has yet to provide a convincing link between variables (Mwangi and Jerotich, 2013). Therefore, this study attempts to contribute in the area and may facilitate more intensive research on CSRD and financial performance, specifically in Malaysia as a developing and emerging country.

The aim of this study is to examine the relationship between CSR disclosure and the company financial performance (CFP) among Malaysian public listed companies (PLCs) specifically focus on consumer product and
Trading and services industry only. The selection of these industries was due to their direct relationship with the society and environment. Furthermore, it is significant to the Malaysian's economy contribution since it contributes more than half of the Malaysian Gross Domestic Product (GDP) (EPU, 2015).

Corresponding author's email: nor_1143@yahoo.com

In particular, the objectives of this study:

a) To examine the extent and quality of CSR disclosure practices between consumer product and trading and services sectors
b) To investigate the relationship between CSR disclosure and financial performance of consumer product sectors
c) To investigate the relationship between CSR disclosure and financial performance of trading and services sectors

DATA AND RESEARCH METHOD

Sample and Data Size

This quantitative study involved in examining the annual reports and stand-alone sustainability reports for trading and services and consumer product sector listed on Bursa Malaysia for the year 2003 – 2013. The sample selection is done by using simple random sampling. The data for this study were obtained from the web page of Bursa Malaysia and Thompson Reuters data stream. The total population listed on Bursa Malaysia comprises 186 companies for trading and services and 126 companies for consumer product. However, after data cleaning of non-availability or missing data, it made up of 38 companies and 31 companies used as samples for trading and services and consumer product respectively. The annual reports and stand-alone sustainability reports were employed to gather information as it is the most source of corporate reporting (Jenkins and Yakovlena, 2006) and the most accessible source of information either in hard copy or electronic publications for Malaysian public listed companies (Rusnah, Mustaruddin and Norhayah, 2006).

Based on the research objectives of the study, the following research questions and hypotheses statements are created. The hypotheses of this study are based on the propositions that good CSR practices will directly affect corporate performance in the companies.

Research questions:

a) What is the extent and quality of CSR disclosure practices between consumer product and trading and services sectors? There is difference between the extent and quality of CSR disclosure practices between consumer product and trading and services sectors.
b) What is the relationship between CSR disclosure and financial performance of consumer product sectors?
c) What is the relationship between CSR disclosure and financial performance of trading and services sectors?

Proxies and Measurement of Variables

Dependent variables

The dependent variable (DV) used in this study is corporate financial performance (CFP) namely; return on asset (ROA) and Tobin's Q (Q).

Corresponding author's email: nor_1143@yahoo.com

a. *Return on asset (ROA)*: ROA indicates the amount of net profit generated by each ringgit invested in assets. It provides investors with an indicator of how effectively the company is converting asset to invest in net income.

ROA = Net profit / Total assets

b. *Tobin's Q ratio (Q)*: Based on Cavaco and Crifo (2009), Q ratio is a neutral indicator which is more sensitive to variations "that may be independent of the operations and social activities of and which can affect the market values.

Tobin's Q = (Equity + Total debt) / Total assets

Independent variables

The independent variables (IV) in this study is measured through CSR disclosure (CSRD), which represents CSR activities disclosure of information in the annual reports or stand-alone sustainability reports. This study applied content analysis to quantify the level of social responsibility information disclosed by selected samples. The disclosure checklist was constructed based on the CSR framework developed by Bursa Malaysia namely, community, environment, marketplace and workplace. The disclosures were rated based on the existence or nonexistence, and the degree of specificity of each item. Haslinda and Lehman (2008) rating approach had been adopted in examining the level of extensiveness in respect of social responsibility disclosure , in which the information was grouped under five categories, namely, general information, qualitative information, quantitative information and a combination of qualitative and quantitative information. These rating scales permit for the measurement of quantified social information (words, sentences or numbers) and also for non-quantifiable disclosures covering graphs, photographs, charts etc.

The total maximum score is set at 288. The maximum for each of the dimensions is 80 for community disclosures, 68 for environment disclosures, 52 for marketplace disclosures and 88 for workplace disclosures. A complete checklist detailing the categories and items of disclosure is attached. Therefore, the development of total CSR disclosure (CSR index),

$$CSRDi = \frac{n}{N} \times 100\%$$

Where CSRDI represents the Corporate Social Responsibility Disclosure Index, n is the number of items which were disclosed by the companies for each of the CSR dimension and N indicates number of items expected disclosed by the companies.

Control Variables

a. *Company size*: It is relevant to consider company size as a control variable because evidence proved that smaller companies may not demonstrate as many socially responsible behaviors as compared to larger companies. According to Waddock and Graves (1997), as they mature and grow, companies attract more attention from external constituents and need to respond more openly to stakeholders demands. For the purpose of this study, the company size is measured through the total assets of company to show the company's liquidity. By including company size as the control variable, it allows the identification of factors other than CSR could influence the CSR performance measure.

Corresponding author's email: nor_1143@yahoo.com

b. *Earnings per share (EPS)*: EPS shows the earnings of how much profit is earned during the period of one year on behalf of each outstanding shares of common stock. Big companies might generate greater income as compared to a smaller company who be doing better in terms per unit of ownership. Thus, EPS is a better performance indicator on how efficient a company in generating earnings from each of the outstanding shares of common stock.

EPS= Net income/Number of outstanding shares of common stock

Statistical Analysis

In lieu to determine the relationship between CSR disclosure and organizational performance in terms of financial performance, a different statistical analysis were used. The Eviews software version 9.0 was used in the analyses of the extent and quality of the information disclosed in the annual reports or stand-alone sustainability reports. The measurement of central tendency provides the mode, median and mean of data collected.

Furthermore, the regression of model to examine the relationship between variables is using panel data which is the observation on cross sectional and time series. For the purpose of this study, Generalized Least Squared (GLS) method is used. The fixed effect (FE) and random effect (RE) model is employed in order to differentiate between models; whether unabsorbed individual effects are correlated with the regressors which is the case of fixed effects, or not in the models, as in the case of the random effects model. In a case of FE model, the intercept in the regression model is allowed to differ among individuals in recognition of the fact that each individual or cross section unit may have some special characteristics of its own. As a result, the Hausman test will be conducted in choosing which model is more precise. The model specification are as follows:

$$\text{Model 1 : } ROA_i = c + COM + ENV + MKT + WORK$$

$$\text{Model 2 : } Q_i = c + COM + ENV + MKT + WORK$$

Where, for company$_i$;
ROA = Return on asset
Q = Tobin's Q ratio
COM = Community disclosure
ENV = Environment disclosure
MKT = Marketplace disclosure
WORK = Workplace disclosure

ANALYSIS AND FINDINGS

Descriptive Findings

Table 4.1 provides the descriptive findings of CSR dimension of CSR disclosure disclosed in the annual reports or stand-alone sustainability reports of the sample. The analysis found that

Corresponding author's email: nor_1143@yahoo.com

the total CSR disclosure, CSR index (CSRD) for trading and services companies was 0.05 while the minimum and the maximum values were 0 and 0.70 respectively. Amongst the four CSR dimension, community involvement had the highest mean score of 0.07, followed by the workplace (0.06) and environment (0.54). The least mean score was related to marketplace (0.04).

CSR Disclosures	Trading & Services					Consumer Products				
	Obs	Min	Max	Mean	SD	Obs	Min	Max	Mean	SD
Community	418	0.00	0.45	0.07	0.09	341	0.00	0.57	0.06	0.09
Environment	418	0.00	0.70	0.05	0.09	341	0.00	0.79	0.06	0.13
Marketplace	418	0.00	0.50	0.04	0.08	341	0.00	0.48	0.05	0.08
Workplace	418	0.00	0.61	0.06	0.09	341	0.00	0.64	0.06	0.11
CSR index	418	0.00	0.44	0.05	0.07	341	0.00	0.56	0.06	0.09

Table 4.1 Descriptive statistics for CSR dimension of CSR disclosure

In respect to descriptive findings for consumer products, the total of CSR disclosure, CSR index (CSRi) was 0.06 while the minimum and the maximum values were 0 and 0.79 respectively. The community involvement, environment and workplace had the highest mean score of 0.06 and the least mean score was related to marketplace (0.05).

These results are consistent with Ibrahim, Zam Zuriyati, Jamal and Norlia (2013) who revealed that in terms of number of sentences, consumer products companies disclosed more information compared with plantation industry of 1,885 sentences and 636 sentences respectively.

Furthermore, Table 4.2 presents the frequency of each of CSR disclosure disclosed in the annual reports or stand-alone sustainability reports. Consistent with the mean score outcomes in Table 4.1, trading and services companies had a high frequency of disclosure in respect to community contribution information (32%), followed by workplace (26%) and environment (22%). The least frequent disclosure was related to marketplace information (20%). These findings are consistent with Mohamad Hafiz, Jamaliah and Nuruu Ain (2015) revealed that the highest frequency score was community disclosure.

CSR Disclosures	Trading & Services		Consumer Products	
	Obs	Total CSRD (%)	Obs	Total CSRD (%)
Community	418	32	341	25
Environment	418	22	341	26
Marketplace	418	20	341	22
Workplace	418	26	341	27

Table 4.2 Frequency of CSRD for each CSR dimension

For consumer products industry, the highest frequency of disclosure was in regard to workplace disclosure (27%), followed by the environment (26%) and community involvement (25%). However, similar with trading and services industry, they disclosed fewer information on marketplace engagement (20%). Such a result implied that the motivation for marketplace disclosure practice in Malaysia was at inconsistency. The evident low level of CSR disclosure on marketplace engagement are due to absence of a appropriate reporting framework, the costs related to the reporting process, doubts of how readers will respond as well as the absence of government and primary stakeholders' pressure (Thompson and Zakaria, 2004).

Corresponding author's email: nor_1143@yahoo.com

A Spearman Rank Order (SRO) test was performed to test the existence high inter-correlation among the independent variables. Based on the Spearman correlation results revealed that none of the correlation results among the independent variables exceed the 0.70 cut-off points as indicated by Sekaran and Bougie (2010). Moreover, the unit root test of Levin, Lin and Chu (LLC) and Philips Pherron (PP) hypothesize that all variables rejected the null hypothesis and are significant at 1 per cent level of confidence. The rejection of null hypothesis indicates that the data series is stationary at first difference. Hence, regression of analysis was conducted by using the transformed data at first difference.

Relationship between CSR disclosure and CFP

The estimation findings of the relationship between CSR disclosure and CFP are revealed in Table 4.3. There are two dependent variables which was used to measure the relationship; return on asset (ROA) and Tobin's Q ratio (T'Q). As a result, for trading and services industry, Model 1 when ROA as the dependent variable, the random effect (RE) is more perfect than the fixed effect (FE) model since the Hausman test is insignificant. It is presumed that the variation across entities to be random and uncorrelated with the predictor variables included in the model. The signs of independent variables have mixed findings except for the total size (LTA) have a negative relationship. The Adjusted R^2 shows that financial performance is explained by total CSR disclosure (CSRD) by 23.7 per cent only. In Model 2 when Tobin's Q ratio (T'Q) used as the dependent variable, it is also found that the random effect (RE) is more appropriate than the fixed effect (FE) model since the Hausman test is insignificant. In general, the CSR disclosure (CSRD) is positive and significantly associated to financial performance at least at 5 per cent level. Adjusted R^2 shows that financial performance is explained by the CSRD and other explanatory variables with estimation at 25 per cent.

	Trading & Services		**Consumer Product**	
Variable	Model 1 ROA	Model 2 T'Q	Model 1 ROA	Model 2 T'Q
	RE Model Coefficient (p-value)	RE Model Coefficient (p-value)	RE Model Coefficient (p-value)	RE Model Coefficient (p-value)
C	**0.08618** (0.0127)**	**0.0086** (0.0000)***	**0.0771** (0.0000)***	**0.0080** (0.0000)***
CSRD	**0.0504** (0.0105)**	**0.0023** (0.0352)**	**0.0300** (0.0315)**	**-0.0014** (0.0373)**
TA	**-0.0060** (0.0285)**	**-0.0002** (0.0891)*	**-0.0214** (0.0140)**	**-0.0002** (0.0477)**
EPS	**5.8476** (0.0177)**	**0.0405** (0.0453)**	**0.0842** (0.0336)**	**0.0005** (0.0377)**
Adjusted R^2	0.2368	0.2500	0.3954	0.2058
F- statistic	14.7886	0.6483	0.9398	0.3222
Durbin-Watson	2.033	1.054	1.4732	1.056

Note: *** Significant at 1% level, ** Significant at 5% level, * Significant at 10% level

Table 4.3 Relationship between CSR disclosure and financial performance using panel data

Corresponding author's email: nor_1143@yahoo.com

The findings showed that correlation between CSR disclosure and financial performance for consumer product industry, Model 1 when ROA as the dependent variable, the random effect (RE) is more precise than the fixed effect (FE) model. The signs of independent variables have mixed findings as for the total size (LTA) have a negative relationship. The Adjusted R^2 shows that financial performance is explained by total CSR disclosure (CSRD) by 39.5 per cent only. In Model 2 when Tobin's Q ratio (T'Q) used as the dependent variable, it is also found that the random effect (RE) is more appropriate than the fixed effect (FE) model since the Hausman test is insignificant. In general, the CSR disclosure (CSRD) is positive and significantly associated to financial performance at least at 5 per cent level. Adjusted R^2 shows that financial performance is explained by the CSRD and other explanatory variables with estimation at 20.6 per cent.

The detail analysis which is based on the dimension of CSRD score is revealed in Table 4.4. The measurement for the financial performance is retain with ROA and Tobin's Q ratio (T'Q) as the dependent variable. In conclusion, the (RE) is more precise than the fixed effect (FE) model since the hypothesis testing accept the null hypothesis.

It is found that CSRD for trading and services industry, Model 1 when ROA the dependent variable, community (COM), environment (ENV) and marketplace (MKT) is positive and significantly related to financial performance whilst workplace (WORK) is negatively significant associated with financial performance. For Model 2, when T'Q ratio used as the dependent variable, community (COM), environment (ENV) and marketplace (MKT) and workplace (WORK) is positive and significantly related to financial performance, whilst size of the company (TA) is negatively significant associated with financial performance.

	Trading & Services		**Consumer Product**	
Variable	Model 1 ROA RE Model Coefficient (p-value)	Model 2 T'Q RE Model Coefficient (p-value)	Model 1 ROA RE Model Coefficient (p-value)	Model 2 T'Q RE Model Coefficient (p-value)
C	**0.1142** (0.0000)***	**0.0086** (0.0000)***	**0.0738** (0.0000)***	**0.0080** (0.0000)***
COM	**0.0184** (0.0081)***	**0.0008** (0.0086)***	**0.0453** (0.0090)***	**0.0014** (0.0305)**
ENV	**0.0191** (0.0055)***	**0.0005** (0.0073)***	**0.1249** (0.0038)***	**0.0002** (0.0066)***
MKT	**0.0138** (0.0069)***	**0.0008** (0.0072)***	**-0.0553** (0.0022)***	**0.0005** (0.0077)***
WORK	**-0.0135** (0.0071)***	**0.0002** (0.0006)***	**0.1533** (0.0021)***	**-0.0004** (0.0154)**
TA	**-0.0116** (0.0000)***	**-0.0002** (0.0133)**	**-0.0185** (0.0607)*	**3.65E-05** (0.0909)*
EPS	**11.9900** (0.0000)	**0.0410** (0.0452)**	**0.0758** (0.0909)*	**8.42E-05** (0.6768)
Adjusted R^2	0.2147	0.2398	0.3103	0.1905
F- statistic	20.1450	0.3222	1.536	0.9699
Durbin-Watson	1.6471	1.056	1.336	1.755

Table 4.4 Relationship between CSR disclosure and financial performance using panel data

Corresponding author's email: nor_1143@yahoo.com

However, the adjusted R^2 at 21.5 per cent and 23.9 per cent is respectively explained the financial performance through CSRD measurement.

In a same time, for consumer product industry, Model 1 when ROA the dependent variable, it shows that community (COM), environment (ENV) and workplace (WORK) is positive and significantly related to financial performance, similar with the trading an services industry. However, dimension for marketplace (MKT) is negatively significant associated with financial performance. For Model 2, when T'Q ratio used as the dependent variable, all the CSR dimensions; community (COM), environment (ENV) and marketplace (MKT) and workplace (WORK) is positive and significantly related to financial performance.

DISCUSSION AND CONCLUSION

This study attempts to examine the comprehensiveness of CSR disclosure disclosed by the companies via the annual report or stand-alone sustainability report. Generally, it can be concluded that public listed companies in Malaysia have started to realize the need and importance of disclosing social responsibility information as a medium to communicate with stakeholders and community. It indicated a development signal for CSR disclosure in Malaysia when there are companies who disclose the CSR information in a special report namely; stand-alone sustainability report. However, many companies were still at low level in disclosing the CSR activities which score below 50 per cent for each dimension for both of the industry. Furthermore, it revealed that there was a gap between the minimum and maximum of CSR disclosure in the report.

From the study it can be concluded that CSRD have an association with the financial performance (CFP). For instance, both of the industry show a positive and significantly relationship between community (COM) and CFP. It shows that there is no negative response from stakeholders when the companies spend specific financial resources to support the society development. The companies believed that in order to maintain the company's reputation and image, it is imperative for them to gain support and being recognize by the society. This includes education sponsorship, rural area development, philanthropic events, national events and many more to mention.

Although it is believed that engagement in environment matters may cause higher cost and a lot of environmental management guidelines to follow, it is found that environment (ENV) and marketplace (MKT) dimension is positive and significantly related to CFP for both industry. It indicate that involvement on environment issues and investing on product development help to add sufficient value to the reputation of the companies.

The workplace (WORK) dimension is positively significant associated with the CFP when ROA as dependent variable for consumer product. It can be concluded that, investing on employee's development and incentives scheme able to increase the morale of employees to put high commitment in their productivity. These include loyalty award, training and development for employees, health and safety and others (Waddock and Graves, 1997). However, the workplace (WORK) dimension have a negative associated with the CFP for trading and services.

Corresponding author's email: nor_1143@yahoo.com

DOI: 10.1051/shsconf/20173600003

In general, the findings confirmed that there is limited evidence of a significant relationship between CSRD and CFP in a long-term period. This is also revealed that there is still limited number of companies consistently in disclosing CSR activities in the reports. Hence, this results may be inconsistent for both industry due to different industry and companies may contribute and disclose their CSR activities in a different way. The involvement of CSR activities may also may be determined by the pressure from the shareholders or authority body and policy makers. The government agencies could also support companies to involve in CSR activities and later disclose them.

The limitation of this study is on the inconsistency of results obtained by using different proxy for the CFP for two different industry. In addition, the sample is taken at randomly regardless the size of the company. This include small size of company in the industry who assume that disclosing CSR activities could affect their cash flows reporting. However, the nature of Malaysia guidelines does not have any specific amount or area for the company to disclose CSR activities and this make content analysis more viable option.

REFERENCES

Auger, P., Devinney, T.M., and Louveniere, J.J. (2003). What will consumers pay for social product features? *Journal of Business Ethics, vol. 43, no.3, pp. 281-304*

Cavaco, S. And Crifo, P (2009). The CSP-CFP missing link: Complementarity between environmental, social and governance practices? *Journal of Economic Literature*

Choi, J., Kwak, Y., & Choe, C., (2010). Corporate Social Responsibility and Corporate Financial Performance: Evidence from Korea. *Australian Journal of Management, vol. 35, no.3, pp. 291-311*

Dalilawati Zainal, Norhayah Zulkifli and zakiah Salleh (2013). Corporate social Responsibility in Malaysia: A Comparison between Shariah and Non-Shariah Approved Firms, *Middle-East Journal of Scientific Research, vol. 15, no. 7, pp. 1035-1046*

Everaert, P., Bouten, L., Van Liedekerke, L., De Moor, L. and Christiaens, J. (2009). Discovering Patterns in Corporate Social Responsibility (CSR) Reporting: A transparent framework based on Global Reporting Initiatives (GRI) Sustainability Reporting Guidelines. *Working papers of Faculty of Economics and Business Administration, Ghent University, Belgium*

Haslinda, Y., and Lehman, G. (2008), International differences on corporate environmental accounting developments: A comparison between Malaysia and Australia accounting and finance in transition, *vol.4, Greenwich University press, pp. 92-124*

Ibrahim Thamby Chek, Zam Zuriyati Mohamad, Jamal Yunus and Norlia Mat Nawawi (2013). Corporate Social Responsibility (CSR) Disclosure in Consumer Products and Plantation Industry in Malaysia, *American International Journal of Contemporary Research, vol. 3, no. 5, pp. 118-125*

Corresponding author's email: nor_1143@yahoo.com

DOI: 10.1051/shsconf/20173600003

Inoue, Y & Lee, S (2010). 'Effects of different dimensions of Corporate Social Responsibility on corporate financial performance in tourism-related industries', *Tourism Management, vol.32, no.2, pp. 790-804*

Ioannuo, I. & Serafeim, G (2012). What drives corporate social performance? *International evidence from Social, Environmental and Governance Scores, HBS Working Paper*

Jenkins, H & Yakovleva, N (2006). 'Corporate Social Responsibility in the mining industry: Exploring trends in social and environmental disclosure', *Journal of Cleaner Production, vol.14, no.3, pp.271-84*

Mahoney, L.S. & Robert, R.W (2007). Corporate Social Performance, financial performance and institutional ownership in Canadian firms, *Accounting Forum, vol. 31, pp. 233-253*

Margolis, J.D., Elfenbein, H. & Walsh, J.P. (2007). Does it pay to be good? A Meta analysis and redirection of reserach on the relationship between Corporate Social and financial performance, *working paper*

McWilliams, A., Siegel, D.S., & Wright. P.M, (2000). Corporate Social Responsibility and Financial Performance: Correlation or misspecification? *Strategic Management Journal, vol. 21, no.5, pp. 603-609*

Miller, J.G. (1994). Cultural Diversity in the morality of caring. Individually oriented versus duty-based interpersonal moral codes, *Cross Cultural Research, vol. 28, no.1, pp. 3-39*

Mustaruddin Saleh (2009). Corporate social responsibility disclosure in an emerging market: A longitudinal analysis approach. *International Business Research*, *vol. 2 no.1, pp. 131-141*

Mustaruddin Saleh, Norhayah Zulkifli & Rusnah Muhamad (2011). Looking for evidence of the relationship between corporate social responsibility and corporate financial performance in an emerging market, *Asia-Pacific Journal of Business Administration, vol. 3 no.2, 2011, pp. 165-190*

Mwangi, C. I., and Jerotich, O.J. (2013). The relationship between corporate social responsibility practices and financial performance of firm in manufacturing, construction and allied sector of the Nairobi Securities Exchange. *International Journal of Business, Humanities and Technology, vol. 3(2)*

Norhawani, Mustaffa & Norashfah (2011). CSR disclosures and its determinants: evidence from Malaysian government link companies, *Social Responsibility Journal, vol. 7, no.2, pp.181 – 201*

Orlitzky, M., James, G., Schmidt, F.L., & Rynes, S.L. (2003). Corporate Social Responsibility and Financial Performance: A Meta-Analysis, *Organization Studies, vol. 24, pp. 403-441*

Othman, S., Darus, F. and Arshad, R. (2011). The influence of coercive isomorphism on corporate social responsibility reporting and reputation. *Socially responsibility Journal, vol.1, no. 1, pp.119-132*

Corresponding author's email: nor_1143@yahoo.com

Ruf, B.M., Muralidhar, K., Brown. R.M., Janney, JJ & Paul, K. (2001). An empirical investigation of the relationship between change in corporate social performance and financial performance: A stakeholder theory perspective, *Journal of Business Ethics, vol. 32, pp. 143-156*

Sheveta Kapoor & Sandhu, H.S (2010). Does it pay to be socially responsible? An empirical examination of impact of corporate social responsibility on financial performance. *Global Business Review, vol. 11, no.85, pp. 185-208*

Thompson, P. and Zurina, Z. (2004). Corporate social responsibility reporting in Malaysia progress and prospects. *Journal of Business Corporate Citizenship*, *vol. 13, pp. 125-136*

Tsoutsoura, M. (2004). Corporate social responsibility and financial performance: In possession of Centre of Responsible Business, *Working paper series, Paper 7, University of California, Berkeley*

Turban, D.B. and Greening, D.W. (1997). Corporate Social Performance and Organizational Attractiveness to Perspective Employees, *Academy of Management Journal, pp. 658-672*

Corresponding author's email: nor_1143@yahoo.com

THE DEVELOPMENT OF AUDITING ARENA AND CORPORATE GOVERNANCE LANDSCAPE IN MALAYSIA: AN INITIATIVE TO COMBAT FRAUD AND CORRUPTIONS

Nur Adura Ahmad Noruddin*, I.K. Norziaton
Faculty of Accountancy, Universiti Teknologi Mara Puncak Alam

ABSTRACT

This research provides two phases of discussions on the evolution in auditing arena and corporate governance (CG) landscape in Malaysia. Phase 1: Pre-Malaysian Code of Corporate Governance (MCCG) (1957-2000) and Phase 2: Post-Malaysian Code of Corporate Governance (2001-2015). MCCG was designed to increase oversight and regulation of the accounting profession and to strengthen CG practices and increase transparency of financial audits. The emergence of audit failures has focused the world's attention on accounting standards and the role of auditors. The cause of corporate collapse was basically due to lack of professionalism in auditing coupled with scandals on fraud and corruptions. Moreover, the quantum of fraud has increased over the past years in Malaysia and has become more sophisticated. Due to this, the Government has put all the efforts to ensure that audit and CG practices are able to comply with stringent and rigorous standards imposed by the statutory bodies as to enhance audit quality and CG practices in line with international benchmarks. Thus, this paper attempts to analyze the evolution of auditing and CG practices as initiatives taken by the Government of Malaysia in order to combat fraud and corruptions.

Keywords: Auditing, Auditors, Corporate Governance, Fraud, Corruptions

INTRODUCTION

This paper provides knowledge on the development of auditing arena and CG landscape as part of the important initiatives taken by Malaysian Government in combating fraud and corruptions. In accordance with a report by KPMG Malaysia Fraud, Bribery and Corruption Survey 2013, the quantum of fraud has increased over the past three years and has become more sophisticated. Due to this reason, the Government of Malaysia has put all the effort to ensure that the audit and CG practices are able to comply with the stringent and rigorous standards imposed by the statutory bodies as to enhance audit quality and CG practices in line with the international benchmarks. Hence, this paper provides an overall view on the evolution of auditing and CG practices as initiatives taken by the Government in order to curb and combat fraud and corruptions.

To better understand the auditing evolution and CG practices, a study was conducted by Azham *et al.* (2006) which examine various episodes in the Malaysian auditing saga. The duration of their study covering the period of the first 40 years post-independence in 1957 until 1997. They divided the analysis into three episodes denoting three different timeframes with reference to the dynamic changes in its political and socio-economic environment. However, this paper on

*Corresponding author's email: nurad185@salam.uitm.edu.my

the other hand, will add on to the literature by extending the timeframe starting from 1957 to 2015. It will focus on the establishment of accounting and auditing statutory bodies and any other significant changes or amendments to the regulations through these years as initiatives taken by the Government in combating frauds and corruptions in Malaysia. This paper is organized into two phases, Phase 1: Pre-Malaysian Code of Corporate Governance (MCCG) (1957-2000) and Phase 2: Post-Malaysian Code of Corporate Governance (2001-2015). Each phase will describe the important events in the development of the auditing profession and CG practices in Malaysia. Besides that, this analysis will also enable one to understand, analyse and even interpret the initiatives taken by the Government in curbing fraud and corruptions.

Objectives of the paper

1. To provide evidences on auditing arena and CG landscape that have changed over the years to meet the needs and expectation of the society and audit profession.
2. To examine the auditing arena and CG landscape experiences in order to assess their significance and to draw reasonable lessons in combating fraud and corruptions.

METHODOLOGY

This paper is produced based on extensive reviews of related literatures including government reports and policy documents, academic papers and other publications. Sources of readings related to scandals and corporate collapses due to wrongdoings of board of directors, the unresolved ethics and independence of auditors and issues pertaining to fraud and corruptions were also referred to.

The Development of Auditing Arena and CG Landscape in Malaysia

The activities of the auditing profession in recent years are considered to be tremendously eventful. There have been many occasions where external auditors have failed in performing their duties to exercise independence and objectivity when reporting the results of their audit findings. In addition, issues of governance in corporations have become central to the public policy agenda. This has been reinforced by the shocking corporate scandals exhibited by renowned conglomerates such as Enron in 2001 and WorldCom in 2002 in the US. Thus, the combination of auditing and CG practices which has contributed to the uniqueness of the corporate environment in Malaysia will be discussed below.

Phase 1: Pre-Malaysian Code on Corporate Governance (MCCG) (1957-2000)

In order to impede the assessment of auditing arena and the emergence of CG in Malaysia, this review will start off with the establishment of the accounting and auditing statutory bodies and regulations, focusing on the post-independence development from 1957 to 1987, as shown in Figure 1 below.

*Corresponding author's email: nurad185@salam.uitm.edu.my

Phase 1: Pre-MCCG (1957-2000)

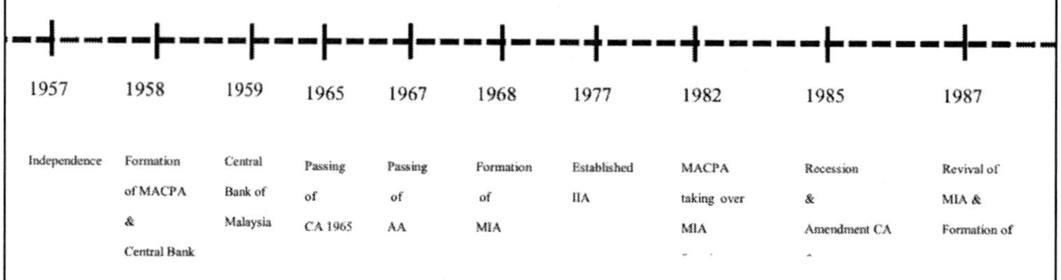

The Development of Malaysian Auditing Arena and CG Landscape from 1957 until 1987

Figure 1 Source: 1957 to 1968, 1985 & 1987 taken from Azham *et al.*, (2006) & 1977 & 1982 from current study

The Malaysian Association of Certified Public Accountants (MACPA) was established in 1958 with the main idea is to promote the status and development of accounting profession in Malaysia. On 23 October 1958, The Central Bank of Malaya Ordinance 1958 (CBO) (Central Bank of Malaya until the formation of Malaysia in 1963) was enacted while the Central Bank of Malaysia was established on 24 January 1959. At the same time, the Banking Ordinance, 1958, which provided for the licensing and regulation of the business of banking in the Federation of Malaya also came into force. Later in 1965, the Companies Act was established which sets out the rules and regulation on accounts and audit that need to be complied with by all companies in Malaysia. The Act specifically spells out the guidelines need to be addressed by auditors in terms of being independent and powers to perform such duties.
In 1967 the Government has approved the Accountants Act 1967 and the establishment of Malaysian Institute of Accountants (MIA) in 1968. It regulates the practice and promotes the interest of the accounting profession. At the same time, MIA's function is also to set on the qualification of members and to register qualified accountants under its membership. However, the statutory body was remained dormant for two decades due to the fact that it limits its role to only registering accountants. Failing to execute its function successfully, in 1982 MACPA has taken over the responsibility.

The Institute of Internal Auditors Malaysia (IIA) was set up in 1977. The main function of IIA is to assist management in ensuring that proper internal control systems are in place and the operation of the company are carried out efficiently, economically and effectively. The world economy was experiencing recessions in 1985 and following the recession, a substantial revision of the Companies Act 1965 was commenced which incorporates best accounting standards and practices, encouraging higher level of disclosures and accountability to minority shareholders including foreigners. The amendments of the Companies Act 1965 have given the auditors further responsibility towards financial statement audit. They are required to perform complete checking of transactions and to prepare a true and fair view of audited accounts.

*Corresponding author's email: nurad185@salam.uitm.edu.my

DOI: 10.1051/shsconf/20173600004

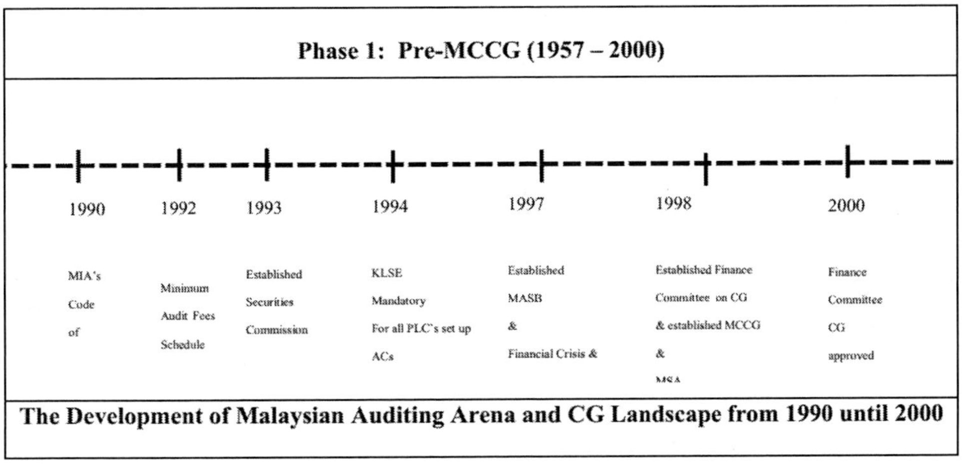

Figure 2 Source: 1990, 1992 & 1997 taken from Azham *et al.*, (2006) & the remaining years are from current study

Figure 2 of the paper still draws attention to the Pre-MCCG (2000) which covers a timeframe from 1990 to 2000, as shown above. In this particular period, the functions of external auditors are to ensure the truthfulness and trustworthiness of the financial statements of their clients. Auditors are expected to enhance the credibility of the financial statements and also to provide value added services such as reporting on irregularities, identifying business risk and advising management on the internal control environment (Boynton & Johnson, 2005). This period begins in 1990 on MIA Code of Ethics. It clearly states the primary duty of an external auditor is to prevent, detect and correcting fraud and at the same time to evaluate and report the results if they manage to recognise fraud and error that may materially affect the financial statements.

MIA under the Accountants Act 1967 was entrusted to promote and regulate the accounting profession. In doing so, the institute has proposed a schedule for determining audit fees in 1992. The intention was to harmonize the fee setting basis and to minimize disparity in fees schedule (*Akauntan Nasional,* 1993). The schedule was developed in response to numerous requests by members in view of maintaining high costs on accountants' staff and offices. They are also in the opinion that the scale has become essential to the auditing profession. The purpose of having a standard scale of fees for audit services are to prevent undercutting and thus protecting the auditing industry as well. As stated by the MIA's Public Practice Committee Chairmans 'in the past, there was no guideline whatsoever to say how much a firm should charge' (*Akauntan Nasional*, 1993). However, the audit fee schedule was only approved by MIA council to be effective beginning of April, 1993 but the schedule was ended up to be as guidelines for audit practitioners in 1994 (*Akauntan Nasional*, 1994). The issue of charging low audit fees is still an issue until now. Even though the MIA has provided guidelines for the audit pricing but it has not been properly adopted by most of the audit firms due to lack of enforcement by the regulatory authority (Lee et al., 2009). As a consequence, the auditors perceived that audit pricing in Malaysia are considered to be at the lower side in comparison with other countries in the region which are at a similar stage of development (Lee et al., 2009).

*Corresponding author's email: nurad185@salam.uitm.edu.my

The Securities Commission (SC) was established in 1993, a regulatory body for the capital market. Its main function is to regulate all matters pertaining to securities and futures market and to ensure enforcements of securities and futures laws (Khoo, 2003). SC identifies and builds up front line regulators for instance the Kuala Lumpur Stock Exchange (KLSE, now Bursa Malaysia) and self-regulatory organizations such as professional bodies which give emphasis on the auditor's responsibilities in financial statement reporting. In 1991, the MIA, in conjunction with the Malaysian Institute of Certified Public Accountants (MICPA) and the Malaysian Institute of Internal Auditors, was proposing that an audit committee to be made mandatory for companies seeking listing at the KLSE. However, later in 1994 the KLSE listing requirements has made it mandatory that every company to form an Audit Committee comprising of a majority of directors that were independent of the company's management.
In 1997, Malaysia became the first Asia country to set up its own accounting standard setting body, the Malaysian Accounting Standards Board (MASB) (*Akauntan Nasional*, 2003). It carries a function of an independent standard setting body for establishing financial accounting and reporting standards in Malaysia. It was set up as to bring harmonization between national and international accounting standards. MASB has been developing standards with reference to the accounting standards issued by the International Accounting Standards Board (IASB). With regards to the convergence with International Financial Reporting Standards (IFRSs) in Malaysia, MASB has taken all necessary steps in promoting high quality and transparent financial reporting.

In making further and better provision for the prevention of corruption and for matters necessary thereto and connected therewith, the Anti-Corruption Act 1997 which formerly known as Act 575 was formed. The Anti-Corruption Agency was then established with the function to investigate and prosecute corruptions in the public and private sectors (The Commissioner of Law Revision, Malaysia, 2006). Affords have been put forward by the Government on the development of CG in Malaysia. The initiative started off with the establishment of Finance Committee on Corporate Governance in 1998 which consists of both government and industry. The recognition on CG was evidenced by the released of the MCCG by the Committee in March 2000. The codes focus on four main areas that are board of directors, director's remuneration, shareholders and accountability and auditing. In March 1998, the High Level Finance Committee on Corporate Governance established the Malaysian Institute of Corporate Governance (MICG). It is a non-profit public company limited by guarantee, with founding members consisting of the Federation of Public Listed Companies (FPLC), MIA, MICPA, Malaysian Institute of Chartered Secretaries and Administrators (MAICSA), and Malaysian Institute of Directors (MID). MICG's mandate was to raise the awareness and practice of CG in Malaysia.
Prior to 1998, Malaysia had no standardised set of approved auditing standards. In this regard, MIA has been adopting most of these standards from the International Federation of Accountants (IFAC) as the main criteria to assess the appropriateness of accounting treatment. Later in July 1998, MIA had approved standards on auditing known as Malaysian Standards on Auditing (MSA) which is strictly adhere to the International Auditing Standards (IASs). These guidelines and recommendations on auditing together with the adopted IFAC standards have form the Generally Accepted Auditing Principles (GAAP) in Malaysia. Thus, professional accountants in public practice are expected to comply with these standards in the performance of their audits under all reporting framework. Transparency International Malaysia (TI-M) was also established in 1998. As an independent, non-governmental and non-partisan organisation committed to the fight against corruption, the mission is to eradicate corruptions and promote transparency, integrity and accountability throughout society.

*Corresponding author's email: nurad185@salam.uitm.edu.my

Through these establishments, it is evidenced that the Government has continuously putting efforts in combating fraud and corruptions in Malaysia.

Minority Shareholder Watchdog Group (MSWG) was another Government's initiative set up in 2000 as a body that looks after the interest of minority shareholders. Act as a non-profit organization, it is intended to protect minority shareholders' right, increase awareness among them of their responsibilities, and ensure they understand their role in an organization. This was due to the fact that minority shareholders were left in the lurch when companies collapses and were often the last to know. Therefore, this shareholder activism is said to positively affecting companies in enhancing the shareholder's value (The Star, July 8, 2009).

In short, the auditing developments that took place after independence basically are more concerned with the establishment of accounting and auditing professional bodies which are mainly MIA and MACPA. Throughout these years there are also some changes made on the accounting and auditing regulations which to certain extend have some implications on the role of external auditor in carrying out their audit engagement. During this period, it can be observed that the audit function is not only focusing on preventing and detecting fraud and errors but also assessing the truth and fairness of the companies' financial statements. Later the practice of auditing in Malaysia highlighted the financial reporting framework which includes the standards, legal responsibilities, codes of conduct and reporting requirements that guide the financial statement auditor's work. It also draws the attention to the professional, regulatory and standard setting bodies that directly impact the auditing profession. This auditing arena and CG landscape are an important introduction to the complex and ever-changing environment involving the elements of fraud and corruptions in which financial statement auditing is performed.

Phase 2: Post-Malaysian Code of Corporate Governance (2001-2015)

Sarbanes Oxley Act (SOX) was released in July 2002, the most comprehensive corporate reform and financial market protection as a reaction to the string of financial scandals in the U.S. It was designed with the intention to increase oversight and regulation of the accounting profession after the collapse of Enron and its auditor Arthur Andersen (Cosgrove & Niederjohn, 2008). The Act aimed to strengthen the CG and increase transparency of financial audits in the U.S. However, it does not only affecting the CG issues in the U.S. but also other Asian countries including Malaysia.

As shown in Figure 3 below, the Post-MCCG covers from years 2001 until 2015. With regards to this timeframe, the role of auditors is expected to converge. They are expected to carry out duties up to the expectations of the public, to maintain high standards of professionalism and to uphold a good reputation in auditing profession (Accountants Today, 2008).

*Corresponding author's email: nurad185@salam.uitm.edu.my

DOI: 10.1051/shsconf/20173600004

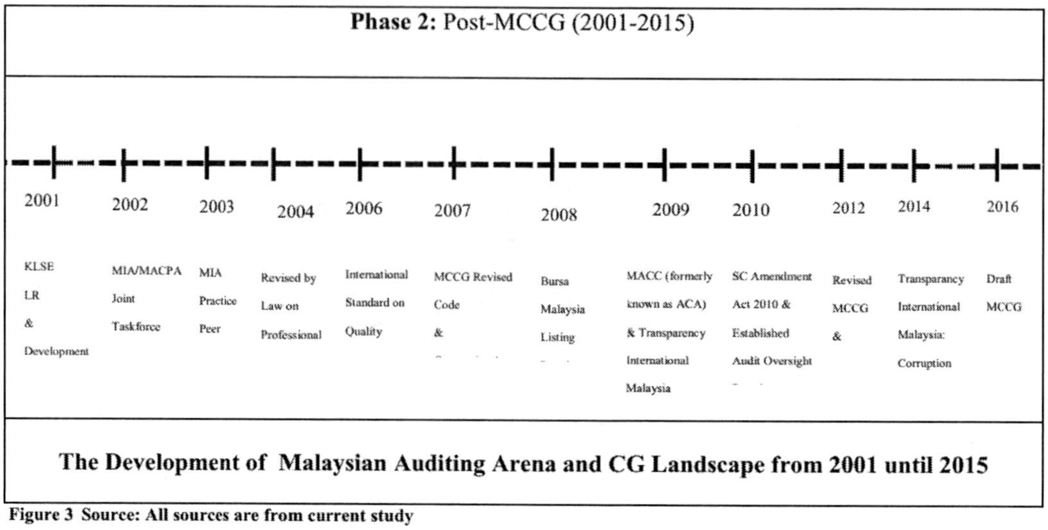

Figure 3 Source: All sources are from current study

The KLSE has revamped its Listing Requirements in year 2001, subsequent to the MCCG's suggestions on the Corporate Governance Best Practices in 2000. The main requirement amongst other requires that all listed companies to disclose a 'CG Statement' in their annual report and its level of compliance with the code. Other than that, companies are also required to disclose a 'Statement on Internal Control' which clearly states the internal control system of the company and to explain the responsibility of the Board in preparing the annual audited accounts. The new Listing requirements also emphasised on the role and composition of audit committee in the company.

In May 2002, the two accounting professional bodies agreed to establish an MIA/MACPA Joint Taskforce on Auditor Independence (MIA 2002). Given that auditor independence and competency is directly unobservable to the Malaysian investing public at the moment, a revised By-Law on Professional Independence was issued in 2004 responding to the SOX issued in the U.S. The main objective of the By-Law is to make it compulsory for all external auditors to be independent of clients for whom audit engagements are conducted (MIA 2004). Whereby, the auditors are expected to separate its non-audit functions and other advisory services since the role of auditors has always been seen as highly associated with such services to audit clients. The MIA's practice review which was introduced in 2003 applies to all member firms that conduct audits. Practice review refers to the evaluation of the work performed by the auditors and of these firms' internal quality controls (Business News, 2014). The work of a practising accountant is to be reviewed by another qualified member of the profession. In other words, the MIA Practice Review Committee is acting as a monitoring mechanism to gauge members' compliance with professional standards and to oversee the implementation of this programme. In areas where the standards are found to be lacking, the committee might give suggestions and recommendations for improvement.

With regards to audit quality, International Standards on Quality Control (ISQC) 1 was introduced in July 2006 by MIA. According to Omar *et. al* (2006), the MIA has urged all audit firms to comply with ISQC 1. Additionally, by implementing ISQC 1 it would enhance the audit quality and preparing those firms to face competitive marketplace. Once again in year

*Corresponding author's email: nurad185@salam.uitm.edu.my

2007, the Securities Commission (SC) released a revised MCCG as to further improve the Malaysia's CG framework. The Revised Code of best practices 2007 intend to intensify the roles and responsibilities of board of directors and audit committee and to ensure that they perform their responsibilities effectively (MCCG, 2007). The revised Code also requires that all public listed companies to carry out their own internal audit functions. The MCCG best practices however became mandatory under the new Bursa Malaysia Listing Requirements (BMLR) in 2008. The revised Listing Requirements was designed to increase oversight and regulation with major provisions on the role of board and audit committees within organization and, the competency of the internal audit function. Following the revision, several important amendments to the Companies Act 1965 have been made. The new provisions are in response to the role and duties of auditors as to promote better auditing practice. The amendment of the Companies Act includes new provision which relates to duty to inform upon ceasing to hold office as auditor (section 172A) and detection of fraud by auditors (section 174 and 174A).

Apart from the above amendments, the whistle blowing law in the Companies Act 1965 has been removed and embodied in the Capital Market and Services Act (CMSA) 2007 (Accountants Today, 2009). The whistle blowing provisions under Section 320 of the CMSA imposes mandatory duty upon auditors of listed corporations to report breaches of the securities law and stock exchange rules to the authority. At the same time, CMSA also contain provisions that prevent victimization of whistle blower. Therefore, the role of statutory auditors is highly controversial with more responsibilities and duties that need to be complied with in ensuring that the company's accounts are trustworthy and reliable.

Another initiative taken by the Government in combating fraud and corruption is through the formation of Malaysian Anti-Corruption Commission (MACC) Act. The Act is formerly known as Anti-Corruption Agency Act 1997 came into effect on 1 January 2009. This has led to the official establishment of MACC as an independent, transparent and professional body effectively and efficiently manage Malaysia's anti-corruption efforts. Through the formation of five panels of independent body with the function to closely and constantly monitor MACC, in order to ensure its integrity and to protect citizens' rights. The 'Check and Balance Mechanisms' consists of the Anti-Corruption Advisory Board, the Special Committee on Corruption, the Complaints Committee, the Operations Review Panel, and the Corruption Consultation and Prevention Panel. The recent passage of the Securities Commission Amendment Act 2010 in Parliament has made the Audit Oversight Board (AOB) is now a reality. AOB in Malaysia was established as to create a new governance structure that will enhance quality of audit services by auditors (Ismail and Mustapha 2015). This is seen as a direct response to the serious irregularities in financial reporting in Malaysia. Therefore, the AOB is said to provide an independent oversight and regulation to external auditors of public listed companies in Malaysia. The board also has the power to reprimand auditors that had done wrong including issuing penalty and deregistering them from the audit practice (Ismail and Mustapha 2015). The Transparency International Malaysian (TI-M) was also introduced in 2009. It is an independent, non-governmental and non-partisan organization committed to the fight against corruption. The Government has commission in the Prime Minister Department, and a consultant, Ethos and Company to engage TI-M in a series of discussions on ways to curb corruptions. By these establishments as mentioned above, it can be seen the Government has taken great initiatives in curbing and combating fraud and corruptions in Malaysia.

On 6 May 2010, the Parliament of Malaysia has passed the Whistleblower Protection Act, which came into force on December 2010. Whistleblowers will be protected under the Act for

*Corresponding author's email: nurad185@salam.uitm.edu.my

DOI: 10.1051/shsconf/20173600004

the report lodged relating to disclosure of improper conducts to any seven of the top government agencies including MACC, the Royal Malaysian Police, Royal Malaysian Customs, Road Transport Department of Malaysia, Immigration Department of Malaysia, Securities Commission of Malaysia and CCM. The Act empowers enforcement agencies to investigate into complaints of detrimental action and allows the court to order damages, compensation, injunction any order remedies against the party who makes the detrimental action. Year 2012 marks another milestone in CG landscape when the MCCG 2012 superseded the MCCG 2007. With the revision of the MCCG in 2012, the BMLR was further revised in November 2012. Although the requirement on audit committee characteristics and internal audit function attributes were not raised, there are now additional requirements on CG such as the establishment of nomination committee (Norziaton 2013).

In 2014, the Transparency International Malaysia (TI-M) which was established in 1998 has launched the results of the Corruption Perception Index (CPI) as well as the Malaysian Corruption Barometer (MCB). Datuk Akhbar Satar, the President of TI-M mentioned that the score and ranking of Malaysia between 2011- 2012 have improved (The Edge Financial Daily, 2014). The CPI ranks countries based on the perceived level of corruptions in the public sector. The score of 0 is for "highly corrupt" and the score of 100 is for "very clean". Those scores can be considered as a positive feedback from experts and laymen on the Government's initiatives in combating fraud and corruptions in those years. The annual report of TI-M 2009 also reported after the launch of 2009 Corruption Perception Index (CPI), Malaysia showed a drastic drop of 5.1 to 4.5 and the country ranking from 47 to 56. On the other hand, the 2014 MCB results reported a decrease in the perceived levels of corruption in Malaysia compared to 2013, along with a decline in the belief that corruption in the public sector in Malaysia is a problem. The perception that Malaysia's government is run by a few big interests looking out for themselves also rated better in 2014 compared to 2013. The reported results by TI-M explained that the initiatives taken by the Government to curb and combat frauds and corruption have contributed to positive outcomes in controlling the level of fraud and corruption in Malaysia.

To end the evolution of auditing arena and CG landscape in Malaysia, this study reveals that in 2016, the SC has issued a Proposed Draft of MCCG. As a key document in Malaysia's corporate governance framework the Code has contributed significantly in improving the corporate governance standards of Malaysian listed companies. As mentioned in the earlier section, the Code has been reviewed and revised twice in 2007 and again in 2012. The effort to consistently revise the Code is seen as part of the serious initiatives undertaken by the Government in combating frauds and corruptions in Malaysia.

CONCLUSIONS

This paper relates to the development of auditing arena and CG landscape in Malaysia which covers a timeframe starting from post-independence in 1957 until 2015. Several key developments have occurred as to guide the auditors in carrying out audit practice more diligently and to enhance the integrity of financial statement audit. Over the years, stringent rules and regulations were imposed by the Government which needs to be conforming by the statutory auditors especially after the release of MCCG 2000. Hence, extra audit procedures are to be undertaken in performing the audit assignment. Based on the discussions on the two phases above, this paper has demonstrated auditing arena and CG landscape in Malaysia that has evolved over the years starting from an establishment of professional bodies and regulations until the divergence of the auditing profession. Hence, the changes that took place

*Corresponding author's email: nurad185@salam.uitm.edu.my

over these periods clearly state that considerable resource have been channeled to the evolution of the auditing arena and CG landscape in Malaysia. The Government's continuous initiative in curbing and combating fraud and corruptions can be seem as great efforts in making sure of fraud and corruptions in Malaysia are under good control. According to the Auditor General, Tan Sri Ambrin Buang the number of fraud cases, forgery and corruption in corporate Malaysia are under control as the government introduces new measures and policies to mitigate the risks (Business Times, 26 July, 2010). It is hoped that Malaysia Government will consistently able to manage fraud and corruptions through the revision of existing procedures and several other new mechanisms which may be introduced in the future.

REFERENCES:

Accountants Today, (2008). Enhancing Audit Quality: Impact of the Audit Oversight Board.

Accountants Today, (2009). Auditors and the Whistleblowing Law from http://www.mia.org.my/at/at/2009/04/04.pdf

Akauntan Nasional, (1993). MIA Sets Standard Audit Fees

Akauntan Nasional, (April 1994). Submission by the Council of the Malaysian Institute of Accountants on Guidelines for Minimum Audit Fees - An Overview

Akauntan Nasional (March 2003). Striving For Quality Financial Reporting

Azham A., Roszaini H., Mohammad H. (2006). Episodes in the Malaysian Auditing Saga, Managerial Auditing Journal, Volume 21, Issue 7, 684-701

Boynton W.C., Johnson R.N. (2005). Modern Auditing: Assurance Services and Integrity of Financial Reporting, 8th Edition, Wiley.

Business News (29 March 2014). Auditing The Auditors

Business Times, (26 July, 2010). Corporate Fraud, Graft 'Under Control'.

Cosgrove, S. B., and Niederjohn, M. S. (2008). The effects of the Sarbanes-Oxley Act of 2002 on audit fees. *Journal of Business Strategies, 25*(1), 31.

http://transparency.org.my/

http://www.mia.org.my/handbook/guide/masa/AI600.htm

Ismail H., Mustapha M. (2015). Auditing the Auditors: The Audit Oversight Board and Regulating Audit Quality in Malaysia, Journal of Modern Accounting and Auditing, March 2015, Vol. 11, No. 3, 138-142

Khoo B. Y. (2003). Review of Corporate Governance in Asia: Corporate Governance in Malaysia, http://adbi.org/file/2003.11.10.corporate.governance.malaysia.pdf

*Corresponding author's email: nurad185@salam.uitm.edu.my

DOI: 10.1051/shsconf/20173600004

KPMG Malaysia (2003). Fraud, Bribery and Corruption Survey

Lee, T., Azham, A., Gloeck, J. (2009). The Audit Expectation Gap in Malaysia: An Investigation Into Its Causes and Remedies. Southern African Journal of Accountability and Auditing Research, 9.

MASB (Malaysian Accounting Standards Board). Retrieved on 21 March 2010 from http://www.masb.org.my.

MCCG (Malaysian Code of Corporate Governance) (2007). Malaysian Law Journal Sdn. Bhd., Malaysia.

MIA (2002) By Laws (on Professional Ethics, Conducts and Practices) of the Malaysian Institute of Accountants from http://www.mia.org.my

MIA (2004) By Laws (on Professional Ethics, Conducts and Practices) of the Malaysian Institute of Accountants from http://www.mia.org.my

Norziaton I.K. (2013). Internal Governance Mechanisms and Audit Fees: The Moderating Role of Political Connection and Regulatory Oversight, Thesis Submitted for the Degree of Doctor of Philosophy, Universiti Malaya

Omar N., Mohd Sanusi Z., Yaakob N.A.,Mohamed S. (2013). Audit Quality Assessment on Audit Firm: Compliance and Challenges, The 5th International Conference on Financial Criminology

Sarbanes Oxley Act (2002) from http://www.sec.gov.

SC (Securities Commission Malaysia). Retrieved on 5 November, 2011 from http://www.sc.com.my/.

The Commissioner of Law Revision, Malaysia, 2006

The Edge Financial Daily (December 4, 2014). Malaysia Moves Up 3 Spots in Global Graft Index

The Star, (July 8, 2009). Shareholder Activism Can Enhance Shareholders Value.

Transparency International Malaysia Annual Report 2009 from http://transparency.org.my/

*Corresponding author's email: nurad185@salam.uitm.edu.my

PERFORMANCE MEASUREMENT AND ACCOUNTABILITY OF *WAQF* INSTITUTIONS IN MALAYSIA

Roshayani Arshad* and Norzaihan Mohd Zain
Accounting Research Institute, Universiti Teknologi MARA, Malaysia

ABSTRACT

It has been seen for the past few years that the revival of *waqf* institutions has been an upsurge interest of Muslim communities around the world. In line with the revitalization, the issues of measuring and managing *waqf* performance are growingly being discussed and concerned by the academicians and constituents. *Waqf* institutions need to demonstrate their performance as whether they have effectively and efficiently managed in order to discharge their accountability to various *waqf* stakeholders. There are many studies conducted to measure the performance of various organizations in the private, public and third sector. However, studies on the performance measurement of *waqf* institution are still limited, and financial ratios become the dominant measurement in those studies. Being a non-profit in nature and religious entity, managing the entrusted *waqf* assets for social and economic development of the society, the performance measurement of *waqf* institutions should also focus on realizing their missions. Therefore, the aim of this paper is to discuss both financial and non-financial measurement that can be adapted by *waqf* institutions to assess their performance in discharging their accountability. Drawing from the experience on performance measurement discussed in the non-profit organizations literatures, this conceptual paper is hoped to provide significant insight on how *waqf* institutions performance may be measured and provide a tool to benchmark the best practices that can guide them to achieve their goals and missions.

Keywords: *Waqf*, Performance Measurement, Accountability

INTRODUCTION

Waqf is regarded as a mechanism that plays an important role in the economic development of Muslim society throughout the Islamic history until today. It is not only a religious matter towards Allah SWT but it is also considered as an economic and social institution which benefits the Muslims. It is also claimed that *waqf* can contribute significantly towards the ultimate goal of Islamic economic system which covers eradication of poverty, socio-economic justice, and equitable distribution of income. However, despite the contribution that can be obtained from *waqf* in fulfilling the needs of Muslim society, the great potential of *waqf* has not been fully realized and accomplished by *waqf* institutions around the world (Mohammad, Iman, Hamid, & Omar, 2005). The growth in scholarly concern on the effectiveness of *waqf* management is reflected in

* Corresponding author's email: roshayani@salam.uitm.edu.my

numerous studies. Researchers have been studying, reviewing, and surveying the performance of *waqf* management and found that in general, substantial improvements need to be done (Akhunov, 2015), (D. Ibrahim & Ibrahim, 2013), (Rashid, 2012), (Chowdhury, Chowdhury, Muhammad, & Yasoa, 2012).

At present, *waqf* institutions in Malaysia have not yet become the bodies that can contribute to solve the economic problems of *Ummah* (society). For example, there are many *waqf* lands being left undeveloped even though they are in very strategic areas. The 2010 statistics issued by the Department of *Waqf*, Zakat, and Haj (JAWHAR) showed that there were 11,511 hectares of *waqf* land worth RM 116,441,667 in Malaysia, but only 0.72% of it was developed (Mohamad, 2014). In the year 2015, the National Audit Department has audited the management of *waqf* by the State of Islamic Religious Councils (SIRCs) , Malaysia for the period of year 2014 and found that in general, *waqf* properties were not being managed properly and effectively by the institutions (Auditor General's Report, 2014). The Auditor General has recommended that the SIRCs should improve the management of *waqf* properties so that the Muslims would be able to gain the socio-economic benefit derived from the properties. The SIRCs were also recommended to continuously take an effort to improve the management and development of *waqf* properties in order to ensure the responsibilities that have been entrusted to them and *waqf* objectives are achieved.

Given this issues, it is important for the SIRCs to demonstrate their performance and exercise transparency in order to discharge accountability to the stakeholders. The institutions need to transpire their performance as whether they have successfully achieved *waqf* missions and objectives as required from them. The level of *waqf* performance will help the institutions to enhance support from the public and gain confidence from the investors to invest in developing *waqf* properties. In this regard, the performance of *waqf* institutions need to be measured based on the most appropriate and suitable measurement. Accordingly, the measurement can be a tool to benchmark the best practices that can guide them to achieve their goals and missions.

However, review on the literature reveals that there are limited studies on performance measurement of *waqf* institutions (Sulaiman, Adnan, & Nur, 2009) especially through the empirical research (Sulaiman & Zakari, 2015). The research to date has tended to focus more on financial ratios (Sulaiman et al., 2009), (Sulaiman & Zakari, 2015), (H Pirasteh, 2011) rather than the non-financial aspect of performance measurement. Hence, this study will seek to address this gap by analysing the literature on performance measurement as practised by the non-profit organizations (NPOs) which can be adapted by *waqf* institutions, as both organizations have similarity in nature that involve philanthropic and charitable activities (Ramli & Muhamed, 2013). Therefore, drawing from the experience on performance measurement discussed in the NPOs literatures, both financial and non-financial perspectives of performance measurement for *waqf* institutions will be proposed. This conceptual paper is hoped to provide significant insight on how *waqf* institutions performance may be measured and improved in the near future, specifically in Malaysia.

AN OVERVIEW OF WAQF

Literally meaning of *waqf* in Arabic is confinement or detention and in the Islamic jurisprudence it means "withholding or immobilizing an asset and mobilizing or releasing its proceeds" (Sabiq, 2004). Withholding means that the assets cannot be sold or disposed in any way, releasing the proceed means that the benefit and *usufruct* can be obtained from the asset without consuming it, need to be spend in the way as permitted by Allah s.w.t. (Kahf, 1998). In other words, *waqf* is a perpetual endowment created in the form of property, money or other items for charitable or religious purposes and the *usufruct* (benefit) is confined to the society as a whole or for a specific segment of society in need (Hassan & Shahid, 2010).

This religious act is one of the commendable forms of charity and repeatedly encouraged and mentioned in the *Quran* and *Sunnah*.

"The parable of those who spend their wealth in Allah's way is as the parable of a grain; it grows seven ears, and in each ear there are a hundred grains. Allah gives manifold increase to whom He wills. And Allah is All-Embracing, All-Knowing". (Al Baqarah: 261)

"You will not attain piety until you expend of what you love; and whatever thing you expend, Allah knows it". (Al Imran: 92)

From a *Hadith* of Abu Hurairah r.a. who reported that Rasulullah s.a.w said:

"When the son of Adam dies, his deeds are cut of except in three: perpetual charitable donation and useful knowledge that is made good use of it and the pious son who prays for him". Narrated by Muslim

Although the word *waqf* is not specifically mentioned, majority of the Islamic jurists are in consensus on the legality of *waqf*, based on the verses in the *Quran* and the *Hadith*. Cizakca (1998) views that the perpetual charitable donation mentioned in the Hadith become the essence of *waqf*.

The foundation of *waqf* began since the time of the Prophet Muhammad s.a.w. whereby the first *waqf*, Masjid Quba was established by the Prophet s.a.w when he arrived in Madinah during the *Hijrah*. Later, Prophet's mosque, Masjid Al Nabawi was built which become the place for worships and also administration centre for the Muslim society. Subsequently, the companions of the Prophet s.a.w continued to establish *waqf*; Abu Bakar r.a donated a piece of land for the society in Mecca, Umar r.a allocated a piece of land for the poor and needy people, and Uthman r.a endowed a water well, Bi'r Raumah for the consumption of the society (Gadot, 2014).

WAQF INSTITUTIONS IN MALAYSIA

The perpetuity of *waqf* implies that *waqf* property needs to be preserved and the benefit can be gained without consuming it (Kahf, 1998) , (Chowdhury et al., 2012). The responsibility to preserve and administer *waqf* property lies in the hand of *mutawalli* that is the administrator, *nazir* or trustees (Mahamood, 2000). Therefore, the *mutawalli* has to be appointed to manage the property in order to ensure that the benefits will continually disseminate to the beneficiaries.

In Malaysia, the institutions that are given the authority to manage *waqf* are the State Islamic Religious Councils (SIRCs) (Siraj, 2012). In order to execute the administration of *waqf*, the SIRCs have established a dedicated *waqf* unit or department in the council or an agency within its jurisdiction (Sulaiman & Zakari, 2015). In Selangor, the State of Islamic Religious Councils has formed Perbadanan Wakaf Selangor (PWS) to manage the overall aspect of *waqf* administration in the state. In Sarawak, the administration of *waqf* is being handled by Tabung Baitulmal Sarawak which was incorporated under the Majlis Agama Islam Sarawak Ordinance (Incorporation) (Amendment) 1984 (Siraj, 2012). The State of Pulau Pinang Islamic Religious Council has also established Perbadanan Wakaf Pulau Pinang, a subsidiary company of the council as an effort to increase the wealth of *waqf* in the state (Ismail, Muda, & Hanafiah, 2014).

Among the functions of the department or agency are to:

- Advise the Council with respect to the policies, methods, measures need to be implemented and taken to encourage the development of *waqf* property and products
- Implement and give effect to such policies, instructions or directives of the Council in respect of property development and endowment products
- Develop, adopt and implement policies, ways and measures on property development and endowment products
- Coordinate the implementation of the *waqf* property and product development rapidly
- Plan, develop, implement and promote the development and implementation of *waqf* property and products
- Manage and run the operation and maintenance of *waqf* facilities, infrastructure and equipment in the *waqf* area
- Promote, coordinate and carry out research and development in all aspects of *waqf* property and products.

Source: Perbadanan Wakaf Selangor Website

The main purpose of these functions is to assist the SIRCs to achieve the mission and objective of *waqf*. Generally, the mission of *waqf* management is to manage and add value to *waqf* properties effectively through cultivating *waqf* culture for the benefit of the society. Meanwhile, the objectives are to enhance the value added in *waqf* property management and cultivate *waqf* practices towards economic development of the society by efficacy and holistic approach in accordance with the Islamic Law.

ACCOUNTABILITY AND PERFORMANCE OF WAQF INSTITUTIONS

Being the sole trustees of *waqf*, as well as religious-based institutions, there are basic requirements that the *waqf* institutions need to fulfil which include accountability to the parties that have an interest in the institutions. Accountability can be defined as "the duty to provide an account (by no means necessarily a financial account) or reckoning of those actions for which one is held responsible" (Gray, Owen, & Adams, 1996). S. H. M. Ibrahim (2000) elaborate further the definition of accountability as "the duty of an entity to use (and prevent the misuse of) the resources entrusted in an effective, efficient, and economical manner, within the boundaries of moral and legal framework of society, and to provide an account of its actions to the accountees who are not only the persons who provide it with financial resources but also to the groups within the society and to society at large". This definition indicates that *waqf* institution has a duty to carry out certain actions in managing the resources entrusted to them by the resources provider in a proper and permissible manner and also to fulfil its obligation towards the society.

In the context of non-profit organizations (NPOs), Edwards and Hulme (1996) pointed out these organizations have multiple accountabilities; "downwards" which refer to the organizations' partners, beneficiaries, staff, and supporters, and "upwards" which refer to their trustees, donors, and host governments. Ebrahim (2003) has suggested that accountability involves many parties, namely the financiers who are normally interested to know how the fund is being used, the beneficiaries who are the receiver of the services, and the internal organization itself. The level of accountability also varies based on who are the interested parties and the type of organization (Ebrahim, 2010).

However, the concept of accountability in Islam is different from the conventional as it goes beyond accountability to mankind and worldly life. In Islam, the ultimate accountability of man is to Allah SWT and then to mankind and nature. All deeds will be counted in this world and in hereafter *(al-falah).* As mentioned in the *Qur'an,* a man is entrusted as the *Khalifah* (vicegerent) to safeguard the earth, which ultimately belongs to Allah SWT.

"To Allah SWT belongs whatever is in the heavens and whatever is in the earth. Whether you show what is within yourselves or conceal it, Allah SWT will bring you to account for it. Then He will forgive whom He wills and punish whom He wills, and Allah SWT is over all things competent". (*Al Baqarah:* 284)

Previous studies by S. H. M. Ibrahim (2000), Yaacob (2006), and Ihsan and Adnan (2009) have proposed dual accountability for *waqf* management. For primary accountability, everybody who is responsible for *waqf* should discharge his or her accountability to Allah SWT (*hablun min Allah*) by fulfilling Allah's command and avoiding His prohibitions. This can be done by following the guidelines as stated in the *sharia* rules. As an administrator of *waqf*, *mutawali* should manage and utilize the *waqf* assets effectively in accordance with the *sharia* rules. Concerning secondary accountability, Yaacob (2006) and Ihsan (2007) have stressed that the *mutawali* is also accountable for various *waqf* stakeholders, namely the donors (*waqif*), *waqf* board, regulators, beneficiaries, and the society.

There are mechanisms that can be used by an organization to discharge its accountability to the stakeholders. Ebrahim (2003) proposed the mechanisms of accountability comprises of reports and disclosure statement, participation, self-regulation, social audits and performance assessments and evaluations. Stewart and Walsh (1992) asserted that accountability is associated with performance measurement as the managers who are managing resources entrusted to them are required to achieve certain objectives and goals and they are accountable for their actions and performance. Therefore, in *waqf* institutions, the *mutawallis* are required to achieve the highest level of performance in *waqf* management in order to discharge their accountability as they are not only accountable to the stakeholders but also to Allah SWT.

PERFORMANCE MEASUREMENT OF WAQF INSTITUTIONS

Over the years, scholars have been studying and proposed several definition and measurement that relate with the performance of an organization. However, it is not an easy task to define, conceptualizing and measuring performance especially for non-profit organizations and particularly for *waqf* institutions. Scholars among themselves have different opinions on the measurement of the performance. This is due to the complexity of the measurement as these organizations focus and emphasize on achieving their missions which the accomplishments are difficult to measure (Forbes, 1998), (Epstein & McFarlan, 2011).

A broad definition of performance as defined by Payer-Langthaler and Hiebl (2013) is "intentional action", thus performance measurement can be described as an assessment of the results from intentional action (Cordery & Sinclair, 2013). The results can be reflected by success or failure of the organizations in accomplishing their objectives (Helmig, Ingerfurth, & Pinz, 2014). According to Kaplan (2001), the success of an NPO should be measured based on the effectiveness and efficiency of the organization's activities in fulfilling the needs and expectation of the stakeholders.

Effectiveness is defined as the possibility of an organization to accomplish the objectives of an activity, policy, other interventions (Scott, Burall, Highton, & Wong, 2008), programmes and operations (Cutt & Murray, 2002). On the other hand, efficiency refers to the utilization of resources for the attainment of desired results (Cutt & Murray, 2002) and maximizing the results of an action in relation to the resources used (Mihaiu, Opreana, & Cristescu, 2010). Hence, efficiency is related to the conversion of inputs into outputs and how inputs are transformed into output successfully, while effectiveness reflects the relationship between output and objectives (Sulaiman & Zakari, 2015). Helmig et al. (2014) and Lecy, Schmitz, and Swedlund (2012) have pointed that efficiency can be considered as a part and a subset of effectiveness. This view is supported by Mouzas (2006) who highlights that effectiveness has broader scope in which it consider quality, creation of value added, employee satisfaction, output interaction with the social and economic environment.

There are measurements that are being used in determining the performance of an organization. Zheng, Yang, and McLean (2010) in their study on assessing organizational effectiveness have adopted the measurement that relates to members' perception of the overall success, profitability, growth rate, and innovativeness. However, since the main objective of NPO is not achieving financial success and profit maximization (Kaplan, 2001), the measurement of management effectiveness for the organization is more complex (Helmig et al., 2014). The measurement should not only be based on monetary factors but also on the extent of NPO reaction to moral values that inspire contributors and beneficiaries (Herman and Renz (1999). Fishel (2004) also argued that the measurement has to take into account other considerations whether the formation of the organization is to achieve social, cultural, welfare or religious objectives.

The management of *waqf* is associated with the concept of accountability and one of the mechanisms to discharge the accountability is through performance assessment. Since the main

DOI: 10.1051/shsconf/20173600005

purpose of *waqf* is to fulfil non-profit purposes and its nature is similar to NPOs (Ramli & Muhamed, 2013), the definition and measurement of performance for the NPO can be adapted to measure the performance of *waqf* institutions. In this study, both financial and non-financial aspects will be considered in measuring the performance of *waqf* institutions in Malaysia. The perspectives of *waqf* institutions performance measurement proposed in the study are depicted in Figure 1, below:

Figure 1: Proposed Perspectives of *Waqf* Institutions Performance Measurement

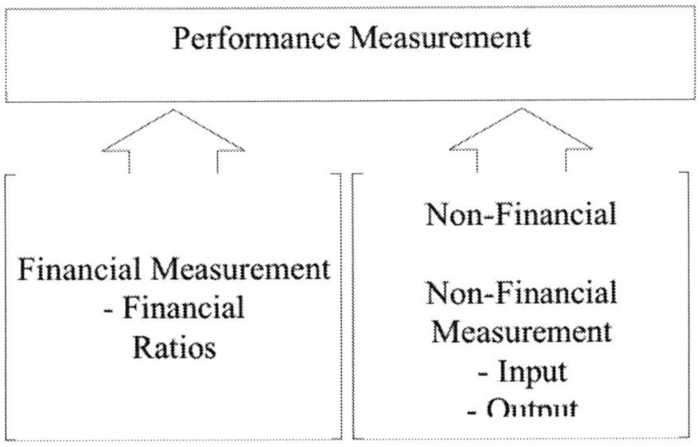

Financial Measurement – Financial Ratios

The effectiveness of an organization depends on the financial condition, vulnerabilities (Keating, Fischer, Gordon, & Greenlee, 2005), and sustainability (Helmig et al., 2014) of the organization. The organizations can enhance their effectiveness if they have sufficient fund and strong financial support from the funders. Previous studies have used financial ratios to indicate the financial performance and vulnerability of an organization. Items in the balance sheet and profit and loss account are commonly used in calculating the ratios (Cohen, 2008). The advantages of using the ratios are they are more comparable to organizations of similar activities, scalable, and collectable (Liket & Maas, 2015).

Several studies on performance of *waqf* institutions have been conducted which concentrated on the financial ratios as the performance measurements. Sulaiman et al. (2009) examined the accounting practices of the International Islamic University Malaysia's *waqf* fund and measured the management efficiency of the fund. The study focused on three ratios that were programme efficiency ratio, return of investment and fundraising efficiency ratio. In another study by H Pirasteh (2011) on the efficiency of *waqf* management by government and private agencies in Iran, the ratios used to measure the efficiency were disbursement to proceeds (objective achieved index) and the balance remaining for the year to total earning (the expected income achieved index). A study was conducted by (Shafii, Yunanda, & Rahman, 2014) which focused on ratios to measure the financial and operational performance of an SIRC and Majlis Ugama Islam Singapore. The ratios used in the measurement were net income, operating income, income growth, administrative efficiency, program efficiency, growth of program, fundraising efficiency and cash availability ratios. A recent study on the efficiency of *waqf* management by the State Islamic Religious

Councils in Malaysia was carried out Sulaiman and Zakari (2015) and the measurement focused on the financial vulnerability and sustainability ratios.

The ratios that were being used as performance measurement of *waqf* institutions are summarized in Table 1. These ratios are applicable and can be adopted to measure the fiscal strength of *waqf* institutions and will enlighten the *waqf* institutions on what information that is required to be disclosed in their reporting in order for them to exercise transparency and enable the stakeholders to evaluate their performance.

Table 1: Financial Ratios as *Waqf* Institutions Performance Measurement

No	Ratios	Description	Indication	Literature
1.	Programme Efficiency	Programme expenses to total expenses	Recommended ratio is 65%. Low ratio indicates that less beneficiaries receive the assistance or the total expenses are much too high. Higher ratio indicates that more expenses are disbursed to run programs	(Sulaiman et al., 2009) (Shafii et al., 2014)
2.	Cash Availability	Cash & Bank Balances to Total Assets	Indicate the amount of idle cash and bank balances. The higher the ratio tells that the organization maintain high amount of idle cash	(Shafii et al., 2014)
3.	Return of Investment	Investment Income To Average Investment	Indicate the return of investment on *waqf* assets.	(Sulaiman et al., 2009)
4.	Fundraising Efficiency	Total Fundraising Expenses To Total Funds Raised	Indicate the extent to which the organization is able to generate surplus from the donation	(Sulaiman et al., 2009) (Shafii et al., 2014)
5.	Operating Income	Core income to Expenses	Indicate the extent to which the income generated from core activities	(Shafii et al., 2014)
6.	Objective Achieved Index	Disbursement (output) To Proceeds (input)	Indicate the degree of achievement that *mutawalli* has fulfilled *waqif* designated objectives. If the ratio is closer to one, this indicates that there is a balance between disbursement and proceeds	(Hossein Pirasteh, 2011)
7.	Expected Income Achieved Index	The Balance Remaining For The Year To Total Earning	Indicate better income acquisition if the ratio is close to one.	(Hossein Pirasteh, 2011)
8.	Equity Balances	Ratio of equity to revenue	Indicate financial sustainability if the ratio is high	(Sulaiman & Zakari, 2015)
9.	Revenue concentration	Square of the percentage share that each revenue source represents of the total revenue	Indicate that organization has equal revenue from diverse sources if each ratio is close to zero. Revenue that has ratio close to one shows that it is dependent on one single source of income which is not healthy to the organization	(Sulaiman & Zakari, 2015)
10.	Administrative Efficiency	Ratio of administrative costs as a percentage of total costs	Determine the ability of the organization to control expenditure and impact of the control on service delivery	(Sulaiman & Zakari, 2015) (Shafii et al., 2014)
11.	Operating margin	Net income (or loss) divided by total revenue	Indicate the organization is financially stable if the ratio is high	(Sulaiman & Zakari, 2015)
12.	Income Growth	Income year n - Income year (n-1) divided by Income year (n-1)	Indicate the extent to which the organization is able to generate income from its activities from time to time	(Shafii et al., 2014)

However, there are a few weaknesses that can be associated with the application of financial ratios alone in measuring the performance of *waqf* institutions. These ratios measure the historical consequences rather than the causes of the event (Balabonienė & Večerskienė, 2015) which provide limited information on how and what need to be resolved. Furthermore, these measures do not reflect other essential strategic matters such as the performance of employees and customer satisfaction (Rahman, Mastuki, & Yusof, 2015). In *waqf* performance measurement studies, most of the researchers had to face the challenges especially in data collection stage due to the inconsistency and inadequate disclosure of *waqf* information in the annual report among the SIRCs (Masruki & Shafii, 2013), (Sulaiman & Zakari, 2015). Therefore, *waqf* institutions need to take into consideration the non-financial aspect of performance measurement as this will provide a wider coverage and useful information for them to discharge their accountability to the relevant stakeholders.

Non-Financial Measurement

Lecy et al. (2012) have conducted a comprehensive review of academic literature on NPO effectiveness over the past 10 years and summarized that one-dimensional measurement is not sufficient for measuring the effectiveness of an organization. They proposed four areas that can guide to a better measurement of organizational effectiveness based on managerial, program, network effectiveness, and legitimacy. Managerial effectiveness can be measured by financial ratios, human resource statistics, and balance scorecard; program effectiveness can be measured by program output and impact; network effectiveness can be measured by mobilizing actors through networks or control resources and the achievement of strategic objectives through participating in networks; legitimacy can be measured by the ability of an organization to raise fund, access to policy process, and gain support from partners and donors. A similar study has been conducted by Helmig et al. (2014) and found that the categories that can depict an organization success are goal achievement, financial performance, social performance, service quality, resource acquisition, survival, efficiency, stakeholder satisfaction, and organization transformation.

A recent study by Lee and Nowell (2015), identified various perspectives that can be adopted in measuring and conceptualizing performance of the NPO. The perspectives are inputs, organizational capacity, outputs, outcomes, public value accomplishment and network/institutional legitimacy. These perspectives offer valuable insights for *waqf* institutions to measure their performance based on both financial and non-financial measurement. The relevant perspectives of *waqf* performance measurement that will be discussed in this study are input, output and outcome.

Input

Input refers to the resources that are dedicated to the operation, and is necessary for producing goods and delivering services of any organization. This perspective focuses on how financial and non-financial resources are being acquired and utilized efficiently in order to ensure sustainability, and support the operation in achieving the objectives of the organization (Helmig et al., 2014), (Mitchell, 2012), (Connolly & Hyndman, 2004), (Cutt & Murray, 2002), (Sillanpää, 2011). There are two approaches to advocate this perspective.

The first approach relates with acquisition and utilization of funds, assets, staff, training and preparation of program and services (Epstein & McFarlan, 2011). Examples of performance measurement for this approach as summarized by Lee and Nowell (2015) are increase in revenue, diversity in revenue stream, ability to acquire and manage human resources, employee satisfaction, motivation, retention. Additional measurements as suggested by Sillanpää (2011) are utilization of facilities and financial resources sufficiency.

In the context of *waqf*, the SIRCs receive *waqf* from the *waqif* in the form of cash, properties or securities. These are the resources that *waqf* institutions are obliged to take care of in accordance with the mission, objectives and functions of the institutions. Among the examples of criteria that are appropriate to measure the performance of the institutions resources or input are all *waqf* immovable properties are registered and gazetted under the SIRCs as required by the *Waqf* Enactment, increase in cash *waqf* collection, diversified method of cash *waqf* collection and the turnover of the employees managing *waqf*.

The second approach emphasizes on the efficiency in managing expenditures that relate with the resources and organization activities (Fishel, 2004). This approach is normally involved financial ratios in measuring the performance.

Output

Generally, output indicates the goods and services that resulted from the organization's activities (Mitchell, 2012), (Cordery & Sinclair, 2013). These perspectives measure the effectiveness in managing inputs that are related with the goals accomplishment (Berman, 2015), and associated with the short term objectives (Mitchell, 2012) of the organization. Based on the criteria of performance measurement summarized by Lee and Nowell (2015), the examples of output are on-time service deliveries, number of participants served, achieve specific goals and quality of services provided. In addition, Sillanpää (2011) focus on financial and quantitative measures such as cost per unit of service, client served and number of client visit.

The output of *waqf* management should also support the mission, objectives and functions of *waqf* institutions. Among the examples of performance measurement that are related with the output are distribution of the *usufruct* (benefit) to the *maukuf alaih* (beneficiaries), development of *waqf* properties, income generated from *waqf* properties and financial investment and fulfillment of *waqif* intention on *waqf* properties.

Outcome

Outcome is defined as the effects and impacts of the good and services delivered by an organization on the beneficiaries, (Cordery & Sinclair, 2013), (MacIndoe & Barman, 2012) and normally related with the long term goal of the organization (Mitchell, 2012). This perspective indicates the behavioral and environmental changes of the intended population as a result of the goods rendered and services received from the organization (Epstein & McFarlan, 2011), (Lee & Nowell, 2015). The impact is measured by comparing the level of beneficiaries attributes or population before and after they received the goods and services in order to know the effectiveness of the organization's activities (MacIndoe & Barman, 2012). Buckmaster (1999) argued that outcome is the most appropriate measurement of an organization performance as it reflects the ultimate achievement of the organization goals and a way to enhance the accountability of the organization

Based on review of NPO performance literatures, Lee and Nowell (2015) identify two broad approaches in measuring outcome of organization activities and these approaches can be complementary to each other or segregated due to different methods in their measurement. The first approach deals with environmental changes approach in which performance of an organization is measured based on whether there is an accomplishment of substantial changes in the environment or targeted group (Cordery & Sinclair, 2013), (Packard, 2010). A few examples summarized by Lee and Nowell (2015) are improve in condition or status, increase in skills and knowledge and modified behavior or attitude the targeted population. The second approach focus on customer satisfaction in which the quality of goods and services rendered by the organization are being assessed based on the perception of customers through satisfaction and complaint survey. Among the elements of customer satisfaction measurement summarized by Lee and Nowell (2015) are accessibility, timeliness, condition of facilities, customer retention and gain of new customer. In the context of *waqf* institutions performance, the measurement can be increase in number of *waqif* and *waqf* properties, improve the awareness on the significant of *waqf* and participation among the society.

CONCLUSION

This paper discusses the conceptualization of *waqf* institutions performance based on financial and non-financial aspect of measurement. Extensive literature review was conducted on the performance measurement practiced by the NPO provide an understanding of the management best practices that can be adaptable by *waqf* institutions. The performance measurement perspectives discussed in the study are financial ratios, input into the organization, output produced by the organization by utilizing the input and outcome generate form the activities of the organization. These perspectives offer valuable guidance for *waqf* institutions to measure their performance comprehensively based on both financial and non-financial measurement. It is hoped that the practical relevance of this study would be able to help *waqf* institutions to improve their performance in order to discharge their accountability to the stakeholders.

REFERENCES

Akhunov, A. M. (2015). Issue of Waqf in the Context of Establishment of Muslim Institutions in the Post-Soviet Tatarstan. *Journal of Sustainable Development, 8*(5), p157.

Auditor General's Report. (2014). *Auditor General's Report (2014) The Activities Of The Departments/ Agencies And Management of State Government's Companies, 3rd Series*. Retrieved from

Balabonienė, I., & Večerskienė, G. (2015). The Aspects of Performance Measurement in Public Sector Organization. *Procedia-Social and Behavioral Sciences, 213*, 314-320.

Berman, E. (2015). *Performance and productivity in public and nonprofit organizations*: Routledge.

Buckmaster, N. (1999). Associations between outcome measurement, accountability and learning for non-profit organisations. *International Journal of Public Sector Management, 12*(2), 186-197.

Chowdhury, M. S. R., Chowdhury, I. A., Muhammad, M. Z., & Yasoa, M. R. (2012). Problems Of Waqf Administration And Proposals For Improvement: A Study In Malaysia. *Journal of Internet Banking and Commerce, 17*, 1-8.

Cizakca, M. (1998). Awqaf In History And Its Implications For Modern Islamic Economies. *Islamic Economic Studies, 6*(1), 43-70.

Cohen, S. (2008). Identifying The Moderator Factors Of Financial Performance In Greek Municipalities. *Financial Accountability & Management, 24*(3), 265-294.

Connolly, C., & Hyndman, N. (2004). Performance Reporting: A Comparative Study Of British And Irish Charities. *The British Accounting Review, 36*(2), 127-154.

Cordery, C., & Sinclair, R. (2013). Measuring performance in the third sector. *Qualitative Research in Accounting & Management, 10*(3/4), 196-212.

Cutt, J., & Murray, V. (2002). *Accountability And Effectiveness Evaluation In Nonprofit Organizations*: Routledge.

Ebrahim, A. (2003). Accountability In Practice: Mechanisms for NGOs. *World Development, 31*(5), 813-829.

Ebrahim, A. (2010). The Many Faces of Nonprofit Accountability. *Jossey-Bass Handbook of Nonprofit Leadership and Management*.

Edwards, M., & Hulme, D. (1996). Too Close for Comfort? The Impact of Official Aid on Non-governmental Organizations. *Word Developmet, 24*(6), 961-973.

Epstein, M. J., & McFarlan, F. W. (2011). Measuring the efficiency and effectiveness of a nonprofit's performance. *Strategic finance, 93*(4), 27-35.

Fishel, D. (2004). *Performance and Accountability in the Non-profit Sector.* Paper presented at the Fourth One-Day Symposium On Accountability, Governance And Performance In Transition.

Forbes, D. P. (1998). Measuring the Unmeasurable: Empirical Studies of Nonprofit Organization Effectiveness From 1977 to 1997. *Nonprofit and Voluntary Sector Quarterly, 27*(2), 183-202.

DOI: 10.1051/shsconf/20173600005

Gadot, N. (2014). *Pembangunan Wakaf: Angkasa*. Paper presented at the Seminar Waqaf Antarabangsa 2014, Kuala Lumpur.

Gray, R., Owen, D., & Adams, C. (1996). *Accounting & Accountability Changes and Challenges In Corporate Social And Environment Reporting*. England: Prentice Hall.

Hassan, A., & Shahid, M. A. (2010). *Management And Development Of The Awqaf Assets*. Paper presented at the Seventh International Conference – The Tawhidi Epistemology: Zakat and Waqf Economy, Bangi.

Helmig, B., Ingerfurth, S., & Pinz, A. (2014). Success and Failure of Nonprofit Organizations: Theoretical Foundations, Empirical Evidence, and Future Research. *Voluntas (Manchester, England), 25*(6), 1509-1538.

Herman, R. D., & Renz, D. O. (1999). Theses On Nonprofit Organizational Effectiveness. *Nonprofit and Voluntary Sector Quarterly, 28*(2), 107-126.

Ibrahim, D., & Ibrahim, H. (2013). *Revitalization Of Islamic Trust Institutions Through Corporate Waqf*. Paper presented at the 4th International Conference On Business And Economic Research (4th Icber 2013) Proceeding, Bandung Indonesia.

Ibrahim, S. H. M. (2000). *The Need For Islamic Accounting; Perceptions Of Its Objectives And Characteristics By Malaysian Muslim Accountants And Accounting Academics*. University Of Dundee, Unpublish Thesis.

Ihsan, H. (2007). *An Exploratory Study of Waqf Accounting and Management in Indonesia Waqf Institutions, The case of Domet Duafa and Universitas Islam Indonesia (UII)*. International Islamic University, Malaysia, Unpublished Thesis.

Ihsan, H., & Adnan, M. A. (2009). Waqf Accounting and The Construction of Accountability *Humanomics, 27*(4), 252-269.

Ismail, C. Z., Muda, S., & Hanafiah, N. J. A. (2014). Challenges and Prospects of Cash Waqf Development in Malaysia. *Journal of Basic and Applied Scientific Research, 4*(2), 340-348.

Kahf, M. (1998). *Financing The Dvelopment of Awqaf Property*. Paper presented at the Seminar on Development of Awqaf Kuala Lumpur.

Kaplan, R. S. (2001). Strategic Performance Measurement And Management In Nonprofit Organizations. *Nonprofit Management And Leadership, 11*(3), 353-370.

Keating, E. K., Fischer, M., Gordon, T. P., & Greenlee, J. S. (2005). Assessing Financial Vulnerability In The Nonprofit Sector.

Lecy, J. D., Schmitz, H. P., & Swedlund, H. (2012). Non-Governmental and Not-for-Profit Organizational Effectiveness: A Modern Synthesis. *Voluntas, 23*(2), 434-457.

Lee, C., & Nowell, B. (2015). A framework for assessing the performance of nonprofit organizations. *American Journal of Evaluation, 36*(3), 299-319.

Liket, K. C., & Maas, K. (2015). Nonprofit Organizational Effectiveness: Analysis of Best Practices. *Nonprofit and Voluntary Sector Quarterly, 44*(2), 268-296.

MacIndoe, H., & Barman, E. (2012). How organizational stakeholders shape performance measurement in nonprofits: Exploring a multidimensional measure. *Nonprofit and Voluntary Sector Quarterly*, 0899764012444351.

Mahamood, S. M. (2000). *The Administration Of Waqf, Pious Endowment In Islam: A Critical Study Of The Role Of The State Islamic Religious Councils As The Sole Trustees Of A Wqaf Assets And The Implementation Of Istibdal In Malaysia With Special Reference To The Federal Territory Of Kuala Lumpur.* (PHD), Birmingham University.

Masruki, R., & Shafii, Z. (2013). The Development of Waqf Accounting In Enhancing Accountability. *Middle East Journal of Scientific Research, 13*, 1-6.

Mihaiu, D. M., Opreana, A., & Cristescu, M. P. (2010). Efficiency, Effectiveness And Performance Of The Public Sector. *Romanian Journal of Economic Forecasting, 4*(1), 132-147.

Mitchell, G. E. (2012). The Construct of Organizational Effectiveness: Perspectives From Leaders of International Nonprofits in the United States. *Nonprofit and Voluntary Sector Quarterly, 42*(2), 324-345.

Mohamad, M. H. (2014, 6 May 2014). Unleashing The Potential Of Waqf. *The Star*

Mohammad, M. T. S., Iman, M., Hamid, A., & Omar, I. (2005). An Ideal Financial Mechanism For The Development Of Waqf Properties In Malaysia: Ministry of Science, Technology, and Innovation, Malaysia.

Mouzas, S. (2006). Efficiency versus effectiveness in business networks. *Journal of Business Research, 59*(10), 1124-1132.

Packard, T. (2010). Staff Perceptions of Variables Affecting Performance in Human Service Organizations. *Nonprofit and Voluntary Sector Quarterly, 39*(6), 971-990.

Payer-Langthaler, S., & Hiebl, M. R. W. (2013). Towards a definition of performance for religious organizations and beyond: A case of Benedictine abbeys. *Qualitative Research in Accounting & Management, 10*(3/4), 213-233.

Pirasteh, H. (2011). *Efficiency Criteria In The Public And Private Waqf Management: The Iranian Experience.*

Pirasteh, H. (2011). Efficiency Criteria In The Public And Private Waqf Management: The Iranian Experience. In Monzer (Ed.), *Essential Readings In Contemporary Waqf Issues* (pp. 269-320). Malaysia: CERT Publications.

Rahman, M. F. B. A., Mastuki, N. a., & Yusof, S. N. S. (2015). Performance Measurement Model of Mosques. *Procedia Economics and Finance, 31*, 26-35.

Ramli, N. M., & Muhamed, N. A. (2013). *Good Governance Framework for Corporate Waqf: Towards Accountability Enhancement.* Paper presented at the Paper accepted for Proceedings Of World Universities' Islamic Philanthropy Conference.

Rashid, S. K. (2012). Measures For The Better Management Of Awqaf. *Iium Law Journal, 20*(1), 103-137.

Sabiq, A. S. (2004). *Fiqh el Sunnah*. Mesir: Al Fath lil I'lam el Arabi.

Scott, A., Burall, S., Highton, N., & Wong, K. (2008). *Assessing Multilateral Organisation Effectiveness*. Retrieved from

Shafii, Z., Yunanda, R. A., & Rahman, F. K. (2014). Financial And Operational Measures Of Waqf Performance: The Case Of State Islamic Religion Council Of Singapore And Malaysia.

Sillanpää, V. (2011). Performance measurement in welfare services: a survey of Finnish organisations. *Measuring Business Excellence, 15*(4), 62-70.

Siraj, S. A. (2012). *An Empirical Investigation Into The Accounting, Accountability and Effectiveness of Waqf Management In The State Islamic Religious Councils (SIRCs) In Malaysia.* (Thesis), Cardiff University.

Stewart, J., & Walsh, K. (1992). Change In The Management Of Public Services. *Public Administration, 70*(4), 499-518.

Sulaiman, M., Adnan, M. A., & Nur, P. N. S. M. M. (2009). Trust Me! A Case Study of The International Islamic University Malaysia's Waqf Fund. *Review of Islamic Economics, 13*(1), 69-88.

Sulaiman, M., & Zakari, M. A. (2015). Efficiency and Effectiveness of Waqf Institutions in Malaysia: Toward Financial Sustainability. *Bloomsbury Qatar Foundation Journals, 1*(43-53).

Yaacob, H. (2006). Waqf Accounting in Malaysia State Islamic Religious Institutions (SIRC) : The Case of Federal Territory SIRC. *Master's Dissertation, International Islamic University.*

Zheng, W., Yang, B., & McLean, G. N. (2010). Linking Organizational Culture, Structure, Strategy, And Organizational Effectiveness: Mediating Role Of Knowledge Management. *Journal of Business Research, 63*(7), 763-771.

AN EXPLORATORY STUDY OF THE RHETORIC IN CSR REPORTS: A COMPARISON OF SHARI'AH COMPLIANT COMPANY AND NON –SHARI'AH COMPLIANT COMPANY IN MALAYSIA

Rafizan Abdul Razak*
Universiti Teknologi MARA, Cawangan Negeri Sembilan

ABSTRACT

This paper advances previous research in corporate social reporting (CSR) by using a taxonomy of five distinct rhetorical strategies to examine any differences in term of type of ethical categories and to identify moral attitudes towards different societal activities between Shari'ah compliant company Bank Islam Malaysia Berhad (BIMB) and Non-Shari'ah compliant company Malayan Banking Berhad (Maybank). The data is analyzed using content analysis. The findings revealed that there are different rhetoric applied in both types of companies such as Agency (profit), Benefit (collective welfare), Compliance (law and contracts), Duty and Ethos (virtue). In term of moral attitudes, both types of companies disclosed relatively similar behaviour and commitment towards CSR. This research is the first of its kind to be conducted in Malaysia by using this new approach and will contribute new insights to existing CSR literatures in Malaysia.

Keywords: Agency, Benefit, Compliance, Duty and Ethos.

INTRODUCTION

Ethics and corporate social responsibility are related to each other in many ways. The behaviour of companies in constantly disclosing their social responsibility shows that they are morally responsible and ethical companies. The stakeholders and the public will normally give preferences and trust to those companies that they perceived as having a good behaviour and/or conduct.

Yunanda and Abd. Majid (2011) highlighted that the erosion of trust by the business's stakeholders is mainly due to the immoral bahaviour of the business's administrators ranging from chief executive officers to lower level managers and its employees. Such act may tarnish the status of the company and subject to lawful approval.

On the other hand, Ditlev-Simonsen and Wenstop (2011) highlighted different dimension on corporate social disclosure research by looking at what ethical perspectives corporation actually had applied in the companies' corporate social responsibility (CSR) report rhetoric. In their study, they developed a taxonomy of rhetorical strategies which consist of five areas.

* Corresponding author's email: rafizanabdulrazak@gmail.com

The taxonomy is used for describing the purpose of CSR and it consist of Agency (profit), Benefit (collective welfare), Compliance (law and contracts), Duty and Ethos (virtue). Their result had shown that different rhetoric is applied in companies' report where ethos had been the most common ethical perspective expressed in the report. †

In Malaysia, the CSR study focuses on the nature, status or trend, quality, quantity and factors of disclosure in both Non- Shari'ah - compliant and the Shari'ah compliant companies' annual reports (Anuar et.al., 2004, Yahya et al., 2005, Amran and Devi,2008; Said et al., 2009; Mohamad et al., 2010 ; Esa and Ghazali., 2012). However, none has been noted to have discussed the type of ethical categories and moral attitude of company in divulging CSR in the annual report. Therefore, by using the taxonomy of rhetorical strategies framework, this study intends to examine the following two research questions:

1. Are there any differences in terms of the type of ethical categories between Shari'ah compliant and Non-Shari'ah compliant companies in Malaysia?
2. To identify any differences in terms of moral attitudes towards different societal activities between Shari'ah compliant and Non-Shari'ah compliant companies in Malaysia.

Thus, this study seeks to contribute to the literatures by reviewing and documenting CSR reporting on the area concerned in Malaysia. The present study will also extend the literatures on CSR in relation to Shari'ah compliant companies as no recent studies in the area have been published in Malaysia. A comparison between Shari'ah compliant and Non-Shari'ah compliant companies would be a meaningful contribution given the fact that in depth studies on this area are limited.

LITERATURE REVIEW

SHARI'AH COMPLIANT AND NON-SHARI'AH COMPLIANT COMPANIES

In Malaysia, the Securities Commission (SC) was established on 1 March 1993 under the Securities Commission Act 1993 to regulate the capital market. In 1996, the SC has set up Syariah Advisory Council (SAC) to provide requirement of specific investors to invest in capital market that abides with Shariah rules and principles. According to Ousama and Fatima (2010) the SAC applies standard criteria in classifying the companies listed in Bursa Malaysia as approved and non-approved companies based on their core activities. Companies with major activities which are not conflicting to the Shariah principles are categorized as Shari'ah compliant whereas Non-Shari'ah compliant are companies involved in primary activities deemed not lawful according to the Shariah.

ETHICS AND CSR

There were many definitions of ethics which had been stated in previous literatures (Freeman and Gilbert, 1988; Raiborn and Payne, 1990; Carroll, 1991). Brenda and Payne (2002) stated that the definitions of ethics and morality are cross-referenced to each other. According to De George (1999) as cited in Brenda and Payne (2002), he defined business ethics as the interaction of ethics and business. Such a definition encompasses a moral evaluation of the

economic system of the free enterprise system in the United States, the businesses which operate in this system, a moral evaluation of individuals and their actions in conducting business and a review of business behavior in the international arena. De George provides further illumination: moral judgements should be universally applicable, they involved serious matters with potential to cause serious results, and moral judgements invoked praise or blame. Additionally, moral judgements can only be made by individuals for themselves: others, including governmental agencies, cannot forced moral judgements on anyone.
Brenda and Payne (2002) stated that businesses will in fact engage in ethical business practices for one of two reasons, one ethical in nature and the other one is for more Machiavellian. The ethical motivation guiding business is related to a desire to do the right thing, without external pressure or governmental constraint. Furthermore, Brenda and Payne (2002) highlighted that the more Machiavellian approach that businesses espouse in their use of ethics has its roots in a desire to convince the stakeholder that the firm is doing the right thing.

On the other hand, according to Carroll (1979) as cited in Dusuki and Tengku Mohd Yusof (2008), the definition of CSR is too wide in that it encompasses economic, legal and voluntary activities. Carroll (1979: 500) defines CSR as "the social responsibility of business encompasses the economic, legal, ethical, and discretionary expectations that society has of organizations at a given point in time". In order for the business to fully address their obligation to society, all these four categories of expectation on the economic, legal, and ethical and philanthropy (discretionary) must be met.
Therefore, we can assert that a company's action is influenced by the desire to evade legal consequences of its actions or to persuade the stakeholders that the company does have their best interests at humanity and seeks to assist their interests rather than their own. This is where we can normally see in companies' annual report under their CSR's section or stand-alone sustainability report.

ETHICS RHETORIC IN CSR REPORTING

Ditlev-Simonsen and Wenstop (2011) defined rhetoric as the art of using language to communicate effectively which is applied to convey message. They anticipated corporation presumptively apply ethics rhetoric in their CSR reports to achieve certain communication goal such as enhancing legitimacy. Furthermore, Ditlev-Simonsen and Wenstop (2011) emphasized that the ethical attitude behind a company's CSR is presumably a determining element for its CSR activities and a relevant place to look for the expression of such attitudes is the rhetoric applied in CSR reports. Therefore, this study is conducted to fill the gap in this area in Malaysia.

DATA COLLECTION AND METHODOLOGY

This study focused on disclosure made in two corporate annual reports. Since this is an exploratory study and due to time constraint, the author focused on Bank Islam Malaysia Berhad (BIMB) to represent Shari'ah compliant company and Malayan Banking Berhad (Maybank) to represent Non-Shari'ah compliant company to make the comparisons. BIMB was selected as a subject because it is the first Islamic bank established in Malaysia and is based on Shari'ah principle. On the other hand, Maybank was selected because the company had won one of the NACRA awards for year 2014. This study embraced the taxonomy of ethics rhetoric applied in Ditlev-Simonsen and Wenstop (2011) which consists of five

DOI: 10.1051/shsconf/20173600006

categories to classify the rhetoric strategies revealed in the two annual reports. The following section will describe the definition of five rhetoric strategies.

AGENCY

Ditlev-Simonsen and Wenstop (2011) described agency in their study as a criteria that linked to the ethical perspective of egoism. However, they looked at wider perspective than individual level of egoism. In their taxonomy, agency is an extension of egoism. According to Ditlev- Simonsen and Wenstop (2011), for example, in a typical business organization, the motives of its owners by extension also become the motives of the agents employed to work on the organization. Therefore, in this study the author used the same notion to identify the ethical commitment made by BIMB and Maybank.

BENEFIT

Ditlev-Simonsen and Wenstop (2011) described benefit as a criteria that linked to benevolent consequentialism and utilitarianism. It expressed a commitment to further interests, welfare or well-being of stakeholders outside the boundaries of the firm. Thus, in this study the author is interested to compare how Shari'ah and Non- Shari'ah companies expressed their concerns to customers and social good and the environment.

CONTRACT

According to Ditlev-Simonsen and Wenstop (2011) contract is a criterion that linked to contractualism where it can be thought as either actual or ideal. For instance, organization will follow existing rules and regulations and not go beyond what is required from them. Therefore, in this study the author would like to know how Shari'ah and Non- Shari'ah companies are different from each other.

DUTY

Next, Ditlev-Simonsen and Wenstop (2011) described duty as criteria relates to ethical duty. They further conferred that there are certain things that organization should engaged in which is independent of laws and regulations. In other words, the organization should have embarked into certain things as other organization did. Hence, in this study the author is concerned about the similarities between Shari'ah and Non- Shari'ah companies as well as their differences.

ETHOS

Lastly, Ditlev-Simonsen and Wenstop (2011) defined ethos as criteria that expressed a commitment to virtues or to avoid, (how one would want to be as opposed to look at what to do). According to them, ethos needs to be separated out as distinct rhetoric strategy because it communicates a distinctive sort of moral commitment. In this study, the author intends to highlight which company used its CSR reporting to express their moral commitment in most extensive manner.

METHODOLOGY

SELECTION OF QUOTES AND INTERPRETATION

The annual reports of BIMB and Maybank were read thoroughly to identify the social responsibility disclosure. Similar with Rahman et.al (2010), this study used a number of CSR sentences to identify ethical commitment. Milne and Adler (1999) conferred that using sentences for coding and measurement would provide complete, reliable and meaningful data for further analysis. Therefore, the number of sentences was chosen to classify ethical rhetoric. Then, the quotes were assigned and grouped to ethical categories based on a checklist of disclosure item designed by Ditlev-Simonsen and Wenstop (2011) above.

Once the sentences or paragraph were group to ethical categories, we also identify the moral attitudes as to why the company disclosed it. As stated in Ditlev-Simonsen and Wenstop (2011) one company may emphasize a commitment to benefit, but also state that it is doing so in order to meet a commitment to agency. They further elaborated that the end motive here is agency and the pattern of moral reasoning is that benefit is an instrument for agency.

FINDINGS

The following discussion encompasses the result of the study.

AGENCY

Both companies disclosed their CSR activities related to agency strategies. This is anticipated as the general objective of doing business is to make profit. Apart from focusing on the bottom line, both companies do cognizant about sustainability of their organization operation and activities as can be seen from their involvement and commitment in CSR. The followings are the quotes on agency strategy in BIMB:

"The *core of the Bank's corporate social responsibility programme is driven by Shariah principles that pertain to uplifting the lives of destitute and poor Muslims through the contribution of Zakat, or obligatory alms. As our business revenue and profit grow, so does our Zakat contribution. In 2014, the Bank's Zakat contribution exceeded RM12 million. A significant portion of this was channeled to fund vital social causes across Malaysia with emphasis on poverty reduction projects, education initiatives, health and welfare support and unity programmes".*

(BIMB Annual report 2014, page 28)

It is clear that BIMB emphasised on the welfare of the community and their obligation as Shari'ah compliant company to share the business income. This is coherent with the Tawhid concept as stated in Yunanda at el (2011) where believing in one and the only God is the paramount essence of Islam. This leads to the conformity of Qur'anic ordinances and Sunnah as the primary rules and man-made rules in particular. The above statement also emphasis that as a Shari'ah compliant company, BIMB moral attitude is driven by their responsibility for distribution of wealth and financial interest as a secondary objective.

On the other hand, the following is the quote pertaining to the agency strategy in Maybank:
"Our sustainability progress is in line with our mission to humanise financial services. In 2014, the Board-approved 20/20 Sustainability Plan took effect with the objective of

implementing an impact-based approach focused on environment, social and governance (ESG) issues. Through the 20/20 Sustainability Plan, we aim to align our sustainability progress with our corporate strategy in order to achieve holistic growth for all of our stakeholders. We want to create enduring meaning and value across the workplace, marketplace, environment and community in ways that are globally significant and yet locally relevant. The achievement of these goals and the realisation of our value proposition contribute to our business competitiveness and are significant to maintaining our social licence to operate. As the region's leading financial institution and to prepare our business for the future, we have to embed sustainability into our operations and conduct our business in away that creates value *for our stakeholders over the short and long term. We also have the responsibility to manage our impacts positively".*

(Maybank Annual report 2014, page 46)

Similar with BIMB, Maybank's attention on CSR effort can be seen on their *20/20 Sustainability Plan.* However, as a Non-Shari'ah compliant company, the motives behind their CSR activities is more on business ethics in general or doing what is right. Non-Shari'ah compliant company perceived that in achieving their business goal, they do belief towards portraying the social contribution activities to be accepted in community and industry.

BENEFIT

The next taxonomy is benefit. From the analysis, we found that both companies disclosed the information about how they are focusing on their benevolents to the stakeholders, environment and community in which they operate.
The following quotes were reported in BIMB:

"The development of future talent for the industry is a core concern of the Bank. Towards this end, the Bank has launched a number of Human Resource initiatives targeting the development of its own 4,242 strong workforce as well as supported government launched programmes to help groom the potential of fresh graduates; namely Skim Latihan 1Malaysia ("SL1M") and the Graduate Employability Management Scheme ("GEMS")".

(BIMB Annual report 2014, page 30)

"The best workplace produces a productive workforce who work towards one goal in unison. In building an effective team who are driven by a common purpose with individual measurable goals, the Bank understands that good management, clear communication and building trust and rapport through a two-way communication is necessary. Employee engagement activities outside the workplace accelerate this strengthening of comradeship. In 2014, activities aimed at promoting healthy lifestyles and building team spirit amongst the employees continue to take centre stage. The HR Division co-organised the Bank Islam Sports Carnival 2014 in Kuching, Sarawak, Bank Islam's Health Day and other sports activities such as the National Banks Sports Council Futsal Tournament, National Banks Sports Council Carrom Tournament and Futsal Challenge, Piala Dato' Sri Zukri. Hari Raya and Buka Puasa gatherings also continued to provide occasions to gather in positive spirit."

(BIMB Annual report 2014, page 150)

The above report revealed that BIMB focused more on development of human capital within the industry as well as the nation too. At the same time BIMB also recognized healthy lifestyles among their employees.

On the other hand, the following quotes were highlighted by Maybank:

"People

We are constantly working towards providing our people with the tools and support in order for them to perform better. In 2014, we created an inclusive working environment through various employee engagement programmes, employee training programmes as well as rewarding our people fairly for their work.

Access to service and product

We are committed to having deep knowledge of our clients, including understanding their social and environmental needs. At the same time, we believe that our business will only thrive when embedded with elements of good governance, social innovation and environmental integrity. Our efforts to-date were made to enhance our services by improving the accessibility and convenience for our customers and to make valuable contributions to the communities where we operate.

Focus on Customers

In 2014, we continued on our mission to humanise financial services regionally with the launch of various products and services aimed at our improving the accessibility and efficiency of our services, reaching out to the financially underserved, and expanding our physical and digital footprint without compromising the security of our customers' data and information.

Access to Finance

As a financial institution, our services provide access to finance that spreads across Asia. We have one of the widest networks in Malaysia and the only bank with branches in 18 rural areas in the country. 16 of our branches in Malaysia are disabled-friendly whilst all 55 Maybank branches in the Philippines are catered to people with disabilities.

Product Stewardship

Our commitment to enhance access to financing for all levels of SMEs was reflected in the Portfolio Guarantee (PG) agreement with the Credit Guarantee Corporation (CGC). A sum of RM1 billion was set aside to provide financing to viable SMEs under the PG schemes, a portion of which will be allocated to the Bumiputera segment, thus making it the largest agreement of its kind to be inked between CGC and any financial institution in Malaysia. We provided further support to SMEs by partnering the Bumiputera Agenda Steering Unit (TERAJU) to provide RM600 million funding to high-performing Bumiputera SMEs. This partnership expanded the Teras Fund to RM2.2 billion. This fund, created by TERAJU, will assist SMEs to obtain working capital to grow their businesses".

(Maybank Annual report 2014, page 204-205)

We can see that there were no differences between Shari'ah compliant and Non-Shari'ah compliant companies in terms of their involvements and roles in giving back to the stakeholders and the society. The only difference is in terms of presentation of the strategies whereby Maybank has remarked their commitments in a more comprehensive and structured manner compared to BIMB. In terms of expression, it has shown that Maybank's reports are more enlightened in winning the heart of the reader. Form the above paragraph, the benefit instrument also act as commitment to achieve agency commitment. In addition, in terms of moral attitude, there were no differences Shari'ah compliant and Non-Shari'ah compliant companies. Both type of companies expressed the same attitude to take a good care of their stakeholders.

CONTRACT

The next taxonomy is contract. The BIMB report has articulated a contract commitment which is bound by the bank to ensure fair and proper distribution of Zakat funds to those who need it (Asnaf). This is consistent with research done by Abdul Rahman, Mohd. Hashim and Abu Bakar (2010) which indicated that Zakat obligations and Sha'riah Supervisory Council (SSC) were disclosed in BIMB reports from year 1992 to 2005. This is due to the fact that payment pertaining to Zakat matters, reports and attestation of SSC, as well as the investment account, are required to be disclosed by paragraphs 71 and 72 of the FRS i-1.
The following quotes were reported in BIMB:

CAPTURING THE SPIRIT OF SOCIAL DEVELOPMENT

"As Malaysia's leading Islamic financial institution, many of our CR efforts are underpinned by Zakat contributions. In line with Islamic principles and laws, it is our utmost priority to ensure fair and proper distribution of Zakat funds for a variety of noble causes and to contribute to the nation's social development. Bank Islam remains the leading corporation in Malaysia in developing and organising Zakat-funded CR projects, playing an important role in disseminating this wealth among the community.

As Bank Islam's asset and performance continue to grow, so has our Zakat fund increased to support our CR initiatives. Our spending over the years in helping to build better lives for those who truly need it is a reflection of Bank Islam's commitment to promote social justice and foster a sense of solidarity in society."

(BIMB Annual report 2014, page 131)

The above statement also highlighted the moral attitude of Shari'ah company that is tied with Islamic principles and laws. This is consistent with research done by Wan Abdullah, Percy and Stewart (2013). On the other hand, in Non-Shari'ah compliant company their contract is linked with the growth and stability of the company in the present and future. The company will look for the opportunity to portray a good business model to environment in which they operate.

The following is the quote pertaining to the contract strategy in Maybank:

"Sustainability at Maybank means conducting business in a way that enhances performance responsibly - today and over the long-term. We continue to serve ASEAN communities while strengthening and expanding our operations regionally. Our sustainability journey is a reflection of Maybank's position as a regional leader in the provision of financial services and one that supports business growth through impact based approaches."

(Maybank Annual report 2014, page 198)

DUTY

Next taxonomy is related to duty. The reports in BIMB and Maybank focused on community and human capital development. Both Shari'ah and Non-Shari'ah compliant companies engaged in many activities that show their concerned and responsibility to fulfil the need of people in which the business operate.

The following are the quotes reported in BIMB:

"Since 2008, our Housing Aid Project has been helping to provide a roof over the heads of those suffering from extreme poverty. We believe that this effort is pivotal not only in providing a basic necessity to those in need, but also in enabling them to gain independence to improve their livelihoods.

The holy month of Ramadan is a time of giving and sharing. While Bank Islam continuously strives to foster the spirit of goodwill throughout the year, the fasting month provides us with an opportunity to intensify our CR efforts to improve the well-being of the communities we operate in.

Bank Islam's affiliation with the PINTAR Foundation in its School Adoption Programme has made it possible for us to organise motivational programmes with the Malaysian army, as part of our efforts in educating and imparting useful skills to the primary school students.

As part of our commitment to the environment in helping create a sustainable future, Bank Islam looked to the ecological paradise that is Sabah in launching its "One Earth, One Quest – Mangroves, Our Priceless Treasure" campaign. By making a contribution of RM50,000 to Sabah Wetlands Conservation Society, we participated in the mangrove trees re-planting effort".

(BIMB Annual report 2014, page 131-136)

The following are the quotes reported in Maybank:

"Since its inception in 1972, Maybank has sponsored more than 1,000 scholars as well as provided them with employment opportunities. Driving education at all levels will enable us to build human capital throughout the region where we operate. Maybank Foundation also offered scholarships to 33 students in Indonesia to pursue their tertiary education in Indonesia, Malaysia and Singapore. The regionalisation of our scholarship awards will be a key driver in our support of educational programmes moving forward.

Under our schools adoption programme, PEERS, we commit to improve the standard of education in rural schools through academic support as well as personal development and capacity building. Together with selected partners, we adopted 15 schools in three areas under the PEERS programme, rewarding achievements, providing ICT facilities as well as promoting language proficiency in English. ICT.

CashUP Programme

The programme aims to encourage low-income families to cultivate the habit of saving money. The CashUp programme provides 1 to 1 ratio matching payment to participating families which essentially doubles their savings.

Maybank Family Protection Programme

The programme has benefited 600 families, insuring the breadwinners of the family by covering work related accidents and injuries. The targeted beneficiaries were chosen from families with total household income of less than RM5,000".

(Maybank Annual report 2014, page 200-201)

From the above reports also, we found that there were similarities in term of moral attitudes between Shari'ah and Non-Shari'ah compliant companies that driven their reporting.

ETHOS

The last ethical strategy is ethos as to how the company would like to be seen by the industry or community. This strategy includes any action taken and long term commitment showed by the companies to contribute to the society in general and how the companies ensure their sustainability.

The following are the quotes reported in BIMB:

"As Malaysia's pioneer Islamic bank, Bank Islam has established a rich history of giving back to the community and environment. In adherence to the Shariah principles that shape the Bank's business, responsible corporate practices make up the very essence of the Bank's operations. With the aim of ensuring the sustainability of our business and long-term well-being of those around us, our Corporate Responsibility ("CR") practices are thus geared not only towards preserving the bank's legacy, but also towards securing its future, and the future of the communities we serve, for years to come. We also seek to cement this sustainability through our efforts to adapt to society's evolving needs and contribute to the overall wellness of local communities.
To this end, our CR programmes focus on creating long-term value for underprivileged communities through education, welfare and enrichment initiatives to help them attain a brighter tomorrow. At the same time, we stand ready to channel financial and medical aid to those who need it the most. We are also firm believers in volunteerism, encouraging our staff and their families to engage with the less fortunate and contribute to the betterment of their quality of life".

(BIMB Annual report 2014, page 130)

From the above, it is reported that BIMB as a pioneer in Islamic bank in Malaysia will continue their legacy as a leading in Shari'ah compliant companies. Therefore, from the report we can see the link between duty and ethos whereby BIMB's commitment in fulfilling its duty would also help them in realizing ethos.

Followings are the quotes reported in Maybank:

"The three pillars serve as a guide in our efforts to achieve the targets set and move purposefully towards the ten commitments which we began tracking in 2015. The ten

commitments, which are to be implemented by different departments, promote Maybank's sustainability focus that cuts across the three pillars.

The 20/20 Sustainability Plan

People: To build the most engaged and inclusive workplace to help us connect with customers and make better, more informed decisions for our business.
Community and Citizenship: To deliver long-term value for the community and environment by being the region's most impactful corporate citizen.
Access to Products and Services: We plan to integrate ESG considerations into our products and services by 2020 to practise responsible financing, promote financial inclusion and help our customers build financial capability".

(Maybank Annual report 2014, page 198- 215)

"In 2014, we continued with the mission to humanise financial services by implementing activities and high-impact initiatives in our efforts to uplift the lives and wellbeing of underprivileged Malaysians, which has brought the bank closer to the community in many ways".

(Maybank Annual report 2014, page 154)

Maybank on the other hand, had showed their exertion through long term vision as highlighted above. The company had a very clear direction and focused group. All engagement activities in CSR in the company basically will comprehend the 20/20 Sustainability Plan aka ethos.

CONCLUSION

This study has been conducted to classify the type of ethical categories between Shari'ah compliant and Non-Shari'ah compliant companies in Malaysia and to identify any differences in terms of moral attitudes towards different societal activities between Shari'ah compliant and Non-Shari'ah compliant companies in Malaysia. We used the taxonomy of ethics rhetoric applied in Ditlev-Simonsen and Wenstop (2011) which consists of five categories to classify the rhetoric strategies revealed in the two annual reports which were BIMB and Maybank. From the above discussion, both companies expressed their CSR reporting in all five categories of the ethical strategies which are Agency, Benefit, Contract, Duty and Ethos. Since CSR disclosures are voluntary in nature, both companies are freely to communicate any types of rhetoric ethical categories.

However, in term of moral attitudes, BIMB, as Shari'ah compliant company are bound to Islamic principles and laws. Therefore, any activities conducted by the BIMB in relation to business matter or CSR initiatives are tied to Qur'anic ordinances and Sunnah. For instance, in this study the Agency and Contract ethical strategies reported in BIMB are clearly rely on Shari'ah principles. On the other hand, Maybank as Non-Shari'ah compliant company concerned on making profit as the first priority and doing good for society at the same time.

The limitation of this study lies in the data. Due to the limited financing and time this study only used two annual reports. One to represent Shari'ah compliant company and the other one for Non- Shari'ah compliant company. Nevertheless, even with this limitation, the contribution is found to be significant knowing that such CSR study using the taxonomy of five rhetoric ethical strategies is still new in Malaysia. Future study can be conducted by using larger sample and industry.

REFERENCES

Abdul Rahman, A., Mohd. Hashim, M.F.A., and Abu Bakar,F. (2010) "Corporate social reporting: a preliminary study of Bank Islam Malaysia Berhad (BIMB)", *Issues in Social and Environmental Accounting*, Vol. 4, No. 1 June 2010, pp. 18-39.

Amran, A. and Devi, S.S. (2008) "The impact of government and foreign affiliate influence on corporate social reporting: the case of Malaysia", *Managerial Auditing Journal*, Vol.23, No.4, pp. 386-404.

Anuar, H.,Sulaiman, M. and Ahmad,N. (2004) "Environmental reporting of Shari'ah approved companies in Malaysia", *Proceeding of International Accounting Conference II, International Islamic University*, Kuala Lumpur.

Bank Islam Malaysia Berhad (BIMB) Annual report 2014.

Brenda E. Joyner and Payne, D. (2002) "Evolution and Implementation: A study of values, business ethics and corporate social responsibility", *Journal of Business Ethics,* 41, pp. 297–311.

Carroll, A. B. (1991) "The Pyramid of Corporate Social Responsibility – Toward the Moral Management of Organizational Stakeholders", *Business Horizons*, 34, 4: pp. 39-48.

Ditlev-Simonsen.C.D. and Wenstop, S. (2011) "Companies' ethical commitment-an analysis of the rhetoric in CSR reports", *Issues in Social and Environmental Accounting*, Vol.5, No1/2 December 2011, pp. 65-81.

Dusuki, A.W and Tengku Mohd Yusof, T.F.M. (2008) "The pyramid of corporate social responsibility model: empirical evidence from Malaysian stakeholder perspectives", *Malaysian Accounting Review*, Volume 7, No. 2, pp29-54.

Esa, E. and Mohd. Ghazali, N.A. (2012) "Corporate social responsibility and corporate governance in Malaysian government-linked companies", *Corporate Governance*, Vol. 12, No.3, pp. 292-305.

Freeman, R. E. and D. E. Gilbert Jr. (1988) "Corporate strategy and the search for ethics", Prentice Hall, Englewood Cliffs, NJ.

List of Shariah-Compliant Securities by the Shariah Advisory Council of the Securities Commission Malaysia.

Malayan Banking Berhad (Maybank) Annual Report 2014.

Milne, M.J. and Adler,R.W. (1999) "Exploring the reliability of social and environmental disclosures content analysis", *Accounting, Auditing & Accountability Journal*, Vol. 12 Iss: 2, pp.237 – 256.

Mohamad, R., Alwi, K. and Jamil, C.Z.M. (2010) "Sustainability disclosure among Malaysia Shari'ah- Compliant listed companies: Web reporting", *Issues in Social and Environmental Accounting*, Vol. 3, No. 2, pp. 160-179.

Ousama, A.A. and Fatima, A.H. (2010) "Factors influencing voluntary disclosure: empirical evidence from Shariah approved companies", *Malaysia Accounting Reviews*, Vol.9, No. 1, pp. 85-103.

Raiborn, C.A. and Payne, D. (1990) "Corporate codes of conduct: a collective conscience and continuum", *Journal of Business Ethics* 9, pp. 897– 889.

Said, R., Zainuddin, Y.H. and Haron, H. (2009) "The relationship between corporate social disclosure and corporate governance characteristics in Malaysian public listed companies", *Social Responsibility Journal*, Vol.5, No.2, pp. 212-226.

Yahya, M.A., Abul Rahman, A. and Tayib, M. (2005) "The relationship between corporate social disclosure and Islamic unit trust shareholding", *Accounting, Commerce and Finance: The Islamic Perspective Journal*, Vol.9, No. 1, pp. 95-63.

Yunanda,R.A and Abd. Majid, N. (2011) "The contribution of Islamic ethics towards ethical accounting practices", *Issues in Social and Environmental Accounting*, Vol. 5, No. 1/2 , pp. 124-137.

Wan Abdullah, W.A., Percy, M. and Stewart, J. (2013) "Shari'ah disclosures in Malaysia and Indonesian Islamic Bank", *Journal of Islamic Accounting and Business Research*, Vol.4, No.2, pp. 100-131.

CORPORATE GOVERNANCE & AUDITOR CHOICE IN MALAYSIA

Wan Asma Wan Nasrudin*, Norhayati Mohamed, Nur Aima Shafie
Universiti Teknologi Mara

ABSTRACT

The aim of this paper is to investigate the determinants of firm's auditor choice in Malaysia in respect of their corporate governance mechanisms. A logit regression model was developed to test the impact of firms' internal corporate governance mechanism on auditor choice decisions made by public listed companies listed on main board of Bursa Malaysia from year 2006 to 2015. Five variables are used to proxy for firm's internal corporate mechanism which are the ownership concentration, the duality of CEO and chairman of BOD, the size of audit committee, the size of BOD and the number of independent directors on the board. All auditors in Malaysia were classified into Big Four and non-Big Four, assuming Big Four auditors can provide higher quality audit services. The final result show that firms with less concentrated ownership, with larger size of audit committee, larger size of the BOD, with lower proportion of independent directors on the board, or in which CEO and BOD's chairman are not the same person are more likely to hire a high-quality auditor. Hence, it suggests that when benefits from lowering capital raising costs are trivial, firms with good corporate governance mechanism are prone to choose a high-quality auditor.

Keywords: Corporate Governance, Auditor Choice, Logit Regression, Firm Risk, Largest Ownership

INTRODUCTION

This paper studies the role of internal corporate governance mechanisms as the deciding factors that determine auditor choice decision in Malaysia. This study is inspired by the growing body of researches that scrutinizes auditor choice decision (Ahmad, Houghton and Yusof, 2006; Beattie and Fearnley, 1995; Chaney, Jeter and Shivakumar, 2003; Guedhami, Pittman and Saffar, 2009; Hope, Kang, Thomas and Yoo, 2008; Johnson and Lys, 1990; Linda Elizabeth, 1982; Mahdavi, Maharlouie, Ebrahimi and Sarikhani, 2011; Pittman and Fortin, 2004; Wang, Wong and Xia, 2008).

Prior study (Fields, Lys and Vincent, 2001), mentions that financial disclosure and reporting are essential in order to address the imperfections and incomplete financial asymmetry within the financial world. Information asymmetries in general are related with the bond between managers and investors in which managers who are better informed tend to disseminate less information to

* Corresponding author' email: wan9513@ns.uitm.edu.my

investors. Nurwati and Wan Nordin (2010) stress that to minimize agency problems, what is needed is to reduce the information asymmetry between management and shareholders, which can be gained through transparent financial disclosure. In addition, the after effect of corporate scandals such as the Enron shocking incident and other current financial crisis further lead the policy makers and regulators to draw more attention to the needs for a better quality and more transparent financial reporting (Bardos, 2011).

LITERATURE REVIEW

CORPORATE GOVERNANCE IN MALAYSIA

Prior study such as that by Alnasser (2012) notes that corporate government scandals have persuaded scholars to give meaning to corporate governance based on their perspectives on problem solving and corporate affairs. As such, a common definition sets down that internal governance mechanisms take into account elements that relates to the firm's internal organizational structure, such as, board of directors and ownership structure while the external governance mechanisms includes taking over the markets and the legal system.

The turning point of corporate governance reformation in Malaysian was after the Asian financial crisis in 1997 (Alnasser, 2012; Kamardin and Haron, 2011; MCCG, 2012). The outbreak of the crisis taught the policy makers valuable lessons and placed their attention, amongst others, on the necessity to raise corporate governance standards. It pointed out the weaknesses in governance practices and made it clear for the need to establish a more responsible, and transparent management in step with international best practices.

Seeing how corporate governance has become an essential part of the Malaysian business environment, it is therefore important to study the relative influence it has on firms' decisions and performances. One particular matter is its implication on auditor choice decision as many prior studies have delved into the association between these two variables (Guedhami, et al., 2009; Lin and Liu, 2009; Mahdavi, et al., 2011; Wang, et al., 2008).

CORPORATE GOVERNANCE, AUDITOR DEMAND AND AUDITOR CHOICE

The interrelations between corporate governance and auditor choice have been the issue of many prior discussions and studies. Mahdavi, et al. (2011) look into the influence of corporate governance mechanisms on auditor choice among Iranian listed firms. They explore the impact of ownership structure, concentration of ownership, board composition and duality of the CEO and chairman roles as the proxies of corporate governance mechanisms on firms' auditor choice decisions. The researchers conclude that the corporate governance mechanisms make a difference to the choice of auditor in firms listed on Tehran Stock Exchange, where the possibility of a high-quality audit firm being chosen will increases in line with the percentage of outside directors.

AUDITOR AND AUDIT QUALITY

Based on previous research paper on Malaysian corporate governance, there are three classifications of people that act as “external enforcers” of good governance identified by The Finance Committee: the auditors, corporate advisers and regulators. In other words, auditors perform a corporate governance role in monitoring a firm's financial reporting process, or they act as “watchdogs” of the firms (Ashbaugh and Warfield, 2003). Audit function plays a crucial role not only to monitor managerial actions but also to create a better information environment as well as to provide a secondary source of assurance against corporate failures (Wallace, 2004). As such, auditors are required to give appropriate assurance through their opinions on whether the firms’ annual accounts have been properly drawn up and in compliance with the approved standards, and if they portray a true and fair view of the firms’ affairs.

CORPORATE GOVERNANCE PROXIES

OWNERSHIP CONCENTRATION AND AUDITOR CHOICE

In Malaysian public listed firms, shareholding is highly concentrated in the hands of small number shareholders (La Porta, Lopez-de-Silanes, Shleifer and Vishny, 2000; Samad, 2002). Samad (2002) reports that the average of the first largest shareholding was 30.30 percent and the five largest shareholdings were 58.84 percent, which accounted for more than half of the voting shares. Furthermore, Kamardin and Haron (2011) mention that around 71.4 percent of firms on the Main Board and Second Board of Bursa Malaysia were controlled by their five largest shareholders, having a shareholding of more than 50 percent and were under majority ownership.

SIZE OF AUDIT COMMITTEE AND AUDITOR CHOICE

The audit committee is fundamental to the concept of corporate accountability and sound governance. It has become an integral part of the corporate framework to help fulfil board’s stewardship accountability to its shareholders and financial stakeholders. An audit committee provides the board with assurance of the quality and reliability of financial information used by the board and of the financial information issued publicly by the firm. Although the board as a whole is responsible for the accuracy and integrity of the firm’s financial reporting, it can be difficult for all directors to keep up with the complexities of financial reporting. That is where the role of the audit committee emerges as it provides an additional and more specialized oversight of the financial reporting process by facilitating the discharge of the board’s responsibility with respect to the timely preparation and issuance of financial statements (MCCG, 2012).

DUALITY OF ROLES OF BOD CHAIRMAN AND CEO, AND AUDITOR CHOICE

The Companies Act, 1965 (CA) states that the business and affairs of a firm must be managed by, or under the direction of, the Board of Directors (BOD). The BOD is obligated to play an active role in directing management. While this does not mean they should be involved in operational matters, it certainly implies that board cannot passively endorse decisions of the management. As such the BOD's role is to provide entrepreneurial leadership of the company within a framework of prudent and effective controls, which enable risk to be assed and managed.

SIZE OF BOARD OF DIRECTORS AND AUDITOR CHOICE

Due to limitations of corporate structure since its inception, shareholders take a precaution step by appointing representative to serve their interests. These shareholders advocates took the form of the "Corporate Board of Directors" whose responsibilities were to meet with and oversee the management of the entity and look out for the interest of the owners. The role of the board is often referred to as the role of steward whose the main responsibility was to ensure the interest of shareholders has been taken care of (Eugene A. Imhoff, 2003). The BOD is the key decision makers in the organization and they have the power to atone the whole decisions that are made by the top management. Furthermore, they said that, in the process of decision-making, the initiation and implementation should be separated from the ratification and monitoring of the decisions in order to make sure the efficient monitoring functions.

NUMBER OF INDEPENDENT DIRECTORS AND AUDITOR CHOICE

Independence of board influences the extent to which the board is independent of company management. A number of studies document a positive relationship between board independence and actions that are in the best interest of shareholders. The BOD is the best control mechanism to monitor actions of management. The study explored board independence based on the agency theory. Outside directors have a motivation to monitor management because the value of their own human capital is partially determined by the effectiveness of their monitoring performance. Thus, independent (outside) directors are more willing to provide effective oversight. They also proposed that the proportion of independent directors on the board was negatively related to the likelihood of financial reporting fraud.

RESEARCH METHODOLOGY

HYPOTHESIS DEVELOPMENT

The objective of this study is to examine the effect of corporate governance mechanisms toward public listed firms' auditor choice decisions. Thus, five main hypotheses have been developed.

Hypothesis 1 (H1): The higher the percentage of total shares held by the largest owner, the less likely a firm will choose a high-quality (Big Four) auditor, *ceteris paribus.*

Hypothesis 2 (H2): The fewer the number of the audit committee members, the less likely a firm will choose a high-quality (Big Four) auditor, ceteris paribus.

Hypothesis 3 (H3): When the same persons hold the positions of CEO and BOD chairman, it is less likely a firm will choose a high-quality auditor (Big Four) auditor, ceteris paribus.

Hypothesis 4 (H4): The bigger the size of BOD, the less likely a firm will choose a high-quality (Big Four) auditor, ceteris paribus

Hypothesis 5 (H5): The fewer the number of the independent directors, the less likely a firm will choose a high-quality (Big Four) auditor, ceteris paribus

SAMPLE SELECTION AND DATA COLLECTION

For the sample, public listed firms on the Main Market of Bursa Malaysia during the period of 2010-2015 were initially considered. The list of the firms was obtained from the Bursa Malaysia web site (http://www.bursamalaysia.com). In total, the sample population contained approximately 810 listed firms (List updated as at 31 August 2016). From there, the final target sample were selected down to 261 firms. Table 1 shows the summary of variables used in this study.

***Table 1*: Summary of Variables**

NO	VARIABLES	ABBREVIATIONS	SOURCE OF DATA Data Stream (DS) / Annual Report (AR)
1	Auditor choice	AUD	AR
2	Largest ownership	LSH	AR
3	No of audit committee member	AC	AR
4	Duality of CEO	DUALITY	AR
5	No of BOD	BOD	AR
6	No of independent director	NED	AR
7	Firm size (Log of total assets)	LNASSET	DS
8	Firm growth (Asset turnover ratio)	ATR	DS
9	Firm profitability (Return on assets)	ROA	DS
10	Firm asset structure (Current assets ratio)	CURR	DS
11	Firm leverage (Debt to asset ratio)	DA	DS
12	Firm risk (Beta)	BETA	DS

MODEL SPECIFICATION AND VARIABLE DEFINITIONS

In this study, following Lin and Liu (2009), a binary classification was used to divide auditors in Malaysia into two categories. This study identifies Big Four Malaysian audit firms as Deloitte, Ernst and Young, KPMG and PricewaterhouseCoopers (PWC).

Since the dependent variable, auditor choice, is binary in nature (Big Four versus non-Big Four), following Lin and Liu (2009), this model uses logit regression to test the hypotheses of the relationship between corporate governance variables and auditor choice. The model is as specified below.

$$AUD = {}_{\beta}0 + {}_{\beta}1LSH + {}_{\beta}2AC + {}_{\beta}3DUALITY + {}_{\beta}4BOD + {}_{\beta}5NED + {}_{\beta}6LNASSET + {}_{\beta}7ATR + {}_{\beta}8ROA + {}_{\beta}9CURR + {}_{\beta}10DA + {}_{\beta}11BETA + \varepsilon$$

Where,

AUD = 1 if the auditor is a Big Four auditor; 0 otherwise

LSH	=	the largest owner's shareholding as a percentage of total shares
AC	=	number of Audit Committee member,
DUALITY	=	1 if the CEO also holds the position of BOD chairman; 0 otherwise
BOD	=	size of BOD
NED	=	number of independent directors
LNASSET	=	log of total assets
ATR	=	asset turnover ratio, calculated as sales divided by total assets
ROA	=	earnings after interest and taxation divided by total assets
CURR	=	current assets divided by total assets
DA	=	total liabilities divided by total assets
ABBETA	=	the absolute value of beta to proxy for firm risk

This model tested the association between the internal corporate governance mechanisms of the public listed Malaysian firms and their auditor choice decision as whether firms with strong internal corporate governance mechanisms would be more likely to choose auditors of high quality. In relation to the five hypotheses, it is predicted that the sign of coefficient for hypotheses "1, 3 and 4" will be negative while for hypotheses "2 and 5" will be positive. This study controls for firm size (log of total asset), growth (asset turnover ratio), profitability (return on asset), asset structure, (current asset divided by total asset), financial leverage (total liabilities divided by total asset) and firm risk (beta) in the model.

EMPIRICAL FINDINGS

DESCRIPTIVE STATISTICS

The data collected are analyzed using Statistical Package for Social Science (SPSS). Descriptive statistic is useful to describe the data or see the characteristics of the data. It provides simple summaries about the sample and about the observations that have been made. Table 2 and Table 3 represent descriptive statistics results and correlation matrix of independent variables consecutively.

Table 2: Descriptive Statistics

	N	Minimum	Maximum	Mean	Std. Deviation	Variance
AUDITOR CHOICE	1566	0	1	.62	.486	.236
LARGEST OWNERSHIP	1566	1.660	73.370	28.43657	16.333257	266.775
NO OF AUDIT COMMITTEE MEMBER	1566	3	6	3.37	.637	.406
DUALITY OF CEO	1566	0	1	.23	.418	.175
NO OF BOD	1566	3	17	7.47	1.989	3.957
NO OF INDEPENDENT DIRECTOR	1566	1	7	3.18	.985	.971
LNASSET	1566	3.809	8.013	5.54765	.590511	.349
ATR	1566	.000	3.693	.75676	.535700	.287
ROA	1566	-1.245	.542	.03011	.099548	.010
CURR	1566	.006	.992	.50333	.200335	.040
DA	1566	.026	2.033	.40569	.207469	.043
BETA	1566	.062	2.094	.96788	.409225	.167
Valid N (listwise)	1566					

Table 2 presents the basic descriptive statistic regression variables. Around 62% (971/1566) of the sample companies were audited by Big Four audit firms. The high concentration rate in the auditing market could be due to the intention of the public listed company to increase investors' confidence due to the increase in corporate scandals over the years.

On the other hand, the largest shareholders can normally have infinite controlling power over the listed firms, holding a mean of 23.44% of the total equity shares of the sample public listed firms. This circumstance of monopolist shareholding has essentially led to poor corporate governance practices and raised serious concerns by investors, professionals, regulators, and the public (DeFond et al., 2000).

The average size of the BOD for the sample firms is 7.47 with a minimum BOD size of three (3) persons and the largest with 17 persons. In addition, about 23% of the sample firms (360/1566) had the same person holding both positions of BOD Chairman and CEO. The results shows that about 77% of the firms (1206/1566) have a separate BOD chairman and CEO which would suggest better monitoring mechanism thus promote better firm performance. Furthermore, the average number of audit committee for the sample firms is 3.37 with minimum audit committee size of three (3) persons and the largest with six (6) persons.

***Table 3:* Correlation Matrix**

		Constant	LSH	AUDIT_ COMMITTEE	DUALITY	BOD	NID	LNASSET	ATR	ROA	CURR	DA	BETA
Step 1	Constant	1.000											
	LSH	-.084	1.000										
	AUDIT_ COMMITTEE	-.244	-.081	1.000									
	DUALITY	-.036	-.050	.059	1.000								
	BOD	-.044	.043	-.066	.046	1.000							
	NID	.120	-.072	-.149	-.035	-.533	1.000						
	LNASSET	-.811	-.020	-.118	-.036	-.161	-.146	1.000					
	ATR	-.201	-.078	.011	-.040	-.131	.110	.163	1.000				
	ROA	.260	-.050	-.006	-.018	-.011	.049	-.281	-.219	1.000			
	CURR	-.263	-.044	-.054	.057	.004	-.006	.106	-.291	-.069	1.000		
	DA	.144	.130	.021	.058	.101	-.020	-.323	-.253	.313	.049	1.000	
	BETA	.025	.063	.006	-.059	.151	-.087	-.249	.177	-.018	-.014	-.130	1.000

Table 3 presents the correlation coefficient matrix. The choice of Big Four auditor is positively related to the audit committee size, largest ownership, firm risk, size of BOD, log of total asset (firm size) and return on asset. Correlation coefficients among the independent variables are not high with only one at the level of 0.360 (between number of independent director and log total asset), thus multicolinearity is low and will not have a significant effect on the relationship between the dependent and independent variables in the test model.

LOGISTIC REGRESSION ANALYSES

Following the study done by Lin and Liu (2009), binary logistic regression was used to test the relationship between firms' internal corporate governance mechanisms and their audit choice decisions. Table 4 presents the results of the logistic regression of the model.

DOI: 10.1051/shsconf/20173600007

Table 4: **Variables in the Equation**

		B	S.E.	Wald	df	Sig.	Exp (B)
Step 1[a]	LSH	.014	.004	16.116	1	.000	1.014
	AUDIT_COMMITTEE	.180	.095	3.581	1	.058	1.198
	DUALITY	-.460	.126	13.430	1	.000	.631
	BOD	.063	.035	3.262	1	.071	1.065
	NID	-.214	.070	9.372	1	.002	.807
	LNASSET	1.235	.133	85.533	1	.000	3.437
	ATR	.572	.118	23.407	1	.000	1.771
	ROA	.307	.620	.245	1	.620	1.359
	CURR	-1.063	.289	13.519	1	.000	.346
	DA	-1.886	.297	40.355	1	.000	.152
	BETA	.508	.147	11.883	1	.001	1.661
	Constant	-6.596	.697	89.531	1	.000	.001
	LSH	B	S.E.	Wald	df	Sig.	Exp (B)
	AUDIT_COMMITTEE	.014	.004	16.116	1	.000	1.014

a. Variable(s) entered on step 1: LSH, AUDIT_COMMITTEE, DUALITY, BOD, NID, LNASSET, ATR, ROA, CURR, DA, BETA.

The results show that duality of CEO, number of independent directors, asset structure (CURR) and firm leverage (DA) does not contribute to the model. The negative B indicates that the target group (audit quality) tends to have more of those coded "0" (non-Big Four) than of these coded "1" (Big Four) but not significantly, after controlling for the other predictors. Meanwhile, audit committee size, largest shareholders, firm risk (Beta), BOD size, firm size (log total asset), growth (asset turnover ratio) and profitability (return on asset) do contribute to the model. The positive B indicates that the target group (audit quality) tends to have more of those coded "1" (Big Four) than of these coded "0" (non-Big Four) but not significantly, after controlling for the other predictors.

The coefficient for the variable of ownership concentration (the largest owner's shareholding) is negatively correlated with audit quality. It advocate that with a larger portion of shares held by controlling owners, public listed firms would be less likely to choose a high quality (Big Four) auditor. The result is consistent with proposed H1 and study done by Lin and Liu (2009). The finding concede that with high concentration of ownership, firms opt for selecting a low level of external audit monitoring in order to profit from more aggressive earnings manipulation and 'tunnelling' behaviours even at the expense of the monitory shareholders or outside investors (Lin and Liu, 2009).

On the other hand, consistence with the proposed H2, the audit committee size is positively associated with audit quality (B2= 0.180, Wald= 3.581). It indicates that, larger audit committee size will be more likely to choose a Big Four (high quality) auditor. The results reveal that an audit committee, which is a subcommittee for the board, plays an important role in the process of selecting external auditor. Effective audit committee, which is associated with audit committees that are active, independent, consist of accounting and industry experts as well as have more members, is a credible indication of strong corporate governance.

The coefficient for the variable of duality of CEO is negatively associated with audit quality (B3= -0.460, Wald= 13.430). It shows that with when a CEO also holds the position as chairman of board, a listed firm is less likely to hire high quality (Big Four) auditors. It is consistent with proposed H3 of the study and in line with the study done by Lin and Liu, (2009). The finding signify that recent regulations in western countries that separate firms' CEO and BOD Chairman positions should be applicable to Malaysian listed firms as well. Such a measure is an effective arrangement for the BOD to oversee the performance of the management. This results contradict with stewardship theory and consistent with agency theory. It seems that, if two different persons hold the position of CEO and Board Chairman, there would be higher chances that a public listed company will select a high quality auditor to oversee and ensure the quality of firm's financial reporting process and management performance (Lin and Liu, 2009).

As indicated by B4, the BOD size is positively related to the selection of a Big Four auditor. The results suggest that firms with larger BOD size are more likely to choose a Big Four (high quality) auditor which supports H4 which proposed that smaller BOD size will be less likely to choose a high quality (Big Four) auditor. Our findings demonstrate that if there is a large BOD that relatively exhibits greater monitoring role; a firm may be more likely to employ a high quality (Big Four) auditor to enhance the supervision or monitoring role. Again, the finding of this study is consistent with those found by Lin and Liu (2009).

As for number of independent directors, our result shows that there is a negative connection between number of independent directors in the board and audit quality. The findings is contrast with proposed hypothesis 5 (H5) which suggest that the fewer the number of the independent directors, the less likely a firm will choose a high-quality (Big Four) auditor. The results advocate that, a large proportion of independent directors on the board would contribute to the firm choosing a lower audit quality (non-Big Four). Though prior study suggest otherwise, the negative association between those two maybe due to the desire of independent directors to protect their image by hiring lower quality auditor to enhance the possibility of earnings management and at the same time to reduce the likelihood of auditors found prove of material misstatement or financial irregularities in the financial statement.

For the control variables, firm size (log of total assets), firm specific risk (beta) and profitability (return on assets) are all positively related to the selection of Big Four (high-quality) auditors. Consistent with prior research in the developed markets, the results suggest that large firms, firms with greater risk and more profitable firms are inclined to hire high-quality (Big Four) auditors in the Malaysian stock market. As for asset structure (current asset divided by total asset) and financial leverage (debt to asset ratio) are negatively associated with the selection of Big Four (high quality) auditors. The finding suggests that high leverage firms and firms with greater asset structure would be more inclined to choose lower quality (non-Big Four) auditors.

SUMMARY OF HYPOTHESES

Table 5 provide overall summary of the hypotheses that has been tested above using descriptive and binary logistic regression.

***Table 5*: Summary of the hypotheses**

Hypotheses	Result
H1	Supported
H2	Supported

H3	**Supported**
H4	**Supported**
H5	**Rejected**

CONCLUSION

DISCUSSION

The findings of the study may create implications for the stakeholders of the Malaysian public listed especially the one who has direct impact, which is the investors as well market regulators. As weak internal corporate governance itself would probably shaken the public interest and could jeopardize a going concern of the firm in the future, it would not be a good practice for listed firms to choose low quality auditors. In doing so, the controlling shareholders of the listed firms can easily capture private profits from exploiting small shareholders. As such, to strengthen the confidence and trust of the market participants, the regulator should promote the reform of corporate governance of the listed firm, enforcing an effective functioning of the BOD as well as independent non-executive directors or subcommittees of the board (i.e. audit committee, risk and management committee, etc.) to continuously improve firms' internal and external supervision mechanisms thus enhancing transparency and information credibility of the listed firms.

LIMITATION OF THE STUDY AND SUGGESTION FOR FUTURE RESEARCH

There are some limitations in this study. First, more accurate results may be derived from more sophisticated simultaneous equation methods for the auditor choice test. Simultaneous equation method could be used to control both the demand and the supply effect of audit service. However, audit fee (a key variable to proxy for the supply effect in the audit market) was not included in this study. Perhaps, future study should explore on this matter. Second, the size of audit committee may not be a good proxy for internal corporate governance mechanisms since the sample chosen consists of public listed companies, which are governed by listing requirement of Bursa Malaysia that require them to have at least three members in the committee. Results found in the study shows that most of the listed firms have three members in the committee with a maximum of six members. Hence, the testing results may be more persuasive if we include the expertise of independent non-executive directors and other characteristics of audit committee.

REFERENCES

Ahmad, A. C., Houghton, K. A., & Yusof, N. Z. M. (2006). The Malaysian market for audit services: ethnicity, multinational companies and auditor choice. Managerial Auditing Journal, 21(7), 702-723.

Alnasser, S. (2012). What has changed? The development of corporate governance in Malaysia. The Journal of Risk Finance, 13(3), 269 - 276.

Ashbaugh, H., & Warfield, T. D. (2003). Audit as a corporate governance mechanism: Evidence from the German market. Journal of International Accounting Research, 2, 1-21.

Bardos, K. S. (2011). Quality of financial information and liquidity. Review of Financial Economics, 20(2), 49-62.

Bartlett, J. E., II , Kotrlik, J. W., & Higgins, C. C. (Spring 2001). Organizational Research: Determining Appropriate Sample Size in Survey Research. Information Technology, Learning, and Performance Journal, 19(1).

Beattie, V., & Fearnley, S. (1995). The Importance of Audit Firm Characteristic and the Drivers of Auditor Change in UK Listed Companies. Accounting and Business Research, 25(100), 227-239.

Chaney, P., Jeter, D., & Shivakumar, L. (2003). Self-Selection of Auditors and Audit Pricing in Private Firms.

DeFond, M. L., Wong, T. J., & Li, S. (1999). The impact of improved auditor independence on audit market concentration in China. [doi: 10.1016/S0165-4101(00)00005-7]. Journal of Accounting and Economics, 28(3), 269-305.

Eugene A. Imhoff, J. (January 2003). Accounting Quality, Auditing and Corporate Governance.

Fama, E. F., & Jensen, M. C. (1983). Separation of Ownership and Control. Journal of Law and Economics, XXVI.

Fields, T. D., Lys, T. Z., & Vincent, L. (2001). Empirical research on accounting choice. Journal of Accounting and Economics, 31(1,Äì3), 255-307.

Guedhami, O., Pittman, J. A., & Saffar, W. (2009). Auditor choice in privatized firms: Empirical evidence on the role of state and foreign owners. Journal of Accounting and Economics, 48, 151-171.

Hope, O.-K., Kang, T., Thomas, W., & Yoo, Y. K. (2008). Culture and auditor choice: A test of the secrecy hypothesis. J. Account. Public Policy, 27, 357-373.

Johnson, W. B., & Lys, T. (1990). The market for audit services: Evidence from voluntary auditor changes. Journal of Accounting and Economics, 12(1,Äì3), 281-308.

La Porta, R., Lopez-de-Silanes, F., Shleifer, A., & Vishny, R. (2000). Investor protection and corporate governance. Journal of Financial Economics, 58(1,Äì2), 3-27.

Lin, Z. J., & Liu, M. (2009). The impact of corporate governance on auditor choice: Evidence from China. [doi: 10.1016/j.intaccaudtax.2008.12.005]. Journal of International Accounting, Auditing and Taxation, 18(1), 44-59.

Linda Elizabeth, D. (1982). Mandated successful efforts and auditor choice. Journal of Accounting and Economics, 4(3), 171-203.

Mahdavi, G., Maharlouie, M. M., Ebrahimi, F., & Sarikhani, M. (2011). The Impact of Corporate Governance on Auditor choice. International Research Journal of Finance and Economics(68).

MCCG. (2012). MALAYSIAN CODE ON CORPORATE GOVERNANCE 2012. Securities Commission Malaysia.

Nurwati, A. A.-Z., & Wan Nordin, W.-H. (2010). Corporate governance and earnings forecasts accuracy. Asian Review of Accounting, 18(1), 50-67.

Pittman, J. A., & Fortin, S. (2004). Auditor choice and the cost of debt capital for newly public firms. Journal of Accounting and Economics, 37(1), 113-136.

Rainsbury, E. A., Bradbury, M., & Cahan, S. F. (2009). The impact of audit committee quality on financial reporting quality and audit fees. Journal of Contemporary Accounting & Economics, 5(1), 20 - 33.

Samad, F. A. (2002). Ownership Structure In The Malaysian Corporation Sector: Its Impact On Corporate Governance, Performance, Financing And Investment Patterns.

Shanthy, R., & Elsa, S. (2009). Improving corporate governance of SMEs in emerging economies: a Malaysian experience. Journal of Enterprise Information Management, 22(4), 468-484.

Wallace, W. A. (2004). THE ECONOMIC ROLE OF THE AUDIT IN FREE AND REGULATED MARKETS: A LOOK BACK AND A LOOK FORWARD. Research in Accounting Regulation, 17(0), 267-298.

Wang, Q., Wong, T. J., & Xia, L. (2008). State ownership, the institutional environment, and auditor choice: Evidence from China. Journal of Accounting and Economics, 46, 112-134.

ACCOUNTING STUDENTS' PERCEPTION AND THEIR INTENTION TO BECOME PROFESSIONALLY QUALIFIED ACCOUNTANTS

Dalilah Abdul Aziz*[1], Mohd Adha Ibrahim[2], Morni Hayati Jaafar Sidik[1], Mazilena Tajuddin[1]
Department of Accounting UniKL Business School Universiti Kuala Lumpur[1], Faculty of Economics and Muamalat, Universiti Sains Islam Malaysia[2]

ABSTRACT

The purpose of this paper is to study accounting students at Majlis Amanah Rakyat (MARA) institutions[1] who are predominantly Bumiputra about their perception of the accounting profession, their grit level, the need for financial assistance and whether or not these are significant in influencing them to become professionally qualified accountants. The research was conducted using a questionnaire survey structured around seven independent variables and translated into 27 statements which was then distributed to accounting students at various MARA institutions. To analyse the data, descriptive and regression analysis were employed and the findings show that all predictor variables explain about 6.3 percent of the variability of students' intention to pursue professional qualification. The research results also show that when studying the independent variables individually, only three independent variables have significant relationship with students' intention to pursue professional qualification. These are job security or stability, financial assistance and grit. The findings have implications for accounting educators as well as the professional bodies and other related organisations in their effort to increase the number of professionally qualified accountants in Malaysia. The study also provides evidence that students' grittiness is a significant factor in influencing them to become professionally qualified accountants.

Keywords: Accounting, Students, Perception, Attitudes, Motivation, Accounting Education, Accounting Profession, Professional Qualification, Professional Qualified Accounta

*Corresponding author's email: dalilahaa@unikl.edu.my

INTRODUCTION

Malaysia is striving to be a high income nation by the year 2020 and one of the main economic drivers of a high income nation is its financial markets. Therefore, in order to attract more investments, especially from foreign investors, the credibility of an entity's financial statements is essential. Useful financial reporting may help to ensure that the process of allocating capital for productive economic purposes will be done effectively. As such, professional accountants are needed to help achieve this goal (The Star, 2013). Malaysia will need to double its number of accountants to 60,000 by year 2020 to be able to sustain the nation's projected economic growth (New Straits Times, 2015). Existingly however, Malaysia is still struggling in attracting its graduates to pursue professional qualifications.

One of the initiatives made by the government of Malaysia to help increase the number of professional accountants is to provide a pool of potential candidates by offering diploma and degree in accounting through colleges and universities. However, most of the accounting graduates eventually opt to not pursue their professional qualifications; choosing to remain in jobs that do not require professional qualifications. It was reported that out of approximately 41,000 accountancy graduates from institutions of higher learning recognised by MIA[2] since 1978 under Part 1 of the First Schedule, only 27% of them became members of MIA (Committee to Strengthen the Accountancy Profession (CSAP), 2014). This percentage is low considering the fact that for these group of graduates to join as a member, they are only required three years of relevant working experience in the accounting field. This also suggests that accounting graduates may have deviated into non-accounting fields.

Additionally, only 8% of total professional accountants are Bumiputra (The Star Business News, 2015; Basaruddin, 2015). This significant under representation of Bumiputras in the accounting profession has sparked concerns amongst various stakeholders and as such, raising the need to determine the rationale behind the shortage.

The purpose of this research is to investigate the perception of the accounting students' towards pursuing professional qualifications; and in specific, students from Majlis Amanah Rakyat (MARA) institutions. Students from these institutions are mainly of Bumiputra ethnicity. In the following sections, a number of relevant previous studies are first presented which then leads into the research question and the conceptual framework. Subsequently, the research approach is dicussed and the results and findings are presented. The paper then concludes with a summary of key points.

LITERATURE REVIEW

In Malaysia, several studies have shown that the accounting profession is still popular amongst university students (Samidi and Tew 1995; Said, Ghani, Hashim and Mohd Nasir, 2004). A study by Samidi and Tew (1995) reported that majority future accounting graduates had chosen accounting as a profession. Said et al. (2004) and Ghani, Said, Mohd Nasir and Jusoff (2008) found that the accounting profession is ranked amongst the most preferred careers by the public and private university students. Unfortunately, despite their interest in the profession, not many especially the Bumiputra graduates have continued on to pursue professional qualifications. As such, this research aims to understand the incoherence.

[2] Malaysian Institute of Accountants.

*Corresponding author's email: dalilahaa@unikl.edu.my

DOI: 10.1051/shsconf/20173600008

Perception, extrinsic and intrinsic motivation

A study conducted by Germanou et al. (2009) found that if a student has a positive perception of the accounting profession, he or she is more likely to pursue a profession in accounting. The finding shows Malaysian students with positive perceptions on the accounting profession are more likely to pursue an accounting career. In the study, perceptions are measured by three categories namely extrinsic, intrinsic prestige and social. Extrinsic category is represented by salary, security of job and physical environment at work place. Prestige is represented by opportunities for promotions, decision making and recognition and status in the eyes of others. Intrinsic is represented by the experience of learning new things, accomplishing something important and a concern for being free from imposed constraints in the work environment. Whilst social relates to items that are affective in nature. All four categories are significantly related with the Malaysian students but only the intrinsic and social categories are found to relate significantly with the English students. This finding suggests that physical retributions such as salary and benefits, opportunities for promotions, job security and status are important motivators to the Malaysian students.

Another study by Jackling, de Lange, Phillips and Sewell (2012) found that Australian students to be intrinsically driven when pursuing a career in accountancy compared to a group of international students which mainly consist of Asian students. Both findings seem to suggest that extrinsic factors are highly relevant to Asian students in pursuit of a career as accountants. Furthermore, Sugahara and Boland (2006) found accounting students value good advancement opportunities, high social prestige and high salary significantly higher than non-accounting students.

Although other studies have shown that intrinsic as well as extrinsic factors may motivate students to be a professional accountant, various studies especially ones involving Asian students found that they are consistently intrinsically motivated. Hence why this research is biased towards extrinsic motivation.

Salary

Abdullah and Zakaria (2006) conducted a study in two public universities in Malaysia and found expected future salary an important attribute for students intending to join public accounting firms. A study conducted by Ghani et al (2008) indicates that first year accounting students perceive salary as a major factor to make an accountant happy. However, this result is not supported in a study by Mustapha and Abu Hassan (2012) and Law (2010) which do not find salary to be a significant factor in motivating a student to pursue CPA career. Since the two earlier studies show support for salary and the two latter studies do not, this study seeks to investigate if salary is a relevant variable for students to be professional accountants.

Job security and stability

Many studies supports the significance of job security and stability to be a factor in students pursuing career as an accountants. A study by Mustapha and Abu Hassan (2012) among final year accounting students at a public university in Malaysia found job security and and stability to be positively and significantly related to the respondents choosing to be professional accountant. Abdullah and Zakaria (2006) also support job security as an important attribute for accounting students to join public accounting firms. In another study by van Zyl and de Villiers (2011), it is found that accounting major students rank availability of employment as one of the motivating factors to become chartered accountants. Similarly, Myburgh (2005) reports

*Corresponding author's email: dalilahaa@unikl.edu.my

that students choose availability of employment as the most important benefit for pursuing a career as chartered accountants.

Prestige or status

Jackling et al (2012) conducted a study on Australian and international students mainly from China, Malaysia and Hong Kong and found that for both groups, prestige is a significant factor for students to continue into accounting profession. This finding is consistent with Mustapha and Abu Hassan (2012) and Germanou and Hasall (2009).

Opportunity for advancement

Abdullah and Zakaria (2006) found evidence that opportunity for advancement as the most important attribute for accounting students to join a public accounting firm especially among the male students. This finding is reinforced by Mustapha and Abu Hassan (2012) who also found that opportunities for advancement to be postively correlated to students' career choice as professional accountant.

Advisory received

Other elements that have been investigated are 'parental influence' (Law, 2010) and parent and teachers' influence (Myburgh 2005) on motivating factors for students to be professional accountants. Law (2010) found that 'parental influence' significantly contribute to predicting the career choice of accounting students in Hong Kong. While Myburgh (2005) study on first year accounting students at University of Pretoria, South Africa shows that advice given by significant family members and influence of school teachers were ranked amongst the top three sources of influence in the respondents' career decision making. Therefore, this study seeks to also examine the significance of family and educators influence in predicting accounting students' intention to pursue professional accounting profession.

Financial assistance

Other than extrinsic and intrinsic factors, another factor that has been reported to influence students' intention to pursue professional qualification is its cost. A report by the CSAP (2014) cites that one of the top reasons for graduates not to pursue professional accountancy qualification is due to its costly expense. Myburgh (2005) findings corroborates this. Both studies imply that cost of qualifying as a professional accountant is a factor perceived negatively by respondents. Therefore, this study tests the same amongst MARA accounting students.

Grit personality

Bacanli (2006) found support that personality characteristics influences career indecisiveness while Gunkel, Schlaegel, Langella and Peluchette (2010) opines personality traits have direct effect on students' career. Duckworth and Peterson (2007) propose one personal quality prominent in successful leaders is grit. Grit is defined as perseverance and passion for long-term goals. The authors developed a self-report questionnaire called the Grit Scale and in total six studies were carried out. Study 1, shows that adults with higher grit achieved higher education. This finding also suggests that the ability to stick to long-range goals makes

*Corresponding author's email: dalilahaa@unikl.edu.my

completion of higher studies possible. Study 2, shows grit to be the only significant predictor of a person less frequently to change jobs. Study 3 tested if high school Scholastic Aptitude Test (SAT) scores are linked to the Cumulative Grade Point Average (CGPA) of undergraduates in an elite university. The finding indeed shows that grit is positively linked with higher CGPA and grit is associated with undergraduates who had lower SAT scores. Studies 4 and 5 selected first year US Military cadets as their respondents and both studies found respondents with higher grit are more likely to complete their first year training. Finally, study 6 proves that among participants of a national level spelling bee competition, children who are grittier achieved better performance by working harder and longer. The results of all six studies suggests that grit is an important factor for an individual to sustain through tough environments and to achieve higher level of success. A predicament which mirrors closely the nature of most professional qualifications.

Thus, this study seeks to provide an insight on the following question:

> Do these seven independent variables together motivate an accounting student to be a professional accountant?

Based on the discussion and research question, the proposed research model is as follows:

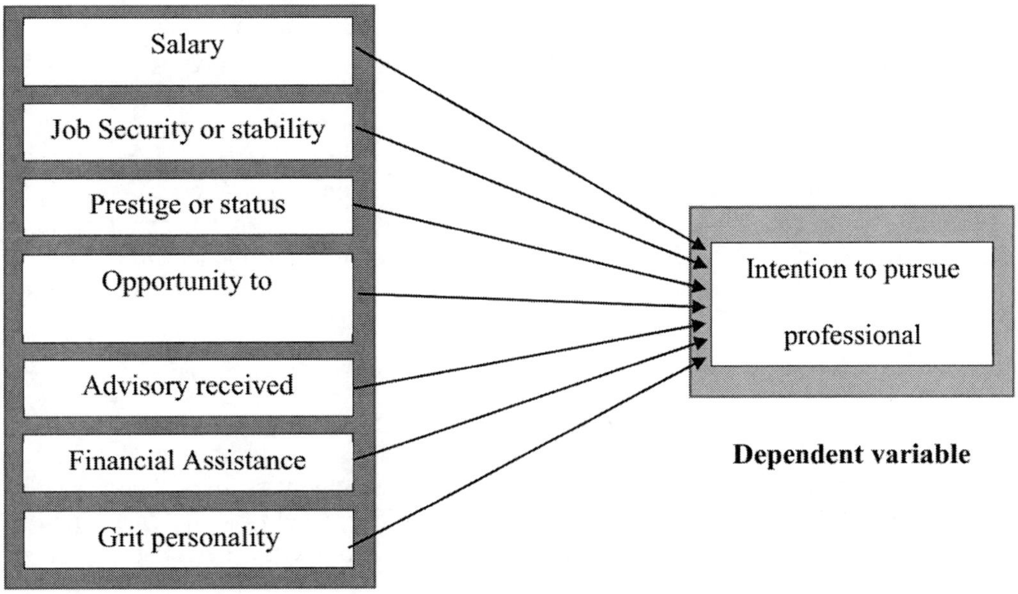

Figure 1: Conceptual Model

The model which has been adapted from the study by Mustapha and Abu Hassan (2012) shows that "Intention to pursue professional qualification" as the dependent variable and items such as salary, job security or stability, prestige or status, opportunity to advancement, advisory received, financial assistance and grit personality to be the independent variables for this study.

*Corresponding author's email: dalilahaa@unikl.edu.my

RESEARCH APPROACH

This research is conducted to study the relationship between students' perception and their intention to pursue professional qualification. The approach taken is quantitative in nature and is in line with the positivism paradigm. The paradigm relies on deductive reasoning following a structured process leading to the identification of causal relationships, logical conclusions and the making of predictions according to various confidence levels (McKerchar, 2008).

The quantitative method was decided to be the most appropriate approach for this study in order to achieve the intended objectives. Although using more than one method may result in more rigorous outcome, other constraints have been considered. Ramenyi, Williams, Momey and Swartz (1998) suggest that the research approach should accommodate the research questions, costs, available budget, available time and target dates for completion, and skills of the researcher.

Survey Research Design

Survey Method and Rationale

There are various research methods in the quantitative approach that may be employed for this study such as the controlled experiment, content analysis, analysis of existing data, participant observation and survey (Babbie, 2013). The present study was undertaken by employing the questionnaire survey method. The main advantage of survey research is the reduced time and costs of carrying out the survey as compared to personal interviews (Babbie, 2013). This in turn allows a wide range of standardized information to be obtained from a large sample (Babbie, 2013). It enables respondents to complete the questionnaire in their own time and in a more convenient way compared to the face-to-face interview (Loo, 2006). Moreover, it is found to be the most reasonable method to cover a wide geographical area, to represent a given population (Loo, 2006). This method is also popular in social science studies and therefore it is also a method of empirical verification (Babbie, 2013) which can produce reliable measures.

Survey Instruments

The primary aim of the questionnaire is to examine the perception of the accounting students at MARA institutions in relation to their intention to pursue professional qualification. The questionnaire has four sections, A, B, C and D. Section A consists of questions on demographic information of the respondents such as institution, age, level of study, year of study, gender, ethnic, current CGPA, parent educational background and Sijil Pelajaran Malaysia (SPM) result for English paper. Respondents were asked to tick a box to answer each question or fill in the appropriate details as requested. This information is important to build the respondents' profile for this study and to determine the representativeness of the respondents in relation to the whole population. Section B is designed to collect information related to students' future plan after graduation. This section consists of four questions, one of which represents the dependent variable for this study[3].

Additonally, sections C and D is developed to collect information in relation to students' perceptions of the accounting profession, advisory received, financial asisstance and their grit personality. There were 27 statements on students' perceptions and these were developed based

[3] Refer to Table 7, item **.

*Corresponding author's email: dalilahaa@unikl.edu.my

on the conceptual model discussed earlier in Section 2. Using five-point Likert scales (1=strongly disagree to 5=strongly agree), respondents were requested to indicate their level of agreement with each of the statements.

Selection of Participants

Ideally, samples are selected to represent the population of interest (Tabachnick & Fidell, 2007). For the purpose of this study, the population of interest is the accounting students at three MARA institutions. The sample is collected from University Kuala Lumpur Business School (UniKL BIS), Kolej Poly-Tech Mara (KPTM) and Kolej Professional Mara (KPM) students. To ensure that the respondents came from the group of respondents, this study employs an appropriate method in selecting the population by individually contacting these institutions. The total population of the current study is derived to be 2,050 as shown in Table 1.

Table 1: Total Population of the Study

	Number of Respondents
Gross Population in UniKL Business School	1,100
Gross Population in Kolej Poly-Tech Mara - Cheras	500
Gross Population in Kolej Professional Mara - Beranang	450
Total Population	**2,050**

Hair, Black, Babin, Anderson and Tatham (2006) highlight in their study that researchers need to be fully aware of the impact of sample size on the statistical test. Krejcie and Morgan (1970), developed a scientific table for determining sample size by taking into account population size. The scientific table suggests a sample size of 274 and 278 for a given population of 950 and 1,000 respectively. Alternatively, Roscoe (1975), argues that a sample size of more than 30 and less than 500 is appropriate for most research.

The views presented above suggest that a sample size of anywhere between 274 and 278 is adequate and the current study considers a sample size of 1,000 (i.e. 49 percent of the total population) to be reasonable. This study uses random sampling technique to obtain a representative sample. The population and sampling frame of the current study is presented in Table 2.

Table 2: Population and Sample Size

	Number of Respondents	**Percentage of Population**	**Sample Size**
Total	2,050	49%	1,000

Survey Execution

The data collection period was approximately one month, commencing in middle of May and ending in the middle of June 2016. Most of the questionnaire were personally administered using face-to-face method to the targeted sample.

*Corresponding author's email: dalilahaa@unikl.edu.my

Measurements

Development of the Determinants of Students' Perception towards Intention to Pursue Professional Qualification model

The conceptual model for this study captures the relationship between the independent and dependent variables. The independent variables for this study are salary, job security or stability, prestige or status, opportunity to advance, advisory received, financial assistance and grit personality. The dependent variable for this study is the intention to pursue professional qualification.

Figure 1, presented in Section 2 shows the conceptual model for this study. This model was derived based on the literature review and problem statement. In this study, based on the conceptual model, there are two types of variables which are dependent variable and independent variables. The dependent variable can be defined as the variable that is observed to change in response to the independent variables. Meanwhile, the independent variables are defined as those variables that are deliberately manipulated to invoke a change in the dependent variables. Changes in the independent variables are expected to give significant result to the dependent variable, which is in this case, relates to 'the intention to pursue professional qualification'.

Measurement of the Determinants of Students' Perception towards Intention to Pursue Professional Qualification

A regression analysis is carried out to examine the determinants of students' intention to pursue professional qualification. The independent variables included are those relating to salary, job security or stability, prestige or status, opportunity to advance, advisory received, financial assistance and grit. The statement and the size of the independent and dependent variables are presented in Table 3 and 4.

Table 3: Statements Measuring the Determinants of Student's Perception and Grit Personality towards Intention to Pursue Professional Qualification (Independent Variables)

Determinant/Factor	Measurement (Statement)	Score (Likert Scale)
Salary	1. Professional graduates are paid higher salary than degree graduates. 2. High salary is the main factor why many choose to pursue professional exam after graduation. 3. An accounting graduate can gain higher salary if he/she has professional qualification. 4. Degree holder are paid less than professional qualification holder.	Higher scores indicate higher agreement about salary.
Job security or stability	1. The career of a professional accountant is highly secured. 2. It is not enough to have degree in accounting for job security. 3. Other professional fields are not as stable compared to accounting field.	Higher scores indicate greater agreement about job security or stability.

*Corresponding author's email: dalilahaa@unikl.edu.my

DOI: 10.1051/shsconf/20173600008

Opportunity to advance	1. Professional accountants will have better opportunity for promotions. 2. Accounting field is the professional field that provides better training opportunities.	Higher scores indicate greater agreement about opportunity to advance.
Prestige or status	1. Professional qualifications are highly valued by the public. 2. Accountants with professional qualification are more respected than accountants with degree in accounting. 3. Professional qualification and non-professional qualification are valued differently by the public.	Higher scores indicate greater agreement about prestige or status.
Advisory received	1. Your parents are supportive of your ambition to be pursue professional examination. 2. Your lecturers or academic advisor encourage you to pursue professional examination.	Higher scores indicate greater agreement about advisory received.
Financial assistance	1. Financial support from various organizations are available to pursue professional examination.	Higher scores indicate greater agreement about financial assistant.
Grit	1. I often set a goal but later choose to pursue a different one. 2. I become interested in new pursuit every few months. 3. My interests change from year to year. 4. I have been obsessed in a certain idea or project for a short time but later lost interest. 5. New idea and new projects sometimes distract me from previous. 6. I have difficulty maintaining my focus on projects that take more than a few months to complete. 7. I have achieved a goal that took years of work. 8. I have overcome setbacks to conquer an important challenge. 9. I finish whatever I begin. 10. Setbacks don't discourage me. 11. I am a hard worker. 12. I am diligent.	Higher scores indicate greater grit about student's intention to pursue professional examination.

Table 4: Statement to link Student's Perception towards Intention to Pursue Professional Qualification (Dependent Variable)

Dependent variable	Measurement
Students intention to pursue professional qualification	Agreement towards selection to intention to pursue professional qualification, was obtained via selection between certainly, probably and unlikely.

Data Analysis

Data are analysed mainly using descriptive statistics and regression analysis. For this purpose, all data are entered into the Statistical Package for the Social Sciences (SPSS) software. Frequency and means analysis are used mainly in analysing the profile of the respondents, namely institution, age, level of study, year of study, gender, ethnic, current CGPA, parent educational background and SPM result for English paper, while variables influencing MARA

*Corresponding author's email: dalilahaa@unikl.edu.my

 DOI: 10.1051/shsconf/20173600008

accounting students' institutions intention to pursue professional qualification are analysed by way of multivariate regressions.

RESULTS AND FINDINGS

The researcher is fortunate to receive an appropriate number of returned questionnaires. Out of 1,000 questionnaires sent, a total of 345 usable questionnaires are returned, which represents 34.5 percent of the net sampling frame (see Table 5).

Table 5: Sample Frame and Response Rate

	Gross Sample Frame	Useable Responses	Useable Response Rate (%)
Number of Respondents	1,000	345	34.5

Demographic Profile of Respondents

A total of nine questions are included in the survey to identify the respondents' demographic profile. These questions seek to gather information such as respondents' institution, age, level of study, year of study, gender, ethnic, current CGPA, parent educational background and SPM result for English paper. The descriptive analysis on the background of the respondents is discussed in this section.

As shown in Table 6 below, analysis of the respondents' institution indicates that about 72.8% are students from UniKL Business School, 15.1% from KPTM and 12.2% from KPM. Most of the respondents (45.5%) have a current CGPA between 3.00 and 3.49. In terms of age, majority of respondents are between 21 and 23 years old, representing 22.6% and 24.3% of the returned sample respectively. In terms of gender, the respondents are 67.8% female and 32.2% male. Malaysia is a multi-cultural country with a main ethnic group, Bumiputra. In this study, 99.1% of the respondents are Bumiputra and the remaining are non-Bumiputra. Information on the respondents' parents' educational background was also collected and the data showed that 51% hold SPM or below certificate.

Table 6: Background of respondent

Item	Description	Frequency	Percentage (%)
Institution	UniKL Business School	251	72.8
	KPTM	52	15.1
	KPM	42	12.2
Age	19	15	4.3
	20	28	8.1
	21	78	22.6
	22	72	20.9
	23	84	24.3
	24	40	11.6
	25	25	7.2
	26	1	0.3

*Corresponding author's email: dalilahaa@unikl.edu.my

	27	1	0.3
Year of study	Second year	122	35.4
	Third year	183	53.0
	Fourth year	40	11.6
Gender	Female	234	67.8
	Male	111	32.2
Ethnic	Malay Bumiputra	335	97.1
	Non-Malay Bumiputra	7	2.0
	Others	3	0.9
Your current CGPA	Below 2.00	4	1.2
	2.00 – 2.49	19	5.5
	2.50 – 2.99	75	21.7
	3.00 – 3.49	157	45.5
	3.50 – 4.00	90	26.1
Parent educational background	SPM or below	176	51.0
	Diploma	66	19.1
	Bachelor Degree	71	20.6
	Postgraduate (master and above)	26	7.5
	Professional Accounting qualification	6	1.8
SPM result for English paper	A+	9	2.6
	A	61	17.7
	A-	16	4.6
	B+	41	11.9
	B	104	30.1
	C+	17	4.9
	C	64	18.6
	D	31	9.0
	E	2	0.6

Table 7 below are the results of descriptive analysis on the second section of the questionnaire. In this section, the respondents are asked about their future plan after graduation. Based on the survey, more than 60.9% respondents choose to work immediately after graduation and 34.2% choose to further study. Additionally the respondents are also asked of their intention to pursue professional accounting course and about 96.3% of the respondents have the intention to at least probably pursue professional qualification. The survey also highlights that ACCA[4] is the

[4] Association of Chartered Certified Accountants.

*Corresponding author's email: dalilahaa@unikl.edu.my

program of choice for these respondents (82.6%) and that more than 63.5% of respondents became aware of the professional programmes from their lecturers.

Table 7: Future plan after graduation

Item	Description	Frequency	Percentage (%)
Planning after graduation	To work immediately	210	60.9
	To further study	118	34.2
	Others	17	4.9
**Intention to pursue professional qualification	Certainly	132	38.3
	Probably	200	58.0
	Unlikely	13	3.8
Preferred professional programmes to pursue	ACCA	285	82.6
	CIMA	14	4.1
	MICPA	10	2.9
	CPA Australia	22	6.4
	ICAEW	3	0.9
	Others	11	3.2
Knowledge of the professional programmes	From lecturers and academic advisor at your institution	219	63.5
	Seminar/workshop/exhibition	31	9.0
	Career guide from counselling unit	18	5.2
	Advertisement from newspaper	11	3.2
	Notice board at university	9	2.6
	From faculty and friends	50	14.5
	Others	7	2.0

Assessment of Reliability and Validity

As presented in Section 3, respondents were asked to state their perception towards 27 attitudinal statements that may or may not influence the students' intention to pursue professional qualification. All these statements are grouped into seven aspects: salary, job security or stability, prestige or status, opportunity to advance, advisory received, financial assistance and grit personality. For the measurements of salary, job security or stability, prestige or status, opportunity to advance, advisory received and grit personality which utilised more than one item for each aspect, both reliability and validity test was performed.

The measurements used in this study are considered reliable and consistent as the alpha coefficients were above the acceptable value of 0.6 (Nunnally, 1978; Hair et al., 2006; Sekaran, 2003) as shown in Table 8.

*Corresponding author's email: dalilahaa@unikl.edu.my

DOI: 10.1051/shsconf/20173600008

Table 8: Reliability Coefficient of Determinants of Students Perception

Determinant/Factor	Number of Items	Alpha Coefficients
Salary	4	0.787
Job Security or stability	3	0.608
Opportunity to advance	3	0.668
Prestige or status	2	0.722
Advisory received	2	0.669
Financial Assistance	1	-
Grit	12	0.606

Analysis of the Variables

Relationship between Salary, Job Security or Stability, Prestige or Status, Opportunity to Advance, Advisory Received, Financial Assistance, Grit Personality and Students' Intention to Pursue Professional qualification

This section provides the analysis of the relationship between independent variables with the students' intention to pursue professional qualifications. A regression analysis was carried out to examine the determinants of students' intention. The independent variables for this analysis consist of seven variables: salary, job security or stability, prestige or status, opportunity to advance, advisory received, financial assistance and grit personality. The equation for the proposed regression model is depicted in Equation 1:

Equation 1: Regression Analysis Model to Examine Student's Intention to Pursue Professional Qualification

$$Y = \beta_0 + \beta_1(X_1) + \beta_2(X_2) + \beta_3(X_3) + \beta_4(X_4) + \beta_5(X_5) + \beta_6(X_6) + \beta_7(X_7) + e$$

Where:
Y = students' intention to pursue professional qualification,
X1 = salary,
X2 = job security or stability,
X3 = opportunity to advance,
X4 = prestige or status,
X5 = advisory received,
X6 = financial assistance,
X7 = grit, and
e = error.

Assessment of the assumptions using the regression model showed that the model did not violate any of the four assumptions of normality, linearity, multicollinearity and heteroscedasticity. A summary of results for the regression analysis to examine students' intention to pursue professional qualification is presented in Table 9 below.

*Corresponding author's email: dalilahaa@unikl.edu.my

Students' Intention to Pursue Professional Qualification

The regression model is statistically significant (p = 0.003) in relation to students' intention to pursue professional qualification. All predictor variables explained about 6.3 percent of the variability of students' intention to pursue professional qualification and were accounted for as the independent variables in this model. Equation 1 also describes the formulation of the multiple regression analysis for the seven independent variables. Out of seven, only three variables, namely job security or stability (Beta = 0.140 and p = 0.038), financial assistance (Beta = -0.119 and p = 0.040) and grit personality (Beta = -0.123 and p = 0.025) were found to be significant determinants of students' intention to pursue professional qualification. As for the other variables, namely salary, prestige or status, opportunity to advance and advisory received; these were not found to be significant determinants toward the students' intention to pursue professional qualifications.

Table 9: Summary of Results of Multiple Regressions – Students' Perception of the Accounting Profession, Advisory Received, Financial Assistance and Grit personality towards their Intention to Pursue Professional Qualification

	B	SE	Beta	F-value	P-value
Salary	-0.078	0.062	-0.085	-1.270	0.205
Job security or stability	-0.119	0.057	-0.140	-2.088	0.038*
Opportunity to advance	0.067	0.064	0.074	1.057	0.291
Prestige or status	0.019	0.063	0.021	0.295	0.768
Advisory received	0.086	0.052	0.100	1.664	0.97
Financial assistance	0.079	0.038	0.119	2.061	0.040*
Grit	0.157	0.0694	0.123	2.254	0.025*

Model Summary: $R^2 = 0.063$; Adjusted $R^2 = 0.043$; F (345) = 3.224; p (0.003) < 0.05
*Significance level is at $p<0.05$.

CONCLUSION

Despite significant interest in the profession, not many especially the Bumiputra graduates have continued to pursue professional qualifications. This is evidenced by the low percentage of Bumiputra professional accountants registered with MIA. As such this research is conducted to better understand this predicament. This study is undertaken specifically to understand how the perception of accounting students' towards accounting profession, advisory received, financial assistance and grit personality would affect their intention to pursue professional qualifications.

With the sample of the study comprising accounting students from MARA institutions who are mainly of Bumiputra ethinicity, the findings show that all predictor variables explain about 6.3 percent of the variability of students' intention to pursue professional qualification. Further analysis also shows that out of seven independent variables included in this study, only three independent variables have direct significant relationship with a student's intention to pursue professional qualification. These are job security or stability, financial assistance and grit personality. Many previous studies supports the significance of job security and stability to be a factor in students pursuing career as accountants (Mustapha and Abu Hassan, 2012; Abdullah and Zakaria, 2006; van Zyl and de Villiers, 2011; Myburgh, 2005). And financial assistance has also been cited as a significant factor in becoming a chartered accountant for black students

*Corresponding author's email: dalilahaa@unikl.edu.my

at a university in South Africa by Myburgh (2005). Grit personality holds a direct significant relationship with a student's intention to pursue professional qualification also as expected.

Other than the main findings, based on the data collected, there are a few other findings and limitations of this study that are worth highlighting. Firstly, virtually all (96.3%) of the respondents stated that they would at least probably pursue professional qualification. However, only 38.3% were certain of it. Majority of the repondents showed interest but are noncommittal.

This study also has its limitations. Firstly, it is recognised that in this study, accounting students of different cohorts are treated as a homogeneous group. As such, the sample under study may not be as accurately representative of its population as it can be. Secondly, although the study finds that all predictor variables explain about 6.3 percent of the variability of students' intention to pursue professional qualification, the much smaller adjusted R-squared of the regression analysis implies room for improvement. Thus, future study may examine other variables such as culture, intrinsic motivation and internship experience of students.

Despite these limitations, it is hoped that the findings of this study will contribute towards further effort and research to improve the percentage of professionally qualified accountants in Malaysia, especially the percentage of professionally qualified Bumiputra accountants.

REFERENCES

Abdullah, M., & Zakaria, Z. (2006). Desired attributes of public accounting firms from accounting students' perceptions: the case of University of Malaya & International Islamic University of Malaysia. *Journal of Financial Reporting and Accounting, 4,* 25-37.

Ahmad, Z., Ismail, H., & Anantharaman, R.N. (2015). To be or not to be: an investigation of accounting students' career intentions. *Education + Training, 57,* 360-376.

Babbie, E. R. (2013). *The basics of social research.* Cengage Learning.

Bacanli, F. (2006). Personality characteristics as predictors of personal indecisiveness. *Journal of Career Development, 32(4)*, 320-332.

Basaruddin, N. (2015). Akauntan bertauliah bumiputera kritikal. [online] Available from: http://www.utusan.com.my/berita/nasional/akauntan-bertauliah-bumiputera-kritikal-1.143460 [Accessed 8th August 2016].

Department of Statistics Malaysia Online Portal (2011). Population Distribution and Basic Demographic Characteristic Report 2010 (Updated: 05/08/2011). [online] Available from:
https://www.statistics.gov.my/index.php?r=column/cthemeByCat&cat=117&bul_id=MDMxdHZjWTk1SjFzTzNkRXYzcVZjdz09&menu_id=L0pheU43NWJwRWVSZklWdzQ4TlhUUT09
[Accessed 8th August 2016].

*Corresponding author's email: dalilahaa@unikl.edu.my

Duckworth, A. L., & Peterson, C. (2007). Grit: Perseverance and passion for long-term goals. *Journal of Personality and Social Psychology, 92,* 1087-1101.

Germanou, E., Hassall, T., & Tournas, Y. (2009). Students' perceptions of accounting profession: work value approach. *Asian Review of Accounting, 17,*136-148.

Ghani, E. K., Said, J., Mohd Nasir, N., & Jusoff, K. (2008). The 21st century accounting career from the perspective of the Malaysian university students. *Asian Social Science, 4,* 73-83.

Gunkel, M., Schlaegel, C., Langella, I. M., & Peluchette, J. V. (2010). Personality and career decisiveness. *Personnel Review, 39,* 503-524.

Hair, J. F., Black, W. C., Babin, B. J., Anderson, R. E. & Tatham, R. L. 2006. *Multivariate Data Analysis. Sixth Edition.* Pearson International Edition, New Jersey.

Jackling, B., de Lange, P., Phillips, J., & Sewell, J. (2012). Attitudes towards accounting: differences between Australian and international students. *Accounting Research Journal, 25,* 113-130.

Krejcie, R. V. & Morgan, D. W. (1970). Determining sample size for research activities. *Educational and psychological measurement. 30.* p. 607-610.

Law, P. K. (2010). A theory of reasoned action model of accounting students' career choice in public accounting practices in the post-Enron. *Journal of Applied Accounting Research. 11,* 58-73.

Loo, E. C. (2006). The influence of the introduction of self assessment on the compliance behaviour of individual taxpayers in Malaysia. Ph.D Thesis, University of Sydney.

McKerchar, M. A. (2008). Philosophical paradigms, inquiry strategies and knowledge claims: applying the principles of research design and conduct to taxation. *Inquiry Strategies and Knowledge Claims: Applying the Principles of Research Design and Conduct to Taxation*, 5-22.

Mustapha, M. & Abu Hassan, M. H. (2012). Accounting students' perception on pursuing professional eaxamination. *International Journal of Education, 4,* 1-15.

Myburgh, J. E. (2005). An empirical analysis of career choice factors that influence first-year accounting students at the University of Pretoria. *Meditari Accountancy Research, 13,* 35-48.

New Straits Times Online (2015). Malaysia needs 60,000 accountants by 2020. [online] Available from: http://www.nst.com.my/news/2015/09/malaysia-needs-60000-accountants-2020
[Accessed 16th July 2016]

Nunnally, J. (1978). *Psychometric methods*. New York: McGraw Hill.

Remenyi, D., Williams, B., Money, A., and Swartz, E. (1998). *Doing research in business and management*. London, Sage Publications.

*Corresponding author's email: dalilahaa@unikl.edu.my

Roscoe, J. T. (1975). *Fundamental research statistics for the behavioral sciences [by] John T. Roscoe*. New York, NY: Holt, Rinehart and Winston.

Samidi, J. and Tew, Y.H., (1995). The study on the awareness of secondary school students towards accounting as a profession. In Accounting Seminar: Past, Present and Future, Universiti Teknologi Mara.

Said, J., Ghani, E.K., Hashim, A., & Mohd Nasir, N. (2004). Perceptions towards accounting career among Malaysian undergraduates. *Journal of Financial Reporting and Accounting, 2,* 17-30.

Sekaran, U. (2003). Research methods for business: A skill approach. *New Jersey: John Willey and Sons, Inc.*

Sugahara, S. and Boland, G., (2006). Perceptions of the certified public accountants by accounting and non-accounting tertiary students in Japan. *Asian Review of Accounting, 14(1/2),* pp.149-167.

Sugahara, S., Hiramatsu, K., & Boland, G. (2009). The factors influencing accounting school students' career intention to become a Certified Public Accountant in Japan. *Asian Review of Accounting, 17*, 5-22.

The Star Online (2013). More accountants needed, says MIA chief. [online] Available from: http://www.thestar.com.my/business/business-news/2013/09/17/more-accountants-needed-says-mia-chief/
[Accessed 16th July 2016].

The Star Online (2015). Govt to boost number of bumiputra accountants. [online] Available from: http://www.thestar.com.my/business/business-news/2015/10/06/govt-to-boost-number-of-bumiputra-accountants/
[Accesed 16th July 2016].

The Committee to Strengthen the Accountancy Profession. (2014). Report on the strengthening of the accountancy profession in Malaysia.

Trump, G.W., & Hendrickson, H.S. (1970). Job selection preferences of accounting students. *The Journal of Accountancy*, June, 84-86.

van Zyl, C., & de Villiers, C. (2011). Why some students choose to become chartered accountants (and others do not). *Meditari Accounting Research, 19,* 56-74.

*Corresponding author's email: dalilahaa@unikl.edu.my

EVALUATING QATAR CENTRAL BANK'S GOVERNANCE ON ISLAMIC BANKS IN QATAR

Hamad Alsayed*
Faculty of Islamic Studies, Qatar University, 120 Al-Qirawan street, Alwakra, Qatar

ABSTRACT

This paper aims to evaluate the governance of Qatar Central bank (QCB) on Islamic banks (IBs). Related-data will be taken from secondary sources. It is expected that QCB efficiently governs IBs in a way fulfills its unique nature. However, evidence suggests a number of issues in QCB governance towards IBs when looking at various dimensions. In Islamic Financial product dimension, QCB has no clear indications for enforcing its tawaruq-related regulations. This results on the extreme usage of tawaruq by IBs. In the social dimension, while it is suggested by early Islamic economists promoting equity products, such products are very slightly used by IBs in Qatar. The absence of QCB's regulations to promote such social aspects has led to the concentration of debt-based products. In terms of Sharia governance (SG) dimension, a number of issues have been raised such as the absence of cross-boarding regulations and the proper definition for the qualifications of sharia scholars. It is recommended, thus, to enforce tawaruq regulations properly, to create social-oriented regulations, to fix CG related issues discussed in this paper, and to unify Islamic financial contracts among IBs.

Keywords: Qatar Central Bank, Sharia governance, Islamic bank

*Corresponding author's email: h.alsayed999@gmail.com

INTRODUCTION

Research Background

It is required for corporate governance (CG) in Islamic financial institutions (IFIs) to have additional layer of governance for ensuring sharia-compliance known as sharia governance (SG). SG is defined by IFSB as "the set of institutional and organisational arrangements through which an IIFS ensures that there is effective independent oversight of *Shari`ah* compliance over each of the following structures and processes" (Dusuki. 2015). This includes a number of principles such as setting relevant sharia resolutions, through juristic opinions, operative experts who monitor sharia-compliance for each transaction on a daily basis, internal sharia reviewing body, annual sharia compliance review. Historically, there was no proper definition for sharia governance before this (Dusuki. 2015). While it is the duty of IBs regulators to ensure having sound sharia governance, the current practice of the regulators of IFIs vary in terms of the extent of regulatory support and the availability of friendly-legal frameworks (Dusuki. 2015).

Hasan (2012) has studied the role of regulatory authorities of both Gulf Corporate Countries (GCC) and Malaysia in sharia governance of IFIs. He provides insights on the different types of regulatory authorities in these countries and to what extent they achieve the efficiency in the IF industry. His findings suggest that the sharia governance practiced in Malaysia is more regulated and transparent than the one in GCC. He added that Malaysia is superior on GCC in having a central sharia body, known as Sharia Advisory Counsel, that is involved in both fatwa issuance and dispute resolving.

Hashim et al. (2015) contributes in studying the CG of IFIs from the conventional perspective on the basis of variables such as the size of the board. It has also considered SSB as a main player in IFIs. In a sample of 82 IFIs in GCC and non-GCC, it has been found out that there is a positive relationship between IFIs and board size, directors' independence, and the aspects of environmental, social, and profit in GCC. On the other hand, the size of SSB happened to be in insignificant relationship with the sustainability practices of IFIs. The implication of this is that there is no impact on the size of SSB of IFIs in GCC while there is such impact in non-GCC countries. This study drew attention to the important role of SSB and that further study need to be conducted in evaluating such manner.

In the Qatari context, Nahar (2015) has studied the regulations of QCB pertaining directing SSBs and their internal auditing bodies as well as the monetary policy of QCB. He evaluates the practice of QCB that using Accounting and Auditing Organization for Islamic Financial Institutions (AAIOFI) standards and sharia fiqhi views uncovering a number of issues will be discussed in details later.

That being shown, there is a need to evaluate further different dimensions of the governance of QCB to detect the extent of best can be achieved by IBs in Qatar.

Research Aim and Objectives

This paper aims to evaluate the governance of QCB in IB in Qatar. The objectives of this research include

1- Detecting the importance of central banks in governing Islamic financial instructions.

2- Identifying the pros and cons of governance practice of Qatar Central banks.

3- Providing suggestions for improving governance of Qatar Central banks

That being said, the research questions of this paper are as follows.

1- What is the regulatory importance of Central Banks in governing Islamic financial institutions in general?

2- What dimensions can be identified in the role of central banks as regulators for Islamic financial institutions?

3- What is the current governance framework for governing Islamic banks in Qatar?

4- What are the strengths and weaknesses of sharia governance framework of Qatar Central bank for Islamic banks?

5- What can be suggested for improving the governance of Qatar Central Bank of IFIs?

Significance of the Research

Although a number of papers discuss the general role of central banks in governing Islamic Financial institutions, a specific study in evaluating the central bank in the Qatari context remains unavailable. Such study is important to provide insights on CG issues occurred between central banks and IFIs in Qatar. The fruits of understanding such issues can be extended to other central banks in either approving or disapproving their practice.

Research Methodology

Secondary sources related to QCB shall be used. Zawya-based data relevant to Qatari Islamic banks also will be collected. Multiple secondary sources pointed out the pros and cons of current governance practice of QCB in comparison with other central banks. Data obtained from Zawya can provide certain details on Islamic financial products, which will be used to show the implications of corporate governance of IFIs.

THE RATIONALIZATION OF THE CRUCIAL ROLE PLAYED BY CENTRAL BANKS IN REGULATING ISLAMIC FINANCIAL INSTITUTIONS

This chapter is to introduce sharia governance as main governance of IFIs. Next, standards setting agencies such as AAIOFI shall be briefly explained as they are used adopted in some times like in Qatar. It then emphasizes the role of central banks in influencing IFIs through the legal and regulatory frame work not only on achieving their financial objectives, but also going beyond that towards socio-economical ones.

Sharia Governance

As stakeholder interests in Islamic financial institutions differ from those in the conventional, the corporate governance structure should be then articulated in a way that fulfills the rights and obligations of IFIs based stakeholders (Wilson 2012). The profit-sharing investment account, for example, which based on mutual relationship between the clients and banks, suggests further disclosure for tackling the conflict of interest issue that may arise (Wilson 2012). It is thus reasonable to state that corporate governance of IFIs is more sophisticated and fruitful, theoretically at least.

SG framework systematically tackles different issues on the heart of governance (Dusuki. 2015). As shown in the figure below, SG concerns on issues such as approving the products, documenting the contracts legally, and reviewing the transactions, which all centers on SSB and internal sharia audit (Dusuki. 2015). Failure to achieve their work results in reputational sharia risk that may lead clients to distrust their banks.

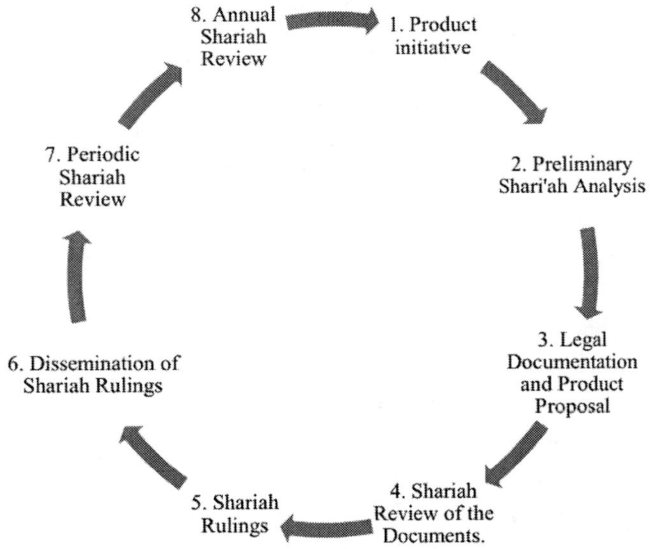

Source: (Dar 2009, cited by Dusuki. 2015, p. 703).

A unique characteristic that lies on the heart of IFIs is SSBs. Their role varies from one board to another depending on their nature and extent of sharia compliance. Ayub (2007) emphasizes that the degree of sharia compliance and the products relationship to islamic norms are important for the integrity of IFIs. The function of sharia scholars is to assist IFIs in approving products and to ensuring sharia compliancy of financial products during their lifecycle (DeLorenzo 2007). Beyond this, it is noteworthy that the regulators either in bank level or country level as in central banks play a role in determining the scope of authority to SSBs, in addition to legal framework in which the country based on (El-Hawary et al. 2007). Thus, the better the regulatory framework, the better SSBs can be, which aimed to be explored in this paper.

Standards setting Agencies

International standards setting institutions such as AAIOFI and IFSB play an important role in providing guidelines in sharia, auditing and corporate governance for best practices expected by IFIs, theoretically at least. NawalKasim et al (2013) contributed in comparing different shariah standards, namely, AAIOFI, IFSB and BNM to examine the pros and cons of each. As can be seen from the table below, while there is an advantage discussed by a standard setting institution, the other two remain silent. They conclude that there is a need for a more comprehensive standards related to shariah governance. It is then expected that the regulators of IFIs, in addition to adopting such standards, are considering further regulations than mentioned by such agencies in order to serve stakeholders of IFIs more professionally.

No	Discussion	AAIOFI	IFSB	BNM
1	Definiton of Shariah Governance	Not provided	Provided	Not provided
2	Independence, competence, responsibility of shariah supervisory board	Discussed	Discussed	Discussed
3	Shariah Review and Shariah Audit	Discussed	Discussed	Discussed
4	Risk management	Not Discussed	Not Discussed	Discussed
5	Research Function	Not Discussed	Not Discussed	Discussed

Source: (Nawal Kasim et al 2013)

Table 1: Comparison between different standards of standards settings agencies

Central Banks

Karim et al. (2013) emphasize the implications the supervisory implications of IF in the sense that the scope of sharia governance can be more difficult depending on the support, the choice, and environment of the regulators. In the United States, for example, there is a constitutional separation of the state from the religion that prevents the involvements of sharia matters. In Sudan or Malaysia, on the other hand, sharia matters face no such restrictions due to the establishments of a central sharia body that is supportive for the existence of Islamic banks in various ways (Karim et al. 2013).

Shedding the light on central banks, they are different types depending on the established framework of them and the countries they are based on. Briefly, known as nominal approach, there are Islamic central banks in which the whole system becomes Islamic, in having friendly Islamic regulations in particular, such as in Iran and Sudan (Ginena et al. 2015, Hasan 2012, Al-Sayed 2013). However, the rest of central banks operate in the conventional system in which such central banks vary in terms of their supportive friendly regulations and regardless of that, IFIs are required to adopt. These can be divided into three. First, as in Malaysia, its proactive jurisdiction is unique in having an active SSB at the regulatory level while there is no external sharia auditing, and the role of frequent sharia audits by the banking supervisor (Ginena et al. 2015, Hasan 2012, Al-Sayed 2013). Second, as in Oman, the committed approach of sharia governance concerns with the necessary legal framework to recognize, regulate, and supervise Islamic banks in addition to having internal sharia governance arrangements and an SSB ant the bank and regulatory level (Ginena et al. 2015, Hasan 2012, Al-Sayed 2013). Third, the engaged approach to sharia governance constitutes of the legal framework's recognition to Islamic banking through the enactment of law, the basis of prudent regulatory and supervisory policies and procedures for the industry (Ginena et al. 2015, Hasan 2012, Al-Sayed 2013). Having said this, each jurisdiction differs on articulating their legal framework because while some established laws that permit Islamic banks, others made amendments in the laws to allow the operation of Islamic banks (Ginena et al. 2015). Examples of this type include UAE and Qatar.

Next chapter is the aim of this paper starting with introducing the legal Qatari environment and general characterstics of QCB in general. It is then followed by evaluating QCB's governance for IBs.

DOI: 10.1051/shsconf/20173600009

EVALUATING QCB IN GOVERNING IBS IN QATAR

In order to examine QCB properly, how supportive Qatari law to IFIs shall be explored first. Next, a general description on the role of QCB towards IFIs shall be overviewed. Finally, going to the aim of this paper, evaluating sharia governance of QCB shall be established through a number of dimensions mentioned.

The legal framework of Qatar

It remains ambiguous to what extent sharia is incorporated in Qatari law concerning Islamic financial contracts. The business transactions are governed by the civil and commercial code, which was modified from the Kuwaiti code (Al-Khatib et al. 2001). The Kuwaiti code goes back to the Egyption civil law, which is originally adopted from French civil code. The whole chain of modifications results in a certain ambiguity of the Qatari law (Ballantyne 2000). This can also explain having elements of both sharia and French civil code in the civil and commercial code of Qatar, as mentioned in a recent study focused in the influence of sharia in Qatari law conducted by Elbarawy (2013). More importantly, the current Qatari law lacks the modern Islamic financial contracts related laws, which create problems for judges in choosing right interpretation (Alsayed 2016) as can be seen from the dispute example discussed later. The absence of such laws led a recent study to conclude that Qatari law is outdated in this regard (Maita 2013).

There are two Qatari regulators for IBS as shown in table (2). Qatar Central bank enjoys high authorities towards IBs because it has been established by the Amiri Decree no. 15/1993 inheriting and replacing Qatar Monetary Agency, the former central banking body (Hasan 2012). In terms of sharia governance, there are two regulatory guidelines provided by QCB and Qatar financial Center (Hasan 2012). The focus here is on the regulations of QCB, known as Instructions to Banks (IB) (Hasan 2012). Part 7 of it is concerned on instructing particularly IFIs (Hasan 2012). Further details will be discussed next.

DOI: 10.1051/shsconf/20173600009

Supervisory Authority	Financial Sectors	Primary Law/Regulations for IFSI	Essential features	Development of IIFS
Qatar Central Bank	Banking and Capital Market	Amiri Decree no. 15 of 1993	The Amiri Decree was issue for establishing the Central Bank as the monetary authority, and the mandates it to formulate monetary and banking and credit policies so as to achieve certain financial and economic objectives. All commercial banks, including IIFS, are governed under this Decree	The two full-fledged IIFS in Qatar are Qatar Islamic Bank and Qatar International Islamic Bank (1991).
Qatar Financial Centre (QFC) Regulatory Authority	Players at the QFC	QFCRA Islamic Finance Rulebook	The Rulebook forms the regulatory framework for the conduct of Islamic financial business in or from the QFC. Other than requiring IIFS to have its own Shari'ah Supervisory Board, the QFCRA also expects IIFS to apply some of the AAOIFI standards	No license issued yet

Source: (Ginena et al. 2015, P166)

Table 2: Qatari banking and Islamic banking related supervisory authorities

General role of QCB

According to Alsayed (2015), the objectives of QCB in supervising IBs in Qatar can summarized in five main points. First, it aims to protect depositors' funds. Although banks are responsible in protecting such funds, it is subject to losses caused by risks, for example, the mismanagement and poor supervision by the bank (speculation) and market fluctuations. QCB, thus, attempts to limit or reduce them. Second, QCB regulates the monetary and credit rxpansion, although IBs are not much concerned on money value as much as the assets that IF contracts are based on. Third, it provides directions for financial activity using means such as quality credit ceilings, types of guarantees, and conditions of borrowing. All of these can be developed in sharia-complaint way. Fourth, it maintains the integrity of the financial position of Banks. Fifth aim is the provisions of aid. This is an important assistance provided by QCB in case of imbalance or failures in stability in the banks.

Although these objectives are in-line with the sharia in fulfilling the financial sector, it is, however, does not provide particular socio-economics objectives to direct IBs towards achieving social objectives that early Islamic economists were concerned about (Asutay 2007). Furthermore, according to Kahf et al. (2014), QCB still lacks IF friendly products to fulfil the need of IFIs in case of financing as last resort as will be discussed later.

QCB has adopted Basal, IFRS and IFSB accounting standards, and AAOIFI governance standards, particularly, for Islamic banks[1]. The focus here will be three dimensions, namely, Islamic financial products, social, and sharia governance dimensions.

Islamic Financial Products dimension

There are different issues related to how supportive QCB for IFIs in sharia-complaint products. From one side, sharia-complaint products from QCB for the financier as last resort are missing. From another, QCB has not unify or restrict the diversity of different Islamic financial contracts among IBs in Qatar leading indirectly to misunderstanding issues in Qatari courts as shown in the dispute example. As the regulatory framework established in QCB is conventional based rather than Islamic-based, QCB is only risk and profit focused rather than ensuring the unity and smoothness of Islamic financial products in different Islamic banks in Qatar.

Kahf et al. (2014) has pointed out that QCB lacks shariah-complaint instruments for improving liquidity management, in particular, for financing as last resort and interbanking. As such, they attempted to develop theoretically such instrument that can be sharia complaint and in-line with national law of Qatar. They also have surveyed the experience of other central banks such Bahrain and Malaysia in developing such instrument that QCB is in serious need. This shows the absence of the role of QCB in supporting sharia-complaint product for IFIs in the case of last resorting.

Second, the unity of IFCs among IBs is a cause of numerous issues. Above all, the hybrid Islamic financial contracts that are characterized as sophisticated contracts are difficult to understand for non-IF specialists including judges in courts[2]. Having such contracts have led to different interpretations made by judges in Qatari courts.

Alsayed (2016) has analyzed a Qatari-based case involves Ijarah muntahiah bittamleek (IMBT) (lease to own) contract, a hybrid Islamic financial contract in which the client defaults due to the failure to fulfill this obligations. In the case he explores, the judge interprets IMBT contract as promise of sale because the contract does not fall into the provisions of Law No. 4/2008 concerning the Rent of Buildings and as such, the judge

[1] This is what I was told in an interview with an expert from QCB during my Master's dissertation.
[2] This is another point that the expert of QCB I interviewed has emphasized strongly.

has refused looking at an issue that is not under the judge's specialty. In supporting this, the judge evidenced that IMBT reflects the prophetic prohibition of two sales on one without referring to a particular Islamic financial code. The judge's' understanding for the contract led to this interpretation, which is contradicting what AAIOFI made lawful on standard no. 9. This suggests inconsistency related issue between IBs in Qatar and Qatari courts. Such conclusion is expected to be avoided if Islamic financial contracts are united and documented in similar way that prevents the confusion of understanding. The unity of Islamic financial contracts is expected to be taken into account under the Qatari law if it comes from Qatar Central Bank. This shows an issue of inconsistency of interpretation between sharia scholars' interpretation and judges under Qatari law, which QCB can be partially blamed for indirectly.

Social dimension

A number of maqasid-oriented literature emphasizes that IF based products should encourage equity based such as musharaka and mudaraba rather than debt-based such as murabaha and tawaruq. Asutay (2007), for example, drew attention to the socio-economic failures of IF in disintegrating Islamic maxims on the products provided by IBs.

However, according to Asutay (2007), the current practice of Qatari IBs seems to be a part of this socio-economic failure. Based on data collected from Zawya, table (3) demonstrates that the total assets of musharaka, mudaraba and murabaha in, as an example, Masraf alrayyan between 2010 and 2014. The total assets of debt-based products are doubled multiple times and superior to equity-based ones. This raises questions over the regulatory role in directing the usage of IF products in socio-economics friendly products.

DOI: 10.1051/shsconf/20173600009

Instrument	Period End Date 12/31/2014	Period End Date 12/31/2013	Period End Date 12/31/2012	Period End Date12/31/2011	Period End Date12/31/2010
Musharaka (USD '000)	992,732	142,832	0	18,946,000	40,848
Mudaraba (USD '000)	0	334,407	0	0	605,785
Murabaha (USD '000)	14,870,770	11,309,664	11,746,558	9,331,297	6,284,310

Source: Zawya

Table 3: Total assets of Musharaka, Murabaha, and Mudarabaha in MAR between 2010 and 2014

The only QCB based regulation available in this regard is limiting the usage of tawaruq except in times of personal necessity. Nahar (2015) and Hasan (2012) criticize the lack of proper definition for necessity in this case. This leads to question the enforability of such regulation on Qatari IBs (Nahar 2015). Such issue can be a major cause for the lack of diverse usage of Islamic financial products in Qatar. In his interview with staff from Masraf Alrayyan, Nahar (2015) has actually confirms that the most used Islamic financial products go to tawaruq.

Having said this, a possible explanation for the missing implications of maqasid or social objectives on the products in Qatari IBs is the high risk of equity-based products as well as the high GDP of Qatari-based citizens. Having less poverty in Qatar seems to affect the choice of IBs products. In this regard, it is expected from QCB or IFIs to disclose the reason behind the choice of the products of IBs not only on the basis of risk, but for achieving optimal Islamic banking practice as well. Further investigation on debt-based products in Qatar is required.

Sharia Governance dimension

Although AAIOFI governance standards are adopted by QCB, there is a number of issues has been raised. First, sharia board members are restricted from receiving credit facilities for commercial purposes, but not personal matters. It is noteworthy that this restriction is not only for SSB members, but also for the members of board of directors (Nahar 2015). It has been questioned why not expanding this restriction further to personal purposes. A possible explanation is that QCB would like to provide a chance for such members to get access for financing in the event of serving in all boards (Hasan 2012, Ginena et al. 2015).

Second, although QCB requires IFIs to have qualified sharia members, it has not defined such qualifications (Nahar 2015, Ginena et al. 2015). Given his sharia background, Nahar (2015) suggested certain qualities in sharia scholars sitting in SSBs. These are the fiqhi ability (malakah fiqhiyyah) in understanding scholarly opinions, the ability of analogical thinking for identifying similar cases in past and future, the quality of smartness, quick noticing and understanding the life of people. These qualities are also mentioned by AAIOFI, but Nahar (2015) has also added the following qualities, namely, having strong personality for not being tricked or forced in non-sharia complaint manner, and learning English language.

Third issue is to have a minimum of two sharia board members (Ginena et al. 2015, Nahar 2015). Nahar (2015) criticizes choosing a minimum of two members rather than three because in the case of dispute in opinions, there will be no third person to settle between the two in choosing the final opinion, which is is required for proper management's point of view.

Fourth, Nahar (2015) raises some issues related to the regulation of having sharia body auditing in IFIs. The issues he raised are the absence of precise definition for the qualifications of sharia auditors and supervisors as well as the lack of identifying a minimum number for sharia auditors. As for the qualifications, Nahar (2015) has referred to the qualifications mentioned for the members of SSBs. Regarding the number of sharia auditors, he suggested that such matter is subject the size of the bank.

That being said, there are a number of non-discussed issues by QCB. One of the issues pointed out by AAIOFI in sharia standard no. 29 on 3/5 but not referred to by QCB is the fact that it is not allowed for IFIs to ask for fatwa issuance from other than their SSBs (Ginena et al. 2015). This can play an important role in determining the choice of products to be more in-line with the maqasid, rather than being market and profit oriented.

Second, QCB has not limited the number of cross-boarding for the members of SSBs, unlike the Malaysian regulator (Ginena et al. 2015). A possible explanation for

DOI: 10.1051/shsconf/20173600009

overlooking this issue by QCB is the slim chance of impact on this (Ginena et al. 2015) given the defined role of the members of SSB. Also, it is possible that in the regulator's view the financing given for the members do not reach the level that creates independence issue (Ginena et al. 2015).

Overall, it can be seen that a number of the issues discussed are due to the usage of abstract language that can lead to weak sharia governance practices.

CONCLUSION

This paper aims to evaluate the governance of Qatar Central bank (QCB) on Islamic banks (IBs). The methodology used is, mainly, the secondary sources and Zawya. It is expected that QCB efficiently governs IBs in a way fulfills its unique nature. This unique nature can be seen from sharia governance of IFIs represented by SSBs. Their role is important in the stability of IFIs and in ensuring the avoidance of reputational and sharia risk. This requires extra consideration in the regulations given to IFIs. However, evidence suggests a number of issues regarding QCB governance towards IBs when looking at various dimensions.

In Islamic Financial product dimension, QCB has not defined clearly the personal necessity for allowing the usage of tawaruq in its regulations. This possibly results on the extreme usage of tawaruq by IBs. As such, there are no clear indications for enforcing its tawaruq-related regulations. In the social dimension, there is lack of diversity of islamic financial products as the concentration is on the usage of tawaruq, rather than promoting for example equity based products.

In terms of Sharia governance (SG) dimension, a number of issues have been raised such as the absence of cross-boarding regulations and the proper definition for the qualifications of sharia scholars. It is recommended, thus, to enforce tawaruq regulations properly, to create maqasid-oriented regulations, to fix the CG issues discussed in this paper and to unify Islamic financial contracts among IBs in Qatar

The limitation of this paper is the lack of quantitative evidence that can be strong indications for the issues raised. Another limitation is restricting the research in the Qatari context with sharia governance focusing. Future research are advised to apply the research of CG of IFIs in MENA region through productive comparisons and optimal practices of CG.

REFERENCES:

AAIOFI, (2015) *Accounting, Auditing and Governance Standards for Islamic Financial Institutions*. Manama, Bahrain, Accounting and Auditing Organization for Islamic Financial Institutions.

Al-Khatib, F. and Al-Abdulla, S. (2001) The State of Qatar: A Financial and Legal Overview. *Middle East Policy*. [Online] 8 (3), 110-125. Available at: doi:doi:10.1111/1475-4967.00031 (Accessed: 31 August 2016).

Al-Sayed, H. (2013) *[i]A Study on The Development and Analysis of Investment Tools in Islamic Banks*[/i]. Ph. D. University of Wales, Lampeter.

Alsayed, H. (2016) *[i]Explorative study on the extent of consistency between Islamic Financial Contracts and Qatari laws through a Qatari-court-based dispute analysis involved an Islamic financial contract*[/i]. MSc. University of Durham.

Asutay, M. (2007) Conceptualisation of the Second Best Solution in Overcoming the Social Failure of Islamic Banking and Finance: Examining the Overpowering of Homoislamicus by Homoeconomicus. *IIUM Journal of Economics and Management*. 15 (2), 167-195.

Ayub, M. (2007) *Understanding Islamic finance*. Hoboken, NJ, John Wiley & Sons.

Ballantyne, W. (2000) *Essays and addresses on Arab laws*. Richmond, Surrey, Curzon.

DeLorenzo, Y. (2007) Shari'ah Compliance Risk. *Chicago Journal of International Law*. 7 (2), 1-13.

Dusuki., A. (2015) *Islamic financial system: Principles and Operations*. 5th edition. Kuala Lumpur, International Shari'ah Research Academy for Islamic Finance (ISRA).

Elbarawy, H. (2013) Ta'athir Alsharia Alislamiah ala Alqanon Almadani AlQatari: Dirasa muqarana (The influence of Islamic Shari'a on Qatar's civil law: A comparative study). *International Review of Law*. [Online] 21-21. Available at: doi:10.5339/irl.2013.cl.2 (Accessed: 31 August 2016).

El-Hawary, D., Grais, W. and Iqbal, Z. (2007) Diversity in the regulation of Islamic Financial Institutions. *The Quarterly Review of Economics and Finance*. [Online] 46 (5), 778-800. Available at: doi:10.1016/j.qref.2006.08.010.

Ginena, K. and Hamid, A. (2015) *Foundations of Shari'ah governance of Islamic banks*.

Hasan, Z. (2012) *Shari'ah Governance in Islamic Banks*. Edinburgh, Edinburgh University Press.

Hashim, F., Mahadi, N. and Amran, A. (2015) Corporate Governance and Sustainability Practices in Islamic Financial Institutions: The Role of Country of Origin. *Procedia Economics and Finance*. [Online] 3136-43. Available at: doi:10.1016/s2212-5671(15)01129-6.

Kahf, M. and Hamadi, C. (2014) An Attempt to Develop Sharīʿah Compliant Liquidity Management Instruments for the Financier of Last Resort: With Reference to Qatar Development Plan. *Islamic Economic Studies*. 22 (1).

Karim, R. and Archer, S. (2013) *Islamic Finance: The New Regulatory Challenge, 2nd Edition*. John Wiley & Sons.

Maita, A. (2013) *[i]Development of a Commercial Arbitration Hub in the Middle East: Case Study - The State of Qatar*[/i]. Ph. D. Golden Gate University School of Law,

Nahar, N. (2015) *Alsairafa Alislamiah fi Dawlat Qatar (Islamic Banking in the State Islamiah fi Dawlat Qatar*

EMPLOYERS' PERCEPTION ON INTERNSHIP PROGRAMME

Nur Jannah Mohaidin*[1], Masniza Supar[1], Mohd Adha Ibrahim[2], Morni Hayati Jaafar Sidik[1]
Universiti Kuala Lumpur Business School[1]
Universiti Sains Islam Malaysia[2]

ABSTRACT

Internship program is compulsory for Bachelor in Accounting (BIA) students in University Kuala Lumpur (UniKL). We believed internship program is useful to facilitate students learning opportunities outside classroom. These experiences provide the opportunity to apply classroom theory into real working environment thus enhancing students' academic and career goals. Constructive comments from supervisor will give us indicator that we must prepare the students with all aspects of accounting wide knowledge. Employer's feedback is important in preparing the students for the industry by developing a better programme structure and subjects offered. It is indirectly improves the lecturers' teaching methods and skills. Therefore, this paper explores the employers' perception towards internship programme for accounting students in Universiti Kuala Lumpur. Data was collected from employer's feedback form using five point-likert scales distributed to employers of the participating companies from Semester January 2013 to Semester January 2015. The evaluation form is used to evaluate the students' performance throughout their 6 months internship period. The statistical results found that student's score is positively associated with employer's feedback. The results also indicate that the employers' perception is important for the students in preparing themselves for the industry and for the university in developing proper programme structure.

Keywords: Internship, Programme Structure, Employer's Perception and Accounting Students

INTRODUCTION

Internship programme is introduced in order to furnish the undergraduate students with the opportunity to attain hands on experience and to provide students with a smooth transition from the academic world to the working environment. One of major rationale for internships is a transition experience from the university to the work environment, where interns learn job-related knowledge and skills, along with employer workplace skills (Maertz, Stoeberl and Marks, 2013). The classroom learning does not provide the practical experience students need upon leaving college. Some students may not fully understand the concepts that the lecturer is trying to illustrate through textbook exercises. Internship programme provides students with direct experience through which they can use analytical skills and reflection to apply new ideas gained from their working experience to classroom learning. Pernsteiner (2015) states employers felt that accounting interns develop significantly in both hard and soft skills over the course of their internship experience.

Universiti Kuala Lumpur, one of the private universities in Malaysia, through UniKL Business School also integrates a structured internship into Bachelor in Accounting (Hons) programme.

*Corresponding author's email: nurjannah@unikl.edu.my

The internship is compulsory for students in their fourth year (semester 7) which carries eight (8) credit hours. The duration of the internship is 24 weeks (1 semester) and the students are compulsory to return to the university and complete their final semester (semester 8) before they graduated. Ebied (2004) claimed that students' academic performance was improved significantly following an internship programme. Furthermore, employers are able to use internship training as a platform for identifying potential young graduates to be hired. Gault, Leach and Duey (2010) suggested that experiential education plays a vital role in enhancing the career preparation and marketability of undergraduates in the entry-level job market.

This programme is coordinated by International, Industrial, Institutional and Partnership Department (IIIP) through Accounting Section of UniKL Business School. The internship coordinators assisted them in preparing resume, interview skills and guidance on how to survive during internship. The students need to find internship placement on their own based on their preferred location. Students are allowed to choose any organisations that are related to their field of study. For students who are considering a professional career in accounting, it is advisable that they pursue an internship programme with a professional firm engaged in providing assurance services.

The first batch of students had completed their internship programme on semester January 2013. To date, there are 5 batches of students who have completed their internship at various organisations namely audit firms, government institution, private sectors, financial institution and multinational corporations. There are a small number of researches carried out to examine the employers' perception on the internship programme. Most of prior research studies involving internship examined the perception of students. Thus the purpose of this research is to examine the employers' perception on internship programme in UniKL Business School.

This study aims at examining the employers' opinion and evaluation on internship programme. The research objectives are as follows:

1. To examine the employers' perception towards accounting students during their internship programme.
2. To investigate whether there are any significant differences on employers' perception for female students and male students (gender analysis).

The findings from this research give us an overview of the internship programme from the employers' perspectives. The findings would give some information to UniKL Business School to prepare the students with adequate accounting knowledge and soft skills prior joining the internship programme.

LITERATURE REVIEW

According to Beck and Halim (2008), internships are considered as a valuable learning experience for students, academics and prospective employers. Students are able to gain experience in real working environment and apply what they have learnt in classroom. Most previous research concentrated on perception of students towards internship programme. However, this study aims to examine the employer's perception towards accounting internship programme. In this section therefore, we will briefly review the literature which discusses various ways on how the accounting internship is valuable to the employer and the accounting students from their perspective. This includes what employers' value about offering

*Corresponding author's email: nurjannah@unikl.edu.my

internships, what employers feel about hiring candidates with experiences, what employer's think the value is for the accounting students and the ability for accounting students to develop their hard and soft skills.

Taylor (1988) defined internships as "structured and career relevant work experiences obtained by students prior to graduation from an academic program". The internship experience and learning gained throughout internship period before graduation can also be critical in the eyes of employers. Internship program is able to prepare the students to be more marketable after graduate by helping them develop soft skills such as communication skill and critical thinking skills that many employers seek for employment.

According to Gault, Redington and Schalger (2000), other than education area, little research has been discovered on the benefit can accrue for educator and their institution. Higher – education researches have continued to focus primarily in improving teaching skills and other pedagogical processes operating within classroom setting. Such learning is more difficult to deliver in classroom, thus internship programme offered to the students assumed to improve the classroom curriculum in preparing the students for a job in future. In addition, the university should build and maintain good relationship with the employer especially audit firms in the list of Big 4, so that they may be able to market themselves not only to the potential students but also to their parent when attracting new students to enrol with them. This shown that all motivated and qualified students can have internship opportunities at their chosen field with those companies. It is interesting if the university can promise the incoming students any possible full time jobs before graduation. In the students' perspective, they will have the sense of loyalty to the university in helping them in job search and therefore promote the university to other prospective students.

Perception of Employers

Extra labor capacity and the opportunity to try out a potential future staff are the most common reason given by employer for using interns (HR Focus, 2005). Hiring interns as full time employee after they graduated can reduce recruitment and selection cost. Other than that it provides lower risk for employers as they already know the performance of the students during internship period. Once hired, the former interns also do not need same degree of training and need less time to adapt with working environment as required by most of new employees. Assuming the interns did well during internship, the interns may be knowledgeable enough to contribute to the company after they are being hired. Interns may also be more loyal to the company and stay longer than the average non-intern hire.

Kelley and Gaedeke (1990) investigate employers' perception of the relevance importance among hiring criteria:

1. Oral communication skill
2. Written communication skill
3. Problem solving
4. Analytical skills
5. Computer applications, and
6. Leadership/ teamwork skills

In this study oral is found to be favoured in the eyes of employer for hiring purpose. Thinking and reasoning skills such as analytical ability, computer application, creative thinking,

*Corresponding author's email: nurjannah@unikl.edu.my

information search and problem solving have been found to be important across a range of field area with the degree of importance varying by industry (Floyd and Gordon, 1998). This is consistent with research done by Andrews and Higson (2008), that employers felt communication skills were important in hiring the interns. Moreover, leadership and teamwork have been found to be the prime importance to recruiters (Boatwright and Stamps, 1988).

According to Gault et al. (2010), they found that the interns must have ten career preparation skills as follows:

1. Reliability
2. Consistency of performance
3. Eagerness to learn new skills
4. Timeliness
5. Effectively prioritising tasks
6. Demonstrating initiative/ self-motivation
7. Exhibiting ethical behaviour
8. Accepting criticism constructively

However, in our study we only focus on few aspects that covered in the questionnaire distributed to employers as below:

1. Written and oral communication skills.
2. Demonstrating initiative in performing work given.
3. Teamwork and leadership.
4. Analytical skills.
5. Accepting criticism constructively.
6. Timeliness.

RESEARCH METHODOLOGY

Questionnaire

Data for this study was collected using employer's feedback form distributed to employers of the participating companies from Semester January 2013 to Semester January 2015. The evaluation form is used to evaluate the students' performance throughout their 6 months internship period. The employer's feedback form is divided into 2 parts. Part A is on the students' performance that required the supervisors to evaluate based on five point Likert-scale and Part B consisted of an open-ended question. This was intended to elicit supervisor's general comments and suggestions regarding students' performance and capabilities.

Research Model

This study attempts to examine the employers' opinion and evaluation on internship programme. Based on the literature, the factors that have been identified are written and oral communication skills, demonstrating initiative in performing work given, teamwork and leadership, analytical skills, accepting criticism constructively and timeliness. The research model is as depicted in the figure 1 below. Employers' perception towards UniKL accounting students is the dependent variable while the six factors are the independent variables. Data from the questionnaire were coded and analysed using the SPSS software. Non-parametric and t-test

*Corresponding author's email: nurjannah@unikl.edu.my

analysis were used to examine the relationship between the independent variables and the dependent variable.

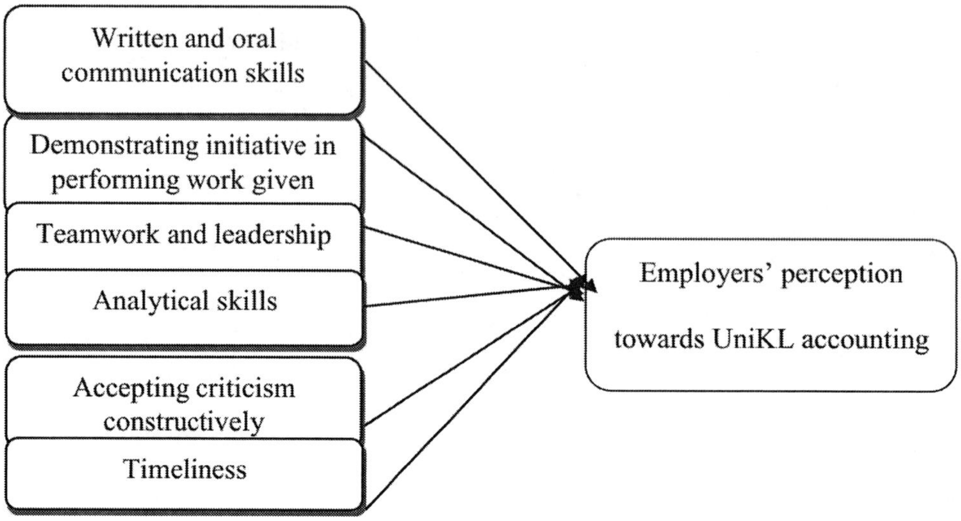

Figure 1: Theoretical Framework

Data Analysis

The survey population involved Semester January 2013 to Semester January 2015 accounting students from UniKL Business School who had completed their internship programme. The actual number of respondents is only 116 which consists about 74.1% female and 25.9% male as presented in Table 1 below.

Table 1: Background of respondent

Item	Description	Frequency	Percentage (%)
Gender	Male	30	25.9
	Female	86	74.1

The questionnaire needs to be completed by supervisor at the attachment place assessing the student's performance throughout their internships period. The majority of interns did their internship attachment in auditing/accounting firms. Such places considered as favourite place among the students because they will be assigned tasks totally related to their study. Based on this research, 79.3% students did their intern in auditing/accounting firm the rest of 20.7% students did their intern at non-auditing firm (Table 2).

Table 2: Internship attachment

Item	Description	Frequency	Percentage (%)
Firm	Auditing/Accounting firm	92	79.3
	Non-Auditing/Accounting firm	24	20.7

*Corresponding author's email: nurjannah@unikl.edu.my

Employers' Perception

Table 3 presented below is the results of the non-parametric correlations that meant to examine the employers' perception towards accounting students during their internship programme. The results indicated that three variables, namely demonstrating initiative in performing work given, analytical skills and accepting criticism were found to be significant determinants of employers' perception towards UniKL accounting students during their internship programme. The result suggests that demonstrating initiative in performing work given, analytical skills and accepting criticism were likely to be factors that will be influence the employers to recruit them as their employee in the future.

Table 3: Employers' Perception towards UniKL Accounting Students

	Spearman's rho Employers' perception		
	Correlation Coefficient	Sig. (2-tailed)	N
Written and oral communication skills	-0.169	0.070	116
Demonstrating initiative in performing work given	-0.214*	0.021	116
Teamwork and leadership	-0.130	0.163	116
Analytical skills	-0.285**	0.002	116
Accepting criticism	-0.229*	0.013	116
Timeliness	-0.117	0.213	116

**. Correlation is significant at the 0.01 level (2-tailed).
*. Correlation is significant at the 0.05 level (2-tailed).

Table 4 presented below is the results of the *t-test* analysis which used to investigate whether there are any significant differences on employers' perception for female students and male tudents (gender analysis). The results indicated that there was no significant difference in scores for males (M = 3.9867, SD = 0.58765) and females [M = 3.9930, SD = 0.60309; t (114) = -0.050, p = 0.960]. Based on the equation 1 below, the magnitude of the differences in the means also was very small (eta squared = 0.004).

Table 4: Independent samples T-test between Gender and Employers' Perception

	t	df	Mean Difference	Std. Error Difference
Employers' Perception towards UniKL Accounting Students	-0.050	114	-0.0064	0.127

CONCLUSION

This study focuses on the employer's perception on internship programme in the case of accounting students in UniKL. The findings of this research concerning employer's perceptions are consistent with previous research suggesting intern must have a very good demonstrating initiative in performing work given, analytical skills and accepting criticism. However, there is no significant difference on the perception of employers towards gender. Therefore, the

*Corresponding author's email: nurjannah@unikl.edu.my

gender factor will not influence employers' decision on whether to recruit the students after completion of their study. Undergraduates with internship placement experience are perceived to be better prepared and more marketable to employers. There are a lot of benefit gained from internship programme, thus educators are wise to promote the advantages of internships to the students. Further research should be done on the students' perception toward internship programme.

References

Andrews, J., Higson, H. (2008), Graduate Employability, Soft Skill Versus Hard Business Knowledge: A European Study, Higher Education in Europe, 33(4), 411-422.

Beck, J.E., Halim, H. (2008), Undergraduate Internships in Accounting: What and How Do Singapore Interns Learn from Experience?, Journal of Accounting Education, 17(2), 151-172

Carl P. Maertz Jr Philipp A. Stoeberl Jill Marks , (2014),"Building successful internships: lessons from the research for interns, schools, and employers", Career Development International, Vol. 19 Iss 1 pp. 123 – 142

Ebied, H. (2004). An Empirical Investigation of the Impact of the Student Accounting Internships on Subsequent Academic Performance: The Case of UAE University. *Journal of Economic and Administrative Sciences*, *20*(2), 33-44.

Floyd, C.J and Gordon M.E (1998), What skills are most important? A comparison of Employer, Student and Staff Perceptions, Journal of Marketing Education, Vol. 20, August, pp. 103-9

Gault, J., Leach, E., & Duey, M. (2010). Effects of business internships on job marketability: the employers' perspective. *Education+ Training*, *52*(1), 76-88.

Gault, J., Redington, J., Schlager, T. (2000), Undergraduate Business Internships and Career Success: are they related?, Journal of Marketing Education, Vol. 22 No. 1., pp. 45-53.

HR Focus, (2005), Plan for the most Effective Internships Programs, pp. 7-11

P. Maertz Jr, C., A. Stoeberl, P., & Marks, J. (2014). Building successful internships: lessons from the research for interns, schools, and employers.*Career Development International*, *19*(1), 123-142.

Pernsteiner, A. J. (2015). The Value of an Accounting Internship: What Do Accounting Students Really Gain?. *Academy of Educational Leadership Journal*, *19*(3), 223.

Taylor, M.S (1998), Effects of College Internships on Individual Participants, Journal of Applied Psychology, Vol 73, pp 393-401.

*Corresponding author's email: nurjannah@unikl.edu.my

THE INTEGRATION OF DISCLOSURE OF ISLAMIC SOCIAL REPORTING (ISR) IN ISLAMIC BANK FINANCIAL STATEMENTS

Sugianto*, Nurlaila Harapan
Faculty of Islamic Economics and Businesses, Universitas Islam Negeri Surabaya

ABSTRACT

Islamic bank has an economy function as well as a social function. So that Islamic banks posses an ethical identity for their social objectives as important if not more important than economic goals due to the fact that the system and its operation is based on the Islamic Shari'ah. This objective is intended to Islamic bank also participate in the improvement of society, so it is expected to describe clear corporate social responsibility in their social reporting practices as evidenced in their annual reports. This study replicated the Haniffa and Hudaib research by examining social reporting practices of Islamic Banks in Indonesia. This examination involves a comparison of social disclosure 12 Islamic banks conducted through their annual reports to the ideal level of social disclosure that Islamic banks should be made, during the years 2014-2015. This comparison is done by using the Ethical Identity Index (EII) developed by Haniffa and Hudaib (2007). The findings revealed that at present, Islamic banks in Indonesia is still poor in the practice of social reporting.

Keywords: EII, Islamic bank, disclosure, social reporting

INTRODUCTION

The institution conducts its business based on sharia is essentially based on a philosophy that comes from al-Qur'an and Sunnah. So for businesses this philosophy became the basis for interacting with the environment and each other. Therefore ties between the institution and its environment in the concept will be stronger than in the conventional concept. Therefore, sharia business institution such as Islamic banking, according Dusuki and Dar (2005) has a social responsibility because of several factors: Islamic banking is based on sharia, operates with a foundation of morals, ethics and social responsibility and the principles upon obedience to the commandments of Allah and caliph. Islamic banking is a unique institution because it does not only focus on economic issues alone but also more importantly the role of social and spritual dimension (Meutia, 2010). Even Islamic banking was established with the mandate to carry out the transaction in accordance with the provisions of sharia (Karim, 1999).

Characteristics of Islamic banking as described above require accountability in implementation. Accountability Islamic banking is partly demonstrated by the submission of an annual report from the economic aspect and social aspects. Disclosure of social responsibility reports of companies known as corporate social responsibility (CSR) disclosure. Thus the information disclosed not only financial information, but

*Corresponding author's email: sugianfa@gmail.com

other information is about the social impacts (externalities) and the environment resulting from the company's activities. Therefore, companies can gain legitimacy by demonstrating social responsibility through CSR disclosure in the media is included in the company's annual report (Oliver, 1991; Haniffa and Coke, 2005).

Disclosure of social responsibility in Islamic banking is not enough to CSR alone. Therefore, AAOIFI (Accounting and Auditing Organization for Islamic Financial Institution) include matters relating to Islamic principles such as zakat, the status of sharia compliance and the transaction is free from elements of usury and gharar and aspects of social as sodaqoh, qardhul hasan, up to the disclosure of worship in a corporate to the disclosure of CSR in Islamic banking. Development of CSR disclosure is known as ISR index (Othman et al, 2009). Islamic Social Reporting (ISR) was first proposed by Haniffa and Hudaib (2007) to develop the concept of "Ethical Identity Index".

This paper attempts to analyze the integration of social responsibility disclosure in the financial statements of Islamic banking is based on the items Islamic Social Reporting (ISR) Index.

THEORETICAL FRAMEWORK KERANGKA TEORI

Theories of Social Reporting

Social reporting by Islamic banking be the main topic in this study.Gray et al. (1987: 4) define social reporting as "The process of providing information designed to discharge social accountability. Typically this act would ... be undertaken by the accountable organization and thus might include information in the annual report, special publications or reports or even socially oriented advertising."

Driven by recent business scandals around the world such as Enron, WorldCom and Parmalat, there has being growing interest in corporate, environmental and social reporting for achieving better corporate accountability (Hess, 2007).

The social reporting provides various theories propounded to explain the motivation for disclosing social information by firms and the particular set of stakeholders for whom such social information is provided for. The more prominent theories are as follows:

Theory of Legitimacy

Today the theory of legitimacy used in the study of accounting in the disclosure of social and environmental responsibility. Legitimacy of the community is a strategy for the company in order to develop the company's future. It can be used as a vehicle to construct the company's strategy, especially related to efforts to position themselves amid increasingly advanced society (Ardianto and Machfudz, 2011: 73). The theory implies that the legitimacy of the activities in the form of corporate social responsibility is a business with respect to the pressure from the surrounding environment, such as political pressure, social or economic. The theory of legitimacy based on the notions implied social contract between society and social institutions and required by institutions to achieve the objectives in line with the broader community (Gray et al, 1995: 47-77).

The point is how companies can synergize with the environment and society by the company can focus on the activities of companies in getting profit and can also help investors in making investment decisions. Then the disclosure of social environment is to gain legitimacy from the people, especially on the continuity of the organization.

In the perspective of an entity's orientation system is affected and the communities where otherwise affect the entity's activity. Corporate disclosure policy is seen as an important matter in which managers can influence the perception of other parties on the organization. The theory of legitimacy has become one of the most frequently used theory especially when it pertains to the area of social and environmental accounting.

Stakeholders Theory

Stakeholder theory states that the company is not the only entity that operates for its own sake, but also must provide benefits to the stakeholders. Stakeholders are all internal and external parties that may influence or be influenced by the company either directly or indirectly (Ardianto and Machfudz, 2011: 73). The company's survival depends on stakeholders, and such support should be sought, so that the activity of the company is to seek the support. The more powerful stakeholders, the greater the company's efforts to adapt. Social disclosure is considered as a communication medium between the company and its stakeholders (Gray et al, 1987).

The existence of these assumptions stakeholder theory which states that the relationship between the entities with groups that have rights, the nature of the relationship in the process of the company and making the managerial desperation. Thus, from the assumption that the company can not be separated from the social environment. This theory is used to explain the behavior of social and environmental disclosure in the hope somewhat satisfied stakeholders to survive and as information for making decisions.

Agency Theory

This theory states that the company is place or intersection point for the contractual relationship that occurs between management, owners, creditors and government. This theory tells of monitoring a wide range of cost and the relationship between these groups. For the example is auditing, he gives confidence to stakeholders about managerial by the management company as agent (Umar, 2008: 46). This theory focuses on determining the most efficient contract between the principal/owner/ shareholders and agents/directors/management.

Agency theory is linked to the process of establishing corporate governance system that will bridge the separation between the interests of owners and managers in the company, especially in the duties, powers and functions of other functions so that each function clearly. Application of corporate governance are expected to entrust the agent (management) manage the wealth of the owners (investors) and the owners have become more confident that the agency will not make a fraud to welfare agencies (Rizkiningsih, 2012: 11).

Social Contract Theory

A term which says that humans are social beings also apply to a company. This theory was born because of the interaction and interrelation in social life, in order that happen harmony, harmony and balance including on the environment. Companies whose objective is part of the community with the wider environment, which affect each of them. For that kind a happened equality it needs a social contract based on an agreement that protects the interests of each (Ardianto and Machfudz, 2011: 76).

Islamic Social Reporting

Social reporting takes place within a framework of social relations. Fundamental to an Islamic perspective on social reporting, is an understanding of the concepts of accountability, social justice and ownership that are central to social relations.

The main objective of corporate reporting is to allow Islamic enterprises to show their compliance with Sharia. Other objectives of corporate reporting may include those known in the Western model, such as assisting decision makers in making economic decisions, but from an Islamic perspective, these are secondary objectives. This view of the primary objective is adopted by AAOIFI when setting out its statement of objectives of financial accounting for Islamic banks and financial institutions.

Islamic Social Reporting is based on the Sharia framework, namely the relationship between the interests of the world and the hereafter which will result in material, moral and spiritual aspects in the ISR enterprise reporting. Within the framework of the shariah, monotheism is the foundation of the teachings of Islam. Tawhid is everything that exists in this world comes from Allah the One and the unity embodied in creed, namely recognition of one God and faith in Him. So, people who *syahadah* will accept the consequences of monotheism that is required to submit and adhere to all laws of God written in the Qur'an, hadith and fiqh and qiyas, ijtihad and ijma 'which all these laws aim to enforce justice social and achieve happiness in this world and the hereafter (*al-Falah*).

Islamic Social Reporting is the concept of CSR with the order of Islamic law was born in 2002 initiated Haniffa and Hudaib on research by the writings were given the title "*Social Reporting Disclosure: An Islamic Perspective*". Then developed by researchers hereinafter as Rohana Othman, Azlan Thani, Raditya more intensively in 2009 in Malaysia and to date ISR concept must still be developed. According pengagas ISR are still many deficiencies in conventional CSR concept, so the birth of ISR makes the provisions of sharia as the basis for the proposed conceptual framework or skeleton known as *sharia* (*sharia* framework). Given that the ISR not only helps stakeholders in decision-making but also help the company fulfill its obligations to Allah and society.

Islam regulates all aspects of human life on earth, encapsulates all aspects of life, both worship and social (muamalah). The basic foundation of the religion of Islam is

Aqeedah (belief and faith), worship, and morality and ethics. Karim (2012: 34) explains that there are five universal values, namely: tauhid (belief), 'adl (justice), nubuwwah (prophethood), khilafah (government) and ma'ad (results) which is the basis of inspiration to compose proportions and theories of Islamic economics. Allah has created human beings as well as possible the shape and power as creatures on earth. As the most perfect creature that Allah created it is appropriate that men always go through all the commands and avoid His prohibitions as a sign of devotion and which are associated with it is damaging the environment. Fromf syariah framework or conceptual framework initiated by Haniffa (2002) can be seen that tauhid is the basic foundation of the teachings of Islam as described above. In this framework also shows the concept of ethics in Islam which consists of ten concepts include faith, taqwa (piety), amanah (trust), ibadah (workship), khilafah (vicegerent), ummah (community), akhirah day of reckoning, ‘adl (justice) and zulm (tyranny), halal (allowable) and haram (forbidden), and I'tidal (moderation) and israf (extravagance). According to the Islamic concept of ethics in the form of accountability in the perspective of Islamic economics ie reporting of corporate social responsibility in accordance with Islamic principles.

In conventional economics, social responsibility reporting known as the reflection of social expectations associated with the role of the community in economic and business activities of companies in the form of financial statements. Related to this, so the conventional system only focuses on the aspects of material and moral. In the spiritual aspect of Islam should also serve as a major focus in the reporting of corporate social responsibility for the decision makers of Muslims have expectations that companies disclose the latest information voluntarily to assist in fulfilling their spiritual needs. Hence the need for a special framework for social responsibility reporting in accordance with the principles of Islam.

The framework is not only useful for decision-makers of Muslim, but also useful for Islamic companies in fulfilling its accountability to Allah and society. This framework known as the Islamic Social Reporting (ISR). Islamic Social Reporting (ISR) using Islamic principles as a fundamental point. Sharia principles in the ISR produce material, moral, and spiritual aspects that became the main focus of the company's social reporting. Islamic Social Reporting (ISR) is an extension of social reporting which is not only a great desire of the whole society on the role of companies in the economy but rather related to spiritual perspective (Haniffa, 2001).

The ISR is more pressure to social justice in its reporting in addition to reporting on the environment, minority interests and employees (Othman, 2010). This involves economic issues related to the welfare of society such as trade practices such as unbalanced and uneven distribution of income. Important factors that form the basis of *sharia* in the formation of Islamic Social Reporting (ISR) is *tauhid* (one God) and not ascribing partners to Him, submit all matters to God and submit to His commandments, believing that Allah belong the kingdom of the heavens and the earth (Qur'an 57: 5), and then to Him you shall be returned (Qur'an 2: 28).

ISR index is disclosure items are used as indicators in reporting social performance of companies. This study uses a framework of Islamic Social Reporting with primary reference Haniffa and Hudaib (2007), which developed with the items on the research Othman et.al. (2009).

Imam (2000), Belal (2001), and Azizul Islam and Mathew (2009) surveyed corporate social disclosure practices in Bangladesh. They found very poor level of disclosure and descriptive in nature. Abu Baker and Naser (2000) also utilized content analysis to examine annual reports of 143 Jordanian companies; their findings appear to mirror the results of prior studies.

Zakimi and Abdul Hamid (2009) investigated the extent of corporate social disclosure by banks and finance companies in Malaysia. They found that social disclosure level among banks and finance companies in Malaysia is still low. In addition, firm size was found to have a positive relationship with the amount of social disclosure made by firms. However, the variables of profitability, scope of business, listing status and audit firm size show insignificant relationship.

Zubairu (2008) examined the social reporting practices of Islamic banks in Malaysia to use the Ethical Identity Index (EII). The results showed that only 1 (one) Bank Islam of 7 (seven) Bank Islam is above average EII. The mismatches related to four factors: 1) a commitment to the community; 2) disclosure of its vision and mission; 3) contributions and management of zakat, charity and qardhul hasan; and 4) information regarding top management. While the results of his research with Sakariyu and Dauda (2011) in the Islamic Bank in Saudi Arabia shows that Islamic banks in Saudi Arabia have much in common with conventional banks.

RESEARCH METHOD

The researchers‟ study adopts as a guide the study carried out by Haniffa and Hudaib (henceforth referred to as Haniffa and Hudaib) in 2007, which sought to explore the ethical identity of Islamic banks as communicated via annual reports. The researchers‟ study sought to discover the social reporting practices of Islamic banks in Indonesia. To achieve this objective, content analysis was utilized in order to examine the annual reports of the selected Islamic banks. Content analysis was selected as research method for the researchers‟ study because it is "a research technique for making replicable and valid inferences from texts to the content of their uses. As a research technique, content analysis provides new insight, increases a researcher‟s understanding of new phenomena, or informs practical action. Content analysis is a scientific tool" (Krippendorf, 2004: 18).

In many previous social disclosure studies, content analysis has been the method of choice for determining the extent of an organization‟s social disclosures through whatever media of communication; whether annual reports (Maali et.al, 2003; Grunning, 2007; Boesso and Kumar, 2007), company websites (Branco and Rodrigues, 2006; Holcomb, 2007) or any other media. This widespread use of this method by previous social reporting literature is another reason why content analysis was chosen as the research method by the researchers. To determine whether or not Islamic banks in Saudi Arabia were fulfilling the Shari'ah requirement of full disclosure, a checklist of items that Islamic banks ought to disclose was developed. This checklist was adopted from two studies, Haniffa and Hudaib (2007) and Maali et.al (2003).

The population in this study are all Islamic banks in Indonesia. Samples are twelve (12) fully-fledged Islamic banks that publish annual financial statements and other information in 2014-2015. The sample of 12 Islamic banks in Indonesia are as seen tabel 1.

No	Name	Abbreviation
1.	PT. Bank Muamalat Indonesia	BMI
2.	PT. Bank Victoria Syariah	BVS
3.	PT. Bank BRISyariah	BRIS
4.	PT. Bank Jabar Banten Syariah	BJBS
5.	PT. Bank BNI Syariah	BNIS
6.	PT. Bank Syariah Mandiri	BSM
7.	PT. Bank Mega Syariah	BMS
8.	PT. Bank Panin Syariah	BPS
9.	PT. Bank Syariah Bukopin	BSB
10.	PT. BCA Syariah	BCAS
11.	PT. Maybank Syariah Indonesia	MSI
12.	PT. Bank Tabungan Pensiunan Nasional Syariah	BTPNS

Tabel 1: Distribution of Research Sample

The researchers adopted, the "ideal ethical identity" checklist developed by Haniffa and Hudaib (2007) in order to determine the social reporting practices of Islamic banks in Indonesia. The reason for adopting Haniffa and Hudaib (2007)'s checklist was because it contained a detailed list of items known as "constructs" which Islamic banks ought to disclose. In addition, Haniffa and Hudaib (2007)'s study was one of two most recent studies on the disclosure patterns of Islamic banks (the other one is Maali et.al 2003), and thus provided the most recent scholarship in the area of Islamic banks social reporting practices. The researchers adopted Haniffa and Hudaib (2007)'s checklist completely.

The annual reports of 12 Indonesia Islamic banks in the sample were obtained on their various websites for the years covered by the study, 2014-2015. These annual reports were then read completely to discover the items that they disclosed as required by the ideal ethical identity checklist.

To measure whether a discrepancy existed between actual disclosure levels and ideal disclosure levels as delineated by the checklist, a disclosure index was utilized which Haniffa and Hudaib (2007) dubbed the Ethical Identity Index (EII). In determining the level of disclosure in the annual reports of the 12 Islamic banks in the sample, a score of 1 was assigned for every item or construct of the checklist that was disclosed in the annual report for each bank, and zero for every item not disclosed. In addition, a company was not given extra marks if it disclosed a construct more than once in its annual report.

Each of the 9 dimensions had a certain number of constructs under it, which Islamic banks had to disclose to fulfill the Shari'ah requirement of full disclosure. To determine how well a bank did in terms of disclosure for each of the 9 dimensions, the EII for each dimension was computed for each annual report examined in every year covered by the study. The resultant computation was referred to as the Dimension EII (DEII). The DEII was calculated as follows:

DEIIy = Xy/Xj ....(1)

Where:

DEIIy = the annual dimension EII for a bank "y"
Xy = number of items disclosed by a bank "y" under dimension "X"
Xj = ideal number of items to be disclosed under thet dimension.

Analysis and Discussion of Results

Based on the results of the ethical ideal of refined identity checklist of 12 Islamic banks in Indonesia period 2014-2015 can be seen in Table 2.

Based on Table 2 above it can be concluded that not all Islamic banks that meet the criteria of all dimensions. Only two banks that cater to the nine dimensions EII above, namely Bank Muamalat Indonesia and BRI Syariah. Two banks are only one dimension unfulfilled, seven banks that do not meet the two dimension, whereas that does not meet the three-dimension is one bank.

In 2014, the "Vision and Mission statements" dimension, all banks disclosed more than half of the required items necessary in the annual report. In 2015, there was one bank increase disclosing from 2014 is BNI Syariah.

The previous dimension result, on "BODs and top management", all banks disclosed more than half of the required items necessary in the annual report in 2014. In 2015, there was one bank that increased disclosure, namely BNI Syariah. And one bank that decreased disclosure, namely Bank Mega Syariah.

DIMENSION	YEAR	BMI	BVS	BRIS	BJBS	BNIS	BSM	BMS	BPS	BSB	BCAS	MSI	BTPNS
Vision and Mission	2014	0.56	0.56	0.67	0.67	0.56	0.56	0.78	0.78	0.78	0.78	0.56	0.56
	2015	0.56	0.56	0.67	0.67	0.67	0.56	0.78	0.78	0.78	0.78	0.56	0.56
BODs and top management	2014	0.77	0.69	0.77	0.69	0.69	0.77	0.77	0.77	0.77	0.77	0.69	0.69
	2015	0.77	0.69	0.77	0.69	0.77	0.77	0.69	0.77	0.77	0.77	0.69	0.69
Product and service	2014	0.40	0.40	0.30	0.30	0.40	0.30	0.30	0.40	0.60	0.30	0.40	0.30
	2015	0.60	0.40	0.30	0.50	0.60	0.30	0.30	0.60	0.60	0.60	0.40	0.30
Zakāt, charity and benevolent loans	2014	0.40	0.40	0.67	0.53	0.33	0.27	0.27	0.40	0.13	0.40	0.13	0.47
	2015	0.40	0.40	0.67	0.53	0.40	0.27	0.27	0.40	0.13	0.47	0.13	0.47
Commitments toward employees	2014	0.89	0.67	0.67	0.67	0.89	0.67	0.78	0.67	1.00	0.67	0.56	0.89
	2015	0.89	0.67	0.67	0.67	0.89	0.67	0.78	0.67	1.00	0.67	0.78	0.89
Commitments toward debtors	2014	1.00	0.00	0.75	0.00	0.00	0.00	0.00	0.00	1.00	0.00	0.00	0.00
	2015	1.00	0.00	0.75	0.00	0.00	0.00	0.00	0.00	1.00	0.00	0.00	0.00
Community	2014	0.86	0.00	0.57	0.71	0.43	0.57	0.57	0.57	0.57	0.57	0.57	0.86

	2015	0.86	0.00	0.57	0.71	0.43	0.57	0.57	0.71	0.71	0.57	0.57	0.86
Environment	2014	0.50	0.00	0.50	0.00	0.00	0.50	0.00	0.00	0.00	0.00	0.00	0.00
	2015	0.50	0.00	0.50	0.00	0.50	0.50	0.00	0.00	0.00	0.00	0.00	0.00
SSB	2014	0.73	0.55	0.64	0.64	0.55	0.64	0.27	0.73	0.64	0.36	0.64	0.45
	2015	0.64	0.55	0.64	0.64	0.55	0.64	0.27	0.73	0.64	0.36	0.64	0.45
Total	2014	0.58	0.40	0.56	0.49	0.44	0.44	0.42	0.50	0.54	0.44	0.40	0.48
	2015	0.59	0.40	0.58	0.51	0.51	0.44	0.41	0.53	0.56	0.49	0.42	0.48

Tabel 2: DEII Result

On the 3rd dimension, "Product and Services", only one bank disclosed up to half of the required items in 2014, namely Bank Syariah Bukopin. In 2015, there are five banks disclosed up to half of the required, namely BMI, BNI Syariah, Bank Panin Syariah, Bank Syariah Bukopin and BCA Syariah.

Under the 4rd dimension, "Zakat, charity and benevolent loan", only two banks disclosed up to half of the required items in 2014 and 2015, namely BRI Syariah and Bank Jabar Banten Syariah. For the dimension, some banks do not manage directly zakat, charity and benevolent loan. They have a special institution that manages the fund, namely amil zakat institution. Bank Muamalat Indonesia for example has Baitul Mal Muamalah (BMM) which manages the fund. Likewise BSM has LAZ BSM.

Under the 5th dimension, "Commitments toward employees", all the banks disclosed more than half of the required items necessary in the annual report both in 2014 and 2015. Only one bank that increased disclosure in 2015, namely Maybank Syariah Indonesia.

On the 6th dimension, "Commitments toward debtors", only three banks disclosed up to half of the required items both in 2014 and 2015, namely Bank Muamalat Indonesia, BRI Syariah and Bank Syariah Bukopin. Other banks that do not meet the dimension.

The 7th dimension, "Community", there are ten banks disclosed up to half of the required items both in 2014 and 2015. However, there is one bank disclosing at least half, namely BNI Syariah and one bank that do not meet the dimension, namely Bank Victoria Syariah. Other banks that do not meet the dimension. In 2015, there are two bank that increased disclosure, namely Bank Panin Syariah and Bank Syariah Bukopin.

On the 8th dimension, "Environment", only three banks disclosed and more to half of the required items both in 2014, namely Bank Muamalat Indonesia, BRI Syariah and Bank Syariah Mandiri. Other banks that do not meet the dimension. In 2015, the banks that disclosing add to four banks, namely BNI Syariah.

The 9th dimension, "Syariah Supervisory Board (SSB)", only three banks disclosed at least half of the required items both in 2014, namely Bank Mega Syariah, BRI Syariah, BCA Syariah and BTPN Syariah. In 2015, there is one bank that decreased disclosure, namely Bank Muamalat Indonesia.

Lastly, overall of nine dimensions only three banks disclosed up to half of the required items in 2014, namely Bank Muamalat Indonesia, BRI Syariah and Bank Syariah Bukopin. In 2015, there are six banks disclosed up to half of the required items, namely Bank Muamalat Indonesia, BRI Syariah, BJB Syariah, BNI Syariah, Bank Panin Syariah and Bank Syariah Bukopin.

Based on the above results show that Islamic Banks in Indonesia had poor implement Islamic social disclosure in reporting. In 2014 only three banks of twelve banks and increased to six banks in 2015. Only on three dimensions all of the Islamic banks disclosed up to half of the required items both in 2014 and 2015. The dimensions are "Vision and Mission statements", "BODs and top management" and "Commitments toward employees".

However, the banks had poor disclosure practice on three dimensions, namely "Products and services", "Zakat, charity and benevolent loan", "Commitment of debtors" and "Environment". For the "Zakat, charity and benevolent loan" dimension, some banks do not manage directly zakat, charity and benevolent loan. They have a special institution that manages the fund, namely amil zakat institution. Bank Muamalat Indonesia for example has Baitul Mal Muamalah (BMM) which manages the fund. Likewise BSM has LAZ BSM.

As Haniffa and Hudaib (2007: 111) put it, the findings are surprising because Islamic banks, as social and economic institutions are expected to disclose more on those dimensions that reflect accountability and justice not only to society, but also ultimately to God. What can be gathered is that Islamic banks in Indonesia have a lot of work to do before they become worthy of the tag they claim of been Islamic. This is not to imply that the development of Islamic banks in Indonesia is not a step in the right direction; it is, but, these institutions have to realign their activities and priorities to make sure they truly reflect the Islamic principles which they claim to follow. This is very important if they are to retain their credibility and reputation in the society in which they operate.

CONCLUSION

This study shows that practice of the Islamic social disclosure in reporting on Islamic Banks in Indonesia is still weak. In fact, Islamic banks are not only aimed at seeking commercial profit but also has a social function as accountability, especially to God.

Limitations of this study are only on the disclosure of the dimensions is not more specific as to how understanding and awareness of the bank's stakeholders, the factors that affect disclosure. Similarly, the public response to the effects of social accountability practices of Islamic banks.

REFERENCES

Abdullah Saeed. 1996. *Islamic Banking and Interest: A Study of the Prohibition of Riba and its Contemporary Interpretation*. Leiden: EJ Brill.

Antonio, Muhammad Syafi'I. 2011. *Islamic Banking, Bank Syariah dari Teori ke Praktik.* Jakarta: Gema Insani.

Ardianto, Elvinaro dan Machfudz, Dindin, M. 2011 *Efek Kedermawanan Pebisnis dan CSR*. Jakarta: PT. Elex Media Komputindo.

Accounting and Auditing Organization for Islamic Financial Institutions. 1999. *Accounting, Auditing and Governance Standards for Islamic Financial Institutions*. Bahrain: AAOIFI.

Belal, A., R., & Owen, D., L. 2007. The Views of Corporate Managers and the Current State of, and Future Prospects for, Social Reporting in Bangladesh. *Accounting, Auditing and Accountability Journal*, 20(3), 472-494.

Boesso, G., & Kumar, K. 2007. Drivers of Corporate Voluntary Disclosure: A framework and Empirical Evidence Italy and the United States. *Accounting, Auditing and Accountability Journal*, 20(2), 269-296.

Brammer, S., & Pavelin, S. 2006. Voluntary Social Disclosures by Large UK Companies. *Business Ethics: A European Review*, 13(2/3), 86-99.

Branco, M., C., & Rodrigues, L., L. 2006. Communication of corporate responsibility by Portuguese banks: A legitimacy theory perspective. *Corporate Communications: An International Journal*, 11(3), 232-248.

Dar, H., & Pressly, J. 2000. Lack of Profit Loss Sharing in Islamic Banking: Management and Control Imbalances. *International Journal of Islamic Financial Services*, 1-27.

Haniffa, R. 2001. *Social Responsibility Disclosure: An Islamic Perspective*. UK: University of Exeter.

Haniffa, R. & Hudaib, M. 2007. Exploring the Ethical Identity of Islamic Banks via Commnication in Annual Reports. Journal of Business Ethics, 76, 97-116.

Hess, D. 2007. Social Reporting and New Governance Regulation: The Prospects of Achieving Corporate Accountability through Transparency. *Business Ethics Quarterly*, 17(3), 453-476.

Holcomb, J., L., Upchurch, R., S., & Okumus, F., O. 2007. Corporate Social Responsibility: what are top hotel companies reporting? *International Journal of Contemporary Hospitality*, 19(6), 461-475.

Karim, Adiwarman Azwar. 2012. *Ekonomi Mikro Islami* Edisi ketiga. Jakarta: PT. Raja Grafindo Persada.

Krippendorf, K. 2004. *Content Analysis: An Introduction to Its Methodology*. London, New Delhi: Sage Publications.

Maali, B., Casson, P. & Napier, C. 2003. Social Reporting by Islamic Banks. *Discussion Papers in Accounting and Finance*, 03(13), 1-46

Othman, Rohana. 2010. *International Business & Economics Research Journal.* Malaysia: Islamic Social Reporting Of Listed Companies.

Othman, R., & Md Thani, A. 2010. Islamic social reporting of listed companies in Malaysia. *International Business and Economics Research Journal*, 9(4), 135-144.

Othman, R., Md Thani, A., & Ghani, E. K. 2009. Determinants of Islamic social reporting among top Shariah-approved companies in Bursa Malaysia. *Research Journal of International Studies*12: 4-20.

Gray, R., et al. 1995. "Corporate social and environmental reporting: a review of the literature and a longitudinal study of UK disclosure", *Accounting, Auditing and Accountability Journal.*

Gray, R., Owen, D., Maunders, K. 1987. "*Corporate Social Reporting: Accounting and Accountability*", Prentice-Hall International, USA

Rizkiningsih, Prayesta. 2012. *Faktor Faktor yang Mempengaruhi pengungkapan Islamic Social Reporting, Studi Empiris pada Bank Syariah Indonesia, Malaysia dan Negara Negara Gulf Cooperation Council.* Jakarta: FE UI.

APPENDIX

Research Instrument

Theme 1: Underlying Philosophy and Values

A. Dimension: vision and mission

No	Construct	Disclosed	
		2014	2015
1	Commitments in operating within Shari'ah principles/ideals		
2	Commitments in providing returns within Shari'ah principles		
3	Focus on maximizing shareholder returns		
4	Current directions in serving the needs of Muslim community		
5	Future directions in serving the needs of Muslim community		
6	Commitments to engage only in permissible investment activities		
7	Commitments to engage only in permissible financing activities		
8	Commitments to fulfill contracts via 'contract (*uqud*) statement'		
9	Appreciation to shareholders and customers		

B. Dimension: BOD and top management

No	Construct	Disclosed	
		2014	2015
1	Names of board members		
2	Positions of board members		
3	Pictures of board members		
4	Profile of board members		
5	Shareholdings of board members		
6	Multiple-directorships exist among board members		
7	Membership of audit committee		
8	Board composition: executive vs non-executive		
9	Role duality: CEO is Chairman of board		
10	Names of management team		
11	Positions of management team		
12	Pictures of management team		
13	Profile of management team		

Theme 2: Interest-Free and Islamically Acceptable Deals

C. Dimension: product

No	Construct	Disclosed	
		2014	2015
1	No involvement in non-permissible activities		
2	Involvement in non-permissible activities - % of profit		
3	Reason for involvement in non-permissible activities		
4	Handling of non-permissible activities		
5	Introduced new product		
6	Approval ex ante by SSB for new product		
7	Basis of *Shari'ah* concept in approving new product		
8	Glossary/definition of products		
9	Investment activities – general		
10	Financing projects – general		

Theme 3: Development and Social Goals

D. Dimension: *Zakāt*, charity and benevolent loans

No	Construct	Disclosed	
		2014	2015
1	Bank liable for Zakāt		
2	Amount paid for Zakāt		
3	Sources of Zakāt		
4	Uses/beneficiaries of Zakāt		
5	Balance of Zakāt not distributed – amount		
6	Reasons for balance of Zakāt		
7	SSB attestation that sources and uses of Zakāt according to *Shari'ah*		
8	SSB attestation that Zakāt has been computed according to Zakāt		
9	Zakah to be paid by individual-amount		
10	Sources of charity (*saddaqa*)		
11	Uses of charity (*saddaqa*)		
12	Sources of *qard al-hassan*		
13	Uses of *qard al-hassan*		
14	Policy for providing *qard al-hassan*		
15	Policy on non-payment of *qard al-hassan*		

E. Dimension: Employees

No	Construct	Disclosed	
		2014	2015
1	Employees appreciation		
2	Number of employees		
3	Equal opportunities policy		
4	Employees welfare		
5	Training: Shari"ah awareness		
6	Training: other		

7	Training: student/recruitment scheme		
8	Training: monetary		
9	Reward for employees		

F. Dimension: Debtors

No	Construct	Disclosed	
		2014	2015
1	Debt policy		
2	Amount of debts written off		
3	Type of lending activities – general		
4	Type of lending activities – detailed		

G. Dimension: Community

No	Construct	Disclosed	
		2014	2015
1	Women branch		
2	Creating job opportunities		
3	Support for org. that provide benefits to society		
4	Participation in government social activities		
5	Sponsor community activities		
6	Commitment to social role		
7	Conferences on Islamic economics		

H. Dimension: Environment

No	Construct	Disclosed	
		2014	2015
1	The amount and nature of any donations or activities undertaken to protect the environment		
2	The projects financed by the bank that may lead to harming the environment		

Theme 4: Reviews by Shari'ah Supervisory Board

I. Dimension: Shari‟ah Supervisory Board (SSB)

No	Construct	Disclosed	
		2014	2015
1	Name of members		
2	Pictures of members		
3	Remuneration of members		
4	Report signed by all members		
5	Number of meeting held		
6	Examination of all business transactions ex ante and ex post		
7	Examination of a sample of business transactions ex ante and ex post		
8	Report defect in product : specific and detailed		
9	Recommendation to rectify defects in products		
10	Action taken by management to rectify defect in product		
11	Distribution of profit and losses comply to shariah		

A COMPARATIVE REVIEW ON COMPANY SPECIFIC DETERMINANTS FOR SUSTAINABILITY REPORTING IN UNITED KINGDOM (UK) AND MALAYSIA

Xue Fa Tong *
Faculty of Business, Multimedia University Melaka

ABSTRACT

This study aims to comparatively assess the effects of company-specific variables on the level of corporate social responsibility (CSR) information disclosed in publicly-traded companies from United Kingdom (UK) and Malaysia. Content analysis was applied to sampled reports from the FTSE 100 Index and FTSE Bursa KLCI against inferred meanings from the Global Reporting Initiative (GRI)-derived coding base to identify similarities and/or differences in CSR disclosure practices. The Spearman's correlation coefficients and multiple linear regressions (MLR) analyses further gauged the associations between the variables and total quantity of CSR disclosure (TQ_{CSR}); and, determined the predictive determinants on sustainability reporting.

The Spearman's correlation has identified a negative association on leverage with TQ_{CSR} for UK companies. In contrast, the TQ_{CSR} in the Malaysian sample was positively associated with directors' CSR-related experiences and profitability but negatively associated with company size. Results from MLR analyses presented company size as a significant determinant on sustainability reporting in the UK model, while directors' experiences were indicated as the crucial determinant in the Malaysian model. This first, direct cross-market sustainability reporting study highlights the importance of board of directors' CSR-relevant experience in influencing the level of CSR disclosures in publicly-traded companies.

Keywords: Sustainability reporting, Global Reporting Initiative (GRI), corporate social responsibility (CSR), board experience

INTRODUCTION

Sustainability reporting has emerged as a powerful tool in gaining long term business growth and competitive advantage. The practice to report high quality non-financial data beyond the financial bottom line helps companies to mitigate potentially material financial impact of environmental and social risk in their businesses. In May 2013, the non-mandatory G4 Sustainability Reporting Guidelines was initiated to commensurate the need for a more relevant, transparent and credible integration of sustainability indicators into sustainability reporting practices. Although the G4 framework is a useful reference in homogenising corporate social responsibility (CSR) disclosures across global market sectors, many sustainable reports are still inconsistent and reactive in nature (Jagd, 2015; Fortainer et al., 2011; PwC's Sustainability Reporting, 2008). The inconsistency in levels of non-financial CSR disclosures is predicated on the polarizing prioritisation of the "stakeholder versus shareholder" approaches in addressing societal issues (Freeman and Moutchnik, 2013), the

*Corresponding author's email: xftong@mmu.edu.my

DOI: 10.1051/shsconf/20173600012

cost of disclosing information voluntarily (Gamerschlag et al., 2011; Kytle and Ruggie, 2005; LaFond and Watts, 2008), and, the varying interpretations of the recommended guidelines.

This study intends to juxtapose the sustainability reporting agenda of publicly-traded companies from the United Kingdom (UK) and Malaysia. While there are substantial empirical researches on the determinants of CSR disclosure practice internationally, cross-market comparison on sustainability reporting activities were scarce and mainly explored the macro-effect of country-level variables and political-legal systems (Ioannou and Serafeim, 2016; Anderson and Gupta, 2009; La Porta et al., 2008). Furthermore, only a few studies on CSR determinants were relevant to the Malaysian context (Mat Nor, 2010; Amran et al., 2007). In order to better understand the readiness of Malaysian companies in pursuing the CSR agenda, a comprehensive comparative study against a developed CSR reporting regime such as the UK is necessary.

For these reasons, the aim of this study is to critically analyse and identify the key drivers for sustainability reporting of the top 30 publicly-traded companies in the UK and Malaysia. This paper extends prior research on the effect of company-level determinants in explaining voluntary CSR disclosure practices (Lan, et al., 2013; Gamerschlag et al., 2011). The study will address two fundamental research questions:

1) What are the company-specific characteristics associated with corporate social responsibility (CSR) disclosures?
2) What are the specific variables predicting the amount of CSR-related information disclosed in the sustainability reports for each country?

Data collected from publicly traded companies operating in the FTSE 100 Index and FTSE Bursa Malaysia KLCI will be subjected to in-depth two-phase analyses to corroborate and quantify the developed hypotheses. Withstanding industry classifications, the research expects a strong positive correlation to exist between the identified internal company-specific drivers (i.e. board experience, profitability, financial leverage) and the level of CSR disclosures in sustainability reporting for both countries regardless of potential cross-market differences. Findings from the study will contribute to a greater understanding on the varying degree of sustainability reporting practices in companies and progress the dimension on CSR-reporting literatures.

LITERATURE REVIEW

The Global Reporting Initiative (GRI) describes sustainable reporting as a form of comparable non-financial reporting, which communicates sustainability-related information on economic, environmental, social and governance performance of a corporation or organisation. On account of a non-existing 'one size fits all' framework, marked variations in the level of CSR disclosures was observed within the Malaysian sustainability reporting practices, which were behind that of in the UK (UNEP, 2013; Chambers et al., 2003). Despite the increased CSR awareness among Malaysian companies and noteworthy efforts to go beyond the standard industry and Bursa Malaysia reporting requirements, the quality of CSR disclosures within Malaysian annual reports is dismal to the normal scope of corporate financial reporting (MaSRA, 2013). Previous studies have suggested the importance of company-specific characteristics as proxies in inducing their disposition and responsiveness in disclosing additional CSR information voluntarily (Lan et al., 2013; Gamerschlag et al., 2011; Siegel and Vitaliano, 2007; Orlitzky et al., 2003).

*Corresponding author's email: xftong@mmu.edu.my

Thus, this study seeks to determine the pre-existing influence of company-level variables on the level of sustainability reporting preceding the transitional implementation period of the G4 Sustainability Reporting Guidelines in 2014 and 2015. In addressing the internal organisational factors of sustainability reporting, extensive literature search indicated dearth of quantified measures on board experience as an antecedent to CSR disclosure levels. This research proposes to establish the role of board of directors with CSR-relevant experiences as a new empirically measurable variable in sustainability reporting. Figure 1 illustrates the identified determinants of sustainability reporting tested in this study.

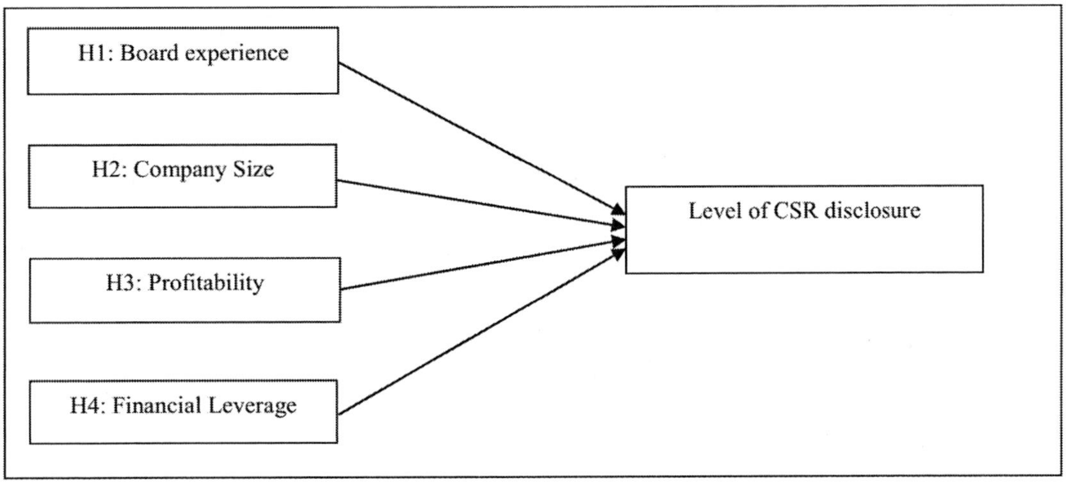

Figure 1: Framework of determinants of CSR disclosure in sustainability reporting

Board experience

Both the UK Financial Reporting Council and Securities Commission Malaysia recommends the composition of the board and its committees to comprise of an appropriate balance of skills, experience and knowledge to adapt to changing circumstances and discharge their responsibilities effectively. A direct inclusion of CSR-relevant experience, expertise, competence and reputation on the boards provide valuable perspectives on sustainability matters in corporate performance and CSR reporting (Mallin and Michelon, 2011; Carpenter and Westphal, 2001). Such presence inadvertently warrants an active incorporation of the CSR agenda into the strategic policies and decisions of companies. This study believes the overall proportion of board of directors with CSR-relevant experience is imperative to the level of CSR disclosure. An improved CSR governance and disclosure practices is achievable with the inputs from all directors during board-level strategic planning. The independence of the directors is irrelevant in considering the boards' advocacy towards CSR activities (Ratna et al., 2016; Gul and Leong, 2004; Kassinis and Vafeas, 2002). For this study, the presence of CSR-relevant board experience is determined from the corporate profiles of the board of directors. Active experience, assignment and participation in CSR-related are inferred from the directors' biographic descriptions provided in the annual report. The following hypothesis is postulated:

H1$_{\textbf{(UK/Malaysia)}}$: *The level of CSR disclosure is positively associated with a CSR relevant board experience.*

*Corresponding author's email: xftong@mmu.edu.my

Company size

Grounded on the association of social disclosure with political visibility, Belkaoui and Karpik (1989) posit a positive association between company size and CSR disclosure. Given the scale of operations and relatively greater economic and social influence, larger companies are often exposed to recurrent public visibility from the media (Fortainer et al., 2011; Gallo and Jones Christensen, 2011; Brammer and Pavelin, 2008). Empirical studies have substantiated that larger companies tend to disclose higher levels of public information readily to ease informational and political costs associated with information asymmetry (Gamerschlag et al., 2011; Chih et al., 2009; LaFond and Watts, 2008; Reverte, 2008; Cormier and Magnan, 2003). With increased public visibility, the likelihood of companies to use a formal channel to communicate CSR activities and improve their corporate image and legitimacy increases equitably (Haniffa and Cooke, 2005). Based on substantial unanimities to reason size as a CSR indicator, company size is predicted to have a positive relationship with the amount of CSR disclosure. For this study, sampled companies will represent the constituents of the largest market capitalisation from the national equity indices of FTSE Bursa Malaysia KLCI and FTSE 100. Using the natural logarithm of total assets of the companies (Lan et al, 2013), it is hypothesised that:

H2(UK/Malaysia): *The level of CSR disclosure is positively associated with company size.*

Profitability

The disclosure of sustainability reporting is dependent on the adequacy of funding made available to companies. Generally, companies have limited available resources to respond to a wide range of stakeholder concerns. Profitable companies are more financially-abled to report socially responsible behaviours than companies with weaker financial performance (Reverte, 2008; Udayasankar, 2008; Campbell, 2007; Haniffa and Cooke, 2005). Two common financial performance indicators typically employed to measure a company's profitability are the return on equity (ROE) and return on assets (ROA). In review, although ROE inherently does not consider the financial risk of leveraging, it is relatively more stable than ROA in comparatively representing the true profitability of companies' performance quality. A higher level of disclosure to correspond with the managerial motivation to assert a company's favourable ROE with executive competencies and to progress managerial compensations (Lan et al., 2013; Gamerschlag et al., 2011). Since leverage will be analysed as a variable independently, this study will observe ROE as the representative determinant for profitability on CSR disclosure levels in maximising shareholder value. For these reasons, the following hypothesis is proposed:

H3(UK/Malaysia): *The level of CSR disclosure is positively associated with profitability.*

Financial leverage

Studies have indicated that a highly leveraged capital structure with substantial debt financing is synonymously associative to a higher level of information disclosure on account of a more demanding debt covenant and contracting cost between the company and its debt suppliers (Broberg et al., 2010; Mohd Ali et al., 2009). Managers are predisposed to disclose more information voluntarily to justify their financial position and project a positive socially responsible image to stakeholders (Cormier and Magnan, 2003). The likelihood for

*Corresponding author's email: xftong@mmu.edu.my

DOI: 10.1051/shsconf/20173600012

companies to disclose supplementary CSR information voluntarily increases to moderate the cost of information asymmetry and to lower the cost of capital (Jagd, 2015; Lan et al., 2013). Nevertheless, Giannarakis (2014) and Reverte (2008) reason an inverse effect of leverage on disclosure levels. A high level of restrictive debt covenants in debt agreements comprehensibly reduces the need for additional disclosure in annual reports (Eng and Mak, 2003). Owing to a higher level of insolvency and debt-servicing obligations, companies are unwilling to incur extraordinary costs to invest in non-value creating CSR activities on CSR disclosure (Barnea and Rubin, 2010). Similarly, findings from Haniffa and Cooke (2005) demonstrated the irrelevance of debt financing structure on CSR disclosures in Malaysia. Using the ratio of total debt to total assets to confirm the implication of leverage on CSR disclosures, the study hypothesises:

H4$_{\textbf{(UK/Malaysia)}}$: *The level of CSR disclosure is positively associated with leverage in capital structure.*

RESEARCH METHOD

Sample and data

For comparative purpose of this study, a purposive sampling strategy is employed on the constituents from the FTSE 100 Index and FTSE Bursa Malaysia KLCI across ten key industries as classified in the FTSE Industry Classification Benchmark (ICB). The sampled constituents for each index represent the top 30 companies by market capitalisation on the Bursa Malaysia and the London Stock Exchange as at June 2014 (FTSE Group, 2014).

Sustainability reports are expected to include the required specific standard disclosures prepared 'in accordance' with the G4 Sustainability Reporting Guidelines regardless of the size, sector or location. This study will only consider electronic and web-based annual reports (both integrated and separated) provided proactively in English on the companies' websites with a reporting period ending between 30 June 2012 and 31 March 2013 as at May 2014. The defined reporting period is to allow a better comprehension on the current standing of the sampled reports prior to the revised G4 framework. Table 1 and 2 represent the subdivided constituents sampled for both countries.

Table 1. FTSE Bursa Malaysia KLCI constituents

Constituent	ICB Subsector	Rank
0001 Oil & Gas		
Petronas Gas Bhd	0533 Exploration & Production	11
Petronas Dagangan Bhd	0537 Integrated Oil & Gas	25
SapuraKencana Petroleum	0573 Oil Equipment & Services	13
1000 Basic Materials		
Petronas Chemicals Group Bhd	1353 Commodity Chemicals	9
2000 Industrials		
Sime Darby Bhd	2727 Diversified Industrials	6
MISC Bhd	2773 Marine Transportation	20
3000 Consumer Goods		
IOI Corporation Bhd	3573 Farming, Fishing & Plantations	10
Kuala Lumpur Kepong Bhd	3573 Farming, Fishing & Plantations	14
Felda Global Ventures Holdings	3573 Farming, Fishing & Plantations	24
UMW Holdings	3353 Automobiles	26
PPB Group Bhd	3577 Food Products	21

*Corresponding author's email: xftong@mmu.edu.my

British American Tobacco (Malaysia)	3785 Tobacco	19
4000 Health Care		
IHH Healthcare	4533 Health Care Providers	18
5000 Consumer Services		
Genting Group	5753 Hotels	7
Genting Malaysia Bhd	5753 Hotels	16
Astro Malaysia Holdings	5553 Broadcasting & Entertainment	28
6000 Telecommunications		
Axiata Group Bhd	6575 Mobile Telecommunications	5
Digi.com	6575 Mobile Telecommunications	8
Maxis Berhad	6575 Mobile Telecommunications	12
Telekom Malaysia Bhd	6535 Fixed Line Telecommunications	15
7000 Utilities		
Tenaga Nasional	7537 Alternative Electricity	4
YTL Corporation	7575 Multi-utilities	22
8000 Financials		
Public Bank BHD	8355 Banks	1
Malayan Banking	8355 Banks	2
CIMB Group Holdings	8355 Banks	3
AMMB Holdings Bhd	8355 Banks	17
Hong Leong Bank	8355 Banks	23
RHB Capital	8355 Banks	27
Hong Leong Financial Group	8355 Banks	29
KLCC Prop&Reits-Stapled Sec	8633 Real Estate Holding & Development	30
9000 Technology		

Table 2. FTSE 100 constituents

Constituent	ICB Subsector	Rank
0001 Oil & Gas		
BP	0537 Integrated Oil & Gas	2
Royal Dutch Shell PLC	0537 Integrated Oil & Gas	3
BG Group	0537 Integrated Oil & Gas	11
1000 Basic Materials		
Rio Tinto	1775 General Mining	9
BHP Billiton	1775 General Mining	10
Glencore Xstrata	1775 General Mining	19
Anglo American	1775 General Mining	25
2000 Industrials		
Rolls-Royce Holdings	2713 Aerospace	24
3000 Consumer Goods		
British American Tobacco (UK)	3785 Tobacco	5
Imperial Tobacco Group	3785 Tobacco	23
SAB Miller	3533 Brewers	20
Diageo	3535 Distillers and Vintners	8
Unilever	3577 Food Products	16
Reckitt Benckiser Group	3724 Nondurable Household Products	15
4000 Health Care		
GlaxoSmithKline	4577 Pharmaceuticals	4
AstraZeneca	4577 Pharmaceuticals	7
Shire	4577 Pharmaceuticals	26
5000 Consumer Services		
Tesco	5337 Food Retailers and Wholesalers	22
WPP	5555 Media Agencies	28
Compass Group	5757 Restaurants and Bars	29

*Corresponding author's email: xftong@mmu.edu.my

6000 Telecommunications		
BT Group	6535 Fixed Line Telecommunications	18
Vodafone	6575 Mobile Telecommunications	6
7000 Utilities		
SSE PLC	7535 Conventional Electricity	30
Centrica	7573 Gas Distribution	27
National Grid	7575 Multi-utilities	17
8000 Financials		
HSBC Holdings	8355 Banks	1
Barclays	8355 Banks	12
Lloyds Banking Group	8355 Banks	13
Standard Chartered	8355 Banks	21
Prudential	8570 Life Insurance	14
9000 Technology		

Source: FTSE Group as at 20 May 2014

The final sample includes 30 companies listed on each of the main index respectively with a spread across 9 major industries. Given market capitalisation as the basis of selection, neither the Malaysian nor the UK top constituents were represented in the technology industry. Table 3 presents the comparative percentage of industry representation of the sampled companies.

Table 3. Comparative industry distribution

	Malaysia		**United Kingdom (UK)**	
Industry	**Number of companies**	**%**	**Number of companies**	**%**
Oil and gas	3	10.00	3	10.00
Basic materials	1	3.33	4	13.33
Industrials	2	6.67	1	3.33
Consumer goods	6	20.00	6	20.00
Health care	1	3.33	3	10.00
Consumer services	3	10.00	3	10.00
Telecommunications	4	13.33	2	6.67
Utilities	2	6.67	3	10.00
Financials	8	26.67	5	16.67
Technology	0	0.00	0	0.00
Total	**30**	**100.00**	**30**	**100.00**

CSR construction and measures

For the first phase of the empirical study, a content analysis is applied to all 60 sampled annual reports to address cross-market similarities and differences in CSR disclosure. The context-sensitive abductive inferencing from content analysis allows a reliable and significantly meaningful informational representation of large volumes of data (Krippendorff, 1980). Thus, content analysis serves as an evocative gauge on the significance of the CSR agenda to companies. By replicating on past studies, this index study attempts to quantify specific CSR information observable in the sampled reports to examine the extent of various items' disclosures.

As stipulated in the Guidelines, the three main indicators analysed in this study are the economic, social and environmental dimensions of CSR disclosures. An aggregate measure of the quantity of disclosure for each indicator within the sustainability reports was constructed using the binary coding system. Inferred meanings from written words, sentences or paragraphs form the coding base for the disclosure indices for a more complete and meaningful dataset for subsequent analysis (Milne and Adler, 1999; Gray et al., 1995). The

*Corresponding author's email: xftong@mmu.edu.my

coding indices are validly and reliably derived from the latest core sustainability indicators outlined in the G4 Sustainability Reporting Guidelines.

In analysing the CSR information content in the sampled reports, the PDF reader's word count function was used (Gamerschlag et al, 2011). The total quantity of CSR disclosures (TQ_{CSR}) from the analysed reports is represented by the average summative amount of the relative total scores awarded for the economic (ECN), environmental (ENV) and social (SOC) indicators:

$$TQ_{CSR\ (UK/Malaysia)} = SCORE_{ECN} + SCORE_{ENV} + SCORE_{SOC}$$

whereby ECN demonstrates the capital flow and economic impact of the company on its stakeholders and economic system; ENV on impacts relating to operational inputs and outputs on living and non-living natural systems; and SOC on the operational impact on the social systems.

The derived indicators used in this study are presented in Table 4. There are 14 indicators representative of the ECN dimension, 25 indicators for ENV and 38 indicators for SOC.

Table 4. Content analysis indicators

Specific Standard Disclosures		
Economic:14 indicators	***Environment:25indicators***	
Economic performance	Water	Awards
Financial performance	Effluent and waste	Biodiversity
Market presence	Spills	Environmental impacts
Defined pension obligations	Greenhouse gas (GHG)	Conservation
Tax relief, credits and subsidies	Hazardous waste	
Financial assistance and incentive	Discharge	
Risk management	Recycle	
Awards	Reclaim	
Government grants	Restore	
Procurement practices	Reuse	
Domestic job support	Recovery	
Infrastructure development	Composting	
Economic development	Preservation	
Research and development	Disposal	
	Emission treatment	
	Pollution control	
	Energy consumption and saving	
	Compliance and policies	
	Protection	
	Environmental programmes	
	Costs (expenses)	

*Corresponding author's email: xftong@mmu.edu.my

Social: 38 indicators		
Employment	Employee share purchase schemes	Stakeholder engagement
Remuneration	Retirement provision / plans	Lobbying and advocacy (public policy)
Trade / labour unions	Insurance	Compliance
Collective agreement / bargaining	Healthcare	Fines / litigation / claims
Organisational communication	Disability and invalidity coverage	Sanctions
Discrimination	Training, education and skills management	Whistleblowing / ethics hotline / helpline
Child labour	Performance / career development and reviews	
Forced / compulsory labour	Corruption	
Work hazards	Personal gift	
Diversity	Anti-competitive / anti-trust behaviour	
Minorities	Product responsibility and labelling	
Equal opportunity	Customer satisfaction	
Human rights	Customer health	
Occupational safety	Customer safety	
Occupational health	Customer data and privacy	
Employee benefits	Community investment / programmes	

Source: Key themes derived from the GRI Specific Standard Disclosures and adapted to the original list used in Gamerschlag et al. (2011)

The indicators provide necessary information on the incorporation of the triple bottom line (TBL) accounting framework in a company, reflecting the performance or impacts of the ECN, ENV and SOC dimensions on stakeholders (GRI, 2013). It is essential to recognise the analysis of ECN's dimension as distinctive from fundamental financial and economic reporting examination. The reported ECN information entails concerns over organisational impact on stakeholders' economic conditions.

Adapting to the scoring measure used in Lan et al. (2013) and Botosan (1997), the relative degree of CSR disclosure for each respective indicator, *i*, is quantified as follows:

$$\text{SCORE}_i = \frac{\text{DSCORE}^i}{\text{MAXSCORE}} \text{ X } \frac{1}{3}$$

In the second phase of the study, three statistical analysis methods were applied to test hypotheses H1 to H6 using the IBM Statistical Package for Social Sciences (SPSS) version 21 software. Table 5 summarises the examined variables.

Table 5. Summary of variables measures

	Variables	**Measures**
	Independent Variables	
H1	Board experience (EXP)	Percentage of CSR-relevant experience among board of directors
H2	Company size (SZE)	Natural log of total assets
H3	Profitability (PRO)	Return on equity (ROE)
H4	Financial leverage (LVG)	Ratio of total debt to total assets
	Dependent Variable	
	Total CSR disclosure index (TQ_{CSR})	

*Corresponding author's email: xftong@mmu.edu.my

The first statistical analysis method applies descriptive statistics analysis to measure the variables' mean, standard deviation, total quantity of CSR disclosures, and other relevant descriptions of the data. Using Pearson's correlation analysis, the relationships between the independent and dependent variables were measured. This analysis measures the strength of the relationships between the drivers of sustainability reporting and the amount of disclosed CSR information. The indicative results specify the significant values and the strength of the relationship is indicated within the range of $\pm$ 1; a strong correlation as between $\pm$ 0.7 to 0.9; a moderate correlation as between $\pm$ 0.4 to 0.6; and, a weak correlation as between $\pm$ 0.1 to 0.3 (Dancey & Reidy, 2004).

In the second method, the Spearman's rank correlation is used to identify the association of each individual independent variable with the total CSR disclosure (TQ_{CSR}) for its monotonic function. The Spearman's rank correlation coefficient analysis is not a measure for linear relationship between the independent and dependent variables. Spearman's coefficient is denoted by r_s and is constrained by the limit of $-1 \leq r_s \leq 1$. This nonparametric correlation analysis permits two variables to be compared for a monotonic function with no requirement of normality for the variables to be measured. A coefficient is considered significant if the p-value is less than 0.05 (1-tailed). The strength of the monotonic relationship between two variables is measured as either very weak (0.00 to 0.19); weak (0.20 to 0.39); moderate (0.40 to 0.59); strong (0.60 to 0.79); or, very strong (0.80 to 1.0) (Hauke and Kossowski, 2011). Assumption was made that there was evidence of non-linearity for secondary data. For these reasons, hypotheses H1 to H4 were tested using Spearman's rank correlation coefficient to confirm the association between the total CSR disclosure and the independent variables.

Lastly, in substantiating the variation in company-specific characteristics on CSR disclosure levels, the multiple linear regressions (MLR) analysis is applied to identify which predictors or independent variables (IVs) of the company-specific characteristics predict the outcome of the CSR disclosure levels. Accordingly, to determine the multiple effects of the independent variables on the dependent variable, the proposed multiple regression equation is surmised as:

$$TQ_{CSR\ (UK/Malaysia)} = \alpha + \beta_1 EXP + \beta_2 SIZE + \beta_3 PRO + \beta_4 LVG + \varepsilon$$

where the DV for this study is the total CSR disclosure (TQ_{CSR}) of the sampled reports subjected to the IVs consisting of the proportion of the board of directors with CSR-relevant experience; the total assets owned by the companies; the profitability and adequacy of funds available; and the leverage in capital structure.

Specifics for the variables are sourced from the provided reports (both the annual and sustainability reports). In analysing the variables, this study adopted measurements from Lan et al. (2013) and Gamerschlag et al. (2011). All figures are logged at the respective companies' reported year-ends.

RESULTS

Data analysis

Content analysis

Based on the data compiled for $SCORE_{ECN}$, $SCORE_{EVN}$ and $SCORE_{SOC}$, the maximum binary scores for the economic, environment and social indicator were 14, 25 and 38 respectively.

*Corresponding author's email: xftong@mmu.edu.my

A comparative review of the total CSR disclosure, TQ_{CSR} for both countries showed variations across different industries. In UK, the highest score of TQ_{CSR} was reported by Diageo (Consumer Goods) at 0.96, while the lowest was at 0.48 from Shire (Health Care). In Malaysia, Petronas Chemicals Group Bhd (Oil & Gas) reported the highest score at 0.69, while Hong Leong Bank (Financials) scored lowest at 0.37. Diageo's near-perfection score was considerably influenced by the company's explicit public assertion and commitment with the GRI guidelines and UN Global Compact (UNGC) to a full reporting practice in their annual audited report.

Expectedly, companies in the UK have a higher economic, environment and social SCORE than in Malaysia (see Table 6). The average total CSR disclosure TQ_{CSR} scores for UK and Malaysia were 0.76 and 0.52 respectively. On closer observation on the specific (i.e. economic, environmental, and social) TQ_{CSR} scores, it shows a well-balanced information disclosure in all three CSR indicators for the UK companies. For the Malaysia sample, the results demonstrated a bias in CSR disclosure efforts on the ECN dimension. Seemingly, the companies emphasised more on the economic aspect of the CSR reporting than the environmental and social disclosures. The worst disclosure SCORE for the Malaysian companies was at 0.13 from Petronas Gas Bhd (Oil & Gas) for the ENV indicator.

Table 6. Average total CSR disclosure ($AVGTQ_{CSR}$)

	$AVGSUM_{ECN}$	$AVGSUM_{ENV}$	$AVGSCORE_{SOC}$	$AVGTQ_{CSR}$
UK	0.28	0.21	0.27	0.76
Malaysia	0.24	0.13	0.15	0.52

The data from the first phase of the study are subjected to further analysis in the next phase to test the developed hypotheses.

Descriptive statistic

The results presented in Table 7 shows the overall descriptive statistics for the independent and dependent variables' means and standard deviations (SD).

Table 7. Descriptive statistics

	United Kingdom (UK)		**Malaysia**	
	Mean	**Std. Deviation**	**Mean**	**Std. Deviation**
ECN	0.2820	0.03089	0.2373	0.05065
ENV	0.2090	0.06830	0.1287	0.06213
SOC	0.2747	0.02980	0.1483	0.03445
TQCSR	0.7650	0.10608	0.5153	0.09325
EXP	31.9520	16.72012	20.8223	13.61406
SZE	25.0303	1.49238	24.4350	1.72155
PRO	17.4533	16.76181	35.8813	85.87423*
LVG	0.6710	0.19308	0.5823	0.28000

In Table 8, the values in the Pearson's correlation matrix indicate some significant linear relationships for the independent and dependent variable for both countries. In UK, only the

*Corresponding author's email: xftong@mmu.edu.my

relationship for PRO and SZE was identified with a moderate negative correlation strength of -0.441. For the Malaysian model, moderate association strengths were observed for LVG and EXP, and, PRO and SZE, of -0.493, -and -0.409 respectively.

Table 8. Pearson's correlation matrix

		UK – Correlations		
EXPUK	1.000			
SZEUK	-0.169	1.000		
PROUK	-0.047	**-0.441****	1.000	
LVGUK	-0.116	**0.354***	**-0.391***	1.000
		Malaysia – Correlations		
EXPM	1.000			
SZEM	0.011	1.000		
PROM	-0.140	**-0.409***	1.000	
LVGM	**-0.493****	0.172	0.294	1.000

Note: ** - Correlation is significant at the 0.01 level (1-tailed)
* - Correlation is significant at the 0.05 level (1-tailed)

Hypotheses testing

Spearman's rank correlation coefficient analysis

The results for hypotheses H1 to H4 for UK companies indicated only financial leverage (LVG) was significant for Spearman's rank correlation coefficient [r_s= -0.345, n = 30, p < .05 (1-tailed)]. This suggests that H4 is supported (see Table 9). This means the level of CSR disclosure is negatively associated with leverage in capital structure. Other hypotheses were not supported.

In the Malaysian model, H1 to H3 were supported. H1, which indicates directors' CSR-relevant experience, has a significant Spearman's coefficient of r_s= 0.413 [n = 30, p < .05]; while H2, for company size, presents with an r_s= -0.315 [n = 30, p < .05]; and H3, for profitability with the r_s= 0.372 [n = 30, p < .05]. H2 interestingly shows a reversed relationship. H4 was not supported. Only H1 has a moderate strength of monotonic relationship between the dependent and independent variables. Other significant variables reflect weak relationships.

Table 9. Spearman's correlation matrix of variables

UK – Correlations					
TQCSRUK	1.000				
EXPUK	-0.062	1.000			
SZEUK	0.127	-0.092	1.000		
PROUK	0.086	-0.070	-0.527	1.000	
LVGUK	**-0.345***	-0.074	0.117	-0.257	1.000
Malaysia – Correlations					
TQCSRM	1.000				
EXPM	**0.413***	1.000			
SZEM	**-0.315***	-0.187	1.000		
PROM	**0.372***	-0.143	-0.414	1.000	
LVGM	-0.185	-0.446	0.291	0.351	1.000

Ordinary regression analysis (OLS) assumptions test

When performing OLS, assumptions test is important to ensure non-normality of the variables. These can be determined by calculating the z-values the variables' skewness and

*Corresponding author's email: xftong@mmu.edu.my

kurtosis. If the z-values are within ±1.96 for skewness and kurtosis, it indicates some form of normality. This can be obtained by dividing skewness and standard error of skewness. Similarly, this applies to measuring kurtosis (Jöreskog, 1999). The result for the UK companies showed only one variable, PRO, has violated the rule of z-values. For the Malaysian companies, three variables (EXP, SZE, PRO) have high skewness and kurtosis values (see Table 10). These variables were transformed to log (10) in the SPSS command icon to adjust the skewness and kurtosis to near normality for analysis (Field, 2009; Bowerman & O'Connell, 1990). Upon transformation of the non-normality variables, the standard deviation spreads were acceptable (see Table 11). Comparison of the two countries using MLR indicated no violations from the assumptions tests.

Table 10. Assumptions test

UK	**TQCSRUK**	**EXPUK**	**SZEUK**	**PROUK**	**LVGUK**
Skewness	-0.466	0.462	0.782	-5.401	-0.157
Std. Error of Skewness	0.427	0.427	0.427	0.427	0.427
Z-Value (±1.96)	-0.981	1.082	1.831	**12.649**	0.368
Kurtosis	0.485	-0.708	0.026	29.437	0.345
Std. Error of Kurtosis	0.833	0.833	0.833	0.833	0.833
Z-Value (±1.96)	0.582	0.85	0.031	**35.34**	0.414
Malaysia	**TQCSRM**	**EXPM**	**SZEM**	**PROM**	**LVGM**
Skewness	0.168	1.215	1.109	4.604	-0.208
Std. Error of Skewness	0.427	0.427	0.427	0.427	0.427
Z-Value (±1.96)	0.930	**2.845**	**2.597**	**10.780**	-0.487
Kurtosis	-1.129	1.9171	2.818	22.458	-1.121
Std. Error of Kurtosis	0.833	0.833	0.833	0.833	0.833
Z-Value (±1.96)	-1.355	**2.301**	**3.383**	**26.96**	1.346

Table 11. Descriptive statistics – non-normality statistics adjusted for log (10)

	UK			**Malaysia**	
	Mean	*Std. Deviation*		*Mean*	*Std. Deviation*
PROUKLOG	0.6422	1.27453	**EXPMLOG**	1.2472	0.24323
			SZEMLOG	1.3870	0.02976
			PROMLOG	1.1390	0.67033

The results in Table 12 on regression coefficients for UK companies shows SZE was significant with unstandardized coefficient *B* at 0.045 ($p < 0.029$). For Malaysia, only EXP was significant with a *B* of 0.179 ($p < 0.026$). Other variables did not predict the total CSR disclosure. The VIF measures for collinearity in the data were within the limit of < 10. Similarly, the Durbin-Watson test for correlated residuals was within the limit of 0 to 4. In Table 13, the model summary for UK has an R^2 value of 0.284 and for Malaysia, an R^2 value of 0.315. The R^2 value for UK indicated that 28.4% of the variance in dependent variable can be explained by the predictor variables; and for Malaysia, at 31.5%.

*Corresponding author's email: xftong@mmu.edu.my

Table 12. Regression coefficients

UK – Coefficients

Model	*Unstandardised coefficients* B	*Unstandardised coefficients* *Std. error*	*Standardised coefficients* β	*t*	*Sig.*	*95.0% confidence interval for β* Lower bound	*95.0% confidence interval for β* Upper bound	*Collinearity statistics* Tolerance	*Collinearity statistics* VIF
(Constant)	-0.309	0.504		-0.935	0.360	-1.764	0.666		
EXPUK	0.000	0.001	0.002	0.012	0.991	-0.002	0.002	0.916	1.092
SZEUK	0.045	0.019	0.634	2.331	0.029	0.005	0.085	0.421	2.373
PROUKLOG	0.000	0.017	0.003	0.014	0.989	-0.034	0.035	0.781	1.281
LVGUK	-0.062	0.147	-0.113	-0.422	0.677	-0.366	0.242	0.434	2.302

Malaysia – Coefficients

Model	*Unstandardised coefficients* B	*Unstandardised coefficients* *Std. error*	*Standardised coefficients* β	*t*	*Sig.*	*95.0% confidence interval for β* Lower bound	*95.0% confidence interval for β* Upper bound	*Collinearity statistics* Tolerance	*Collinearity statistics* VIF
(Constant)	-0.235	2.197		-0.107	0.916	-4.775	4.313		
EXPMLOG	0.179	0.075	0.466	2.379	0.026	0.023	0.334	0.775	1.291
SZEMLOG	0.339	1.547	0.109	0.221	0.827	-2.857	3.540	0.122	8.180
PROMLOG	0.071	0.069	0.510	1.035	0.311	0.071	0.213	0.123	8.150
LVGM	-0.038	0.114	-0.113	-0.329	0.745	-0.274	0.199	0.253	3.960

Table 13. Model summary for the UK and Malaysia

UK – Model summary

R	*R^2*	*Adjusted R^2*	*Std. Error of the Estimate*	*Change Statistics* R^2 Change	*Change Statistics* F Change	*Change Statistics* df1	*Change Statistics* df2	*Change Statistics* Sig. F Change	*Durbin-Watson*
0.533	0.284	0.097	0.10089	0.284	1.519	6	23	0.216	1.784

Malaysia – Model summary

R	*R^2*	*Adjusted R^2*	*Std. Error of the Estimate*	*Change Statistics* R^2 Change	*Change Statistics* F Change	*Change Statistics* df1	*Change Statistics* df2	*Change Statistics* Sig. F Change	*Durbin-Watson*
0.561	0.315	0.137	0.08664	0.315	1.765	6	23	0.151	1.093

Therefore, the model equations for the two countries can be presented as follow:

$TQ_{CSR\ (UK)} = -0.309 + 0.002\ EXP + 0.634\ SZE + 0.003\ PRO - 0.113\ LVG + \varepsilon$

$TQ_{CSR\ (Malaysia)} = -0.235 + 0.466\ EXP + 0.109\ SZE + 0.510\ PRO - 0.113\ LVG + \varepsilon$

The ANOVA results indicate whether the overall models did predict the total CSR disclosure. In the UK model, the ANOVA has F(6, 0.234) = 1.510 [F(df effect, df error) = F-value] with p = 0.219; and in Malaysia, F(6, 0.173) = 1.765, p = 0.151. The non-significant values of the

*Corresponding author's email: xftong@mmu.edu.my

ANOVA indicated the two regression models overall did not predict the total CSR disclosure (TQCSR). However, the non-normality variables from the UK and Malaysia data may have affected the p-values of the F-test as this test is sensitive to non-normality variables (Markowski and Markowski, 1990). Moreover, the ANOVA results from MLR do not indicate the individual contribution of the variables (Field, 2009). Since this exploratory study aims to review the contribution of specific regression coefficient, the non-significant value in F becomes secondary (Dallal, 2000).

Summary of hypotheses testing

The overall hypotheses results are summarised in Table 14. Results with significant values of less than 0.05 are supported and vice versa.

Table 14. Hypotheses results

Hypotheses		United Kingdom (UK)	Malaysia
H1	TQ_{CSR} vs EXP	Not supported	Supported
H2	TQ_{CSR} vs SZE	Not supported	Supported
H3	TQ_{CSR} vs PRO	Not supported	Supported
H4	TQ_{CSR} vs LVG	Supported	Not supported
Multiple regression results		Supported: SZE	Supported: EXP

FINDINGS AND DISCUSSION

This study aims to compare and assess the effects of company-specific variables on the level of CSR disclosure in United Kingdom (UK) and Malaysia's public-listed companies.

Using Spearman's correlations analysis, the total CSR disclosure TQ_{CSR} for the UK sample was found to be negatively associated with financial leverage ($H4_{UK}$). The result suggests that an increase in leverage would reduce the extent of CSR disclosure. This finding is consistent with recent studies by Giannarakis (2014) and reaffirms earlier literatures from Barnea and Rubin (2010) and Haniffa and Cooke (2005). Interestingly, the size of company ($H2_{Malaysia}$) was found to have a negative association with total CSR disclosure. This finding contrasted with previous studies whereby larger firms with increased public visibility tend to invest more in CSR reporting activities (Lan et al., 2013; Gamerschlag et al., 2011; Udayasankar, 2008).

Comparatively, the TQ_{CSR} in Malaysian companies was positively associated with the board of directors' CSR-relevant experience ($H1_{Malaysia}$) and company's profitability ($H3_{Malaysia}$). A higher balance of board of directors with the appropriate CSR knowledge and experience are able to provide better advice and advocacy on matters pertaining to corporate sustainability agendas (Financial Reporting Council, 2012; Securities Commission Malaysia, 2012; Carpenter and Westphal, 2001). Referring to Table 6, evidences from the descriptive statistics on average TQ_{CSR} (UK/Malaysia) proves an overall better quality of CSR disclosure in the UK than in Malaysia. A higher mean value for UK companies than Malaysia indicates a larger composition of board members with CSR-relevant experiences. This implicates the importance of board experiences in CSR and this factor could contribute to the academic knowledge for sustainability reporting.

*Corresponding author's email: xftong@mmu.edu.my

In terms of profitability, this finding is consistent with most preceding studies. The results consolidate with Haniffa and Cooke (2005) s' associative study on CSR disclosures and financial performance of Malaysian companies. In signalling the discretionary non-financial commitments through reporting, companies instinctively reduces potential agency costs and gains long-term competitive advantage over their future stock liquidity (Setyroni and Ishak 2012; Cormier and Magnan, 2003).

Based on the multiple linear regression (MLR) models, the size (SZE) of company in Malaysia is found to be significant in affecting the overall level of CSR disclosure. This relationship affirmed past studies on the explanatory factor of SZE as a consistent predictor for TQ_{CSR} in both developed and developing countries (Giannarakis, 2014; Gallo and Jones Christensen, 2011; Amran et al., 2007; Haniffa and Cooke, 2005).

As a final point, in the Malaysian model, relevant CSR board experience (EXP) is a key determinant for CSR-disclosure levels. The significant EXP variable is also found to be consistent with the Spearman's correlation coefficient. While board of directors may recognise the strategic value of sustainability reporting in reducing agency costs, the incorporation of social and environmental priorities onto the core business function requires adequate CSR-relevant board experience to effectively manage CSR governance.These results undoubtedly confirmed the importance of a composition of board members with CSR-relevant experiences or skills on sustainability reporting.

Limitation and future studies

The limitations from this research are associated with industry classification and sample size. The incentive to disclose CSR information is driven by the associated cost of lobbying and stakeholder pressure. Companies from industries with a higher level of impactful environmental and social involvements are inclined to have a higher level of CSR disclosures. Due to the time constraints of the research period, a complete analysis on all industries from the entire list of companies to distinguish the effect of industrial differences is not feasible.

In regards to the limitation concerning sample size, this study was conducted using a purposive sampling strategy whereby 30 secondary datasets from the FTSE 100 Index and FTSE Bursa Malaysia KLCI each were selected for analyses. The 30 samples from each index with 95% confidence level have a confidence interval (margin of error) of ±15.05. This means the sample answer is true between 34.95 % [50-15.05] and 65.05% [50+15.05] (Creative Research System, 2012). For this reason, a larger sample size from each country would possibly lower confidence level and yield a higher R2 values and identify more determinants influencing CSR disclosure levels. At this level of exploratory study, the findings are not generalisable. Future research should increase the sample sizes from each country to 60 for more reassuring analyses to compare the results.

Potential implications

Implication for practice

Since the board of directors' experience was found to be the key predictor for total CSR disclosure levels, companies should attempt to increase the board's composition with

*Corresponding author's email: xftong@mmu.edu.my

directors having CSR-relevant experiences or skills. Merely officiating CSR events or receiving CSR awards would not lead the board members' awareness on the importance of sustainability reporting. In order to endorse the importance of CSR accountability explicitly, companies could also assign a specific appointment of at least one director in the board to be responsible in CSR activities or trained in CSR-related programmes. Additionally, the appointed board member should oversee the content of the annual CSR voluntary disclosure in annual or sustainability reports.

Implication for methodology

This study has provided useful results by comparing both the developed and developing countries' sustainability reporting practices in the top 30 publicly listed companies. It provides a direct comparison in terms of the quantity and quality of CSR disclosure in both the integrated or separated sustainability reports. The comparative method allows company representatives in both countries to improve on their reporting weaknesses. Academic researchers who are interested in comparative studies would find this research thought-provoking and possibly extend the analysis further.

CONCLUSION

This research conclusively affirms the existence of specific sustainability reporting drivers in motivating the commitment on voluntary corporate social responsibility (CSR) disclosure practices for companies in the United Kingdom (UK) and Malaysia. Results from the study empirically highlighted substantially higher total quantity of CSR disclosure (TQ_{CSR}) for companies in the UK than in Malaysia. More importantly, this study has established board experience (EXP) in CSR initiatives as a pivotal determinant in the variation of results on sustainability reporting levels. Malaysian companies have much to emulate from the UK companies to improve on the CSR disclosure levels in their annual (both integrated and separated) reports.

As globalisation and trade propagates continuously, organisational stakeholders will increasingly place importance on the non-financial operational impact from companies. Sustainability reporting is no longer merely viewed as a tool to satisfy the companies' moral obligations to stakeholders. Efforts to adopt and institutionalise international sustainability reporting frameworks from the likes of the Global Reporting Initiative (GRI) or UN Global Compact helps companies to improve their operating efficiencies and CSR stewardship. In order to gain competitive advantage and achieve long-term corporate sustainability, strong affirmations from companies, particularly those from the emerging markets should exploit the adoption and gradual integration of sustainably reporting into their reporting efforts.

REFERENCES

Amran, A., Lim, L.L. and Sofri, Y. (2007). A Study of Corporate Philanthropic Traits among Major Malaysian Corporations, *Social Responsibility Journal*, 3 (4), 21 – 30

Anderson, A. and Gupta, P.P. (2009). A cross-country comparison of corporate governance and firm performance: Do financial structure and the legal system matter? *Journal of Contemporary Accounting & Economics*, 5, 61-69.

*Corresponding author's email: xftong@mmu.edu.my

Association of Chartered Certified Accountants (ACCA) (2014). *Report of the judges,* ACCA Malaysia Sustainability Reporting Awards (MaSRA). Malaysia: ACCA.

Barnea, A. and Rubin, A., (2010). Corporate social responsibility as a conflict between shareholders. *Journal of Business Ethics,* 97, 71-86.

Belkaoui, A. and Karpik, P.G., (1989). Determinants of the corporate decision to disclose social information. *Accounting, Auditing and Accountability Journal*, 2 (1), 36-51.

Botosan, C.A. (1997). Disclosure level and the cost of equity capital. *Accounting Review*, 72 (3), 323-50.

Bowerman, B.L. and O'Connell, R.T. (1990). Liner statistical models: An applied approach (2nd ed.), Belmont, CA: Duxbury.

Brammer, S. and Pavelin, S. (2008). Factors influencing the quality of corporate environmental disclosure. *Business Strategy and the Environment*, 17 (2), 120-36.

Broberg, P., Tagesson, T. and Collin, S., (2010). What explains variation in voluntary disclosure? A study of the annual reports of corporations listed on the Stockholm Stock Exchange. *Journal of Management & Governance,* 14, 351-77.

Campbell, J.L., (2007). Why would corporations behave in socially responsible ways? An institutional theory of corporate social responsibility. *Academy of Management Review*, 32 (3), 946-67.

Carpenter, M.A. and Westphal, J.D. (2001), "*The strategic context of external network ties: Examining the impact of director appointments on board involvement in strategic decision making*", *Academy of Management Journal*, 44 (4), 639-60.

Chambers, E., Chapple, W., Moon, J. and Sullivan, M. (2003). CSR in Asia: A seven country study of CSR website reporting. *International Centre for Corporate Social Responsibility (ICCSR) Research Paper Series,* 09-2003. Nottingham: Nottingham University Business School.

Chih, H.L., Chih, H.H. and Chih, T.Z. (2009). On the determinants of corporate social responsibility: International evidence on the financial industry. *Journal of Business Ethics,* 93, 115-35.

Cormier, D., Magnan, M. (2003). Environmental reporting management: a continental European perspective. *Journal of Accounting and Public Policy,* 22, 43-62.

Creative Research System (2012). The Survey System: Sample size calculator. Available at: <http://www.surveysystem.com/sscalc.htm > [Accessed 31 august 2014].

Dallal, G.E. (2000), How to Read the Output From Multiple Linear Regression Analyses. Available at: < http://www.jerrydallal.com/lhsp/regout.htm> [Accessed 31 august 2014].

Dancey, C., & Reidy, J. (2004). *Statistics without maths for psychology: Using SPSS for Windows*, London: Prentice Hall.

*Corresponding author's email: xftong@mmu.edu.my

Eng, L.L. and Mak Y.T. (2003). Corporate governance and voluntary disclosure. *Journal of Accounting & Public Policy*, 22, 325-345.

Field, A., (2009). Discovery statistics using SPSS (3rd ed.), SAGE Publications Ltd. London.

Financial Reporting Council (2012). UK Corporate Governance Code. Available at: http://www.slc.co.uk/media/78872/uk-corporate-governance-code-september-2012.pdf [Accessed 18 June 2014]

Fortainer, F., Kolk, A., Pinkse, J. (2011). Harmonization in CSR reporting: MNEs and global CSR standards. *Management International Review,* 51, 665-96.

Freeman, R.E., Moutchnik, A., 2013. Stakeholder management and CSR: questions and answers. *UmweltWirtschaftsForum: uwf,* 21, pp.5-9. Available at: 10.1007/s00550-013-0266-3 [Accessed 18 June 2014].

FTSE Group (2014a). FTSE 100 UK Index Series Quarterly Data – May 2014. Available at: http://www.ftse.com/analytics/factsheets/Home/ConstituentsWeights [Accessed 15 June 2014]

FTSE Group (2014b). FTSE Bursa Malaysia KLCI Index Series Review Changes Overview – June 2014. Available at: http://www.ftse.com/analytics/factsheets/Home/ConstituentsWeights [Accessed 15 June 2014]

Gallo, P.J., Jones Christensen, L. (2011). Firm size matters: an empirical investigation of organizational size and ownership on sustainability-related behaviors. *Business & Society,* 50, 315-49.

Gamerschlag, R., Moller, K. and Verbeeten, F. (2011). Determinants of voluntary CSR disclosure: Empirical evidence from Germany. *Review of Managerial Science*, 5 (2-3), 233-62.

Global Reporting Initiative (GRI), 2013. G4 Sustainability Guidelines. Available at: https://www.globalreporting.org/resourcelibrary/GRIG4-Part1-Reporting-Principles-and-Standard-Disclosures.pdf [Accessed 7 April 2014].

Gray, R., Kouhy, R. and Lavers, S. (1995). Methodological Themes: Constructing a research database of social and environmental reporting by UK companies. *Accounting, Auditing & Accountability Journal*, 8 (2), 78-101.

GRI Reporting Framework, 2013. Available at: https://www.globalreporting.org/reporting/reporting-framework-overview/Pages/default.aspx [Accessed 14 March 2014]

Giannarakis, G. (2014) "Corporate governance and financial characteristic effects on the extent of corporate social responsibility disclosure", Social Responsibility Journal, 10 (4), 569-590.

*Corresponding author's email: xftong@mmu.edu.my

Gul, F., A. and Leong, S. (2004). Board leadership, outside directors' expertise and voluntary corporate disclosures. *Journal of Accounting and Public Policy*, 23 (5), 351-79.

Haniffa, R., Cooke, T.E. (2005). The impact of culture and governance on corporate social reporting. *Journal of Accounting and Public Policy,* 24, 391-430.

Hauke, J. and Kossowski, T. (2011). Comparison of values of Pearson's and Spearman's correlation coefficients on the same sets of data. *Quaestiones Geographicae,* 30 (2), 87-93.

Jöreskog, K., G. (1999). Formula for Skewness and Kurtosis. Available at: http://www.ssicentral.com/lisrel/techdocs/kurtosis.pdf [Accessed 27 August 2014].

Kassinis, G. and Vafeas, N. (2002). Corporate boards and outside stakeholders as determinants of environmental litigation. *Strategic Management Journal,* 23 (5), 399-414.

Krippendorff, K. (1980). *Content analysis: An introduction to its methodology.* Beverly Hills: Sage Publications.

Kytle, B. and Ruggie, J. R. (2005). Corporate social responsibility as risk management: A model for multinationals. *Corporate Social Responsibility Initiative Working Paper 10.* Cambridge: John F. Kennedy School of Government, Harvard University.

LaFond, R. and Watts, R.L. (2008). The information role of conservatism. *Accounting Review,* 83 (2), 447-78.

La Porta, R., Lopez-de-Silanes, F. and Shleifer, A. (2008). The Economic Consequences of Legal Origins. *Journal of Economic Literature,* 46 (2), 285-332. Available at: http://www.aeaweb.org/articles.php?doi-10.1257/jel.46.2.285 [Accessed 4 July 2014]

Lan, Y., Wang, L. and Zhang, X., (2013). Determinants and features of voluntary disclosure in the Chinese stock market. *China Journal of Accounting Research*, 6, 265-85.

Markowski, C.A. and Markowski, E.P. (1990). Conditions for the Effectiveness of a Preliminary Test of Variance, *The American Statisticia,* Vol. 44, No. 4 (Nov.), 322-326

Mat Nor, F., Mohd. Shariff, F. and Ibrahim, I. (2010). The effects of concentrated ownership on the performance of the firm: Do external shareholdings and board structure matter? *Jurnal Pengurusan,* 30, 93-102.

Milne, M.J. and Adler, R.W. (1999). Exploring the reliability of social and environmental disclosures content analysis. *Accounting, Auditing and Accountability Journal*, 12 (2), 237-56.

Mohd Ali, M., Ibrahim, M.K., Mohammad, R., Zain, M.M. and Alwi, M.R. (2009). Malaysia: Value relevance of accounting numbers. In Idowu, S.O. & Filho, W.L. (Eds.), *Global Practices of Corporate Social Responsibility,* 201-230. Berlin: Springer.

Orlitzky, M., Schmidt, F.L. and Rynes, S.L. (2003). Corporate social and financial performance: A meta-analysis. *Organization Studies*, 24 (3), 403-441.

*Corresponding author's email: xftong@mmu.edu.my

Ratna, N., Grantley, T., Rusminm R., Greg T., Bikram, C. (2016). Factors determining social and environmental reporting by Indian textile and apparel firms: a test of legitimacy theory. *Social Responsibility Journal,* 12 (1), 167-89.

Reverte, C. (2009). Determinants of corporate social responsibility disclosure ratings by Spanish listed firms. *Journal of Business Ethics,* 88, 351-66.

Securities Commission (SC) Malaysia (2012). Malaysian Code on Corporate Governance (MCGG). Available at: <http://www.mia.org.my/new/downloads/circularsandresources/circulars/2012/21/MCCG_2012.pdf> [Accessed 18 June 2014]

Siegel and Vitaliano (2007). An empirical analysis of the strategic use of corporate social responsibility. *Journal of Economics & Management Strategy*, 16 (3), 773-92.

Udayasankar, K. (2008). Corporate social responsibility and firm size. *Journal of Business Ethics,* 83, 167-75.

United Nations Environment Programme (UNEP), 2013. Carrots and Sticks: Sustainability reporting policies worldwide – Today's best practice, tomorrow's trends. Available at: <https://www.globalreporting.org/resourcelibrary/Carrots-and-Sticks.pdf> [Accessed 14 March 2014]

*Corresponding author's email: xftong@mmu.edu.my

THE DEVELOPMENT AND APPLICATION OF WATER GOVERNANCE MATRIX: A CASE OF MALAYSIA

Atina Rahmah Mohd Remali, Inaliah Mohd Ali, Bakhtiar Alrazi*, Norhayati Mat Husin
Department of Accounting, Universiti Tenaga Nasional, Sultan Haji Ahmad Shah Campus, 26700 Muadzam Shah, Pahang, Malaysia

ABSTRACT

The study proposes a framework (in the form of a matrix) to map organisations on their level of water-related risk, taking into consideration 'industry intensity' and 'corporate commitment towards water protection and preservation'. On the basis of Ceres (2011), 'industry intensity' is measured as either low, medium, or high. Furthermore, 'corporate commitment' - measured along the same classification (low, medium, or high) – is based on the extent of corporate disclosure of water policies, initiatives, and performance. These indicators are then aggregated and used to develop a matrix - called Water Governance Matrix (WaGM) - which measures the level of riskiness of an organisation towards water - high (red), moderate (yellow), and low (green). An exploratory study of 30 top public listed companies in Malaysia reveals that only one company falls under the low-risk category, while more than 50% falls under the high-risk category; mainly due to low corporate water disclosure. This innovative matrix is of importance for organisations to monitor and manage their water risk and for stakeholders to make informed investment and other decisions about the organisations.

Keywords: water governance, industry intensity, content analysis, Malaysia.

INTRODUCTION

In December 2015, Malaysia suffered from the El Nino phenomenon, causing the reduction in surge monsoons and changing the weather condition into hot and dry condition with inconsistent rainfall patterns. The rainfall inconsistency, in turn, has affected the climate modeling and at the same time induced greater awareness among the stakeholders and regulators of the negative impacts brought about by organisational activities especially those originating from the industrial sectors. These sectors (e.g., mining, oil and gas production, forestry and electricity generation), more often than not, are relying on water for economic viability (de Loë et al., 2016). Hence, they encounter significant water-related challenges, either through sizeable demand or through their wastewater discharges (Ceres, 2011). While ensuring a stable economic growth for business maturity, good water governance needs to be adopted consistently. According to Montgomery et al. (2016), water governance could be a tool to achieve specific outcomes that are essential in resolving water issues through creating structures and processes that allow for equitable negotiation between stakeholders. Ultimately, good water governance will lead to good firm performance, financially and non-financially. Thus, the presence of water governance model is of paramount importance.

*Corresponding author's email: Bakhtiar@uniten.edu.my

In the absence of any specific model for water governance, this paper aims to develop a matrix that could be used for companies to outline strategies to manage their water risk and for stakeholders to help them making informed economic and other decisions about the companies. The matrix considers both the industry intensity (Ceres, 2011) and corporate commitment towards water protection and preservation. The outcome of the matrix is the classification of companies into high, moderate or low water risk. It then applies the matrix on a sample of public listed companies in Malaysia.

This study contributes to the dearth of literature in water governance through the development of a unique, yet comprehensive, matrix that considers both the background of the industry and the extent of corporate commitment on water. Unlike the issue of carbon which has received considerable attention by both the practice and academia, water-related issue has not been widely researched. We address this gap by providing the most recent evidence of the issue in one of the world's emerging economies, Malaysia. The matrix could be a guideline for companies in terms of how far they need to go in assessing their water risk. By doing so, it assists the companies to understand their strengths and weaknesses in their existing policies and practices for improvements in the future. The findings of this study also will help in creating the awareness among the public of the importance of water issue. Considering the low commitment towards water protection and preservation shown by companies through their corporate reporting, regulators and other authoritative bodies might consider making it mandatory for public listed companies to disclose any initiatives, policies, and performance on water-related aspects. This would ensure greater accountability, comparability, and consistency from the companies.

The remaining sections of this paper are structured as follows. Section 2.0 provides the review of water governance. Section 3.0 presents the development of Water Governance Matrix (WaGM). Section 4.0 discusses the findings from the analysis of 30 Malaysia public listed companies. Section 5.0 concludes the paper.

WATER GOVERNANCE

The Institute of Water Policy (IWP) broadly defines water governance as the set of water laws, policies, programmes and projects adopted by a country or a state to develop and manage its water resources to meet the current and future needs of its population (Mohamad et al., 2008). Water governance is even more essential nowadays than in the past due to accelerating climate change problem. According to the Intergovernmental Panel on Climate Change (2007), climate change has resulted in the increase in global air and ocean temperatures, the widespread melting of ice and snow, and increase in sea levels. If left unattended, the problems will lead to negative health consequences for humans due to shortage of food supply and fresh water.

Water scarcity exists when there is imbalance in the supply and demand of water. The world water supply is inadequate, deteriorating in quality and low flow of surface water (Saimy and Yusof, 2013) but the demand for water is higher nowadays (Braadbaart and Braadbaart, 1997). The problem persists not only in Malaysia but also in other countries as well. According to Saimy and Yusof (2013), good water governance is needed in Malaysia to handle water problems complexity, for a better or efficient water use and management.

The important tool in implementing a good water governance is the legislation, where the legal instrument will support the water policies, programmes or projects to ensure economic, social and environmental sustainability so that the water scarcity problem will be given continuous attention and prevention (Shah et al., 2009: Mohamed et al., 2009). The need for strict regulatory intervention could improve the management of water resources, supplies and services, so as to be continuously aware of the water quality and quantity, to prioritise water-related issues, and to protect the natural resource from being declining drastically. In Malaysia, issues pertaining to water are under the purview of the Ministry of Energy, Green Technology and Water. Both Water Service Industry Act 2006 [Act 655] and National Water Services Commission Act [Act 654] become the framework governing the issues.

Other than governance through legislations, companies could play important role in protecting and preserving water. They use and discharge water during day-to-day operations hence how they manage and control water consumption and discharges are of significantly importance. However, as compared to carbon emissions (which is also associated with climate change), the issue of water has received scant attention by researchers. In the field of corporate reporting literature, to the best of our knowledge, a study by Mohd Remali et al. (2016) is the only study investigating this issue. Based on the assessment of annual reports of 10 Malaysian companies for the year 2014, the disclosure was described as fairly low. This could be due to the fact that this issue is still considered as less critical as compared to other the attention and exposure in other developed economies.

Considering the scant attention paid on the issue of water reporting and the need for a matrix or model to assess companies' risks towards water, hence this study is conducted to address these issues.

THE WATER GOVERNANCE MATRIX (WaGM)

The stages in developing and applying WaGM involve: (i) identification of industries' water intensity; (ii) assessment of corporate commitment towards water issues; and (iii) mapping the companies on WaGM.

Identification of Industries' Water Intensity

For the purpose of this study, the industry profile is determined based on the classification developed and used by the World Wide Fund for Nature (WWF) and German investment bank DEG and published by Ceres' "The Ceres Aqua Gauge: A Framework for 21st Century Water Risk Management". They classified the industries into high, medium, and low priority sectors based on the degree of use and pollution of water. Figure 1 shows the water risk profile by industry.

Industry	Risk Profile
Agriculture (plant & animal products)	High
Beverage producers	High
Biomass power production	High
Chemicals	High
Clothing & apparel	High
Electric power production (gas, coal, oil, nuclear)	High
Food producers (incl. tobacco)	High
Food retailers	High
Forestry & paper	High
Freshwater fishing & aquaculture	High
Hydropower production	High
Mining	High
Oil & gas	High
Pharmaceuticals & biotechnology	High
Technology hardware & equipment, semiconductors	High
Water utilities and services	High
Construction & materials	Medium
Gas distribution & multi-utilities	Medium
Manufacturing of: industrial goods, household goods, home construction, personal leisure goods	Medium
Media (printed)	Medium
Real estate	Medium
Transportation (industrial and personal)	Medium
Travel & leisure (non-transportation)	Medium
Financial services, banks, insurance	Low
General retailers (non-food) & storage (warehousing)	Low
Health care & services	Low
Industrial support services, professional service firms, administration, wholesale, trade, education, arts	Low
Media (non-printed)	Low
Renewable power production (e.g. wind, solar)	Low
Salt water fishing and aquaculture	Low
Software & computer services	Low
Telecommunications	Low

Figure 1: Water risk profile by industry (Source: WWF/DEG in Ceres, 2011)

Despite being useful, this classification does not consider any initiatives undertaken by the individual companies to mitigate their water risks. For example, it could be that a company operates in a 'dirty' industry, but it has policies and system in place to manage and control the impacts of its activities on water. In this case, they should be a better way classifying the individual companies. WaGM addresses this issue by considering the level of corporate commitment towards water protection and preservation using corporate disclosure as the basis.

Assessment of Corporate Commitment towards Water Issues

In the absence of locally established guidelines in the area of water reporting, this research has developed a disclosure index - water disclosure index (WDI) - based on the GRI's Sustainability Reporting Guidelines (specifically focusing on the water-related indicators, i.e., EN8 until EN10) and several information requested by CDP 2015 Water Information Request. In this regard, previous literature (see, for example, Clarkson et al., 2008; Van Staden and Hooks, 2007) have suggested that disclosure could be a proxy for corporate commitment on environmental issues when they found a positive relationship between environmental performance and environmental reporting.

The GRI's guidelines have been developed through a series of stakeholder consultations and past researchers have used the guideline to evaluate the extent of sustainability reporting among corporate sectors (see for example Dumay et al., 2010; Clarkson et al., 2008; Alazzani and Wan-Hussin, 2013). As for the CDP Water information Request 2015, it contains suggested questions that work to catalyse action on corporate water stewardship to safeguard water resources and address the global water crisis (CDP, 2015).

DOI: 10.1051/shsconf/20173600013

Based on GRI and CDP, there are eight items related to water. The items were modified to ensure their suitability in the context of Malaysia. To avoid subjectivity, each item is scored on a dichotomous basis in which 1 (one) is awarded if the item is being disclosed and 0 (zero) if otherwise. Table 1 presents the final WDI consisting of eight items with a total maximum score of eight.

No	Items		Details
A	GRI - G4 EN8: Total water withdrawal by source	1.0	Report the water withdrawn and used from the surface water/ground water; the waste water in and outside organisation; the collected rainwater in organisation; the total volume of rainwater collected
B	GRI - G4 EN9: Water sources significantly affected by withdrawal of water	2.0	Report the number of water sources affected by withdrawal of water; the size of water sources held by organisation; whether the source is designated as a protected area; the benefit or water sources for communities; the biodiversity value of the water sources
C	GRI - G4 EN10: Percentage and total volume of water recycled and reused	3.0	Report the total volume of water recycled in organisation; the source and the volume of water recycled; the total volume of water reused in organisation; the source and the volume of water reused; the total volume of water recycled as percentage of water withdrawal reported under G4 EN8; the total volume of water reused as percentage of water withdrawal reported under G4 EN8.
D	CDP's 2015 water information request	4.0	Water risks: Report the water risk that could generate substantive change in operations, revenue, and expenditure of the organization
		5.0	Facility level water accounting: Report the total volume of water discharge by the organisation
		6.0	Governance & strategy: Report the management in organization involved in water responsibilities
		7.0	Compliance, Complaints and Senses: Report whether the organization has been imposed by any fines or civil penalties regarding water or environmental regulations
		8.0	Targets and initiatives: Report the organisation target and goals related to water

Table 1: Water Disclosure Index

Based on the score obtained by the companies, they are classified into high, medium, or low in terms of commitment. In essence, if the score obtained is between 0 to 2, the companies will be regarded as having 'low' commitment to water protection and preservation; if between 3 to 5 as 'medium', while a score of 6 to 8 is considered as 'high'.

Mapping the Companies on WaGM

The final step is to map companies on WaGM which assesses the levels of water risk faced by companies. They are divided into high (red boxes), moderate (yellow), and low (green). Such a matrix allows companies from the high intensity industry to be flagged as 'moderate' if they have policies, systems, and initiatives to mitigate their impact on water. Similarly, companies from the low intensity industry could also be flagged as 'moderate' if they do not take necessary actions to manage and control their impact.

DOI: 10.1051/shsconf/20173600013

CORPORATE COMMITMENT / INDUSTRY INTENSITY	Low Industry	Medium Industry	High Industry
Low Commitment	**Moderate Risk**	High Risk	High Risk
Medium Commitment	Low Risk	**Moderate Risk**	High Risk
High Commitment	Low Risk	Low risk	**Moderate Risk**

Figure 2: Water Governance Matrix (WaGM)

This matrix enables companies to understand their water-related risks hence to institute appropriate mechanisms to manage those risks. For stakeholders, this will help them to make informed economic and other decisions about the companies. For example, shareholders might want to avoid investing in companies in the high intensity industry with low commitment towards water as they are considered as highly risky.

APPLICATION OF WaGM IN MALAYSIA

As of 1 April 2016, there were 814 companies listed on Bursa Malaysia. Since this study is exploratory, the selection of companies to be analysed is limited to top 10 companies for each category of industry intensity (based on market capitalisation). This makes the sample consists of 30 companies. Table 2 depicts the list of companies included in the sample.

No	Intensity	Companies	Industries
1	High	Tenaga Nasional Berhad	Electric power production
2		Sime Darby Berhad	Diversified industrial
3		Petronas Chemical Group	Chemicals
4		IOI Corporation Berhad	Food producers
5		YTL Corporation Berhad	Gas, water, and multi-utilities
6		Kuala Lumpur Kepong Berhad	Food producers
7		British American Tobacco	Tobacco
8		UMW Holdings Berhad	Automobiles & parts
9		Nestle (M) Berhad	Food producers
10		PPB Group Berhad	Food producers
11	Medium	IHH Healthcare Berhad	Healthcare & services
12		Genting Berhad	Leisure
13		Hong Leong Industries Berhad	Manufacturing
14		KICC Prop&Reits-Stapled Sec	Real estate
15		Hap Seng Consolidated Berhad	Real estate
16		IJM Corporation Berhad	Construction
17		Gamuda Berhad	Construction
18		S P Setia Berhad	Real estate

No	Intensity	Companies	Industries
19		Lafarge Malaysia Berhad	Materials
20		Malaysia Airports Holdings Berhad	Transportation
21	Low	Malayan Banking Berhad	Banking
22		Public Bank Berhad	Banking
23		Axiata Group Berhad	Communication
24		Maxis Berhad	Telecommunications
25		Digi.Com Berhad	Communication
26		CIMB Group Holdings Berhad	Banking
27		Telekom Malaysia Berhad	Telecommunications
28		RHB Capital Berhad	Banking
29		Astro Malaysia Holdings Berhad	Media
30		AMMB Holdings Berhad	Banking

Table 2: List of sample companies

Content analysis of annual reports for the year 2014 was conducted to determine the corporate commitment towards water issues. Annual reports were chosen since they are published publicly on a consistent basis due to regulatory requirements (Mohd. Remali et al., 2016), while the year 2014 was chosen as the year of analysis since it is the latest report produced by the time this study is conducted.

Overall, the disclosure by companies is fairly low. The highest score was obtained by Nestle (M) Berhad with 6 (out of 8), followed by Sime Darby Berhad, YTL Corporation Berhad, Lafarge Malaysia Berhad and Malaysia Airports Holdings Berhad with 4. There were 6 companies (20 percent) that did not provide any disclosure on water. Such a low disclosure indicates that companies are not aware of the need to disclose such information. This is not surprising since the disclosure at present is voluntary in nature. Additionally, since it is voluntary, disclosure beyond what is required by the standards may be deemed as detrimental to companies' competitive advantage and it increases the cost for companies in compiling and preparing such report. However, without disclosure, public will never know if companies do take actions on mitigating their water-related risks.

Table 3 presents the findings for WaGM in Malaysia. Based on the table, more than one-half of the companies fall under the high-risk profile (red), while only one is considered as low-risk profile (green). Even though Nestle (M) Berhad obtained a score of 6 for commitment, stakeholders still need to consider the risk associated with the business that it operates which consumes voluminous amount of water. The majority of companies in the low water intensive category are considered as having moderate risk since the evidence on commitment which is assessed through corporate reporting is very minimal.

No	Companies	Commitment	Risk
	HIGH WATER INTENSIVE		
1	Tenaga Nasional Berhad	3	HR
2	Sime Darby Berhad	4	HR
3	Petronas Chemicals Group	3	HR
4	IOI Corporation Berhad	2	HR
5	YTL Corporation Berhad	4	HR

DOI: 10.1051/shsconf/20173600013

6	Kuala Lumpur Kepong Berhad	2	HR
7	British American Tobacco Berhad	0	HR
8	UMW Holdings Berhad	0	HR
9	Nestle (M) Berhad	6	MR
10	PPB Group Berhad	2	HR
	MEDIUM WATER INTENSIVE		
11	IHH Healthcare Berhad	0	HR
12	Genting Berhad	2	HR
13	Hong Leong Industries Berhad	2	HR
14	KICC Prop&Reits-Stapled Sec	3	MR
15	Hap Seng Consolidated Berhad	0	HR
16	IJM Corporation Berhad	2	HR
17	Gamuda Berhad	2	HR
18	S P Setia Berhad	2	HR
19	Lafarge Malaysia Berhad	4	MR
20	Malaysia Airports Holdings Berhad	4	MR
	LOW WATER INTENSIVE		
21	Malayan Banking Berhad	0	MR
22	Public Bank Berhad	2	MR
23	Axiata Group Berhad	2	MR
24	Maxis Berhad	1	MR
25	Digi.Com Berhad	0	MR
26	CIMB Group Holdings Berhad	2	MR
27	Telekom Malaysia Berhad	3	LR
28	RHB Capital Berhad	1	MR
29	Astro Malaysia Holdings Berhad	2	MR
30	AMMB Holdings Berhad	2	MR

HR = high risk, MR = moderate risk, LR = low risk

Table 3: Application of WaGM

CONCLUSION

This study develops a matrix which map companies based on their water-related risks drawing upon the industry intensity classification and organisational commitment towards water protection and preservation. The matrix was derived on the basis of Ceres (2011)'s industry classification and the level of corporate disclosure on water policies, initiatives, and performance. Companies are regarded as either high, moderate, or low risk. It then applied the matrix on 30 public listed companies and found a predominance of the sample in the high risk category. The findings of this study could be used by companies and their stakeholders to have a better view of corporate water-related risk. The lack of disclosure, as found in the analysis of top Malaysian companies, could become the basis for authoritative bodies to revisit the need to have a mandatory disclosure for better accountability, consistency, and comparability.

There are several limitations inherent in this study. First, the water disclosure index used in the matrix might not be a good proxy for corporate commitment towards water protection and preservation. Alternatively, researchers might use primary data collection method such as

 DOI: 10.1051/shsconf/20173600013

questionnaire survey or interview to measure corporate commitment. Additionally, the items included in the index might be too brief and the use of dichotomous scoring system might have ignored the richness of information being reported. Second, we only analysed annual reports but not other corporate media. It has been argued in the past that the medium of reporting influenced the nature and type of information reported therein. The use of other media including company websites and stand-alone sustainability reports might overcome this limitation. Finally, the analysis of Malaysian companies suffers from small observations. Increasing the sample size will improve the generalizability of the findings.

REFERENCES:

Alazzani, A. & Wan-Hussin, W. N. (2013). Global Reporting Initiative's environmental reporting: A study of oil and gas companies. *Ecological indicators*, 32, 19-24.

Braadbaart, O. & Braadbaart, F. (1997). Policing the urban pumping race: Industrial groundwater overexploitation in Indonesia. *World Development*, 25 (2), 199-210.

CDP (2015). 2015 Water Information Request. Retrieved from https://www.cdp.net/CDP%20Questionaire%20Documents/CDP-Water-Information-request-2015.pdf

Ceres (2011). The Ceres Aqua Gauge: A framework for 21st century water risk management. Retrieved from https://www.ceres.org/resources/reports/aqua-gauge

Clarkson, P. M., Li, Y., Richardson, G. D., & Vasvari, F. P. (2008). Revisiting the relation between environmental performance and environmental disclosure: An empirical analysis. *Accounting, Organizations and Society, 33*, 303-327.

de Loë, R. C., Murray, D., & Brisbois, M. C. (2016). Perspectives of natural resource sector firms on collaborative approaches to governance for water. *Journal of Cleaner Production*, 135, 1117-1128.

Dumay, J., Guthrie, J., & Farneti, F. (2010). GRI sustainability reporting guidelines for public and third sector organizations: A critical review. *Public Management Review*, 12 (4), 531-548.

GRI. (2013). G4 *Sustainability reporting guidelines*. Amsterdam, Netherlands: Global Reporting Initiative. Retrieved from https://www.globalreporting.org/standards/g4/Pages/default.aspx

Mohamad, J. M., Firuza-Begham & Wan-Sobri, W. M. A. (2008). Water governance in Peninsular Malaysia: Strategies for reform.". In *International Sustainable Development Research Conference.* Hong Kong, pp. 1-10.

Mohamed, A.F., Wan Yaacob, W. Z., Taha, M.R., & Samsudin, A.R. (2009). Groundwater and soil vulnerability in the Langat basin Malaysia. *European Journal of Scientific Research*, 27 (4), 628 - 635.

Mohd. Remali, A. R., Husin, N. M., Mohd. Ali, I., & Alrazi, B. (2016). An exploratory study on water reporting among top Malaysian public listed companies. *Procedia Economics and Finance*, 35, 64-73.

Montgomery, J., Xu, W., Bjornlund, H., & Edwards, J. (2016). A table for five: Stakeholder perceptions of water governance in Alberta. *Agricultural Water Management*, 174, 11-21.

Saimy, I. S., & Yusof, N. A. M. (2013). The need for better water policy and governance in Malaysia. *Procedia-Social and Behavioral Sciences*, 81, 371-375.

DOI: 10.1051/shsconf/20173600013

Shah, T., Burke J. & Villholth, K. (2009). Groundwater: A water assessment of scale and significance. Proceedings of the Groundwater Management in Malaysia Status and Challenges, Putrajaya, 25-26 March, 15-47.

Van Staden, C. J., & Hooks, J. (2007). A comprehensive comparison of corporate environmental reporting and responsiveness. *The British Accounting Review,* 39, 197-210.

THE ENERGY DISCLOSURE AMONG ENERGY INTENSIVE COMPANIES IN MALAYSIA: A RESOURCE BASED APPROACH

Nabila Elyana Tasrip, Norhayati Mat Husin, Bakhtiar Alrazi*
**Department of Accounting, Universiti Tenaga Nasional, Sultan Haji Ahmad Shah Campus, 26700 Muadzam Shah, Pahang, Malaysia*

ABSTRACT

This study examines the extent of, and factors influencing, energy disclosure in Malaysia. We content analysed the annual reports of top 100 energy intensive companies for the year 2014. Using resource-based theory, we analysed if companies' tangible and intangible resources as well as human capabilities determine the extent of energy disclosure. Overall, the disclosure is low with an average 'occurrence' of energy-related information of 3.53. We also find that the 'Top Brand Value' companies and companies with a larger number of directors serving on the board tend to provide greater energy disclosures. However, other variables - in particular tangible resources - have no significant influence. The findings provide tentative evidence to support the arguments of resource-based theory.

Keywords*:* energy reporting, resource-based theory, Malaysia.

INTRODUCTION

The world economic and industrial development brings along negative environmental impacts such as climate change and global warming (Akbas and Canikli, 2014). The rising use of fossil fuels continues to increase energy-related carbon-dioxide (CO_2) emissions. A report by International Energy Agency (2013) indicates that the Southeast Asia's energy-related CO_2 emissions had almost doubled, from 1.2 gigatonnes (Gt) in 2011 to 2.3 Gt in 2035, or from 3.7% to 6.1% of global emissions. Additionally, the total electricity net consumption had increased from 109.834 billion kWh in the year 2010 to 115.338 billion kWh in the year 2011 (International Energy Outlook, 2014).

Malaysia is an example of developing country in the East Asian region as there is rapid growth in the economic environment (Arshad et al., 2012). It is known as the world's second-largest exporter of liquefied natural gas and the second-largest oil and natural gas producer in Southeast Asia as reported in the website of U.S Energy Information and Administration. In year 2012, the electricity production in Malaysia stood at 118 billion kWh and consumption at 112 billion kWh. The growing electricity demand has spurred the use of fuels such as coal, oil, and gas. Commercial and residential buildings alone accounted for about 13% of total energy consumption and 48% of electricity consumption in Malaysia (Al-Mofleh et al., 2009). The increase in the consumption of energy and fossil fuels will eventually lead to emissions released to the atmosphere. Hence, estimation of energy consumption for operating household appliances, savings of energy under policy intervention, and emission of poisonous gases in a fast developing country deserve academic attention (Saidur et al., 2007).

*Corresponding author's email: Bakhtiar@uniten.edu.my

Among the mechanisms that have been instituted by the government of Malaysia to mitigate the emissions build up are energy efficiency and the use of renewable energy. For example, Sustainability Achieved via Energy efficiency (SAVE) was introduced in 2011 to encourage the purchase of energy efficient equipment to initiate market development of energy efficient appliances by giving rebates. Moreover, the establishment of Sustainable Energy Development Authority Malaysia (SEDA) and Green Technology Financing Scheme (GTFS) are examples of government initiatives regarding renewable energy. In a similar vein, companies are expected to undertake initiatives to reduce their environmental impacts from energy-related activities and to report them to the public. The reporting of such information can be done through annual reports, stand-alone sustainability reports, and the websites.

Unlike the literature on generic environmental reporting and later on carbon reporting, there are limited studies focusing on the reporting of energy information. To date, only Maneschijn et al. (2014) focused specifically on energy reporting; however it is not conducted on Malaysia. Additionally, most of prior literature in social and environmental reporting had relied on legitimacy theory to explain the reporting practice. The use of alternative theoretical lens is thus needed.

The objectives of this study are twofold. Firstly, it aims to explore the extent of energy disclosure made by energy intensive companies in Malaysia. In this regard, a checklist instrument is constructed to measure the quantity of energy related items being reported in the annual reports, stand-alone reports and also the corporate websites. Secondly, using resource based theory, it aims to examine the influence of tangible and intangible resources as well as human capabilities on the extent of energy disclosure.

The study is pertinent for several reasons. This study will be the first attempt in providing evidence of energy reporting in Malaysia. In so doing, it adds to the body of knowledge which had placed greater emphasis on the reporting of general environmental information and carbon information. It also contributes in using resource-based theory as an alternative explanation for the disclosure practice. The constructed checklist instrument could be used by researchers for conducting further studies and for companies to identify the information they might consider for reporting. Finally, the findings will provide indication for the policy makers to revisit the requirements for energy reporting in Malaysia hence to improve the existing energy-related policy frameworks.

The remaining sections of the paper are structured as follows. Section 2.0 provides the discussion of related literature and hypotheses development. Section 3.0 presents the research methods. Section 4.0 discusses the findings. Section 5.0 concludes the paper.

LITERATURE REVIEW AND HYPOTHESES DEVELOPMENT

Literature review

Malaysian government has embarked on various strategies to enhance its' energy-based initiatives. The most prominent strategy is the introduction of various energy based policies. Among those are the introduction of National Energy Policy (1979), National Depletion Policy (1980), National Biofuel Policy (2006), National Green Technology Policy (2009), and National Renewable Energy Policy (2010). Currently, a final draft of National Energy Efficiency Action Plan is introduced in January 2014. The plan outlines strategic actions that are needed for an effective and efficient implementation of energy efficiency programs. The introduction of National Green Technology Policy (2011-2015) provide details on the strategy to restructure the energy subsidies on natural gas and electricity used by industry sector as well as discounted energy pricing mechanism for selected industrial users. With several energy policies being implemented by the government that acts on behalf of the public, this means a company is now being bequeathed with legitimacy based on its ability to operate within the bounds imposed by the society in order to enjoy continued access to products and resource markets (Campbell et al., 2003). Corporate reporting, therefore, has become one of the strategies used by companies to educate the public or to communicate the changes that have been made, in response to the pressure made by the government/public to contribute to energy-based initiative. Nonetheless, with financial reporting being the center of attention since the existence accounting, the reporting of non-financial based information in Malaysia has only been made compulsory in year 2007 with the mandatory requirement of Corporate Social Responsibility (CSR) reporting to all Malaysian public listed companies (PLCs) (Sulaiman et al., 2014).

Prior studies on climate change reporting (see for example De Aguiar and Bebbington, 2014; Ahmad and Hossain, 2015), CSR reporting (see for example Othman and Ameer, 2009; Salleh et al., 2010) and environmental reporting (see for example Ienchu et al., 2012; Sulaiman et al., 2014) are among studies that have evaluated companies disclosure, in response to pressure resulting from social contract, in different forms of corporate reports. Most of these studies have also incorporated energy based information as part of the information being evaluated. A CSR reporting study conducted by Ahmad et al. (2003), for example, reported only small minority (1 percent) of companies in industrial sector disclose information related to on energy. Nonetheless, with energy-based information being incorporated with other types of sustainability information, the analysis can be regarded as being very limited. To the knowledge of this study, there is only one study that put a specific focus on energy reporting and it is conducted by Maneschijn et al. (2014). This study which is based in South African developed a guideline for the development and implementation of effective energy reporting. Unfortunately, Maneschijn et al. (2014) only focuses on preparing a special energy report for a company' energy management system that may not be able to represent overall energy initiatives implemented by the company. Taking into consideration the importance of energy conservation issue in Malaysian and the lack of focus on energy reporting per se, it is the objective of this study to provide a close examination of energy information reported by top energy intensive companies in Malaysia.

Theoretical Framework and Hypotheses Development

Resource-Based Theory

The resource based theory (RBT) serves as the theoretical foundation of this study. The general argument for this theory is that companies need to have resources for them to implement any strategies and to achieve competitive advantage. In this regard, previous scholars have attempted to categorise the resources. Russo and Fouts (1997) classified resources into (1) physical assets and technology, (2) human resources and organisational capabilities and (3) intangible resources such as reputation and organisation's political acumen. Das and Teng (2000) identified two major types of resources: (1) property-based resources and (2) knowledge-based resources. Branco and Rodrigues (2006) suggested the following categories of resources: (1) tangible resources (financial assets and physical assets); (2) intangible resources and capabilities (intellectual property assets, organisational assets, and reputational assets).

In this study, we consider the following: (1) tangible resources (Russo and Fouts, 1997), (2) intangible resources, including corporate reputation (Russo and Fouts (1997), (3) capabilities, including human resources (Wright and McMahan, 1992; Barney, 1991; Russo and Fouts, 1997). In essence, companies with tangible resources (profitability, leverage, and financial slack), reputation, and human capabilities (board size, CEO with international experience, and board with international experience) tend to provide greater extent of energy disclosures.

Hypotheses Development

Profitability

Sulaiman et al. (2014) expected that a company with good profitability is perceived to have more resources to provide better quality environmental disclosure. Accordingly, companies with higher profitability may increase the quality of environmental disclosure primarily as they have better means and opportunities as compared to companies with lower profitability. A number of previous studies revealed negative findings on the relationship between profitability and energy reporting. Sulaiman et al. (2014) concluded that profitability had no significant relationship with the quality of environmental reporting. Toppinen et al. (2012) also found no statistically significant association between profitability and voluntary disclosures for the sample companies. One explanation could be the use of single profitability indicator (ROCE) and the cross-sectional nature of data in the study.

Nevertheless, the following researchers reached opposing conclusions (Qiu et al., 2014; Luo et al., 2013). Luo et al. (2013) who studied on carbon disclosure argued that profitable firms are less constrained by financial resources in making green decisions and directly enhance their economic performance as compared to less profitable firms. Qiu et al. (2014) also found being profitable firms have the resources and the willingness to commit to investments in the social arena. The study concludes that profitable firms make higher social disclosure compared to environmental disclosure and higher combined social and environmental disclosure. Based on these arguments, we develop the following hypothesis:

H_1: Profitability is associated with the extent of energy disclosure

Leverage

Clarkson et al. (2008) suggested that firms with high leverage may not be able to absorb the adverse financial impact of disclosing energy information, due to high financial risk inherent in such disclosures. Luo et al. (2013) revealed that firms in developing countries with higher leverage are less likely to disclose carbon information. Firms with poor financial condition tend to less disclose and bear the consequences of revealing proprietary carbon information. The level of average is negatively associated with disclosure, due to a heavier liability for debt and interest repayment would restrict a firm's ability to undertake strategies for carbon reduction and disclosures and high leverage firms would be more sensitive to carbon-prevention expenditures.

Thus, the following hypothesis is developed:
H_2: Leverage is associated with the extent of energy disclosure

Organisational Slack

Nohria and Gulati (1996) refer slack as the pool of resources in an organisation that is in excess of the minimum necessary to produce a given level of organisational output. When an excess of available resources exists, businesses are able to invest in socially responsible activities, such as energy reporting, which may enhance the corporate social and environmental performance (Elsayed, 2006). Companies with more slack resources have a wider variety of strategic options compared to a company that have limited resources (Yol Lee and Rhee, 2007). Bansal (2005) recommended that organizational slack allows organisations to engage in creative practices. Whereas inadequate of slack resources limits business flexibility, thus hindering its performance and response to institutional pressures on timely issues such as climate change.

Yol Lee and Rhee (2007) found that slack resources have positive relationship with corporate environmental strategies. Bansal (2005)'s finding shows a positive relationship between organisational slack and corporate sustainability. Recent scholars however found contrary evidence. Amran et al. (2016) indicate that organisational slack is inversely related to climate change business strategy. Darus et al. (2014) posits negative relationship between organisational slack on corporate social responsibility (CSR) of financial institutions in Malaysia. Accordingly, Arshad et al. (2012) find inverse relationship between organisational slack and two types of environmental disclosure; annual report and corporate websites.

The following hypothesis is developed:
H_3: Organisational slack is associated with the extent of energy disclosure

Corporate Reputation

Corporate reputation is arguably one of the most important intangible assets that provide competitive advantage (Toms, 2002). A good corporate reputation is important to firm performance and such resource is more likely to be valuable and inimitable when society demands a cleaner environment (Russo and Fouts, 1997).

Cao et al. (2012) provided evidence that companies with higher reputations are less likely to misstate their financial statements. Zeng et al. (2012) argue that a company which aims to convey a better image and value its public reputation among stakeholders is more likely to voluntarily disclose environmental information. On the other hand, Steverman (2010) gave an example of British Petroleum (BP), receiving high marks from many influential CSR and sustainability rating agencies between 2001 and 2009, was responsible for the Deepwater Horizon oil spill in the Gulf of Mexico, the worst environmental disaster in the U.S. history.

Therefore, we develop the following hypothesis:
H_4: Corporate reputation is associated with the extent of energy disclosure

Board size

The board size affects the monitoring ability of the boards and that larger board sizes are often believed to be more capable of monitoring the actions of top management (Amran et al.

2016) as it will increase the communication problems and cost of poor decision making Mohamad Taha, 2013) . A larger board size is assumed to consist of directors with different backgrounds such as education, industrial, knowledge and skills. The combination of various knowledge and skills could help to improve the quality of actions or decisions made. Thus, the hypothesis is developed as follow:

H_5: Board size is associated with the extent of energy disclosure

CEO/Board of directors with international experience

CEO represents one of the unique resources who lead the organisation by formulating business strategy and implement strategies towards strategic competitive advantage (Slater and Dixon-Fowler, 2009). Wei and Ling (2015) opined that CEOs/boards who have working experience at foreign firms possess higher levels of corporate entrepreneurship than those with CEOs/boards who have worked only at domestic firms. As international assignments mostly hold more responsibility, executives experience higher levels of autonomy, which may instill higher confidence levels (Suutari and Mäkelä, 2007). The international experience which they have gained enabled them to develop a set of best practice based on their collective learning in prior working assignment (Bansal, 2005).

Bansal (2005) further argued CEO/boards with international experience could build up the best practice based on their knowledge obtained from different countries. This is supported by other past studies that have indicated that CEOs' international experience positively impacts business performance (Carpenter et al., 2001; Slater and Dixon-Fowler, 2009). Such experience is considered as tacit knowledge, represent the strategic value for the business and directly leading to higher business performance (Carpenter et al., 2001). Arshad et al. (2012) found that international experience of CEO /boards is significantly positively related to the extent of environmental reporting in corporate websites.

This is however not the case in Amran et al. (2016). The authors found a negative association and argued that CEO/board with international experience does not have any advantage over locally experience CEO in formulating climate change business strategy. Additionally, Darus et al. (2014) found a negative relationship between foreign exposure of top management on corporate social responsibility (CSR) of financial institutions in Malaysia.

Hence, the hypothesis is developed as follow:
H_6: CEO with international experience is associated with the extent of energy disclosure
H_7: Board of directors with international experience is associated with the extent of energy disclosure

RESEARCH METHODS

Sample selection

The sample for this study is public listed companies which are among the energy producers and energy intensive industries. The industries classified as energy intensive industries are energy production, wood, glass, rubber, pulp and paper, food, cement, ceramic, iron, steel, chemical, textile, aluminum, and transportation. Table 1 below provides the classification of industries based on their energy intensity according to prior literature.

Industry	References
Iron & steel	Song & Oh (2015); Liddle (2012); Saygin et al.(2011)
Cement	Li et al. (2014); Napp et al. (2014); Campana et al. (2013)
Chemical	Liddle (2012); Bassi & Yudken (2011); Saygin et al. (2011)
Pulp & paper	Song & Oh (2015); Liddle (2012); Saygin et al., (2011)
Textile	Bassi et al. (2012); Palamutcu (2010)
Aluminum	Li et al. (2014); Liddle, (2012); Bassi & Yudken, (2011)
Oil & gas	Song & Oh (2015); Li et al. (2014); Campana et al. (2013)
Electricity & heat	Li et al. (2014); Bassi et al. (2012)
Ceramic	Li et al. (2014)
Glass	Campana et al. (2013)
Wood	Paramonova et al. (2015)
Coal mining	Levesque et al. (2014)
Transport	Lin & Xie (2014)

Table 1: Energy intensive industries

Since the classification of the industry based on Bursa Malaysia is different, we contacted the officer at Bursa Malaysia to provide a list of companies based on their core activities according to market capitalisation. From the list, we selected the top 100 companies based on the arguments in the literature that companies that are large are more visible to the public hence expected to provide greater disclosures.

Data Collection Procedures

In conducting this research, the data were extracted from the annual reports, stand-alone reports and also corporate web reports. The annual and stand-alone reports of these companies were accessed through website of Bursa Malaysia and the companies' websites. We used all these media since energy disclosure is considered a new area – both in literature and practice – and to ensure that the disclosure is comprehensive (Unerman, 2000). Data for the year 2014 were used because they are the most recent data available at the commencement of this research.

Energy Disclosure Instrument

To achieve the first objective in this study, energy disclosure checklist instrument is constructed. The development of this instrument involved scrutinising preliminary indexes of energy efficiency and renewable energy through a review of previous literature (Kansal et al., 2014; Abdullah et al., 2011; Buniamin, 2010) and Global Reporting Initiative (GRI) indicators that are relevant to energy efficiency and renewable energy

Table 2 below shows a self-developed checklist instrument which suits the Malaysian context on energy.

No.	Items		Indicators
1	Energy consumption of organisation	1	Report total energy consumption within the organisation in joules or multiples
		2	Report energy consumed outside of the organisation, in joules or multiples
		3	Report on energy usage by types (Coal, electricity, etc)
		4	Report on energy use by facility or segment level
		5	Report the energy intensity ratio
2	Initiative to reduce energy consumption	1	Development / Usage of energy-efficient products and services
		2	Procedures related to training and raising awareness in relation to the society aspects
		3	Have a clear target on energy conservation
		4	Disclose relevant technologies (energy management system)
		5	Participation in voluntary environmental initiatives endorsed by Ministry of Energy, Green Technologies & Water (MEGTW) or other organisation
3	Energy governance & ethics	1	Disclosing company's policy on energy
		2	Disclosing governance structure related to energy including committees [Sustainability and Transformational Management Office] (TMO)
		3	Disclose the organisation's values, principles, standards, and norms of behavior related to energy
4	Energy based achievement & awards	1	Receiving an award for an energy conservation programme (Environment, etc)
5	Reduction of energy consumption	1	Comparison of energy consumption with previous year or reduction target level
		2	Report on total energy saving due to direct conservation activities
		3	Report the reductions in the energy requirements of sold products and services
6	Renewable energy	1	Disclosure on the use of renewable energy sources
		2	Disclosure on investment in RE
		3	Acquisition of renewable energy certificates
7	Other disclosure	1	Any other energy disclosures not fitting the items above

Table 2: Energy disclosure checklist instrument

Measurement of variables

For this research, the dependent variable is the extent of energy reporting, while the independent variables consist of variables representing the resources available to the companies. We also included two additional variables asset newness and capital intensity as control variables since these two variables have been used in the past to proxy for environmental performance. High asset newness and capital intensity indicate greater proportion of investment in new technologies that might be more environmentally friendly (see Clarkson et al., 2008). Table 3 below summarises the measurement of variables.

DOI: 10.1051/shsconf/20173600014

Variables	Measurement of variables	References
Dependent Variables		
Extent of Energy Reporting	Occurrence (how many times the energy related information being mentioned)	• Michelon and Parbonetti (2012) • Buniamin (2010)
Independent Variables ***Resource Based Variables***		
Profitability	Return on asset (ROA) = Net Income : Total Asset	• Qiu et al. (2014) • Sulaiman et al. (2014) • Luo et al. (2013)
Leverage	Debt to equity ratio [Ratio of total debt divided by total assets]	• Luo et al. (2013)
Organisational Slack	Current ratio [Ratio of current assets to current liabilities] (CA:CL)	• Amran et al. (2016) • Darus et al. (2014) • Arshad et al. (2012) • Yol Lee & Rhee (2007) • Bansal (2005)
Corporate Reputation	Corporate Brand Value Companies listed in the Top 100 companies of brand value = 1 Companies not listed in the Top 100 companies of brand value = 0	• Zeng et al., (2012)
Board size	Number of board members	• Buniamin et al. (2008)
CEO with International Experience	International experience CEO / Chairman = 1; No international experience = 0	• Darus et al. (2014) • Amran, et al. (2016) • Arshad et al., (2012)
Board of directors with International Experience	Percentage of board members with international experience of the total board of directors	• Amran et al. (2016) • Darus et al. (2014) • Arshad et al. (2012)
Control Variables		
Asset Newness	Ratio of net properties, plant and equipment divided by the gross properties, plant and equipment	• Clarkson et al., (2008)
Capital Intensity	Ratio of capital spending divided by total sales revenues	• Clarkson et al., (2008)

Table 3: Summary of variables

Regression model

The following regression model is used to examine the relationship between the extent of energy reporting and resource-based theory elements which comprise profitability, leverage, organizational slack, corporate reputation, board size, CEO with international experience, boards of directors with international experience, and control variables.

$$EREP = \beta_0 + \beta_1 ROA + \beta_2 LEV + \beta_3 SLACK + \beta_4 REPUT + \beta_5 BSIZE + \beta_6 CEOEXP + \beta_7 BEXP + \beta_8 NEWNESS + \beta_8 CAPIN + \varepsilon$$

Where:

EREP	: Level of Energy Reporting
β_0	: Intercept
ROA	: Profitability
LEV	: Leverage
SLACK	: Organisational Slack
REPUT	: Corporate Reputation
BSIZE	: Board size
CEOEXP	: CEO / Chairman with international experience
BEXP	: Board of directors with international experience
NEWNESS	: Asset Newness
CAPIN	: Capital Intensity
ε	: Disturbance term

FINDINGS AND ANALYSIS

This section presents the findings from the descriptive statistics, correlation analysis, and regression analysis.

Descriptive statistics

Table 4 depicts the descriptive statistics for the variables used in estimation. Based on the table, the extent of energy disclosure among the sample companies is low with an average occurrence of 3.530. Additionally, the majority of the sample companies consist of profitable (*ROA* = 0.415), highly leveraged (*LEV* = 1.398), and positive working capital (*SLACK* = 2.734) companies. Top Brand companies make up only 19% of the sample (*REPUT*). In regards to human capabilities, the average number of board members is 8, with 59% of the CEO and 44.7% of the board members have international experience. For control variables representing the degree of environmental performance, *NEWNESS* and *CAPIN*, due to the nature of the companies which are high energy intensive, the sample can be considered as having high impact to the environment, with means of 0.655 and 0.807, respectively.

Panel A: Descriptive statistics of continuous variables used in estimation

Variable	Mean	Median	Q1	Q3	Std dev
EREP	3.530	1.000	0.000	4.750	5.290
ROA	0.415	0.060	0.025	0.100	0.207
LEV	1.398	0.465	0.138	0.878	7.121
SLACK	2.734	1.655	1.113	2.453	6.409
BSIZE	8.490	8.000	7.000	10.000	2.1630
BEXP	0.447	0.500	3.000	6.000	0.2138
NEWNESS	0.655	0.637	0.498	0.793	0.2632
CAPIN	0.807	0.426	0.248	1.106	1.4129

Panel B: Distribution of categorical variables

	1		0	
Variable	*N*	%	*N*	%
REPUT	19	19	81	81
CEO_EXP	59	59	41	41

This table presents descriptive statistics of the dependent and independent variables used in correlation and regression analyses. Statistics are presented for the full sample of 100 companies.

Table 4: Descriptive statistics of continuous variables

Correlation analysis

Correlation analysis was conducted to determine whether there is a multicollinearity problem that could influence the regression analysis. Table 5 provides the results of the correlation statistics with Pearson above the diagonal and Spearman's rho, below. This is a two-tailed test of significance. There appears to be no serious multicollinearity issues among the independent variables. According to Field (2013), a coefficient correlation of greater than, or equal to, (≥) 0.80 indicates there is cause for concern. Based on the table below, the highest coefficient is 0.505 (p = 0.000) which is between *ROA* and *SLACK*.

	EREP	*ROA*	*LEV*	*SLACK*	*REPUT*	*BSIZE*	*CEOEXP*	*BEXP*	*NEWNESS*	*CAPIN*
EREP	-	0.040	-0.060	-0.010	0.431**	0.318**	-0.013	0.093	0.060	-0.043
ROA	-0.080	-	-0.074	0.008	0.025	0.214	-0.120	-0.193	0.041	0.042
LEV	0.071	-0.433**	-	0.136	-0.036	-0.108	0.095	0.193	0.082	-0.055
SLACK	-0.097	0.505**	-0.467**	-	-0.079	-0.201*	-0.092	-0.147	0.083	-0.097
REPUT	0.356**	-0.112	0.144	-0.118	-	0.174	0.093	0.195	-0.005	0.015
BSIZE	0.356**	0.056	0.081	-0.139	0.193	-	0.029	-0.014	0.015	0.160
CEOEXP	0.008	-0.102	0.099	0.030	0.093	0.063	-	0.471**	-0.034	0.093
BEXP	0.158	-0.232*	0.108	-0.071	0.191	-0.020	0.477**	-	-0.095	0.046
NEWNESS	0.074	-0.123	0.185	-0.047	-0.002	0.004	-0.002	-0.074	-	0.124
CAPIN	0.089	-0.128	0.216*	-0.145	0.131	0.370**	0.048	0.006	0.025	-

This table presents the correlation statistics for the dependent and independent variables used in multivariate analyses. Statistics are presented for the full sample of 100 companies. * and ** represent significance levels (two-tailed) at 5 percent and 1 percent, respectively.

Table 5: Correlation statistics for dependent and independent variables

Regression analysis

Table 6 presents the results of the linear regression (OLS). Each variable's unstandardised coefficients and p-values (two-tailed) and each model's adjusted R^2 and F statistic are provided. Based on the table, the only significant variables are *REPUT* and *BSIZE*. These indicate that Top Brand companies and companies with larger board size provided greater energy disclosures. This is consistent with the argument of resource-based theory. However, other variables are not statistically significant. The adjusted R^2 indicates that the variables included in the model are able to explain 19 % of the variation in the extent of energy reporting.

	β	p-value
Constant	-4.449	0.106
ROA	-0.629	0.802
LEV	-0.307	0.603
SLACK	0.059	0.459
REPUT	5.038	0.000**
BSIZE	0.682	0.007**
CEOEXP	-0.813	0.475
BEXP	2.083	0.449
NEWNESS	1.581	0.407
CAPIN	-0.365	0.307
N	100	
Model *F* stat	3.470	
Adj. R^2	0.190**	

*Significance level (two-tailed) at the 5% level.
**Significance level (two-tailed) at the 1% level

Table 6: Results of the OLS regression

CONCLUSION

This study examines the extent of energy disclosures among the largest energy intensive companies in Malaysia and the factors influencing the disclosure practices. Overall, there is a low level of disclosure on energy among the companies. This provides several implications. First, considering that the sample companies are from the energy intensive industries, this situation is alarming. In the age of climate change, public would like to know what initiatives companies have taken to mitigate, and adapt to, this problem. Such a low disclosure among the sample companies might indicate the overall level of disclosure of companies in Malaysia. Second, the findings of this study also suggest that, if left alone, companies in Malaysia would not disclose beyond what is required by the laws and accounting standards. The disclosure of environmental information (including energy related), although is mandatory, the volume, nature and type are very much the discretion of the companies. This might suggest for more refined guideline for reporting to ensure a more comprehensive, comparable and consistent information.

The findings should be interpreted with caution. Firstly, the sample size is small and the industry is limited. Future studies might consider increasing the sample size and including companies from low energy intensive industries to enhance the generalizability of the findings. Secondly, the measure for energy reporting only considers the quantity of information being reported (i.e. occurrence). The examination of quality of such information, either quantitative or qualitative, might provide better understanding of the disclosure practice. Finally, we have not considered the influence of physical resources (e.g., total assets) and human capabilities (e.g., board age, board tenure). The inclusion of these additional variables is expected to improve the ability of the regression model to explain the variation in the disclosure.

REFERENCES:

Abdullah, S. N., Mohamad, N. R., & Mokhtar, M. Z. (2011). Board independence, ownership and CSR of Malaysian large firms, Corporate Ownership & Control, 8(3), 417-431.

Ahmad, N. N. N., Sulaiman, M., & Siswantoro, D. (2003). Corporate social responsibility disclosure in Malaysia: An analysis of annual reports of KLSE listed companies. International Journal of Economics, Management and Accounting, 11(1), 51-86.

Akbas, H. E., & Canikli, S. (2014). Corporate environmental disclosures in a developing country: An investigation on Turkish listed companies, International Journal of Economics and Finance, 6(2), 50-61.

Al-Mofleh, A., Taib, S., Mujeebu, M. A., & Salah, W. (2009), Analysis of sectoral energy conservation in Malaysia. Energy, 34(6), 733-739.

Amran, A., Ooi, S. K., Wong, C. Y., & Hashim, F. (2016). Business Strategy for Climate Change: An ASEAN Perspective, Corporate Social Responsibility and Environmental Management, 23(4), 213-227.

Arshad, R., Omar, H., & Othman, R., (2012). Corporate resources and environmental reporting practices in Malaysia, 3rd International Conference on Business and Economic Research Proceeding, 12 - 13 March.

Bansal, P. (2005). Evolving sustainably: a longitudinal study of corporate sustainable development, Strategic Management Journal, 26(3), 197-218.

Barney, J. (1991). Firm resources and sustained competitive advantage, Journal of Management, 17(1), 99-120.

Bassi, A. M., & Yudken, J. S. (2011). Climate policy and energy-intensive manufacturing: A comprehensive analysis of the effectiveness of cost mitigation provisions in the American Energy and Security Act of 2009, Energy Policy, 39(9), 4920-4931.

Bassi, A. M., Tan, Z., & Mbi, A. (2012). Estimating the impact of investing in a resource efficient, resilient global energy-intensive manufacturing industry, Technological Forecasting and Social Change, 79(1), 69-84.

Branco, M. C., & Rodrigues, L. L. (2006), Corporate social responsibility and resource-based perspectives, Journal of Business Ethics, 69(2), 111-132.

Buniamin, S. (2010). The quantity and quality of environmental reporting in annual report of public listed companies in Malaysia. Issues in Social & Environmental Accounting, 4(2), 115-135.

Buniamin, S., Alrazi, B., Johari, N.H., & Abd. Rahman, N.R. (2008). An investigation of the association between corporate governance and environmental reporting in Malaysia, Asian Journal of Business and Accounting, 1(2), 65-88.

Campana, F., Bianchi, M., Branchini, L., De Pascale, A., Peretto, A., Baresi, M., & Vescovo, R. (2013). ORC waste heat recovery in European energy intensive industries: Energy and GHG savings, Energy Conversion and Management, 76, 244-252.

Campbell D., Craven B., & Shrives P., (2003), Voluntary social reporting in three FTSE sectors: A comment on perception and legitimacy, Accounting, Auditing, and Accountability Journal, 16, 558-591.

Cao, Y., Myers, L. A., & Omer, T. C. (2012). Does company reputation matter for financial reporting quality? Evidence from restatements, Contemporary Accounting Research, 29(3), 956-990.

Carpenter, M. A., Sanders, W. G., & Gregersen, H. B. (2001). Bundling human capital with organizational context: The impact of international assignment experience on multinational firm performance and CEO pay, Academy of Management Journal, 44(3), 493-511.

Clarkson, P. M., Li, Y., Richardson, G. D., & Vasvari, F. P. (2008). Revisiting the relation between environmental performance and environmental disclosure: An empirical analysis, Accounting, Organizations and Society, 33(4), 303-327.

Darus, F., Mad, S., & Yusoff, H. (2014), The importance of ownership monitoring and firm resources on corporate social responsibility (CSR) of financial institutions, Procedia-Social and Behavioral Sciences, 145, 173-180.

Das, T. K., & Teng, B. S. (2000). A resource-based theory of strategic alliances, Journal of Management, 26(1), 31-61.

De Aguiar, T. R. S., & Bebbington, J. (2014), Disclosure on climate change: Analysing the UK ETS effects, Accounting Forum, 38, 227-240.

Elsayed, K. (2006). Reexamining the expected effect of available resources and firm size on firm environmental orientation: An empirical study of UK firms, Journal of Business Ethics, 65(3), 297-308.

Field, A. (2013), Discovering Statistics using IBM SPSS Statistics (4th ed.), SAGE Publications: London.

Ienciu, I. A., Popa, I. E., & Ienciu, N. M. (2012). Environmental reporting and good practice of corporate governance: Petroleum industry case study, Procedia Economics & Finance, 3, 961-67.

International Energy Outlook, (2014). World energy and economic outlook. Retrieved at: http://www.eia.doe.gov/oiaf/ieo/world.html

Kansal, M., Joshi, M., & Batra, G. S. (2014). Determinants of corporate social responsibility disclosures: Evidence from India, Advances in Accounting, 30(1), 217-229.

Levesque, M., Millar, D., & Paraszczak, J. (2014). Energy and mining–the home truths. Journal of Cleaner Production, 84, 233-255.

Li, L., Wang, J., Tan, Z., Ge, X., Zhang, J., & Yun, X. (2014). Policies for eliminating low-efficiency production capacities and improving energy efficiency of energy-intensive industries in China. Renewable and Sustainable Energy Reviews, 39, 312-326.

Liddle, B., (2012). The importance of energy quality in energy intensive manufacturing: Evidence from panel cointegration and panel FMOLS. Energy Economics. 34, 1819–1825.

Lin, B., & Xie, C. (2014). Reduction potential of CO_2 emissions in China' s transport industry. Renewable and Sustainable Energy Reviews, 33, 689-700.

Luo, L., Tang, Q., & Lan, Y. C. (2013). Comparison of propensity for carbon disclosure between developing and developed countries: A resource constraint perspective, Accounting Research Journal, 26(1), 6-34.

Maneschijn, R., Vosloo, J.C., Pelzer, R. (2014). Effective multi-level energy reporting in South African industry, International Conference on the Industrial and Commercial Use of Energy (ICUE), 19-20 Aug.

Michelon, G., & Parbonetti, A. (2012). The effect of corporate governance on sustainability disclosure. Journal of Management & Governance, 16(3), 477-509.

Mohamad Taha, M.H. (2013). The relationship between corporate social responsibility disclosure and corporate governance characteristics in Malaysian public listed companies. Retrieved from http://papers.ssrn.com/sol3/papers.cfm?abstract_id=2276763

Napp, T. A., Gambhir, A., Hills, T. P., Florin, N., & Fennell, P. S. (2014). A review of the technologies, economics and policy instruments for decarbonising energy-intensive manufacturing industrie, Renewable and Sustainable Energy Reviews, 30, 616-640.

Nohria, N., & Gulati, R. (1996). Is slack good or bad for innovation? Academy of Management Journal, 39(5), 1245-1264.

Othman, R. & Ameer, R. (2009). Corporate social and environmental reporting: Where are we heading? A survey of the literature. International Journal of Disclosure and Governance, 6, 298-320

Palamutcu, S. (2010). Electric energy consumption in the cotton textile processing stages. Energy, 35(7), 2945-2952.

Paramonova, S., Thollander, P., & Ottosson, M. (2015). Quantifying the extended energy efficiency gap-evidence from Swedish electricity-intensive industries, Renewable and Sustainable Energy Reviews, 51, 472-483.

Qiu, Y., Shaukat, A., & Tharyan, R. (2014). Environmental and social disclosures: Link with corporate financial performance. The British Accounting Review, 48(1), 102-116.

Russo, M. V., & Fouts, P. A. (1997). A resource-based perspective on corporate environmental performance and profitability, Academy of management Journal, 40(3), 534-559.

Saidur, R., Masjuki, H. H., Jamaluddin, M. Y., & Ahmed, S. (2007). Energy and associated greenhouse gas emissions from household appliances in Malaysia, Energy Policy, 35(3), 1648-1657.

Saleh, M., Zulkifli, N., & Muhamad, R. (2010), Corporate social responsibility disclosure and its relation on institutional ownership: Evidence from public listed companies in Malaysia, Managerial Auditing Journal, 25, 591-613.

Saygin, D., Worrell, E., Patel, M. K., & Gielen, D. J. (2011). Benchmarking the energy use of energy-intensive industries in industrialized and in developing countries. Energy, 36(11), 6661-6673.

Slater, D. J., & Dixon-Fowler, H. R. (2009). CEO international assignment experience and corporate social performance, Journal of Business Ethics, 89(3), 473-489.

Song, C., & Oh, W. (2015). Determinants of innovation in energy intensive industry and implications for energy policy. Energy Policy, 81, 122-130.

International Energy Agency (2013). Southeast Asia Energy Look. Retrieved at https://www.iea.org/publications/freepublications/publication/SoutheastAsiaEnergyOutlook_WEO2013SpecialReport.pdf

Sulaiman M., Abdullah N., & Abdul Hamid, F. (2014), Determinants of environmental reporting quality in Malaysia, International Journal of Economics, Management, & Accounting, 22, 63-90.

Suutari, V., & Mäkelä, K. (2007). The career capital of managers with global careers. Journal of Managerial Psychology, 22(7), 628-648.

Toms, J. S. (2002). Firm resources, quality signals and the determinants of corporate environmental reputation: some UK evidence, The British Accounting Review, 34(3), 257-282.

Toppinen, A., Li, N., Tuppura, A., & Xiong, Y. (2012). Corporate responsibility and strategic groups in the forest-based industry: Exploratory analysis based on the Global Reporting Initiative (GRI) framework. Corporate Social Responsibility and Environmental Management, 19(4), 191-205.

Unerman, J. (2000). Methodological issues-reflections on quantification in corporate social reporting content analysis. Accounting, Auditing & Accountability Journal, 13(5), 667-681.

Wei, L. Q., & Ling, Y. (2015). CEO characteristics and corporate entrepreneurship in transition economies: Evidence from China, Journal of Business Research, 68(6), 1157-1165.

Wright, P. M., & McMahan, G. C. (1992), Theoretical perspectives for strategic human resource Management, Journal of Management, 18(2), 295-320.

Yol Lee, S., & Rhee, S. K. (2007). The change in corporate environmental strategies: a longitudinal empirical study, Management Decision, 45(2), 196-216.

Zeng, S. X., Xu, X. D., Yin, H. T., & Tam, C. M. (2012). Factors that drive Chinese listed companies in voluntary disclosure of environmental information, Journal of Business Ethics, 109(3), 309-321.

BOARD CHARACTERISTICS, COUNTRY OF ORIGIN, AND THE DECISION TO PARTICIPATE IN THE CDP'S WATER PROGRAMME

Bakhtiar Alrazi*, Inaliah Mohd Ali, Norhayati Mat Husin
**Department of Accounting, Universiti Tenaga Nasional, Sultan Haji Ahmad Shah Campus, 26700 Muadzam Shah, Pahang, Malaysia*

ABSTRACT

This study examines the propensity of the world's largest companies to participate in the Carbon Disclosure Project (CDP) water programme. Drawing upon stakeholder theory, we assess the influence of board characteristics and country of origin on the decision of 748 largest companies from 42 countries to respond to the 2015 CDP water programme. Based on the binary logistic regression, we found that companies which participated in the survey have boards dominated by the independent directors, no specific environmental committee, and were domiciled in countries which placed greater emphasis on the information needs of the shareholders. We also found the evidence of companies operating in high water intensive industries to be more inclined towards participating in the survey. The findings provide tentative evidence to support the arguments of stakeholder theory.

Keywords: CDP water programme, board characteristics, country of origin, stakeholder theory.

INTRODUCTION

Water security is one of the most tangible and fastest-growing social, political and economic challenges faced today. A survey among of the World Economic Forum's global multi-stakeholder community in year 2014 ranked water crises as among the top 10 global risks likely to happen within 10 years and as the single most impactful risk on the society (WEF, 2015). This situation is alarming as prior to 2012, water crises did not even feature in the top 20 list. It is expected that, the demand for water in every sector to increase and analysis suggests that the world will face a 40% global shortfall between forecast demand and available supply by 2030. At present, more than one-third of the world's population—roughly 2.4 billion people—live in water-stressed countries, and by 2025 that proportion is expected to rise by two-thirds (Ceres, 2011). Ironically, the effects of climate change will exert further pressure on this problem causing either floods or droughts which in turn disrupt the supply of food and the availability of fresh water for consumption.

Recent years have seen a growing interest among the investors on environmental, social, and governance issues. One example is the establishment of Carbon Disclosure Project (CDP), a London-based not for profit organisation, which represents the interest of 827 investors with USD100 in assets. Over the years, CDP has built the most comprehensive collection of self-reported environmental data pertaining to climate, water, and forest encompassing over 5,600

*Corresponding author's email: Bakhtiar@uniten.edu.my

DOI: 10.1051/shsconf/20173600015

companies and 533 cities. Investors requested for these information to make financial and other decisions about the companies. To achieve this, CDP has been inviting the world's largest companies, by size of market capitalisation, to participate in its annual surveys.

Unlike climate change, the interest in water and how companies are dealing with this issue is a recent phenomenon. CDP, for example, has started to invite companies to provide information voluntarily since year 2002. The water programme, however was first launched in year 2010. In a similar vein, there has been abundance of research focusing on carbon disclosure (see, for example, Alrazi et al., 2016; Luo and Tang, 2016; Liao et al., 2015; Luo et al., 2013; Luo et al., 2012; Rankin et al., 2011). However research into why companies engage in water reporting receives scant attention.

Thus, the aim of this research is to examine the factors influencing companies' decision to voluntarily report on their water policies, initiatives and performance to CDP. In particular, we assess the influence of board characteristics (independence, diversity, and environmental committee) and country of origin (legal system and strength of law enforcement) on the propensity of companies participating in the CDP water programme.

The study is pertinent for several reasons. First, it adds to the body of knowledge, in particular, the dearth of literature on water disclosure. In so doing, it enhances the discussion in the area of corporate governance (board characteristics) and this is done on an international setting (country of origin). Second, it helps us to understand the possible reasons for companies' voluntary participation in an environmental activity. It could also inform the investors of the characteristics of board of directors that promote environmental agenda. Finally, the results of this research address one of the major social concerns of our time, namely that of water scarcity, and the related company disclosure responses. Therefore, they will be of interest to regulators, managers, accountants, environmental groups, and researchers.

The remaining sections of this paper are structured as follows. Section 2.0 provides the review of related literature and hypotheses development. Section 3.0 presents the methods used for this research. Section 4.0 discusses the findings. Section 5.0 concludes the paper, highlights the limitations, and provides recommendation.

LITERATURE REVIEW AND HYPOTHESES DEVELOPMENT

Water scarcity imposes enormous challenges to businesses. These include disruption to operations from drought or flooding, declining water quality that required costly onsite pretreatment, increases in water prices, and fines and litigation relating to pollution incidents. In essence, for businesses, these problems impact their bottom line. Due to these escalating business impacts, water-related risks and opportunities are becoming a bigger focus among global investors. Thus, they have been seeking increased disclosure from companies on their water use and associated risks.

Businesses respond to these challenges in many ways. At present, there are over 130 companies worldwide that have committed to advance water stewardship, sanitation, and the Sustainable

Development Goals – in partnership with the United Nations, governments, peers, civil society, and others. This is made through the CEO Water Mandate under the UN Global Compact in which signatories are committed to the six elements of water stewardship: direct operations, supply chain and watershed management, collective action, public policy, community engagement, and transparency. Businesses also have worked with World Business Council for Sustainable Development (WBCSD) to develop tools to help companies to better understand and manage their water-related challenges, and communicate effectively what they are doing about them. Some of the tools are industry-specific which are suitable for oil and gas, power utilities, and cement sectors. In 2011, Ceres, a U.S.-based coalition of investors, environmental groups, and other public interest organisations, partnered with WBCSD, Irbaris, and the IRRC Institute to publish a framework for water risk management. The framework emphasises four key areas, namely measurement, management, engagement, and disclosure. Additionally, there is a growing number of companies using CDP to guide their water management efforts. According to CDP (2015), the number of responding companies in 2010 was 150 and had increased to 1226 in 2015 (an increase of 700 percent). This trend is commensurate with the increase in the number of CDP Water programme signatories and the fact that water risk being regarded as one of the top three issues imposing the greatest impact to the society.

One common theme emerged from the abovementioned initiatives is disclosure. In essence, businesses are not only required to protect and preserve water, they also need to demonstrate their water stewardship. This can be done through corporate disclosure. Apart from participating in the CDP water programme, companies could disclose water-related information in their annual reports, stand-alone sustainability reports, or their own websites. There are also guidelines available for the companies to help them prepare the reports in a more consistent and comparable manner. The guidelines include CEO Water Mandates' Corporate Water Disclosure Guidelines and Global Reporting Initiative (GRI)'s sustainability reporting guidelines. According to KPMG (2015), GRI remains widely used by the world's largest companies, with three quarters (74 percent) of the G250 using the GRI framework (including in reporting on water-related information). Moreover, according to publication by KPMG International, GRI, UNEP and the Stellenbosch Business School (2016), a total of 400 sustainability reporting instruments in 64 countries were identified in year 2016 (as compared to 180 instruments identified in 44 countries in 2013 report) with 65% of them are considered as mandatory. This is particularly evident in Europe, Asia Pacific and Latin America.

Water disclosure is a sub-set of broader environmental disclosure. Other components include the disclosure related to materials, energy, biodiversity, emissions, and effluents and waste (GRI, 2013). In the past, the literature on environmental reporting has shown an increasing trend (Hahn and Kühnen, 2013). Recently, the increased awareness on the impact of climate change has brought about the interest in the disclosure of carbon and greenhouse gas information (see Ascui, 2014; Stechemesser and Guenther, 2012; and Hahn et al., 2015; for a review of literature on carbon disclosure/accounting). Of these studies, several have examined factors influencing the decision and quality of the disclosure to CDP Climate Change programme (Ben-Amar and McIlkenny, 2015; Ben-Amar et al., 2015; Depoers et al., 2016; Gonzalez-Gonzalez and Ramírez, 2016; Grauel and Gotthardt, 2016; Li et al., 2016; Liao et al., 2013, 2015; Luo et al., 2012; Peters and Romi, 2014; Reid and Toffel, 2009: Tang and Luo, 2014, 2016).

DOI: 10.1051/shsconf/20173600015

Unlike carbon/climate change disclosure, the literature on water disclosure has received scant attention. Mohd. Remali et al. (2016) has conducted an exploratory research on the extent of water disclosure among top 10 Malaysian public listed companies based on market capitalisation and water risk profile. Overall, they found that the disclosure of such information in the annual report is still low. However, the findings are not generalisable as the sample is too small. Moreover, CDP since 2010 has conducted a survey among the world's largest companies and published the results on its website. However, it is still unclear as to what factors contributed to the decision of the companies to participate in the surveys since there is no statistical analysis conducted to test the relationships. It is of interest to examine whether factors found to be significant for CDP climate change programme in the previous literature are applicable for the water programme. Particularly relevant to this research is a study conducted by Liao et al. (2015) which examined the influence of board diversity, board independence, and the presence of environmental committee on the decision of 329 largest companies in the United Kingdom to participate in the CDP Climate Change programme for the year 2011. They found all the factors to be positive and significant.

Due to the limited attention being paid on water disclosure, this study intends to address this gap. Consistent with Liao et al. (2015) and other several studies (Depoers et al., 2016; Gonzalez-Gonzalez and Ramírez, 2016; Luo et al., 2013), we use stakeholder theory to explain the relationship among the variables in this research.

Stakeholder theory

Clarkson (1995, p. 106) defines stakeholders as "[all] persons or groups that have, or claim, ownership, rights, or interests in a corporation and its activities, past, present, or future". He then distinguishes stakeholders into primary and secondary, with primary stakeholders (including shareholders, employees, customers, suppliers, government and communities) perceived as more critical to the survival of the business. In essence, the stakeholder groups are not only different in terms of the degree of impact they may have on the business and vice versa, but also in terms of the contents and format of the environmental reports addressing their concerns, and in the criteria they use to assess the appropriateness of corporate environmental behaviour.

Stakeholder theory attempts to articulate a fundamental question in a systematic way: which groups are stakeholders deserving or requiring management attention, and which are not?" (Mitchell et al, 1997). It acknowledges the dynamic and complex relationships between organisations and their stakeholders and that these relationships involve responsibility and accountability (Gray et al., 1996). Stakeholder analysis enables identification of those societal interest groups to whom the business might be considered accountable, and therefore to whom an adequate account of its activities would be deemed necessary. In this regard, if water information is important, then stakeholders would request for such information and companies in turn would provide the information (see also O'Higgins and Morgan, 2006 and Gago and Antolìn, 2004 for further discussion on stakeholder salience).

Stakeholder theory is relevant in the context of this research since the role of directors is to ensure that efforts and resources are spent towards the best interest of the company and its shareholders while ensuring that the interests of other stakeholders are not compromised (SC,

2012). Therefore, the structure and composition of the board would determine the effective execution of this role. In essence, the more structured and diverse the board is, the greater the breadth and width of stakeholders be covered. Furthermore, previous studies have argued that the pressure faced by companies would be different across countries. Differences in the social, cultural, political, and economic settings of a country would shape the nature and intensity of stakeholders' expectations (Alrazi et al., 2016).

Hypotheses development

Following Liao et al. (2015), we examine the effect of board's gender diversity, board independence, and the presence of environmental committee on the companies' participation in CDP water programme. Furthermore, since this is an international comparative research, we include countries' legal origin and strength of law enforcement as the country variables. This is consistent with Simnett et al. (2009).

Board independence

One of the roles of board of directors is to ensure that the company's strategies promote sustainability (SC, 2012). Since independent directors do not hold any executive position in the company nor do they hold significant shareholding which will compromise their independence, it is expected that their presence on the board is to represent the interests of other stakeholders, they will have more influence on environmental reporting (Haniffa and Cooke, 2005). They will also improve the monitoring and controlling ability of the board over management (Fama and Jensen, 1983).

The extant literature found that companies with a higher proportion of independent directors tend to participate voluntarily in CDP climate change programme (Liao et al., 2015), have better environmental performance (De Villiers et al., 2011), and provide higher quality of community information. Based on these arguments, we hypothesise that,

H_1: Board independence is associated with the company's participation in the carbon disclosure project (CDP) water programme.

Board's gender diversity

Boards' gender composition is an important dimension of corporate governance because women and men are traditionally, culturally, and socially different. According to Hofstede's classification of national culture, masculinity "represents a preference in society for achievement, heroism, assertiveness and material rewards for success; while femininity stands for a preference for cooperation, modesty, caring for the weak and quality of life" (Hofstede, 2016). Gray (1988) further argued that as people are more caring and quality of life, people and the environment were given more emphasis, societies will tend to be more transparent including with socially related information.

Additionally, women are more committed, involved, and diligent (Huse and Solber, 2006) and less self-interest oriented (Coffee and Wang, 1998) that would create a good atmosphere in the

DOI: 10.1051/shsconf/20173600015

board and improve the decision making process which ultimately lead to enhance board effectiveness enhancement. Due to their roles in society, it is also argued that women more concerned than men with environmental issues (Diamantopoulos et al., 2003) and more inclined to take actions to reduce perceived environmental risks (Bord and O'Connor, 1997). This gives a positive impact on organisation's socially responsible behaviour (Barako and Brown, 2008). Liao et al. (2015) and Ben-Amar et al. (2015) found that higher representation of female directors on the board tend to voluntarily participate in the CDP climate change programme. Thus, the following hypothesis is developed:

H_2: Board's gender diversity is associated with the company's participation in the carbon disclosure project (CDP) water programme.

Environmental committee

The presence of an environmental committee enables the organisation to monitor corporate environmental policy (Vafeas and Nikolaou, 2001), oversee its ecological impacts (Bansal and Roth, 2000; McKendall et al., 1999), and monitor and manage its relationship with key stakeholders (Elijido-Ten, 2007; Kent and Chan, 2009). It is also logical to anticipate that the committee would be more likely to see the importance of environmental reporting to their public, government, and other stakeholders (Rankin et al., 2011) and thus demand accountability from management (on environmental matters) so as to protect its reputation (McKendall et al., 1999).

Prior literature found the presence of an environmental committee to be associated with higher propensity of participating in CDP climate change programme (Liao et al., 2015; Peters and Romi, 2014) and more extensive and greater quality environmental reporting (Kent and Chan, 2009) and human resource reporting (Cowen et al., 1987) Therefore, the following hypothesis is developed:

H_3: Environmental committee is associated with the company's participation in the carbon disclosure project (CDP) water programme.

Legal system

The way businesses are being controlled and managed is partly determined by a country's legal system. This, in turn, defines the company-stakeholders relationship (Van der Laan Smith et al., 2005) and shapes the business culture (Simnett et al., 2009). Business culture can be broadly divided into stakeholder or shareholder orientated. In a stakeholder-orientated culture, a broad spectrum of stakeholders is assumed to have legitimate interests and influence on corporate activities (Simnett et al., 2009). Companies are considered to have social responsibility that goes beyond achieving economic efficiency (Kolk & Perego, 2010; Van der Laan Smith et al., 2005). As such, corporate disclosure needs to consider the information needs of stakeholders other than shareholders. The nature of information can be both financial and non-financial. However, in a shareholder-orientated business culture, the influence of stakeholder groups, other than shareholders, is minimal, as the main purpose of a company is seen as shareholder wealth maximisation (Kolk & Perego, 2010; Simnett et al., 2009; Van der Laan Smith et al., 2005).

Following prior literature, it is assumed that companies domiciled in code law countries are more stakeholder-orientated, while those domiciled in common law countries are more shareholder-orientated. Water, in general, is a stakeholder issue and at present information on consumption and discharges is not subjected to any disclosure requirement or accounting standard. Hence, companies publish it on a voluntary basis to satisfy the information needs of the stakeholders. Consistent with stakeholder theory and the findings of Kolk and Perego (2010) and Simnett et al. (2009) who found significant influence of countries' legal system on sustainability reporting, it is hypothesise that;

H_4: Countries' legal system is associated with the company's participation in the carbon disclosure project (CDP) water programme.

Strength of law enforcement

As stated earlier, environmental disclosure has remained largely voluntary across the world. Mandatory reporting, if any, usually pertains to a single issue with limited disclosure requirements (KPMG et al., 2015). Nevertheless, Walden and Schwartz (1997) argue that the degree of public pressure (for environmental responsibility) is greater in an environment where there is a greater likelihood of additional legislations and sanctions, increased regulatory oversight, and dissatisfaction of the general public or a group of public. Here, regulators, customers, and non-governmental organisations promoting sustainability become the key stakeholders.

Prior literature has also found that the extent of regulatory enforcement has also been associated with lower incidence of corporate environmental lawsuits (Kassinis and Vafeas, 2002), company decisions to have sustainability reports assured by a third party organisation (Simnett et al., 2009), and voluntary participation in the CDP climate change programme (Reid and Toffel, 2009). Consistent with the argument of stakeholder theory, it is expected that companies domiciled in countries with a strong law enforcement environment will be more likely to voluntarily participate in the CDP water programme. Thus, we hypothesise that;

H_5: Countries' strength of law enforcement is associated with the company's participation in the carbon disclosure project (CDP) water programme.

RESEARCH METHODS

Population and sample

The population of this research consists of all companies invited to participate in the 2015 CDP water programme. According to CDP (2015), a total of 1,073 world's largest public listed companies were invited to participate in its survey. The companies were deemed to have high water vulnerability hence expected to provide disclosure as to how they manage risks, and seize opportunities, related to water. A further assessment was made on these companies to ensure the availability of data for the purpose of testing the hypotheses. The final sample comprises 748 companies from 42 different countries. Other companies (i.e. 325) were removed from the sample due to unavailability of complete and/or English-based annual reports and there was no other means of obtaining the required data. Although not tabulated here, the majority of the sample are from US (26.3%), Japan (15%), UK (8.7%), Australia (6.6%), and South Africa (5.9%). Additionally, materials (23%), consumer directory (17%), and industrials (14%) represent the industries with the highest representatives in the sample.

Data collection methods

Data were collected using the secondary method. The dependent variable is represented by the decision of companies to participate in the 2015 CDP water programme which is extracted from CDP (2015). Data for board characteristics were collected from the annual reports for the year 2014 or the company websites. In regards to country of origin, data were collected from the CIA (2016) and World Bank (2016). The classification of industry is based on CDP (2015).

Measurement of variables

Table 1 below summarises the measurement of variables used in this research.

No	Variables	Measurement	Source
1	CDP_{WATER}	1 = if the company responded to the CDP water programme survey; 0 = if otherwise	CDP (2015)
2	*BIND*	The percentage of independent non-executive directors out of total directors	Annual reports
3	*BDIV*	The percentage of female directors out of total directors	Annual reports
4	*BCOMM*	1= if there is board committee for environmental issues; 0 if otherwise	Annual reports
5	*LEGAL*	1 = companies domiciled in code law countries; 0 = companies domiciled in common law countries	CIA (2016)
6	*LAW*	Countries' rule of law	World Bank (2016)
7	*IND*	1 = Companies operating in high water intensive industries; 0 if otherwise	CDP (2015); Ceres (2011)

Table 1: Measurement of variables

Regression model

Hence, the regression model for this research is as follows:
$CDP_{WATER} = \beta_0 + \beta_1 BIND + \beta_2 BDIV + \beta_3 BCOMM + \beta_4 LEGAL + \beta_5 LAW + \beta_6 IND + \varepsilon$

The definition for each variable is available in Table 1. Logistic regression is used to test the hypotheses.

FINDINGS AND ANALYSIS

This section presents the findings from the descriptive statistics, correlation analysis, and logistic regression analysis.

DOI: 10.1051/shsconf/20173600015

Descriptive statistics

Table 2 depicts the descriptive statistics for the variables used in estimation. Based on the table, 44 percent of the sample companies participated in the CDP water programme. Additionally, about one-half of the directors in the sample companies are independent non-executive directors (mean *BIND* = 46%). However, female directors only constituted 16% of the directors. The majority of the companies did not have any specific board level committee to monitor their environmental performance (*BCOMM* = 37%).

Panel A: Descriptive statistics of continuous variables used in estimation

Variable	Mean	Median	Q1	Q3	Std dev
BIND	46.20	46.41	20.36	71.07	29.115
BDIV	16.39	16.67	7.14	25.00	13.154
LAW	1.41	1.62	1.47	1.89	0.669

Panel B: Distribution of categorical variables

	1		0	
Variable	*N*	%	*N*	%
CDP_{WATER}	329	44.0	419	56.0
BCOMM	278	37.2	470	62.8
LEGAL	392	52.4	356	47.6
IND	143	19.1	605	80.9

This table presents descriptive statistics of the dependent and independent variables used in correlation and regression analyses. Statistics are presented for the full sample of 748 companies. All variables are defined in Table 1.

Table 2: Descriptive statistics of continuous variables

In terms of country of origin, there is a balanced composition between companies domiciled in code law countries and common law countries (with 52% and 48%, respectively) and, on average, they are from a strong law enforcement environment in which the average *LAW* of 1.41 is higher than the world's average of 0.00 (World Bank, 2016). Finally, about 20% of the sample companies were operating in the high water intensity industries. In this research, following the classification by Ceres (2011), utilities and energy industries are regarded as high water risk industries.

Correlation analysis

Correlation analysis was conducted to determine whether there is a multicollinearity problem that could influence the regression analysis. Table 3 provides the results of the correlation statistics with Pearson above the diagonal and Spearman's rho, below. This is a two-tailed test of significance. There appears to be no serious multicollinearity issues among the independent variables. According to Field (2013), a coefficient correlation of greater than, or equal to, (≥) 0.80 indicates there is cause for concern. Based on the table below, the highest coefficient is -0.461 (p = 0.000) which is between *LEGAL* and *LAW*.

	CDP_{WATER}	*BIND*	*BDIV*	*BCOMM*	*LEGAL*	*LAW*	*IND*
CDP_{WATER}	-	0.053	0.068	0.166**	0.062	0.039	-0.129**
BIND	0.053	-	0.085*	0.003	-0.207**	0.108**	0.193**
BDIV	0.073*	0.085*	-	0.018	-0.155**	0.090*	-0.044
BCOMM	0.166**	0.003	0.030	-	0.007	-0.023	0.027
LEGAL	0.062	-0.215**	-0.205**	0.007	-	-0.391**	-0.081*
LAW	-0.021	0.229**	0.205**	-0.066	-0.461**	-	-0.177**
IND	-0.129**	0.192**	-0.037	0.027	-0.081*	-0.086*	-

This table presents the correlation statistics for the dependent and independent variables used in multivariate analyses. Statistics are presented for the full sample of 748 companies. All variables are defined in Table 1. * and ** represent significance levels (two-tailed) at 5 percent and 1 percent, respectively.

Table 3: Correlation statistics for dependent and independent variables

Regression analysis

The results of the logistic regression are presented in Table 4. The reported pseudo- R^2, namely Cox & Snell R^2 and Nagelkerke R^2 are 0.062 and 0.083, respectively. Based the table, only *BIND* shows finding that is consistent with the hypothesis. It indicates that companies with higher proportion of independent directors had greater proclivity to respond to the CDP water programme. As argued earlier, the presence of independent directors on the board is as a proxy for the interest of stakeholders other than shareholders. *BCOMM* is significant at 0.01 level, however in the opposite direction. This is inconsistent with the findings of Liao et al. (2015) and Peters and Romi (2014) who found that the presence of environmental committee increases the likelihood companies participating in CDP climate change programme.

In terms of country of origin, *LEGAL* is significant (p=0.015). The negative coefficient is suggesting that companies from the common law countries tend to participate in the CDP water programme. Even though it is code law countries that are argued to be stakeholder-oriented, the fact that the raison d'etre of CDP establishment is to serve investors (CDP, 2015) may have induced greater pressure for companies from the common law countries to be more forthcoming in this issue. This is consistent with the findings of Luo et al. (2012), Luo et al. (2013), and Graul and Gotthardt (2016).

Variables	Expected sign	β	Wald Statistics	Sig.
BIND	+	0.06	5.583	0.018*
BDIV	+	0.10	2.962	0.085
BCOMM	+	-0.733	21.774	0.000**
LEGAL	+	-0.414	5.901	0.015*
LAW	+	0.165	1.676	0.195
IND	+	0.758	12.973	0.000**
Constant		-0.913	9.492	0.002**
Observations		748		
Likelihood		978.116		
Cox & Snell R^2		0.062		
Nagelkerke R^2		0.083		

All variables are defined in Table 1. * and ** represent significance levels (two-tailed) at 5 percent and 1 percent, respectively.

Table 4: Logistic regression analysis

Consistent with the prediction, *IND* is positive and significant (p=0.002) implying that companies operating in the high water intensive industries have greater proclivity to engage in water reporting. This is consistent with Luo et al. (2012). Additional analysis was also conducted to re-classify the industries to include industrials, materials, consumer directory, and consumer staples as high water intensive (they are considered 'medium' in CDP, 2015). Although not tabulated here, *IND* remains positive and significant (p=0.000).

BDIV and *LAW* are not significant, despite the signs show the predicted direction. Gender of the board members do not influence the companies' decision on CDP water programme. This is inconsistent with the findings of Ben-Amar et al. (2015). As shown in Table 2, the average of *BDIV* is 16%, indicating that male directors still dominate the board. The result for *LAW* suggests that the strength of law enforcement in a country does not have significant influence on companies' decision to participate in CDP water programme. This is consistent with Alrazi et al. (2016).

CONCLUSION

This study examines the propensity of 748 global largest companies to respond to CDP water programme in year 2014. The number of responding companies is considered high with 44%. Using stakeholder theory as the underlying framework, we found that companies that tend to participate in the survey have greater proportion of independent directors and no specific board level committee on environmental issues, domiciled in common law countries, and operated in the high water sensitive industries.

There are several limitations inherent in this study. First, the participation in the CDP water programme as the measure for dependent variable might be too crude. It does not capture the quantity and quality of the information being disclosed in such a survey. Additionally, we did not consider cases in which companies participated but did not allow the survey to be made publicly available on the CDP website. Hence, future studies may want to consider these issues (see Ben-Amar and McIlkenny, 2015).

Second, we only examined three board characteristics and two country of origin variables. As evidenced by the R^2 values, they are low suggesting that there are other variables that could explain the variation in reporting decision. Liao et al. (2014) for example included board size, board meeting, CEO compensation, and directors' ownership. Ben-Amar and McIlkenny (2015) used overall board effectiveness as a proxy for board characteristic (see also Rankin et al., 2011). Country factors that can be tested include environmental performance/regulations (Alrazi et al., 2016; Graul and Gotthardt, 2016; Luo et al., 2012), culture (Luo and Tang, 2016), and economic development (Luo et al., 2013).

Moreover, we only examined industry as company characteristic but not company size, multinational orientation (Graul and Gotthardt, 2016), financial risk, listing status, ownership concentration (Gonzalez-Gonzalez and Ramírez, 2016), degree of resource availability (Luo et al., 2013), and internal organisation system factors (Rankin et al., 2011).

REFERENCES:

Alrazi, B., De Villiers, C., & Van Staden, C.J., (2016), The environmental disclosures of the electricity generation industry: a global perspective, Accounting and Business Research, 46(6), 665-701.

Ascui, F. (2014). A review of carbon accounting in the social and environmental accounting literature: what can it contribute to the debate? Social and Environmental Accountability Journal, 34(1), 6-28. DOI: 10.1080/0969160X.2013.870487

Bansal, P. & Roth, K. (2000), Why companies go green: A model of ecological responsiveness, Academy of Management Journal, 43(4), 717-736. DOI: 10.2307/1556363

Ben-Amar, W. & McIlkenny, P. (2015). Board effectiveness and the voluntary disclosure of climate change information, Business Strategy and the Environment, 24(8), 709-714. DOI: 10.1002/bse.1840

Ben-Amar, W., Chang, M. & McIlkenny, P. (2015), Board gender diversity and corporate response to sustainability initiatives: Evidence from the Carbon Disclosure Project, Journal of Business Ethics, July, 1-15. DOI: 10.1007/s10551-015-2759-1

CDP (2015). Accelerating action: CDP Global Water Report 2015, Carbon Disclosure Project: London. Retrieved from: https://www.cdp.net/CDPResults/CDP-Global-Water-Report-2015.pdf

Ceres (2011). The Ceres Aqua Gauge: A Framework for 21st Century Water Risk Management: Retrieved from: https://www.ceres.org/resources/reports/aqua-gauge

CIA (2016). The World Factbook: Legal System. Central Intelligence Agency: Washington. Retrieved from: https://www.cia.gov/library/publications/the-world-factbook/fields/2100.html

Clarkson, M.B.E., (1995), A stakeholder framework for analyzing and evaluating corporate social performance, Academy of Management Review, 20(1), 92-117. DOI: http://www.jstor.org/stable/258888

Cowen, S. S., Ferreri, L. B., & Parker, L. D. (1987), The impact of corporate characteristics on social responsibility disclosure: A typology and frequency-based analysis, Accounting, Organizations and Society, 12(2), 111-122. DOI: 10.1016/0361-3682(87)90001-8

De Villiers, C., Naiker, V., & Van Staden, C.J., (2011), The effect of board characteristics on firm environmental performance, Journal of Management, 37(6), 1636-1663. DOI: 0.1177/014920631141150

Depoers, F., Jeanjean, T. & Jérôme, T. (2016), Voluntary disclosure of greenhouse gas emissions: Contrasting the Carbon Disclosure Project and corporate reports, Journal of Business Ethics, 134(3), 445-461. DOI: 10.1007/s10551-014-2432-0

Elijido-Ten, E. (2007), Applying stakeholder theory to analyze corporate environmental performance: Evidence from Australian listed companies, Asian Review of Accounting, 15(2), 164 – 184. DOI: 10.1108/13217340710823378

Fama, E.F. & Jensen, M.C., (1983), Separation of ownership and control, The Journal of Law & Economics, 26(2), 301-325. DOI: http://www.jstor.org/stable/725104

Field, A. (2013), Discovering Statistics using IBM SPSS Statistics (4th ed.), SAGE Publications: London.

Gago, R.F., & Antolìn, M.N., (2004), Stakeholder salience in corporate environmental strategy, Corporate Governance, 4 (3), 65-76. DOI: 10.1108/14720700410547512.

Gonzalez-Gonzalez, J.M. & Ramírez, C.Z. (2016), Voluntary carbon disclosure by Spanish companies: an empirical analysis, International Journal of Climate Change Strategies and Management, 8(1), 57 - 79. DOI: 10.1108/IJCCSM-09-2014-0114

Graul, J. & Gotthardt, D. (2016). The relevance of national contexts for carbon disclosure decisions of stock-listed companies: a multilevel analysis, Journal of Cleaner Production, 133, 1204-1217. DOI: 10.1016/j.jclepro.2016.05.182

Gray, R., Owen, D., & Adams, C., (1996). Accounting & Accountability - Changes and Challenges in Corporate Social and Environmental Reporting. Prentice Hall Europe: Hertfordshire.

Gray, S. (1988). Towards a theory of cultural influence on the development of accounting systems internationally, ABACUS, 24(1), 1-15. DOI: 10.1111/j.1467-6281.1988.tb00200.x

GRI (2013). G4 Sustainability Reporting Guidelines: Reporting Principles and Standard Disclosures, Global Reporting Initiative: Amsterdam. Retrieved from: https://www.globalreporting.org/resourcelibrary/GRIG4-Part1-Reporting-Principles-and-Standard-Disclosures.pdf

Hahn, R. & Kühnen, M. (2013), Determinants of sustainability reporting: a review of results, trends, theory, and opportunities in an expanding field of research, Journal of Cleaner Production, 59, 5-21. DOI: 10.1016/j.jclepro.2013.07.005

Hahn, R., Reimsbach, D. & Schiemann, F. (2015). Organizations, climate change, and transparency: Reviewing the literature on carbon disclosure, Organization & Environment, 28(1), 80-102. DOI: 10.1177/108602661557554

Haniffa, R.M. & Cooke, T.M. (2006). The impact of culture and governance on corporate social reporting, Journal of Accounting and Public Policy, 24(5), 391-430. DOI: 10.1016/j.jaccpubpol.2005.06.001

Hofstede, G. (2016), National culture, itim International: Helsinki. Retrieved from: https://geert-hofstede.com/national-culture.html

Kassinis, G., & Vafeas, N. (2002), Corporate boards and outside stakeholders as determinants of environmental litigation, Strategic Management Journal, 23, 399-415. DOI:10.1002/smj.230

Kent, P., & Chan, C. (2009), Application of stakeholder theory to corporate environmental disclosures, Corporate Ownership & Control, 7(1), 399-414. DOI: http://epublications.bond.edu.au/business_pubs/258/

Kolk, A., & Perego, P. (2010), Determinants of the adoption of sustainability assurance statements: an international investigation, Business Strategy and the Environment, 19, 182-198. DOI:10.1002/bse.643

KPMG (2015). The KPMG Survey of Corporate Responsibility Reporting 2015. KPMG International Cooperative: Netherlands. Retrieved from: https://assets.kpmg.com/content/dam/kpmg/pdf/2015/11/kpmg-international-survey-of-corporate-responsibility-reporting-2015.pdf

KPMG, GRI, UNEP, & USB (2016). Carrots & Sticks: Global Trends in Sustainability Reporting Regulation and Policy. Retrieved from: https://assets.kpmg.com/content/dam/kpmg/pdf/2016/05/carrots-and-sticks-may-2016.pdf

Li, D., Huang, M., Ren, S., Chen, X., & Ning, L. (2016). Environmental legitimacy, green innovation, and corporate carbon disclosure: Evidence from CDP China 100, Journal of Business Ethics, May, 1-16. DOI: 10.1007/s10551-016-3187-6

Liao, L., Luo, L., & Tang, Q. (2015), Gender diversity, board independence, environmental committee, and greenhouse gas disclosure, The British Accounting Review, 47(4), 409-424. DOI: 10.1016/j.bar.2014.01.002

Luo L., Lan Y.C., & Tang, Q., (2012), Corporate incentives to disclose carbon information: Evidence from the CDP Global 500 Report, Journal of International Financial Management and Accounting, 23, 93-120. DOI: 10.1111/j.1467-646X.2012.01055.x

Luo, L. & Tang, Q. (2016). Determinants of the quality of corporate carbon management systems: An international study, The International Journal of Accounting, 51(2), 275-305. DOI: 10.1016/j.intacc.2016.04.007

Luo, L., Tang, Q., & Lan, Y.C., (2013). Comparison of propensity for carbon disclosure between developing and developed countries: A resource constraint perspective, Accounting Research Journal, 26(1), 6-34. DOI: 10.1108/ARJ-04-2012-0024

Luo, L.L. & Tang, Q. (2016), Does national culture influence corporate carbon disclosure propensity?, Journal of International Accounting Research: Spring, 15(1), 17-47. DOI: 10.2308/jiar-51131

McKendall, M., Sanchez, C. & Sicilian, P., (1999), Corporate governance and corporate illegality: The effects of board structure on environmental violations, The International Journal of Organizational Analysis, 7(3), 201-223. DOI: 10.1108/eb028900

Mitchell, R.K., Agle, B.R., and Wood, D.J., (1997). Toward a theory of stakeholder identification and salience: defining the principle of who and what really counts, Academy of Management Review, 22 (4), 853-886. DOI: http://www.jstor.org/stable/259247

Mohd. Remali, A.R., Mohd. Ali, I., Mat Husin, N., & Alrazi, B., (2016), An exploratory study on water reporting among top Malaysian public listed companies, Procedia Economics and Finance, 35, 64-73. DOI: 10.1016/S2212-5671(16)00010-1

O'Higgins, E.R.E. & Morgan, J.W. (2006), Stakeholder salience and engagement in political organisations: Who and what really counts? Society and Business Review, 1(1), 62-76. DOI: 10.1108/17465680610643355

Peters, G.F. & Romi, A.E. (2014). Does the voluntary adoption of corporate governance mechanisms improve environmental risk disclosures? Evidence from greenhouse gas emission accounting. Journal of Business Ethics, 125(4), 637-666. DOI: 10.1007/s10551-013-1886-9

Rankin, M., Windsor, C., & Wahyuni, D., (2011). An investigation of voluntary corporate greenhouse gas emissions reporting in a market governance system: Australian evidence, Accounting, Auditing & Accountability Journal, 24 (8), 1037-1070. DOI: 10.1108/09513571111184751

Reid, E.M. & Toffel, M.W. (2009), Responding to public and private politics: Corporate disclosure of climate change strategies, Strategic Management Journal, 30, 1157-1178. DOI: 10.1002/smj.796

SC (2012). Malaysian Code on Corporate Governance. Securities Commission: Kuala Lumpur. Retrieved from: https://www.sc.com.my/wp-content/uploads/eng/html/cg/cg2012.pdf

Simnett, R., Vanstraelen, A., & Chua, W. F. (2009). Assurance on sustainability reports: An international comparison. The Accounting Review, 84(3), 937-967. DOI: 10.2308/accr.2009.84.3.937

Stechemesser, K. & Guenther, E. (2012). Carbon accounting: a systematic literature review, Journal of Cleaner Production, 36, 17-38. DOI: 10.1016/j.jclepro.2012.02.021

Vafeas, N. & Nikolaou, V. (2001). The association between corporate environmental and financial performance, Advances in Public Interest Accounting, 8, 195-214. DOI: 10.1016/S1041-7060(01)08010-5

Van der Laan Smith, J., Adhikari, A., & Tondkar, R. H. (2005), Exploring differences in social disclosures internationally: A stakeholder perspective, Journal of Accounting and Public Policy, 24, 123-151. DOI:10.1016/j.jaccpubpol.2004.12.007

WEF (2015). Global Risk 2015, 10th Edition, World Economic Forum: Geneva. Retrieved from: http://www3.weforum.org/docs/WEF_Global_Risks_2015_Report15.pdf

World Bank (2016). Worldwide Governance Indicators. The World Bank Group: Washington. Retrieved from: http://info.worldbank.org/governance/wgi/index.aspx#home

PREDICTING THE PERFORMANCE AND SURVIVAL OF ISLAMIC BANKS IN MALAYSIA TO ACHIEVE GROWTH SUSTAINABILITY

Noraina Mazuin Sapuan, Suzaida Bakar, Hamidah Ramlan*
Department of Finance and Economics, Universiti Tenaga Nasional, Muadzam Shah, Pahang, Malaysia

ABSTRACT

In Malaysia, the growth of the Islamic financial industry has increased tremendously in line with the Government's ambition to make Malaysia as an international hub for Islamic finance since 2010. With the increasing number of foreign players in this industry plus with the increasing demand from domestic and foreign customers would further enhance the possibility for Malaysia to achieve this ambition. Currently, according to the Economic Transformation Programme, 2012 Malaysia is the world's third largest market for Shariah assets that cover Islamic banks, Takaful, and sukuk. Malaysia as one of the main contributors to the global Islamic financial assets with Islamic assets in Malaysia grew by 23.8% in 2011 from RM350.8bil to RM434.6bil. The issues of predicting the performance and the survival of Islamic Banks in Malaysia become amongst crucial issues in academic research. By employing multi – layer perceptron neural network and pooled regression, we found that total assets/ size of the Islamic banks (GROWTH) have high weightage and significantly influence in predicting the performance and the survival of Islamic banks in Malaysia. With the increasing number of Islamic banking institutions in Malaysia, this study can give insight on the sustainability of the Islamic banking system in Malaysia for the benefit of the investors, shareholder and depositors.

Keywords: Performance, Survival, Islamic Banks, Multi – layer Perceptron Neural Network and Pooled Regression and Growth Sustainability

INTRODUCTION

The Islamic financial industry has grown tremendously since it first emerged in the 1970's. Globally, Islamic finance assets grew at double-digit rates in the past decade to reach an estimated $1.8 trillion at the end of 2013, with further growth expected (Ernst & Young, 2014; IFSB, 2014). This growth reflects demand from large and relatively Muslim and non-Muslim populations seeking to deposit money or invest in sharia-compliant banks and financial products—those that are acceptable under Islamic law. Furthermore, by emphasizing the need for transactions to be supported by genuine trade or business related activities, Islamic banking sets a higher standard for investments and promotes greater accountability and risk mitigation.

*Corresponding author's email: Hamidah@uniten.edu.my

Basically, with the rapid growth of Islamic finance industry worldwide, especially Islamic banking industry, many studies has studied the importance of this industry. The earliest study on Islamic banking system is done by Nienhaus (1983) examined the usage of profit loss sharing (PLS) instruments in Islamic bank and the alternative method to measure bank profit through profit sharing ratio. Meanwhile, Samad and Hassan (2000) are the pioneer of the study of determinants of Islamic bank's profitability which is linked to the efficiency and profitability of Islamic banks.

In terms of bank profitability, previous studies have utilized the income before tax and income after tax over total assets, total equity and total deposits calculate the profitability ratio, namely ROA, ROE and ROD as well as bank net margin as the dependant variables. Especially on the determinants of Islamic bank's profitability.

Most of the previous researcher concentrates their study on the determinants of conventional banks' profitability in either cross-country or individual countries' banking system. Only a few similar studies are carried out, particularly for Islamic bank. The earliest study on the determinants of Islamic bank's profitability which is linked to the efficiency and profitability of Islamic banks is done by Samad and Hassan (2000). With the increasing demand of the Islamic banks instruments' worldwide, especially in the 1990s, many researchers have taken the initiative to do the same research in this field. How et al. (2005) and Rosly and Abu Bakar (2003) focus on the performance of Islamic banks in Malaysia, while Turen (1995) on Bahrain Islamic Bank and Bashir (2000) on Middle Eastern countries. bank profitability is determined by internal and external factors. Internal factors can be viewed as factors that can be controlled by the bank management and can be also classified as bank specific variables such as bank size, leverage, assets and liability portfolio mix, overhead expenses, liquidity ratio, capital ratio, ownership and capital adequacy. Meanwhile, the external factors could be expressed as variables which are not associated with bank management. Basically, external factors can be divided into two types, i.e. industry specific determinants and macroeconomic variables.Among the widely discussed industry specific variables are competition, regulation, concentration, market share, ownership and scarcity of capital. As far as the internal factors are considered, Bashir (2000), Hassan and Bashir (2003) and Sanusi and Mohammed (2004) discovered that Islamic banking profitability responds positively to the increases in capital ratios. The results revealed that larger equity to total asset ratio leads to more profit margins. In terms of external factors, Bourke (1989) found that concentration has moderate and positive related to pre-tax returns on assets. Molyneux and Thornton (1992) further extended Bourke's methodology and examine bank profitability in eighteen European countries between 1986 and 1989. There is a significant positive relationship between return on capital and concentration and a positive relationship for nominal interest rates. Concentration shows a positive, statistically significant correlation with pre-tax return on assets. Athanasoglou et al. (2008) found that industry-specific i.e. ownership and concentration are insignificant in explaining profitability. Meanwhile, the latest studies done by Sapuan & Roly (2015) shown that bank profitability, bank specifics variables and financial structure are cointegrated in long run.

In view of the above, this paper will discuss on the performance of the Malaysian Islamic bankings and its survival in the future especially with the increasing competition from internaltional participants whether in Malaysia or from foreign market.

*Corresponding author's email: Hamidah@uniten.edu.my

Empirical studies on the prediction in finance had utilize various methods of soft computing system such as multi layered feed forward neural network trained with backpropagation (MLFF-BP), a probabilistic neural network (PNN), multivariate discriminant analysis (MDA) and support vector machine (SVM)(Ravi et al, 2007). Tam and Kiang (1992), Gunay and Ozkan (2007) and Celik and Karatepe (2007) have used neural networks to predict bank failure and bankruptcy.

Meanwhile, Odom and Sharda (1990), Koster et al. (1990), Coats and Fant (1993), Tsukuda and Baba (1994), and Altman et al. (1994) have used the similar tool to predict company bankruptcy. Dutta and Shekhar (1988) and Surkan and Singleton (1990) predict the ratings of corporate bonds and their expected profitability. Yoon and Swales (1990) predicts stock price performance. White (1988), Kamijo and Tanigawa (1990), and Kryzanwski et al. (1993) predict common stock returns. Mitchell and Pavur (2002), Handzic et al. (2003), Molharta and Molharta (2003) and Eletter and Yassen (2010) on loan application decision.

Tam and Kiang (1992) examined bank bankruptcy prediction using ANN and compare it with linear classifier, logistic regressions, kNN and ID3. The data sample consists of Texas banks that failed in the period 1985-1987. The data are comprised of bank data one year and two years prior to failure. As a control measure, a failed bank was matched with a non-failed bank in terms of (1) asset size, (2) number of branches, (3) age, and (4) charter status. In each period, 118 banks (59 failed and 59 non-failed) were selected as the training sample. Each bank is described by 19 financial ratios that have been used in previous studies. The empirical results show that neural network is a promising method of evaluating bank conditions in terms of predictive accuracy, adaptability, and robustness compared to other forecasting method.

In terms of prediction on bank performance for Islamic banks, Al-Osaimy (1998) used a data sample of twenty six Islamic banks worldwide has been collected for the period 1991-1993. Kohonen neural network was used to group the Islamic banks into high and low performance groups using the seven financial ratios. Twelve banks have been indicated as a high performance group and fourteen banks as a low performance group. The three layers Backpropagation neural network was built to predict the Islamic bank's performance. The network was trained on the one year prior to the performance year data set while using the performance year data as a test set. Meanwhile, the two years prior to the performance year data set was used as a holdout sample to check the network prediction abilities. When the network was applied, the results showed high prediction abilities and show a significant method as a useful early warning signal for various types of performance whether bankruptcy, insolvency or failure.

Additionally, Abu Bakar and Mohd Tahir (2009) further extend the study by Al-Osaimy (1998) by making the prediction the performance of Islamic banks in Malaysia base on thirteen banks from 2001 to 2006 by comparing multiple linear regressions (MLR) and artificial neural network (ANN) in order to search for accurate forecasting results. ROA was used to measure bank performance, meanwhile seven variables, including liquidity, credit risk, cost to income ratio, size, concentration ratio, inflation and GDP were used as

*Corresponding author's email: Hamidah@uniten.edu.my

 DOI: 10.1051/shsconf/20173600016

independent variables. The prediction of bank performance is measured using the mean square prediction error and the R-squared value which indicates the proportion of variation explained by the validation model. The study concludes that ANN gives extremely accurate results and outperformed the MLR in predicting bank performance.

Meanwhile, Gunay and Ozkan (2007) and Celik and Karatepe (2007) studied on prediction of bank failure in the Turkish banking sector using ANN. The results of both studies are in line with the finding by Tam and Kiang (1992) and Abu Bakar and Mohd Tahir (2009) whereby ANN is the best method to predict bank crises and can become an alternative early warning system that can be used along with other common method used in measuring bank performance such as financial ratio, comprehensive bank assessment and econometrics model.

RESEARCH DESIGN

The data were collected from 16 Islamic banking annual reports from the period of 2005 to 2014. The selected banks were Islamic commercial banks and the commercial banks that have been classified under the Islamic banking scheme and whereby the operation recently been categorized as subsidiaries Islamic banking. All related data were collected from the IFS database. For the data analysis, we employed supervised machine learning neural network method (Odom and Sharda, 1990). Multilayer perceptron algorithm has been used in order to detect the highest possible predictors that influence the performance and the survival of the Islamic banks in Malaysia. One hidden layer have been use and the activation function for input layer is hyperbolic tangent. In order to see the direction of the relationship between the inputs (predictors) and output (ROA), pooled regression analysis (ordinary least square) have been employed.

Dependence Variables

The endogenous factor of this study is bank profitability that measured through the balance sheet ratios that represented by Return on Assets (ROA) as a measurement of asset utilization.

ROA = Income after tax or zakah /total assets.

Independence Variables

We identified the exogenous factors as bank characteristics that consist of variables as listed as follows:

GROWTH = log of total asset
Credit risk (LTA)= total loans / total assets
Bank efficiency (NIE)= Non interest income /total assets

Estimation Model

The regression model can be specified as follows:

*Corresponding author's email: Hamidah@uniten.edu.my

$$ROA_{it} = \alpha_0 + \alpha_1 GROWTH_{it} + \alpha_2 LTA_{it} + \alpha_3 NIETA_{it} + \varepsilon_{it} \quad (1)$$

Where;

ROA= bank profitability
GROWTH = bank size represents by log of total assets,
LTA = ratio between total loan over total assets,(credit risk)
NIETA = ratio between non-interest earning over total assets (Bank efficiency)
ε = the error term.

In machine learning, artificial neural networks (ANNs) are a family of statistical learning algorithms inspired by biological neural networks (the central nervous systems of animals, in particular the brain) and are used to estimate or approximate functions that can depend on a large number of inputs and are generally unknown. Artificial neural networks are generally presented as systems of interconnected "neurons" which can compute values from inputs, and are capable of machine learning as well as pattern recognition thanks to their adaptive nature.

Like other machine learning methods - systems that learn from data - neural networks have been used to solve a wide variety of tasks that are hard to solve using ordinary rule-based programming, including computer vision and speech recognition.

IBM SPSS Neural Networks 19 has been used to run the data. Below are an architecture or framework for this study in neural network architecture:

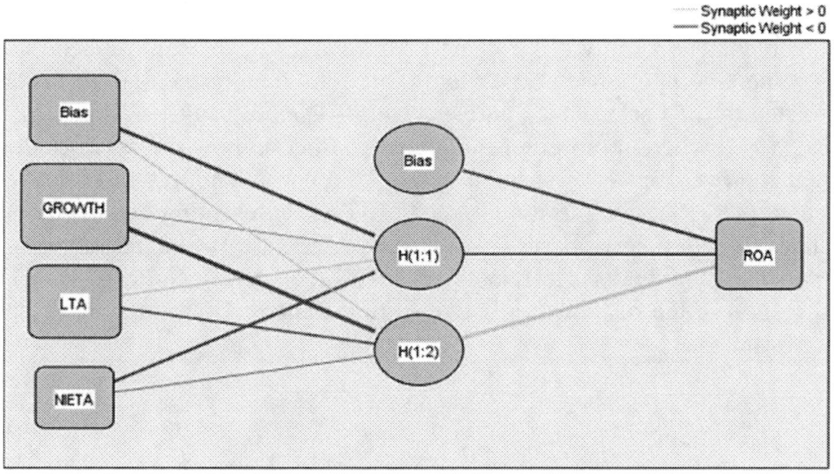

Hidden layer activation function: Hyperbolic tangent

Output layer activation function: Hyperbolic tangent

FINDINGS

Neural Network Analysis

*Corresponding author's email: Hamidah@uniten.edu.my

Table 1 Case processing summary, shows that from the total of 112 cases, 82 cases were assigned to the training sample and 30 to the testing (holdout) sample and no cases were excluded from the analysis.

Table 1: Case processing summary

Sample	N	Percent
Training	82	73.2%
Testing	30	26.8%
Valid	112	100.0%
Excluded	0	
Total	112	

Based on Table 2, the importance of an independent variables as measured of how much the network's model-predicted value changes for different values of the independent variable. Normalized importance is simply the importance values divided by the largest importance values and expressed as percentages.

Table 2: Independent Variable Importance

Input	Importance	Normalized Importance
Log Total Asset(GROWTH)	0.565	100.0 %
Loans to Total Asset(LTA)	0.328	58.1 %
Non Interest Earning to Total Asset(NIETA)	0.107	18.9 %

The importance chart is simply a bar chart of the values in the importance table, sorted in descending value of importance. Based on Figure II, it appears that the most important predictors related to the performance and survival of Islamic banks in Malaysia is the bank growth (GROWTH) where it represent the greatest effect on how the network classifies bank survival; then followed by loans to total asset (LTA) which represent the credit risks and finally the non interest earning to total assets(NIETA) representing bank efficiency; but one of the drawback of neural network it cannot tell the "direction" of the relationship between all predictors and the predicted performance or survival of the Islamic banks. Therefore, to reveal the direction of the relationship between inputs and output, pooled regression analysis has been employed.

*Corresponding author's email: Hamidah@uniten.edu.my

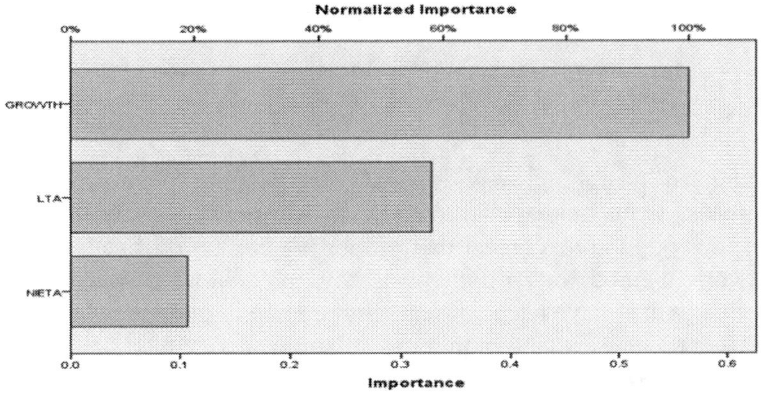

Figure 2. Importance Chart

Pooled Regression Analysis

In Table 3, the results indicates a negative significant relation between GROWTH and credit risks with the performance and the survival of Islamic banks in Malaysia. Theoretically, it is confirmed that Islamic banks will find difficulties when the loan loss provision(credit risks) increases where it will affects earnings before zakat and depreciation. When banks in this condition for two consecutive years (where the EBZDA is negative) (Asquith, 1994), these particular bank will be considered as financially distressed banks. In this case, there is possibility the bank will plan for merger and acquisition with others dominance banks; as happened during Asian financial crisis in 1997. From our findings, it shows that GROWTH which represent by total assets of the banks have negative significant influence with the performance and the survival of the Islamic banks. If Islamic banks have so many branches it might increase the operating cost and it will reduce the profits of the Islamic banks; where it might contribute to the deterioration in Islamic bank growth sustainability. The findings is consistent with the real situation in banking industry nowadays such for example the closing of some CIMB branches all over Malaysia, it is due to the high cost in operating expenses and it will reduce the profits of the banks. The arguments are consistent with Yap et.al (2011) and Alifiah et.al (2013), which shows that the likelihood of firms' financial distress is linked with working capital, cash flow of the firms, the levels firms' debt and not with efficiency ratio. Our findings also consistent with Abu Bakar and Mohd Tahir (2009),

Table 3: Pooled Regression analysis

DV	**ROA**	
IV	β	t
Log Total Asset(GROWTH)	-0.6142***	-2.7491
Loans to Total Asset(LTA)	-1.4250**	-1.7595
Non Interest Earning to Total Asset (NIETA)	-0.1045	-0.0940
C	7.2859***	4.1631
	Adj $R^2 = 0.0566$	
	F-stat = 3.2202**	

Model Equation:

*Corresponding author's email: Hamidah@uniten.edu.my

DOI: 10.1051/shsconf/20173600016

$$ROA = 7.2859 - 0.6142*GROWTH_{it} - 1.4250*LTA_{it} - 0.1045*NIETA_{it} + \varepsilon_{it}$$

CONCLUSION

The hybridization of neural networks was a very popular phenomenon in investigated business disciplines. Even though that specific hybrid networks might work well only for particular tasks, our opinion to suggest that proper integration of metaheuristic methods into the neural network methodology might be a key for achieving the optimal findings. To conclude, neural networks have been successfully applied in wide range of business tasks and were able to detect complex and nonlinear relationships without requiring any specific assumptions about the distribution or characteristics of the data. As from our findings, we able to predict the survival predictors for the Islamic banks and able to identify the directions for each predictors. As a conclusion our findings are consistent with Yap et.al (2011) and Alifiah et.al (2013), which shows that the likelihood of firms' financial distress is linked with the levels credit risks and not with efficiency ratio. While for GROWTH it is supported by Abu Bakar and Mohd Tahir (2009). As for the recommendation, the future development on the findings may also integrate with non-linear predictors such as input from survey that pertaining to the characteristics of the CEO and board of directors and measure whether there is any possible integration with the likelihood of Islamic banks survival in Malaysia.

ACKNOWLEDGEMENTS

A special thanks to Universiti Tenaga Nasional as for being our sponsor for this Internal Grant Scheme and this is a part of the findings from this research grant.

REFERENCES

Abu Bakar, N.M. and Mohd Tahir. I. (2009), "Applying Multiple Linear Regression and Neural Network to Predict Bank Performance", International Business Research, vol 2, no 4, pp.176-181.

Alifiah, M. N., Salamudin, N., & Ahmad, I. (2013), "Prediction of Financial Distress Companies in the Consumer Products Sector in Malaysia", Jurnal Teknologi, 64(1).

Al-Osaimy, M.H. (1998), "A Neural Networks System for Predicting Islamic Banks Performance", Journal of King Abdul Azis University: Islamic Economic & Administration, vol.11, no.1, pp.33-46.

Altman, E., G. Marco and F. Varetto, (1994), "Corporate Distress Diagnosis: Comparisons Using Linear Discriminant Analysis and Neural Networks (the Italian Experience)", *Journal of Banking and Finance*, Vol. 18, pp. 505-529.

*Corresponding author's email: Hamidah@uniten.edu.my

Athanasoglou, P P, Brissimis, S N, and Delis, M. D. (2008), "Bank-specific, industry-specific and macroeconomic determinants of bank profitability". *International, Financial Markets, Institution and Money,* vol. 18, 21–136.

Bashir, A. (2000), "Assessing the performance of Islamic banks: Some evidence from the Middle East". Paper presented at the ERF 8th meeting in Jordan.

Bourke, P. (1989), "Concentration and other determinants of bank profitability in Europe, North America and Australia", *Journal of Banking and Finance*, vol.13, pp.65–79.

Celik A.E. & Karatepe Y. (2007), "Evaluating and forecasting banking crises through neural network models: An application for Turkish banking sector". *Expert Systems with Applications*,33 pp: 809–815.

Coats, P.K. and Fant, L.F. (1993), "Recognizing financial distress patterns using a neural network tool", *Financial Management*, Vol. 22 No. 3, pp. 142-56.

Demirguc-Kunt, A. and Huizinga, H. (2000), "Financial structure and bank profitability". Policy Research Working Paper Series 2430. The World Bank.

Dutta, S. and S. Shekhar, (1988), "Bond Rating: A Non-Conservative Application of Neural Networks", *Proceedings of IEEE International Conference on Neural Networks*, IEEE Press, Alamitos, CA, Vol. 2, pp. 443-450.

Ernst and Young, (2014), World Islamic Banking Competitiveness Report 2013–14: The Transition Begins (London).

Eletter, S.F. & Yaseen, S. G. (2010), "Applying Neural Networks for Loan Decisions in the Jordanian Commercial Banking System", International Journal of Computer Science and Network Security, VOL.10 No.1, pp: 209-214

Gunay, O. E. N. and Ozkan, M. (2007), "Prediction of bank failures in emerging financial markets: an ANN approach". *The Journal of Risk Finance*, Vol. 8 No. 5, pp. 465-480

Hassan, M. and Bashir, A. (2003), "Determinants of Islamic banking profitability". Paper presented at the ERP 10th Annual Conference in Morroco.

Haron, S. (2004), "Determinants of Islamic Bank Profitability", *Global Journal of Finance and Economics. USA,* vol 1, No 1.

Handzic, M.T, Rawibawa F. and Yeo J. (2003). "How Neural Networks Can Help Loan Officers to Make Better Informed Application Decisions", Informing Science Insite

How, J. C., Melina, A. K. and Verhoeven, P. (2005), "Islamic financing and bank risks: the case of Malaysia", *Thunderbird International Business Review,* 47(1), pp.75-94.

Islamic Financial Services Board (IFSB), (2014), Islamic Financial Services Industry Stability Report (Kuala Lumpur).

*Corresponding author's email: Hamidah@uniten.edu.my

Kamijo, K. and T. Tanigawa, (1990), "Stock Price Pattern Recognition: A Recurrent Neural Network Approach", *Proceedings of IEEE International Conference on Neural Networks*, IEEE Press, Alamitos, CA, Vol. 1, pp. 215-221.

Koster, A., N. Sondak and W. Bourbia, (1990), "A Business Application of Artificial Neural Network System", *The Journal of Computer Information Systems*, Vol. XI, pp. 3-10

Kryzanwski, L., M. Galler and D. Wright, (1993), "Using Artificial Neural Networks to Pick Stock", *Financial Analysis Journal*, Vol. 12, No. 1, pp. 21-27.

Malhorta R. And Malhorta D.K., (2003), Evaluating Consumer Loans Using Neural Networks, Elsevier Science Ltd.

Mitchell D. and Pavur R., (2002), "Using Modular Neural Networks for Business Decisions", *Management Decision*, Vol. 40 Issue 1 pp. 58 - 63

Molyneux, P. and Thornton, J. (1992), "Determinants of European bank profitability: a note", *Journal of Banking and Finance*, vol.16, pp.1173–1178.

Nienhaus, V. (1983), "Profitability of Islamic PLS Banks Competing with Interest Banks-Problems and Prospects", *Journal of Research in Islamic Economics*, vol. 1 (1), pp. 37-47.

Odom, M. and Sharda, R. (1990), "A neural network model for bankruptcy prediction". Proceedings of the IEEE International Conference on Neural Networks 2, 163-168.

Ravi, V., Kurniawan, H., Thai, P.N.K. and Kumar, P. R. (2008), "Soft computing system for bank performance prediction. Applied Soft Computing", Volume 8, Issue 1, January 2008, Pages 305–315

Rosly, SA and Abu Bakar MA (2003), "Performance of Islamic and mainstream banks in Malaysia". *International Journal of Social Economics*, 30(12), pp.1249-1265.

Samad, A., and Hassan, M. K. (2000), "The performance of Malaysian Islamic Bank During 1984 - 1997: An Explanatory Study". Thoughts on Economics, 10(1&2), 7-26.

Sanusi, N. and Mohammed, N. (2005), "Profitability of an Islamic Bank: Panel Evidence from Malaysia", in Sanusi, N. Harun, M and Samsudin, S.(Ed.), *Reading in Islamic Economics and Finance*, Penerbit Universiti Utara Malaysia, Kedah, pp: 97-116.

Sapuan, N.M. and Roly, M.R. (2015), "Bank Profitability and Bank-Specific Variables in Malaysia: A Panel Cointegration and Error Correction Model". *Journal of Islamic Finance and Business Research*, Vol. 3. No. 1. March 2015 Issue, pp: 50 – 61

Surkan, A. and J. Singleton, (1990), "Neural Networks for Bond Rating Improved by Multiple Hidden Layers", *Proceedings of IEEE International Conference on Neural Networks*, IEEE Press, Alamitos, CA, Vol. 2, pp. 157-162.

*Corresponding author's email: Hamidah@uniten.edu.my

Tam, K.Y. and Kiang, M.Y. (1992), "Managerial applications of neural networks: the case of bank failure predictions", Management Science, Vol. 38, pp. 926-47.

Tsukuda, J. and S. Baba, (1994), "Predicting Japanese Corporate Bankruptcy In Term of Financial Data Using Neural Network", *Selected papers from the 16th Annual Conference on Computers and Industrial Engineering*, Elsevier Science Ltd., Vol. 27, Nos 1-4, pp. 445-448.

Turen, S (1995), "Performance and risk analysis of the Islamic Banks: The Case of Bahrain Islamic Bank". *Journal of Islamic Economics*, vol. 7, pp.3 -13.

White, H., (1988), "Economic Prediction Using Neural Networks: The Case of IBM Daily Stock Returns", *Proceedings of IEEE International Conference on Neural Networks*, IEEE Press, San Diego, CA, pp. 451-459.

Yoon, Y. and G. Swales, (1990), "Predicting Stock Price Performance", *Proceedings of the 24th Hawaii International Conference on System Sciences*, IEEE Press, Alamitos, CA, Vol. 4, pp. 156-162

*Corresponding author's email: Hamidah@uniten.edu.my

EXAMINING THE FACTORS OF TAX NON-COMPLIANCE: A CASE STUDY OF SMALL AND MEDIUM ENTERPRISES IN METRO AREA

Marziana Mohamad*, Mohd Asri Mohd Ali
**Accounting Department, College of Business Management and Accounting*
Universiti Tenaga Nasional

ABSTRACT

Hidden income is a business income and individual income should be recorded and taxed but taxpayers are not reported in full or not reported at all. In this paper we examine the determinants of tax non-compliance among Small and Medium Enterprises (SMEs) in Metro area. 150 questionnaires were distributed to the three sectors namely, distribution, food and beverages, and service providers; only 86 agreed to participate (57.3%). Kaiser-Meyer-Olkin, descriptive and Pearson correlation test were used to analyse the data. The results indicate that there was a number of significant relationship between tax incentive and company size towards tax non-compliance. Therefore, the importance of this phenomenon is the potential serious revenue losses to government and resulting in the possible under funding of public's service and also an unfair burden on honest citizens.

Keywords: Tax Compliance, Tax Knowledge, Fiscal Psychology Theory

INTRODUCTION

Tax compliance is referring to the willingness of taxpayers to act in accordance with both the spirit and the letter of the tax law and administration. Roth, Scholz and Witte (1989) defined tax compliance as filling all required tax return in proper of time and accurately report their tax liabilities with the tax law and regulation. In term of tax noncompliance, have two categories which are tax avoidance and tax evasion. Tax avoidance is a legal activity because tax laws are used to reduce the tax amount payable which denotes the taxpayers' ingenuity to arrange his affairs in a proper manner and any attempt to minimize tax is acceptable.

Tax evasion on the other hand is an illegal activity which involves a taxpayer's deliberate contravention of the tax law in order to eliminate tax liability by entails not filing income tax returns altogether or misrepresenting the tax liability amount. Tax evasion is the application of fraudulent practices in order to minimize or eliminate tax liability by making a false return of income by omitting or understating income or overstating expenses, making a false statement in a return affecting tax liability, giving false information on any matter affecting tax liability and other. Thus, tax avoidance and tax evasion will lead to the reduction or elimination of tax liability resulting into government revenues loss (www.taxation n-tz.com 2011).

*Corresponding author's email: marziana@uniten.edu.my

Based on the macroeconomic perspective, tax avoidance and tax evasion have similar negative effects on the national budget of a country. Kirchler, Maciejovsky and Scheider (2001) and Cross and Shaw (1982) reported that no longer found discrimination between tax evasion and tax avoidance. However, based on the psychological perspective, due to legal differences and moral considerations, Etzioni (1988) indicated that taxpayers perceive tax avoidance and tax evasion differently. The psychological dimension of tax avoidance and tax evasion has been analyzed from many perspectives, starting as early as the 1960 by "Cologne school of tax psychology" (Song and Yarbrough, 1978).

Otusanya (2011), also agreed that the tax avoidance and tax evasion reduced the government revenue. He found that the evidence from the cases examined has implicated MNCs (Multi-National Companies) in adopting a variety of tax avoidance and tax evasion schemes such as under-reporting their taxable profits and manipulating their accounting reports. Tax evasion has been a general and persistent problem throughout history with serious economic consequences such as hidden economy, not only in transition economics but also in countries with developed tax system.

Therefore, this study explores the perception of Small Medium Enterprises (SMEs) taxpayers in term of their level of tax noncompliance. This paper is organised as follows, review of literature, methodology and results. The last section includes some recommendations and direction for future research.

REVIEW OF LITERATURE

Tax knowledge is the level of awareness or sensitivity of the taxpayers to tax legislation. Tax knowledge also as an element in a voluntary compliance tax system, particularly in determining an accurate tax liability (Kasipillai, 2000; Palil, 2005). More recent studies investigate in Malaysia (Loo, McKerchar, Hansford, 2008; Loo, 2006; McKerchar, 2005) suggested that tax knowledge to be the most influential factor to determine taxpayers' compliance behavior under the self-assessment system (SAS). Kirchler and Niemirowski (2006) also stated that tax knowledge would lead the higher compliance rates.

Based on the economists per say, they have established a relationship between tax rates and the amount of tax evasion (Frey and Pommerehne, 1984; Trandel and Show, 1999). The higher the level of taxations shows the greater the incentive to participate in escape taxes. In a study done by Noor Suhaila and Sarifah (2008), majority of taxpayers from Small Medium Enterprises (SMEs) have low level of tax knowledge. They found that the lower of tax knowledge, it is resulting leads to tax noncompliance. Eriksen and Fallen (1996), also studies the level of tax knowledge taxpayers. The finding shows that the taxpayers have greater tax knowledge would comply the tax more efficiently compared to those with lower level of tax knowledge. It is means, higher level of education exposed more to tax matters and more knowledgeable about tax system in their countries.

In term of company size, there is a mixed evidence of finding between the company size and tax noncompliance level. According to Hanlon, Mills and Slemrod (2007), they found that have

DOI: 10.1051/shsconf/20173600017

positive relationship between tax noncompliance and the company size. On the other hand, previous studies indicated that company size have a negative relationship with tax noncompliance. They found that the bigger the firms are, the more the firm complaint with the tax laws due to proper internal control, better accounting system and greater tax knowledge as compared to small-scale firms. Besides, Marinov (2008) stated that the socioeconomic reason for taxpayers to choose the decision to go underground or hidden income is influencing by the government policy in term of tax and regulation measures.

There are three theories of tax compliance, namely an economic deterrence model, fiscal psychology model and social psychology model. Firstly, economic deterrence model interpreted on the taxpayers compliance views such as perfectly moral, risk-neutral or risk averse individual who seek to maximize their utility and chose to evade tax. Secondly, for fiscal psychology model, consist economic deterrence model and social psychology model which is individuals are not simply independent utility maximizers but they are correlate with attitude and beliefs to respond of tax compliance (Hasseldine and Bebbington, 1991). Thirdly, social psychology model is examined the attitude and beliefs of taxpayers in order to understand and predict the human behavior. In this study, fiscal psychology model has chosen to examine the determinants of tax non-compliance. Hence, this study would like to investigate the factors of tax non-compliance among SMEs taxpayers in metro area, namely in Kuala Lumpur and Selangor. Therefore, the hypothesis are as follows:

H_1 : There is significant relationship between tax incentives towards tax noncompliance for SMEs.

H_2 : There is significant relationship between company size towards tax noncompliance for SMEs.

METHODOLOGY

Data are collected from respondents consist of taxpayers who are doing small medium enterprises business (SMEs) in metro area namely in Kuala Lumpur and Selangor. One hundred fifty (150) SMEs taxpayers were selected randomly to distribute the questionnaires. Only 86 agreed to participate in this study which contributes to 57.3% response rate. The selected sectors are distribution trade, food and beverages, and services providers. The questionnaires is adapted from previous study which is Abdul Jabbar (2009), who studied on the tax compliance cost on noncompliance behavior of SMEs. The questionnaires are divided into four sections, namely Section A; respondents background, Section B; Knowledge on Tax Incentives, Section C; Perception of Company size and Section D; perception on tax non-compliance.

RESULTS AND DISSCUSION

Reliability Test

Cronbach's alpha has been used to test the reliability of questioner's which has set by researcher. Reliability test has conducted on 3 dimensions which respectively are tax incentives, company size and tax noncompliance. The finding shows that all variable is above 0.7 which is acceptable in this study (Table 1).

Reliability test	Cronbach's Alpa	No item
Tax Incentives	0.697	6
Company size	0.770	8
Tax Non-compliance	0.720	5

Table 1: Reliability Analysis

Factor Analysis

Factor analysis has been applied to explore the underlying factors associated with 19 items by using Principal Component Analysis (PCA). The value of KMO varies from 0 to 1, and KMO overall should be 0.60 or higher to perform factor analysis. Result for the Bartlett's Test of Sphericity and the KMO reveal that both were highly significant and concluded that this variables was suitable for the factor analysis which is 0.713 (Table 2).

Kaiser-Meyer-Olkin Measure of Sampling Adequacy		0.713
Bartlett's Test of Sphericity	Approx. Chi-Square	534.731
	df	171
	Sig.	0.000

Table 2: KMO and Bartlett's Test

Descriptive Analysis

Analysis of the profile of the respondents indicates that an equal percentage between female and male respondents which is 50% (n=43) respectively. About 43% (n=37) of the respondents are senior position, 25.6% (n=22) are junior position, 22.1% (n=19) are managers and 9.3% (n=8) are partners.

Table 3 shows the information about the respondents' perception on tax non-compliance. The results indicate that the average of SMEs perception towards tax non-compliance is 2.9767 which is 'neutral'. It is means, some of SMEs taxpayers either agree or disagree with the following statement.

No.	Measures	Mean	Std. Deviation
1	Tax non-compliance is not a serious offense.	2.6744	1.12136
2	Tax non-compliance does not compromised one's integrity.	2.5814	0.87401
3	Tax non-compliance should be allowed if the reason to do so can be justified.	3.2558	1.03112
4	Tax non-compliance is more towards tax evasion (a criminal offense) than tax avoidance (not a criminal offense).	3.0814	1.15009
5	Tax non-compliance is more towards tax avoidance (not a criminal offense) than tax evasion (a criminal offense).	3.2907	1.03885
	Average	2.9767	1.04309

Table 3: Perception on Tax Non-Compliance

Hypotheses Testing

Table 4 presents correlation results of the study. The p value results shows that all variable have significant relationship with tax noncompliance. In term of tax incentive, there is a significant relationship with tax noncompliance which is p value 0.046. Previous studies have evidence that general tax knowledge has a very close relationship with taxpayers' ability to understand the laws and regulations of taxation (Singh, 2003). Therefore, this hypothesis (H_1) is supported and consistent with previous studies.

Then, for company size also shows significant relationship towards tax noncompliance. The p value is less than 0.05 which is the p value is 0.035. Hanlon, Mills and Slemrod (2007) indicate that there is a significant relationship between business size and tax non-compliance of corporate taxpayers. Therefore, this hypothesis (H_2) is supported and consisted with previous studies.

		CS	TI
Tax	Pearson Correlation	**.227***	**.216***
Noncompliance	Sig. (2-tailed)	**.035**	**.046**
(TNC)	N	86	86

* Correlation is significant at the 0.05 level (2-tailed)

Table 3: Perception on Tax Non-Compliance

CONCLUSION AND RECOMMENDATION

The objective of this study is to examine the determinants of tax non-compliance among SMEs and their perception on the tax law and regulations. The results indicate that tax incentive and company size are significantly related to the tax noncompliance. This study shows that strong desire of taxpayers to comply the tax obligations determined by the tax incentive and company size towards tax non-compliance. This suggest that one of the target of this paper is to explain the factors that affect compliance in taxation through Fiscal Psychology Model can be realized.

In order to improve the tax compliance, it is very important for SMEs taxpayers to enroll in at least one of the tax education. This is because general level of education is not sufficient enough. In a study by Loo and Ho (2005), investigate the group of salaried individuals in Melaka, Malaysia, they found that the taxpayers competency level is quite low even though most of them have tertiary education. Previous studies also have evidenced that tax knowledge play an important role in increasing tax compliance. However, this study cannot be generalized to all taxpayers, since it is only being conducted to the SMEs in metro area namely Kuala Lumpur and Selangor. Besides, tax evasion and tax avoidance are an important phenomenon that affects to countries such as potential serious revenue losses to government.

REFERENCES:

Abdul Jabbar, H. (2009). Income Tax Non-compliance of Small and Medium Enterprises in Malaysia: Determinants and Tax Compliance costs. Unpublished doctoral dissertation, Curtin University of Technology, Perth, Australia

Cross, R.B., and Shaw, G.K. (1982). Tax Evasion-avoidance choice: A Suggested Approach. *National Tax Journal*, 34,489-491

Eriksen, K. and fallan, L. (1996). Tax Knowledge and Attitudes towards Taxation: A Report on a Quasi-experiment, *Journal of Economic Psychology*, 17, 387-402

Etzioni, A. (1988). The Moral Dimension. Towards a new Economy, New York: Free Press

Hanlon, M., Mills, L., and Slemrod, J. (2007). An Empirical Examination of Corporate Tax Non-compliance. Cambridge University Press. 171-210

Hasseldine, D., and Bebbington, K. J. (1991). Blending Economic Deterrence and Fiscal Psychology Models in theDesign of Responses to Tax Evasion: The New Zealand Experience. . *Journal of Economic Psychology, 12*(2), 299-324. http://dx.doi.org/10.1016/0167-4870 (91)90018-O

Kasipillai, J. (2000). A Practical Guide to Malaysian Taxation-Current Year Assessment, Kuala Lumpur, McGraw-Hill, New York

Kirchler, E., Maciejovsky, B., and Scheider, F. (2001). Everyday representations of Tax Avoidance, Tax Evasion and Tax Flight: Do not legal differences matter? *Journal of Economic Psychology*, vol. 24, 535-553

Kirchler, E., Niemirowski, A. (2006). Wearing, Shared Subjective views, intent to cooperate and Tax Compliance: Similarities between Australian Taxpayers and Tax Officer, *Journal of Economic Psychology*, 27(4), 502-517

Loo, E.C. (2006). The influences of the introduction on Self Assessments on Compliance Behavior of individual taxpayers in Malaysia, PhD thesis, University of Sydney

Loo, E.C. and Ho, J.K. (2005). Competency of Malaysia Salaried Individuals in Relation to Tax Compliance under Self-Assessment. http://www.austlii.edu.au/au/journals/eJTR/2005

Loo, E.C., Mckerchar, M. and Hansford, A. (2008). Tax Compliance Behavior: Finding derived from a mixed method design, Proceeding of 8th International Tax Administration Conference, Sydney

Marinov, A. (2008). Hidden Economy in the Rural Regions of Bulgaria. International Review on Public and Nonprofit Marketing. 5, 71-81

Noor Suhaila, S., and Sarifah, I. (2008). Factors influencing the level of Tax Compliance: A Case Study of Small Business Entrepreneurs Hentian Nilai, Negeri Sembilan, Selangor International Islamic University College.

Otusanya, O.J. (2011). The role of multinational companies in tax evasion and tax avoidance: The case of Nigeria. *Critical Perspectives on Accounting*. V22. 316-332

Palil, M.R. (2005). Does Tax Knowledge matter in Self-Assessment System? Evidence from the Malaysian Tax Administration, *Journal of American Academy of Business*, 6 (2), 80-85

Roth, J.A., Scholz, J.T. and Wittee, A.D. (1989). Taxpayers Compliance: An Agenda for Research, University of Michigan Press, Philadelphia, PA

Sekaran, U. (2003). Research Methods for Business: A Skill Building Approach, New York: John Wiley & Sons

Singh, V. (2003). Malaysian Tax Administration, 6th Ed. Longman, Kuala Lumpur

Song, Y.D. and Yarbrough, T.E. (1978). Tax Ethics and Tax Attitudes: A Survey. *Public Administration Review*, 38. 442-452

AN ASSESSMENT OF ENVIRONMENTAL IMPACTS ASSESSMENT (EIA) IN MALAYSIA

Masdiah Abdul Hamid *, Kweh Qian Long
Department of Accounting, Universiti Tenaga Nasional, Pahang, Malaysia.

ABSTRACT

The effectiveness of Environmental Impact Assessment (EIA) in evaluating the planning project is a debatable issue among academics and practitioners, since EIA has been claimed to be unable to eliminate the environmental issues. Focusing only on technical improvements is not sufficient for rectifying the problems of EIA; the process of EIA should be clearly identified instead to maximise the effective use of EIA. It is important to note that the effective use of EIA, particularly on process-related issues could significantly minimise bad environmental effects. In summary, this study aims to explore and identify the effectiveness of EIA in the planning process and barriers to evaluate the environmental performance in Malaysia. The findings of this study could be a baseline for organisation to minimize emission, avoid the risk of prosecution and fines arising from potential environment breaches and cost reduction within the organisation.

Keywords: Environmental Impact Assessment (EIA), Environment, Society and Sustainability

INTRODUCTION

Rapid development and urbanisation have been leaving significant side effects on the environment since decades ago. In Malaysia, the Department of Statistics reported that the increasing population number and the level of urbanisation in Malaysia significantly contributed bad impacts towards the environment especially the quality of air. That is, air pollution is one of the major problems due to rapid growth of industrial sectors, increased traffic volume, expansion in land exploration, forest degradation, burning of fuels in industrial processing, open burning etc. It has been reported that air surrounding us is contaminated by any number of substances such as gaseous and particulates not normally found in the air. With these phenomena, the air pollution in a few states in Malaysia are said to have reached the danger zone. Overall, a number of environmental problems such as air pollution, forest destruction, acid rains, global warming, ozone depletion, hazardous waste and over-pollution have caused numerous effects to the country resulting from environmental disaster. Due to the magnitude of these nuisances and their possible effects on businesses, intensive remedial efforts have been taken by countries, business corporations, non-profit organizations and even individuals. However, the afore-mentioned situations remain unchanged but getting worse. Therefore, a comprehensive action should be taken to improve the environmental performance.

To that end, using Environmental Impact Assessment (EIA) would enhance greater awareness and understanding of environmental issues and its benefits. While EIA is widely used, it has been subject to criticism to improve and addressing the issues emerging from the EIA application. EIA has been critiqued for its failure in identifying and determining the process-related issues and interaction with key elements in Environmental Management System (EMS)

*Corresponding author's email: masdiah@uniten.edu.my

which caused the effectiveness of EIA to drop. Moreover, EIA not only identifies the best solution among all possible alternatives, it also might raise confusions among planners or other professionals (Dendena & Corsi, 2015) and thus EIA seems as hard to practise. Also, focusing only on the technical improvements is not sufficient for rectifying the problems of EIA, and therefore function EIA should be clearly identified to maximise the effective use of EIA. Lack of discussion on EIA effectiveness has motivated this study to provide in-depth knowledge about the process issues and enrich the discussion in Malaysia. In response to environmental issues, comprehensive EIA is an imperative tool to eliminate the adverse effects on environment. Effectiveness of EIA which is perceived as the back bone tool to improve environmental performance may thereby indirectly assist businesses, save government money, public funds and ultimately contribute to a healthier environment.

That is, related parties can rely on the establishment of Environmental Impact Assessment (EIA), which aims to protect the environment and eliminate adverse impacts of any numbers of substances by recommending the appropriate mitigation measures prior to the realisation of a particular activity (Pavlickova & Vyskupova, 2015). Hillary (1998) defines EIA as a systematic process to document periodic and objective checking process of company's environmental performance against pre-set standard and objectives in evaluating the compliance status to environment regulatory requirements, the Environment Management System (EMS) and overall environmental risk of an organization (Haslinda & Fuong, 2010; Mustafa, 2011). Traditionally, EIA is used to assess company's activities and likely implications for the environment. Thus, EIA has been practised aggressively throughout the world among organisations because of the ability of EIA in the identification of the major environmental impacts resulting from company's operations and to document suggestions of mitigation measures to minimise the environmental effects. EIA is highly associated with nature of the organisation's operations, which requires methodical review of operations according to specified criteria (Stanwick & Stanwick, 2001) for internal management use and verified the accuracy of estimates in environmental impact statements. Moreover, EIA is also used as an internal control mechanism with the purpose to ensure that organisational operations adhere to environmental laws and regulations (Yusoff, 2013). It is important to note that EIA is used as environmental tool to ensure that available resources had been used economically, effectively and accomplishes the ultimate objective of an organisation.

Taken together, EIA is considered as a technique commonly used to improve environmental performance and ultimately to achieve sustainable development. In Malaysia, environmental protection is governed by Malaysian Environmental Quality Act 1974 which is regulated pertaining to prevention, reduction, control of pollution, and enhancement of environmental performance. To sustain healthy environment for the benefit of mankind and enable all people enjoy the beauty of earth, it is our shared responsibility to preserve our environment to assure the quality of life to reduce adverse effects on the well-being of the people.

ENVIRONMENTAL IMPACT ASSESSMENT (EIA)

Environmental Impact Assessment (EIA) captures the idea of assessing proposed actions of organisation from projects to policies, with respect to their likely implications for the environment (Dendena & Corsi, 2015). EIA considered as one mechanism used to measure the positive and negative impacts prior the proposed planning take place by an organisation. Pavlickova and Vyskupova (2015) suggested that the users seem to be well suited for the comprehensive EIA process because the process of EIA has a positive effect on the

*Corresponding author's email: masdiah@uniten.edu.my

identification of potential cumulative effects in Malacky city. An effective EIA process is helpful in giving a better understanding of the current environmental and socio-economic conditions of the affected area and the specific effect of each potential predicted impact. Soria-Lara, Bertolini, and te Brömmelstroet (2015) remarked about gap between the content related and process related with EIA. The study argued that existing process of EIA only focused on the measurement of environmental effects regardless the roles of EIA in the planning process and the interaction between key actors. On the other hand, Rozema and Bond (2015) posit on the effectiveness of impact assessment tools due to challenge to future practice of impact assessment which seem as ineffective and controversy generated amongst stakeholders in decision making. They further argued about the meaning of effectiveness in impact assessment tools as an inevitable consequence of increased participation in environmental decision making. The analysis on EIA however appears that the EIA was established to support decision making through a better understanding of the implications on the environment. Thus, it is important to map the process of EIA in different decision context and also the role of impact assessment in those processes. A study done on the value of independent EIA follow-up verifiers suggested that the value of those checker significantly add value when the consultant getting involved with screening EIA requirements of new project, allocation of financial and human resources, checking legal compliance, influencing implementation, reporting conformance results, community and stakeholders engagement, integration with self-responsibility programs such as Environmental Management System (EMS) and controlling records, in South Africa context (Wessels, Retief, & Morrison-Saunders, 2015).

CONCLUSION

The lack of environmental related rules and regulations imposed on decision makers makes EIA possibly an ineffective one. Besides, EIA has been proven to not able to address the environmental impacts of developments. Any improving modifications on the EIA procedure in either its structure or applications have to be continuously developed and executed. Measures that should be in the policy makers' considerations include ensuring that the impact of any developments is minimised or even to the extent that no any protective measure is needed afterwards. If the impact is moderate to severe, parties involved should have taken protective or corrective measures to restore the initial environmental health.

ACKNOWLEDGEMENTS

We appreciate the research support given by the Ministry of Higher Education (MOHE) Malaysia under the Fundamental Research Grant Scheme (FRGS/1/2015/SS01/UNITEN/03/1).

*Corresponding author's email: masdiah@uniten.edu.my

REFERENCES

Dendena, B., & Corsi, S. (2015). The Environmental and Social Impact Assessment: a further step towards an integrated assessment process. *Journal of Cleaner Production, 108*, 965-977.

Haslinda, A., & Fuong, C. C. (2010). The implementation of ISO 14001 environmental management system in manufacturing firms in Malaysia. *Asian Social Science, 6*(3), 100.

Hillary, R. (1998). Environmental auditing: concepts, methods and developments. *International Journal of Auditing, 2*(1), 71-85.

Mustafa, M. (2011). The Environmental Quality Act 1974: A significant legal instrument for implementing environmental policy directives of Malaysia. *IIUM Law Journal, 19*(1), 1.

Pavlickova, K., & Vyskupova, M. (2015). A method proposal for cumulative environmental impact assessment based on the landscape vulnerability evaluation. *Environmental Impact Assessment Review, 50*, 74-84.

Rozema, J. G., & Bond, A. J. (2015). Framing effectiveness in impact assessment: Discourse accommodation in controversial infrastructure development. *Environmental Impact Assessment Review, 50*, 66-73.

Soria-Lara, J. A., Bertolini, L., & te Brömmelstroet, M. (2015). Environmental impact assessment in urban transport planning: Exploring process-related barriers in Spanish practice. *Environmental Impact Assessment Review, 50*, 95-104.

Stanwick, P. A., & Stanwick, S. D. (2001). Cut your risks with environmental auditing. *Journal of Corporate Accounting & Finance, 12*(4), 11-14.

Wessels, J.-A., Retief, F., & Morrison-Saunders, A. (2015). Appraising the value of independent EIA follow-up verifiers. *Environmental Impact Assessment Review, 50*, 178-189.

Yusoff, H. N. (2013). Understanding environmental audit in the public sector: Malaysian perspective. *Jurnal Teknologi, 62*(1).

*Corresponding author's email: masdiah@uniten.edu.my

SHARI'AH AUDITING: A REVIEW OF SHARI'AH AUDIT PRACTICES IN ISLAMIC FINANCIAL INSTITUTION (IFIs)

Azwan Abdul Rashid, Masdiah Abdul Hamid*, Ahmad Subhi Mohamad Sidek, Wan Mohammad Taufik Wan Abdullah, Inaliah Mohd Ali
Department of Accounting, Universiti Tenaga Nasional, Pahang, Malaysia.

ABSTRACT

With its increasing number of Islamic Financial Institutions (IFIs) in the country, Malaysia has emerged as a leading hub when benchmarked against the other Islamic countries of the world. Unlike its conventional counterpart, the Islamic financial system focuses on the achievement of societal justice as evaluated within its own framework and uses its own criteria in order to achieve the objective of the *Maqasid* Ash-*Shari'ah*. To help achieve this objective, the Islamic finance industry is in need of *Shari'ah* auditors who are not only knowledgeable and competent in ensuring that the IFIs operate in accordance with *Shari'ah* principles, but also adequate in supply, in order to cater for the increasing number of IFIs in the country. Therefore, a major aim of this study is to conduct a comprehensive review of the auditing process performed by *Shari'ah* auditors in determining the achievement of the *Maqasid Ash-Shari'ah* by the IFIs. Another aim of this study is to assess the existing *Shari'ah* auditing framework and standards practiced by IFIs in Malaysia. The results of this study could serve as a reference point for the regulatory and professional bodies in assessing the implementation of a comprehensive *Shari'ah* auditing framework.

Keywords: *Shari'ah* Auditing, *Maqasid Ash-Shari'ah*, *Shari'ah* Auditing Framework and Islamic Financial Institution

INTRODUCTION

The introduction of the *Shari'ah* Governance Framework (SGF) for Islamic Financial Institutions by Bank Negara Malaysia (BNM) on 1 January 2011 has brought immediate attention to both academicians and practitioners alike on the importance of *Shari'ah* governance. A major challenge for *Shari'ah* auditors toward the implementation of the SGF is the requirement for the auditor to possess adequate knowledge of *Shari'ah* to ensure that Islamic Financial Institutions (IFIs) operate in accordance with *Shari'ah* principles. According to Yaacob and Donglah (2012) the Islamic finance industry has grown at a rate of between 15%-20%, a rate easily exceeding the world economic growth rate. It is therefore of crucial importance that *Shari'ah* auditors are able to play their role in ensuring that activities and operations carried out by the Islamic finance industry are subject to proper checks and balances and that their reporting is in line with *Shari'ah* principles. The presence of an auditor is perceived as fundamental for the quality of financial reporting and is an important element in the process of securing corporate accountability. From a conventional

Corresponding author's email: masdiah@uniten.edu.my

auditing perspective, the auditor is required to critically examine and evaluate whether the financial statements are prepared in accordance with GAAP and any relevant financial reporting framework. On the other hand, *Shari'ah* auditing of IFIs requires the *Shari'ah* auditor ensure that management is not only fully adhering with GAAP and with the relevant reporting framework, but also ensure that management follow the *Shari'ah* framework to ensure a sound and effective internal control system accordance with *Shari'ah* principles in achieving the ultimate objective of *Shari'ah* (Yaacob & Donglah, 2012). The *Shari'ah* auditor is not only required to conduct a review of IFI activities in accordance with *Shari'ah* principles; he is also responsible for expressing his opinion on the IFI financial statements (AAOIFI, 2004; Mohammed Sarea & Mohd Hanefah, 2013). To perform his auditing role effectively, a *Shari'ah* auditor should possess adequate and appropriate Islamic knowledge (AAOIFI, 2004). Yaacob and Donglah (2012) argues that a *Shari'ah* auditor has a similar role to that of the *Muhtasib* (Hisbah member), a role that is related to the concept of *Taklif* (accountability and responsibility) in ensuring that the IFIs strictly follow *Shari'ah* guidelines.

The existence of modern Islamic institution globally is over 50 years (Yaacob, 2012). The development of IFIs is purpose to provide benefit and value for society. In line with conventional development, people have realised that traditional method is not enough for them to capture the real activities of the business. Contradict with traditional system; Islamic system has captured a new dimension for society to perceive the value of accountability. Thus, proper auditing system in Islamic view is a basis for the IFIs to achieve their main objective in achieving the *Maslahah* to the *Ummah*. However, the absence of proper *Shari'ah* guidelines and standards is a major problem facing the current *Shari'ah* auditing framework (Kasim, Ibrahim, Hameed, & Sulaiman, 2009). Additionally, lack of system pertaining to *Shari'ah* auditing also contribute to the failure of IFIs system in Malaysia. Therefore, regular independent *Shari'ah* audit in IFIs are necessary as people are now experiencing a movement along a continuum from a society that trust everything and audit nothing to a society that trust nothing and audit everything. In other words, regular independent *Shari'ah* audit in IFIs perceived as a vital part in order to achieve the ultimate objective of *Shari'ah*. Thus, this study aims to evaluate the *Shari'ah* auditing practices of *Shari'ah* auditors in IFIs in achieving the objectives of the *Shari'ah*.

PROBLEM STATEMENT

The rapid growth of the Islamic finance industry has led to an increase in demand for high quality *Shari'ah* auditing services from IFIs and the Muslim public who do business with them. Unlike its conventional counterpart, the Islamic finance industry is based on interest-free (riba-free) lending and borrowing activities where risks are shared between lender and borrower. Thus, the Islamic system is one that focuses on the promotion of social justice in order to achieve the objective of *Maqasid Ash-Shariah* on which its formation is based (Kassim, 2009). Towards the fulfilment of this objective, it is therefore vital for the *Shari'ah* auditors who audit the IFIs to be able to provide assurance and attestation that the IFIs are operated in accordance with *Shari'ah* principles. Compliance with *Shari'ah* principles, whether real or perceived, is therefore seen as an important consideration for the IFIs in order to earn the Muslim's trust and confidence in its products and services (Rahman, 2008). Therefore, it is important for IFIs to achieve its ultimate objective which

Corresponding author's email: masdiah@uniten.edu.my

is not contradicted to the *fatwa*, guidelines and instruction issued by the *Fatwa* and *Shari'ah* Supervisory Board (SSB) of the institution. Successful of IFIs operations with full compliance to the *Shari'ah* rules heavily depends on the structure of management control in the organisation. A high value of integrity and accountability to achieve *Shari'ah* objectives by reflecting with its economic activities is a key for the Islamic finance successful through proper check and balance. In other words, fully adhered with the *Shari'ah* rules and fatwa with the principle for the society justices are not easy task for IFIs to be achieved. Kassim (2009) argues that the practice of *Shari'ah* auditing has not been seriously taken by IFIs in Malaysia and resulting to the failure of IFIs in achieving their objectives. This may indicate that the management of IFIs lacked an adequate understanding of the *Shari'ah* auditing framework. Additionally, Yahya and Mahzan (2012) reported the *Shari'ah* practices of IFIs in Malaysia are still at the lower stage and that it is necessary for regulators to promote a *Shari'ah* audit.

A big challenge facing the external auditor when asked to carry out work on *Shari'ah* auditing is their current lack of knowledge and expertise on matters related to the *Shari'ah*. The role of *Shari'ah* auditor does not only include providing assurance to management and shareholders that the IFIs are run in accordance with the *Shari'ah*, but that they also have a responsibility to the *Muslim* (society as a whole) in helping to maintain the upkeep of Islamic values. Importantly, the *Shari'ah* auditor could help to play a role in influencing others because his religious knowledge of the *Shari'ah* is highly regarded and is expected to be in accordance with the *Al-Quran, Sunnah* and *Ijma'*. Thus, with his high Islamic knowledge of the *Shari'ah*, the *Shari'ah* auditor could assist IFIs in achieving their objectives. However, recent studies observe that there is a gap between the desired and the actual practices of *Shari'ah* auditing which suggest that IFIs in Malaysia do not seriously implement and fulfill the *Shari'ah* requirement in their business operations. A possible reason for this gap is that the management of IFIs in Malaysia is not clear on the scope of the *Shari'ah* auditing framework. Another possible reason for this gap is that the *Shari'ah* auditors are themselves lacking in the knowledge required to successfully carry out an audit in accordance with *Shari'ah* principles. Thus, adequate knowledge of the *Shari'ah* is a prerequisite for the successful performance of the *Shari'ah* auditing process. This indicates that the implementation of *Shari'ah* auditing is not an easy task for the *Shari'ah* auditor even studies find that the implementation of *Shari'ah* audit framework is considered relevant in order to cater the growing numbers of IFIs in Malaysia. The implementation of *Shari'ah* audit framework is very important because it's covered all aspects including governance, competency, processes and reporting system requirements of IFIs. However, without proper structure in *Shari'ah* auditing process, it will be difficult for IFIs to fulfill the *Maslahah* to the *Ummah* (benefit to the people) through socio-economic justice (Kasim et al., 2009) and further achieving of *Maqasid Ash- Shari'ah.* Hence, following the problems discussed above, the research question of this study as follow; to what extent the *Shari'ah* auditors do examination on the financial reporting and business operation of IFIs?

Shariah Audit Practices

Over the past decades, the auditing profession has come under increased scrutiny about the credibility of the auditor in examining the financial reporting. The biggest collapse and failure of giant corporation in USA have increased attention of many people including investors on the relying of financial reporting, decision making and more concern of corporate governance issues.

Corresponding author's email: masdiah@uniten.edu.my

 DOI: 10.1051/shsconf/20173600019

Thus, public start to evaluate the auditor's work in fraud detection, reducing the manipulation issues and re-assess the level of trust they put on audit to provide assurance and ensure the company's activities reflect with the economic activities. Consequently, public has reduce confidence on the work of auditors and financial reporting. Accordingly, auditing profession has comes out with more guidelines and framework on professional skepticism in order to re-attract a public confident to the profession as a whole. Haniffa (2010) viewed the auditor as important person who live in central of human activities to protect and improve the condition of human life in all dimension. Therefore, the limited scope of conventional auditing framework which focused on the assurance and attestation services is realized and the movement to the Islamic based are necessary and viewed from different perspectives. Islam is concerned on the Islamic values because it reflects to the accountability and justice not only for society *(Ummah),* but also ultimately to Allah. As Allah said in the Holy Quran:

> *"O you who believe, why do you say that which you do not do? Grievously odious is it in the sight of Allah that you say that which you do not do".*
>
> *(Surah As-Saff, 61:2-3)*

The existence of IFIs seemed as a new dimension for the auditor since the wider of scope on the audit process including the compliance with Islamic guidelines is the key element for IFIs to achieve *Maqasid Ash-Shari'ah* instead of relying on the conventional auditing framework for audit purposes. The presence of *Shari'ah* auditor in the IFIs enables them to assist *Muslims* as the whole through their socio-economic justice and to the IFIs itself throughout the compliance of *Shari'ah* requirement. Kasim et al. (2009) claimed that the *Shari'ah* auditor capabilities in auditing IFIs because of wider scope should be focused. Therefore, a *Shari'ah* auditor should knowledgeable in *Shari'ah* and competent in order to reflect with the significant growth in Islamic finance. Apparently, the *Shari'ah* auditor is not only knowledgeable in accounting, auditing and finance, but must have a good knowledge in *Shari'ah* and *Fiqh* (Yaacob, 2012).

CONCLUSION

It is important to note that the processes of Islamic Financial Institutions (IFIs) and the practices of *Shari'ah* auditing clearly identified to minimise the gap and evaluate the obstacles in conducting the *Shari'ah* audit. Apart from that, the competency of Shariah auditor seems as important factor to assess their ability and successful of Shariah auditing. Additional skills and knowledge especially in Fiqh' Muamalat could enhance the independence and reliability of the shariah auditor in performing the shariah audit in IFIs.

ACKNOWLEDGEMENTS

We appreciate the research support given by the Universiti Tenaga Nasional (UNITEN), under the UNITEN Research Grant (UNITEN Internal Grant, Job No. J510050568).

Corresponding author's email: masdiah@uniten.edu.my

REFERENCES

AAOIFI. (2004). Accounting, Auditing and Governance Standards for Islamic Financial Institutions: Accounting and Auditing Organization for Islamic Financial Institutions Bahrain.

Haniffa, R. (2010). Auditing Islamic financial institutions. *Islamic Finance: Instruments and Markets (QFinance), Bloomsbury, London.*

Kasim, N., Ibrahim, M., Hameed, S., & Sulaiman, M. (2009). Shariah auditing in Islamic financial institutions: exploring the gap between the" desirable" and the" actual". *Global Economy & finance journal, 2*(2), 127-137.

Kassim, N. (2009). Dynamics of Shariah Auditing in Islamic Institutions: a study of the Malaysian Islamic financial sector. *Unpublished thesis, International Islamic University Malaysia.*

Mohammed Sarea, A., & Mohd Hanefah, M. (2013). The need of accounting standards for Islamic financial institutions: evidence from AAOIFI. *Journal of Islamic Accounting and Business Research, 4*(1), 64-76.

Rahman, A. R. A. (2008). *Shari'ah audit for Islamic financial services: the needs and challenges.* Paper presented at the ISRA Islamic Finance Seminar.

Yaacob, H. (2012). *Issues and challenges of shari'ah audit in Islamic financial institutions: a contemporary view.* Paper presented at the Proceedings of 3rd International Conference on Business and Economics Research.

Yaacob, H., & Donglah, N. K. (2012). Shari'ah Audit in Islamic Financial Institutions: The Postgraduates' Perspective. *International Journal of Economics and Finance, 4*(12), 224.

Yahya, Y., & Mahzan, N. (2012). *The role of internal auditing in ensuring governance in Islamic financial institution (IFI).* Paper presented at the 3rd International Conference on Business and Economic Research (3rd ICBER 2012), Bandung, Indonesia.

Corresponding author's email: masdiah@uniten.edu.my

CARBON EMISSION DISCLOSURE AND THE COST OF CAPITAL: AN ANALYSIS OF MALAYSIAN CAPITAL MARKET

Noor Raida Binti Abd Rahman*[1], Nor Hasimah Binti Johari[1], Nor Edi Azhar Binti Mohamad[2]
Accounting Department, Universiti Tenaga Nasional[1]
Economic and Finance Department, Universiti Tenaga Nasional[2]

ABSTRACT

The main purpose of this study is to examine the relationship between voluntary disclosure and cost of capital by exploring the impact of voluntary carbon emission disclosure (VCED) on the firm's weighted-average cost of capital. A carbon disclosure index is used to evaluate the quality of carbon emission disclosure in 2013 and 2014 annual reports of 247 Malaysian public listed companies. By using content analysis, the result highlights a significant increase in the level and quality of carbon emission disclosure practice from 2013 to 2014. In addition, the finding from regression analysis indicates insignificant relationship between VCED quality and weighted-average cost of capital. Overall, our findings suggest that the carbon emission disclosure is still low, as such, the quality of VCED do not have an impact on firm's cost of capital. The results of the study allow the government to measure progress toward achieving its target to reduce carbon emission and will add weight to the call by accounting regulation body such as Malaysian Accounting Standard Board for a specific standard on carbon reporting.

Keywords: Climate Change, Carbon emission disclosure, Cost of Capital, CAPM, WACC

INTRODUCTION

Climate change is becoming one of the most important issues of the twenty-first century (Peng, Sun and Luo, 2015) and it is widely recognised as the most significant environmental issue facing the global economy (Freedman, Park and Stagliano, 2015; Renukappa, Akintoye, Egbu and Goulding, 2013). Majority of the scientist agreed that carbon emissions (CO_2) is the most prominent factor responsible for climate change (Solomon et al., 2009). A serious threat of climate change and CO_2emission on the environment (Freedman et al., 2015), people (Al-Amin, Jaafar and Siwar, 2010) and business (Besio and Pronzini, 2014) are already observable. Evidence does suggest that the failure of the business to manage these impacts can expose them to considerable risk (Linnenluecke, Birt and Griffiths, 2015). Nowadays, many businesses aware and accept that they need to address the issue of climate change in order to survive (Renukappa et al., 2015).

Concerns about emissions have prompted the participants in capital markets, both shareholders and creditors to incorporate carbon-related considerations in their risk analysis to assess investment options and lending decision. Taking into this consideration, there is a growing

*Corresponding author's email: noraida@uniten.edu.my

demand for business to publicly report information about their climate change-related business practices (Global Reporting Initiative and KPMG 2007) and it is expected that firms will report this issue in a comprehensive, transparent, and accountable manner (Belal et al., 2010).

Although, there is a demand from investors for the disclosure, a relatively little evidence exists as to the impact voluntary disclosures have on financial markets (Bachoo, Tan and Wilson, 2013). In fact, the benefit from increased disclosure via a lower cost of capital remains a controversial issue among managers, policy makers as well as researcher (Souissi and Khlif, 2012). Much of the previous and current research closely examined the relationship between financial disclosure and cost of capital. In recent years, academic researchers have increasingly investigated the impact of voluntary disclosure on cost of capital. Nevertheless, most of the previous studies in this area have concentrated on environmental disclosures. There is still a lack of substantial empirical evidence on the association between carbon disclosure and cost of capital. One of the reasons is that, carbon emission disclosure is largely voluntary in certain country and unregulated nature of carbon emission disclosure has led to substantial managerial discretion in terms of disclosure choice and content (Merkl-Davies and Brennan, 2007). This motivates us to fill the void by examining the impact of carbon disclosure on firm value from the perspective of cost of capital, particularly in Malaysia. Malaysia provides a good setting to investigate the relationship between VCED quality and cost of capital. Malaysia is among the countries vulnerable to the adverse impacts of CO_2 emissions and has well developed capital market (Embong, Mohd-Saleh and Sabri Hassan, 2012). The raising population reaching approximately 27 million has ranked Malaysia the 26th largest greenhouse gas (GHG) emitter in the world (Murad, Molla, Mokhtar and Raquib, 2010). This posed a great challenge to Malaysia since the growth rate of GHG emissions seems likely to increase rapidly and become a threat to the environment and Malaysian economy.

The aim of this study is to assess the quality and the trend of carbon disclosure of 247 Malaysian public listed companies. Additionally, by using regression analysis, this study aims to examine the potential impact of carbon disclosure on the cost of capital of these companies. This study contributes to the extent literature by provide empirical evidence between the voluntary disclosure and cost of capital literature by focusing narrowly on carbon disclosure and the weighted-average cost of capital.

The findings of this study have several important policy implications. First, through carbon reporting practices of Malaysian companies, it allows the government to measure progress toward achieving its target to reduce carbon emission. Second, the result of the study also will add weight to the call by accounting regulation body such as Malaysian Accounting Standard Board for a specific standard on carbon reporting. Without the use of a defined reporting standard, the credibility of a company's reporting will probably be at stake.

The remainder of this paper is organised as follows: The following section synthesises relevant literature and develops the research hypotheses. Section 3 elaborates on the research design and Section 4 reports the empirical result. The final section summarises the study.

LITERATURE REVIEW AND HYPOTHESES DEVELOPMENT

*Corresponding author's email: noraida@uniten.edu.my

Corporate disclosure is potentially a vital tool of communication between a firm's management and outside investors. Signaling theory is suitable in explaining the behavior of individuals or organization when have access to different information (Connelly, Certo, Ireland and Reutzel, 2011). The sender of the information typically determine whether and how to communicate (or signal) the information, whereas the receiver need to determine how to interpret such signal (Connelly et al., 2011). Signaling theory posits that, if the party with superior information signals to others, information asymmetries can be reduced (Bazine and Vural, 2011). This theory assume that managers have superior information as compared to outside investors on companies' expected future performance, even with the assumption of an efficient capital market, and managers may enhance the quality of their financial reporting by voluntarily providing additional disclosures (Healy and Palepu, 2001).

The impact of disclosure on cost of capital has been investigated in recent years by several theoretical and empirical studies. The effects of voluntary disclosure on the firm's cost of capital are mixed. One stream of analytical research suggests that information might have a negative value as the investors may perceive themselves to be worse off if they consider that the firm is disclosing information which might be exploited to their detriment (Hassan, Romilly, Giorgioni and Power, 2009). Additionally, as the firms are required to release their emission information, this will produce more uncertainty about the value of the equity. In such circumstances, equity owners have to require higher rates of return on these firms. Thus the cost of equity of these companies will be increased.

The other stream of research provides empirical results on the disclosure which are generally consistent with finance-theory predictions in which more public information enhances firm value by reducing the firm's cost of capital. It is found that the voluntary disclosure can improve investor base, which in turn improved risk sharing and decrease the firm's cost of capital. As consequences, the disclosure provides investors with better opportunities to assess the prospects of the company.

Richardson and Welker (2001) examined the influence of social disclosure on cost of capital of Canadian firm and reported a positive and significant relationship. Lambert, Leuz and Verrecchia (2007) investigated the direct and indirect effects of disclosure quality on the cost of capital and conclude that the direct effect occurs because higher-quality disclosures affect the firm's assessed covariance with other firms' cash flows which attributed to a reduction in the estimation of information risk. Further, Li and Yang (2011) also suggested that greater voluntary disclosure is associated with lower market cost of capital. In Malaysia, Embong et al. (2012) examined the relationship between disclosure and cost of equity of firm listed on Bursa Malaysia for the year 2004-2006. Based on 675 observations, the author found that significant negative relationship between disclosure and cost of equity, indicated that higher disclosure is good for the market and the economy in general.

With regard to environmental and carbon emission disclosure, Plumlee, Brown, Hayes and Marshall (2015) investigated how quality of voluntary environmental disclosure is related to the firm value. The author used cost of capital and future expected cash flows as the proxy of firm values and found a positive association between both variables. Dhaliwal, Li, Tsang and Yang, (2011) found that environmental disclosure reduced asymmetry and uncertainty, thus decrease cost of capital. Most recently, Bonetti, Chi, Michelon and Tanaka (2013) found that firm that providing more precise, verifiable and hard environmental information on CO2

*Corresponding author's email: noraida@uniten.edu.my

emissions levels experience a less severe shock to the cost of capital than the firms that not providing such information. Consistent with signaling theory and these studies, the following hypothesis is developed lower market cost of capital.

He, Tang and Wang (2013) also found evidence on the inverse association between cost of capital and carbon disclosure, which is consistent with voluntary disclosure theory. By analyzing firms that participated in the Carbon Disclosure Project (CDP) in 2010, the author concluded that firms make a rational choice in exhorting the voluntary carbon disclosure in reducing the pressure exerted by legitimacy threats and to lower the cost of capital.

H1: There is a negative association between the quality of voluntary carbon emissions disclosure and the cost of capital.

SAMPLE SELECTION AND DATA COLLECTION

Sample Selection

All companies listed on main market of Bursa Malaysia as at 31 December 2014 is select as a population of the study. The final sample comprises 247 observations of annual report disclosure for the year 2013 and 2014 after remove those firms belonging to the finance and insurance sectors and companies with unavailable annual reports and data. Data on carbon emission disclosure is extract using the content analysis method from the annual reports of these companies. Content analysis is utilised since it has been commonly used in the areas of climate change, GHG and carbon emission disclosure. It is defined as "a research technique for making replicable and valid inferences from text (or other meaningful matter) to the contexts of their use" (Krippendorff, 2004). To search for the carbon related information in the annual report of these companies, the keyword outlined by previous researcher is be used. The annual report is choose as the medium of reporting because it is the most likely place that stakeholders and parties interested in environmental disclosure (carbon emission is part of it) would seek and obtain information. For cost of capital data, the information is collected from the Bloomberg and Thompson DataStream.

Dependent Variable

As for this study is concerned, the weighted-average cost of capital (WACC) will be used as the measurement for firm's cost of capital. According to Modigliani and Miller (1958), the firm's overall cost of capital was computed based on weighted average of the cost provided for each capital's components used by the particular company. Acquainted as WACC, it's practically used in assessing the firm's cost of financial structure. Thus, adopted from Modigliani and Miller (1958, 1963), the cost of capital WACC of a firm after corporate taxes (but before personal taxes) is given by the formula as below:

$$WACC = \frac{D}{V}(1 - T_C).r_d + \frac{E}{V}.r_e \qquad (1)$$

The equations denotes Vas representation for the market value of the firm; D as the market value of debt while E is the market value of equity. In addition, Tc represents the corporate tax imposed to that company; r_d is the cost of issuing the debt; while r_e is the cost of equity.

*Corresponding author's email: noraida@uniten.edu.my

To quantify the market value of the debt (D), its equal to the number of bonds outstanding multiplied by the current market price of the bond. While, the total market value of the common stock (E) is equal to the current market price of one share of stock multiplied by the number of shares outstanding. Both total market value of bonds and the total market value of common equity will be add in calculating the total market value of the firm (V). The cost of debt is the yield to maturity of the current price of the bonds relative to the expected future coupon payments and the face value of the bond at maturity. Followed Modigliani and Miller (1958), we will used the cost of debt by estimate the yield to maturity by using firms current bond rating provided by Rating Agency of Malaysia (RAM).The cost of equity r_e can be estimated using expected required return based on Capital Asset Pricing Model (CAPM) framework developed by Sharpe (1964). The equation can be displayed as follows:

$$R_e = R_F + \beta_a (R_M\text{-}R_F) \quad (2)$$

Where, β_a is the company's beta; $R_F$is the risk free rate and R_M is the market risk. Beta was used to measure the systematic component of risk for the common stock. It was calculated as the slope coefficient of the characteristic line, and also can be expressed as:

$$\beta_a = \frac{\mathrm{Cov}(r_a, r_p)}{\mathrm{Var}(r_p)} \quad (3)$$

Where r_a measures the rate of return of the asset, r_p measures the rate of return of the portfolio, and Cov (r_a,rp) is the covariance between the rates of return.

Independent Variable

The quality of voluntary carbon emission disclosure (VCEDQ) is computed based on hand-collected data for the fiscal year 2014 annual report. Although the firms have an alternative way to present corporate information, annual report is one of the main resources of corporate information and the most likely place that stakeholders and parties interested in environmental disclosure (carbon information is part of it) would seek and obtain information.

This study employs content analysis as it is widely used in the studies in the areas of climate change, GHG and carbon emission disclosure. The content analysis index developed by Choi, Lee and Psaros (2013) is employed in this study to measure the quality of voluntary carbon emission disclosure. The author has constructed the checklist based on the factors identified in information request sheets by the Carbon Disclosure Project (CDP). The index consists of 18 disclosure items related to 5 areas of information. All items are treated as equally important and assigned the weights of 0 to 1 to the items disclosed. A dichotomous variable has been used to identify items that can appear in the firm's annual report. As such, 1 is awarded if the firms disclosed information related to the items and 0 is otherwise. The score was then divided by the total number of items included in the information checklist (18) to achieve the overall reporting quality score.

Control Variables

*Corresponding author's email: noraida@uniten.edu.my

DOI: 10.1051/shsconf/20173600020

To control for the effect of other determinants of the cost of capital, we employed firm characteristics variables such as company size, leverage, profitability, market to book to book value and capital intensity. These variables are used as they are consistently found to be related to cost of capital and significant determinants of cost of capital (Jung, Herbohn and Clarkson, 2014).

$SIZE_t$	=	Firm size, measured as the natural log of total assets in year t
LEV_t	=	leverage, measured as total debt divided by total assets in year t
ROA_t	=	profitability, measured as earnings before interest and taxes divided by of total assets in year t value of debt, divided by total assets in year t
MB_t	=	market-to-book ratio, measured as the market value of equity plus the book
CAP_INT_t	=	capital intensity, measured as capital expenditures divided by sales revenues in year t

Empirical Model

The study predicted that cost of capital is influenced by quality of VCR. Thus, the models for this study are presented as follows:

$$WACC = \beta_0 + \beta_1 VCEDQ + \beta_2 LnSIZE + \beta_3 LEV + \beta_4 ROA + \beta_5 MBV + \beta_6 CAP\_INT_t \ \varepsilon$$

WACC is firm's cost of capital; VCEDQ is the quality of voluntary carbon emissions reporting; FSIZE is the natural log of total assets; LEV is the ratio of total debt to total assets ROA is the ratio of return to total asset; CAP_INT is the ratio of capital expenditure to sales revenue and ε is the estimate error.

RESULT AND INTERPRETATION

Descriptive Analysis

Table 1epresents the findings on the number of reporting companies from year 2013 to 2014. Overall, it is found that there is an increase of number of reporting companies in 2014 as compared to 2013. The result indicates that 53% of the total companies in 2014 have disclosed their information on carbon emission while only 35% of the total companies disclosed such information in 2013.

Table 1: Number of reporting company

	2013	2014
Number of disclosing companies	86	131
Number of non-disclosing companies	161	116
Total	247	247

Table 2 reports descriptive analysis for variables used in this study. The mean WACC estimate is 7.6079 in 2013 and 7.2869 in 2014. The finding in Table 2 reveals that the quality of

*Corresponding author's email: noraida@uniten.edu.my

DOI: 10.1051/shsconf/20173600020

reporting is relatively low with an average score of 3.7120% in year 2013 and to 5.8287% in year 2014.

Table 2: Descriptive Statistics of Variables

	2013				2014			
	MEAN	**STD DEV**	**MIN**	**MAX**	**MEAN**	**STD DEV**	**MIN**	**MAX**
WACC	7.6079	2.4229	1.7013	16.8164	7.2869	2.45351	1.6903	19.7688
VCEDQ	3.7120	7.1088	0.00	50.00	5.8287	8.50581	0.00	57.14
SIZE	2271855	8431486	9980	99999300	2328364	8738939	28322	1110000000
LEV	0.37887	0.1941	0.0062	0.9835	0.3777	0.1977	0.0059	1.4604
ROA	3.9345	8.94078	-38.85	58.32	3.6953	11.2965	-59.59	67.70
MB	1.9328	5.31699	.08	58.34	1.7530	5.2637	-0.3522	69.9157
CAP_INT	-0.1002	0.2270	-1.9427	-0.0001	-0.1036	0.3319	-0.5213	-0.0009

Nevertheless based on paired samples test shown in Table 3, there is a significant increase in the quality of voluntary carbon emission disclosure from 2013 to 2014. Consistently, the maximum disclosure as in Table 2 is slightly increased to 57.14% as compared to only 50% in 2013. The increase is might be due to increase in the companies' awareness on the environmental issue particularly on carbon emission.

Table 3: Two paired sample t-test

	COEFFICIENT	T-STAT	P-VALUE
VCEDQ		-3.249	0.001

Correlation Analysis

Since the data in not normally distributed, the Pearson product-moment correlation is used to quantify the severity of multicollinearity in model selected. The correlations between the variables were reported in Table 4 and Table 5 for year 2013 and 2014 respectively. The correlation results indicate there is no problems of multicollinearity within the selected variable since the correlations were relatively low for both years. Gujarati (2004) illuminate that multicollinearity problems exist when the correlations value exceeded 0.80. The correlations results for WACC signify negative coefficient with LEV (-0.337); positive significant coefficient with ROA (+0.253); and MBV (+0.357) all at 5 % significant level for the year 2013. The correlations results for the year 2014 also exhibit the same relations as the previous year. The correlations results for WACC signify negative coefficient with LEV (-0.281); positive significant coefficient with ROA (+0.267); and MBV (+0.331) all at 5 % significant level.

Table 4: Pearson's Correlation of Variables for the year 2013

	WACC	**VCEDQ**	**SIZE**	**LEV**	**ROA**	**MBV**	**CAP_INT**
WACC	1.000						
VCEDQ	.044	1.000					
SIZE	.102	.043	1.000				
LEV	-.337**	.059	.219**	1.000			
ROA	.253**	.020	.309**	-.168**	1.000		
MBV	.357**	.007	.194**	.119	.462**	1.000	
CAP_INT	-.045	-.041	-.221**	-.038	-.065	-.211**	1.000

Table 5: Spearman Correlation of Variables for the year 2014

	WACC	**VCEDQ**	**SIZE**	**LEV**	**ROA**	**MBV**	**CAP_INT**
WACC	1.000						

*Corresponding author's email: noraida@uniten.edu.my

VCEDQ	.051	1.000					
SIZE	-.026	.139**	1.000				
LEV	-.281**	.048	.312**	1.000			
ROA	.267**	.160**	.254**	-.200**	1.000		
MBV	.331**	.195**	.148**	.066**	.426**	1.000	
CAP_INT	-.116	-.039	-.281**	-.069	-.139**	-.204**	1.000

Particularly, there is no rule in determining the strength of association to particular values towards the magnitude of the Pearson's correlation coefficient. Hitherto, there are some general guidelines provided by Cohen, Cohen, West and Aiken (2003) indicating a coefficient with r < 0.3 does indicates a small coefficient; while coefficient's in range of .3 < | r | < .5 indicates medium or moderate correlation as exhibit by the of correlations' result in Table 4 and 5. Though the estimated correlation coefficient does indicates positive and negative correlated within the variable selected however it still considers low; therefore, it is not large enough to cause any concern in the regression model.

Multiple Regressions Analysis

To test the hypothesis that firm's level of WACC is a function of six tested variables which are VCEDQ, SIZE, LEV, ROA, MBV and CAP_INT, the multiple regression analysis was performed using 247 observations for the year 2013 and 2014. The results of the regression analysis provided in Table 6 do conveyed the confirmation for the research hypothesis for both years. For year 2013, the WACC signify positive significant relations with SIZE (+2.480) and MBV (+3.078) at 5% and 1 % significant level respectively. While for leverage ratio (LEV) the coefficient value indicating negative significant relations with cost of capital (-6.845) and (-4,454) at 1 % significant level respectively for year 2013 and 2014. This in line with the study done by Li et al. (2014) which reports leverage is negatively and significantly associated with the cost of debt.

Yet the quality of VCED towards cost of capital cannot be confirms in this study since the results for VCEDQ for both years are insignificant. This results are contradict towards a few study done on these issue (i.e; Bonetti et al. 2013; He et al., 2013, Lambert et al., 2007; Li et al., 2014 and Plumlee et al. 2015) which reported a significant relations with cost of capital. Nonetheless, most of the study used separate indictor for each cost of obtaining the capital which are cost of debt and cost of equity instead of the weighted cost of both debt and equity's capitals as per current study. From the Malaysian experience, the finding of the current study fails to confirm significant relations between the VCED quality with cost of capital.

Table.6. Multiple Regressions

	2013			2014		
	COEFFICIENT	**T-STAT**	**P-VALUE**	**COEFFICIENT**	**T-STAT**	**P-VALUE**
Constant		5.051	.000		6.167	.000
VCEDQ	.022	.378	.706	.025	.405	.686
SIZE	.154	2.480	.014*	.023	.350	.726
LEV	-.433	-6.845	.000**	-.315	-4.454	.000**
ROA	-.011	-.153	.878	.084	1.130	.259
MBV	.213	3.078	.002**	.065	.894	.372
CAP_INT	-.005	-.080	.937	-.086	-1.314	.190
R Square			0.185			0.105

*Corresponding author's email: noraida@uniten.edu.my

Adjusted R Square	0.165	0.082
F statistic	9.109	4.669
Anova	0.000	0.000

Even so, the ANOVA results thus support hypotheses 1 as depicted in Table 6, the F statistics is substantiated at the 1% significant level for year 2013 (9.109) and 2014 (4.669). Thus entail the null hypotheses that the regressions coefficients are all zeros can be rejected at 1% level of significant. Consequently denoting that the estimated regressions for hypotheses 1 is efficient for predictions, and the hypotheses can be accepted. Indicating that there are associations between selected variable with cost of capital for listed companies in Bursa Malaysia is concern.

CONCLUSIONS

This study focuses on the relationship between the quality of voluntary carbon emission disclosure and weighted cost of capital. Overall this study concludes that the voluntary carbon emission disclosure in Malaysia is still low, as such, the quality of carbon emission disclosure do not have an impact on firm's cost of capital. However, the result of the study shows a significant increase in the level and quality of carbon emission disclosure practice from 2013 to 2014. This indicates that level of awareness on carbon emission reporting among companies in Malaysia has increased between those years. The results of the study allow the government to measure progress toward achieving its target to reduce carbon emission and will add weight to the call by accounting regulation body such as Malaysian Accounting Standard Board for a specific standard on carbon reporting.

The findings of the study should be interpreted in light of several limitations. Firstly, only two year of observations are considered in the current study. Thus, it would be interesting to conduct a longitudinal study which involves more than two year of observation as it may help to trace the trend of carbon emission disclosure and the impact towards firms' cost of capital. Secondly, the study focuses merely on the quality of carbon emission reporting. In addition, the aspects focus in this study is quite limited and the area focus should be extended to cover wider aspects in the annual report. Lastly, it would be interesting for future research to consider the development of a comprehensive framework, as to assess the quality of corporate governance and examine its influence on voluntary disclosure of carbon emission.

REFERENCES

Al-Amin, A. Q., Jaafar, A. H., & Siwar, C. (2010). Climate change mitigation and policy concern for prioritization. *International Journal of Climate Change Strategies and Management*, *2*(4), 418–425. doi:10.1108/17568691011089936

*Corresponding author's email: noraida@uniten.edu.my

Bachoo, K., Tan, R., & Wilson, M. (2013). Firm value and the quality of sustainability reporting in Australia. *Australian Accounting Review*, *23*(1), 67–87. http://doi.org/10.1111/j.1835-2561.2012.00187.x

Bazine, E., & Vural, D. (2011). *Voluntary disclosure of financial targets*. Working Paper.

Belal, A. R., Kabir, R., Cooper, S., Dey, P., Khan, N. A., Rahman, T., & Ali, M. (2010). *Corporate environmental and climate change disclosures: Empirical evidence from Bangladesh. Research in Accounting in Emerging Economies* (Vol. 10). Elsevier. http://doi.org/10.1108/S1479-3563(2010)0000010011

Besio, C., & Pronzini, A. (2014). Morality, Ethics, and Values Outside and Inside Organizations: An Example of the Discourse on Climate Change. *Journal of Business Ethics*, *119*(3), 287–300. http://doi.org/10.1007/s10551-013-1641-2

Bonetti, P., Chi, C. H., Michelon, G., & Tanaka, Y. (2013). *Environmental disclosure and the cost of capital: Evidence from the Fukushima nuclear accident*.

Choi, B. B., Lee, D., & Psaros, J. (2013). An analysis of Australian company carbon emission disclosures. *Pacific Accounting Review*, *25*(1), 58–79. http://doi.org/10.1108/01140581311318968

Cohen, J., Cohen, P., West, S., & Aiken, L. (2003). Applied Multiple Regression / Correlation Analysis for the Behavioral Sciences. *2nd Ed Hillsdale NJ Lawrence Erlbaum Associates*, *Third Edit*, 703 S. http://doi.org/10.2307/2064799

Connelly, B. L., Certo, S. T., Ireland, R. D., & Reutzel, C. R. (2011). Signaling Theory: A Review and Assessment. *Journal of Management*, *37*(1), 39–67. http://doi.org/10.1177/0149206310388419

Dhaliwal, D. S., Li, O. Z., Tsang, A., & Yang, Y. G. (2011). Voluntary nonfinancial disclosure and the cost of equity capital: The initiation of corporate social responsibility reporting. *Accounting Review*, *86*(1), 59–100. http://doi.org/10.2308/accr.00000005

Embong, Z., Mohd-Saleh, N., & Sabri Hassan, M. (2012). Firm size, disclosure and cost of equity capital. *Asian Review of Accounting*, *20*(2), 119–139. http://doi.org/10.1108/13217341211242178

Freedman, M., Park, J. D., & Stagliano, a. J. (2015). Mandated Climate Change Disclosures: A Study of Large US Firms That Emit Carbon Dioxide. *Advances in Public Interest Accounting*, *18*, 99–121. http://doi.org/10.1108/S1041-706020150000018003

Gujarati, D. N. (2004). *Basic Econometrics. New York*. http://doi.org/10.1126/science.1186874

Hassan, O. A. G., Romilly, P., Giorgioni, G., & Power, D. (2009). The value relevance of disclosure: Evidence from the emerging capital market of Egypt. *International Journal of Accounting*, *44*(1), 79–102. http://doi.org/10.1016/j.intacc.2008.12.005

*Corresponding author's email: noraida@uniten.edu.my

He, Y., Tang, Q., & Wang, K. (2013). Carbon disclosure, carbon performance, and cost of capital. *China Journal of Accounting Studies*, *1*(3–4), 190–220. http://doi.org/10.1080/21697221.2014.855976

Healy, P. M., & Palepu, K. G. (2001). Information asymmetry, corporate disclosure, and the capital markets: A review of the empirical disclosure literature. *Journal of Accounting and Economics*, *31*(1–3), 405–440. http://doi.org/10.1016/S0165-4101(01)00018-0

Jung, J., Herbohn, K., & Clarkson, P. (2014). *The impact of a firm ' s carbon risk profile on the cost of debt capital : Evidence from Australian firms The impact of a firm ' s carbon risk profile on the cost of debt capital : Evidence from Australian firms*.

Lambert, R., Leuz, C., & Verrecchia, R. E. (2007). Accounting information, disclosure, and the cost of capital. *Journal of Accounting Research*, *45*(2), 385–420. http://doi.org/10.1111/j.1475-679X.2007.00238.x

Li, Y., & Yang, H. (2011). Disclosure and the Cost of Equity Capital: An Analysis at the Market Level, 1–52. Retrieved from http://papers.ssrn.com/sol3/papers.cfm?abstract_id=1965663

Linnenluecke, M. K., Birt, J., & Griffiths, A. (2015). The role of accounting in supporting adaptation to climate change. *Accounting & Finance*, *55*, 607–625. http://doi.org/10.1111/acfi.12120

Merkl-Davies, D. M., & Brennan, N. N. (2007). Discretionary Disclosure Strategies in Corporate Narratives: Incremental Information or Impression Management? *Journal of Accounting Literature*, *26*, 116–196. http://doi.org/10.3386/w19846

Modigliani, F., & Miller, M. H. (1958). The Cost of Capital, Corporation Finance and the Theory of Investment. *American Economic Review*, *48*(3), 261–297. http://doi.org/10.4013/base.20082.07

Murad, M. W., Molla, R. I., Mokhtar, M., & Raquib, M. A. (2010). Climate change and agricultural growth: an examination of the link in Malaysia. *International Journal of Climate Change Strategies and Management*, *2*(4), 403–417. http://doi.org/10.1108/17568691011089927

Peng, J., Sun, J., & Luo, R. (2015). Corporate Voluntary Carbon Information Disclosure: Evidence from China's Listed Companies. *World Economy*, *38*(1), 91–109. http://doi.org/10.1111/twec.12187

Plumlee, M., Brown, D., Hayes, R. M., & Marshall, R. S. (2015). Voluntary environmental disclosure quality and firm value: Further evidence. *Journal of Accounting and Public Policy*, *34*(4), 336–361. http://doi.org/10.1016/j.jaccpubpol.2015.04.004

Renukappa, S., Akintoye, A., Egbu, C., & Goulding, J. (2013). Carbon emission reduction strategies in the UK industrial sectors: an empirical study. *International Journal of Climate Change Strategies and Management*, *5*(3), 304–323. http://doi.org/10.1108/IJCCSM-02-2012-0010

*Corresponding author's email: noraida@uniten.edu.my

Richardson, A. J., & Welker, M. (2001). Social disclosure, financial disclosure and the cost of equity capital. *Accounting, Organizations and Society*, *26*(7–8), 597–616. http://doi.org/10.1016/S0361-3682(01)00025-3

Sharpe, W. F. (1964). Capital asset prices: A Theory of market equilibrium under conditions of risk. *The Journal of Finance*, *19*(3), 425–442.

Souissi, M., & Khlif, H. (2012). Meta-analytic review of disclosure level and cost of equity capital. *International Journal of Accounting and Information Management*, *20*(1), 49–62. http://doi.org/10.1108/18347641211201072

*Corresponding author's email: noraida@uniten.edu.my

LINKING WORKING CAPITAL POLICY TOWARDS FINANCIAL PERFORMANCE OF SMALL MEDIUM ENTERPRISE (SME) IN MALAYSIA

Nor Edi Azhar Binti Mohamad*[1], Noor Raida Binti Abd Rahman[2], Noriza Binti Mohd Saad[1]
Finance and Economic Department, The National Energy University UNITEN
Accounting Department, The National Energy University UNITEN

ABSTRACT

Despite the fact that working capital management (WCM) is vital to businesses of any size that operated in developed and emerging countries, WCM is of particular importance to the small business firms operating in emerging markets. The importance of WCM to small and medium-sized enterprises (SMEs) stems from the limited financial resources available and heavily reliance of SMEs on WCM as a main source of finance. This study aims to provide empirical evidence on the effects of working capital investment policy on firm's financial performance for a sample of 103 small and medium-sized firms listed with the SME Corporation of Malaysia. Data for the period from 2008 to 2013 are analysed to examine if investment policy improves firms' return on total asset. By using correlation and pooled ordinary least square regression, the result provides a significant relationship between the level of aggressiveness of investment policy and SME's financial performance. The findings of this study not only contribute to the scant WCM literature in Malaysia but throw light on the importance of efficient WCM to the policy makers and regulators in motivating and encouraging relevant parties to pay more attention on working capital through improving investors' awareness and improving transparency.

Keywords*:* Working Capital Management, Small and medium-sized enterprises, Working Capital Investment Policy, Return on Total Asset

INTRODUCTION

Small and medium-sized enterprises (SMEs) are the backbone of the resilient national Economy in boosting domestic ultimatum. SMEs have been identified as a significant contributor of the economy particularly those from an emerging countries (Hashim, 2015). In Malaysia, there was a total of 662,939 SMEs operating their businesses, represents over 97% of the total business establishments and contributes for more than one-third of the total Malaysian GDP (SME Annual Report, 2013). Notwithstanding the remarkable achievement of a number of SMEs, nevertheless, many more still struggle to ensure long-term growth, profitability and survival (Lyngstadaas and Berg, 2016). Even for a firm with favorable long-run prospects, inattention to efficient working capital management (WCM) may cause severe difficulties and losses due to adverse short-run developments (Pais and Gama, 2015).

*Corresponding author's email: NorEdi@uniten.edu.my

The implication of efficient WCM for the betterment of the firm has been discussed in the previous studies (Park, 1951). It is considered as one of the key indicators of the firms' financial health (Altman, 1968) and also proven as one of the significant factors causing industrial sickness (Yadav, 1986). Past researches depicted that poor WCM and inadequate long-term financing are the main cause of failure among small businesses (Ramiah, Zhao and Moosa, 2014). Inefficient WCM might lead to tying ups the important sources of funds in an unproductive assets which contribute towards reducing the firm's liquidity and profitability (Panigrahi, 2013).

Given the limited financial resources available to SMEs and rely heavily on WCM as a vital source of finance as compared to its larger counterparts (Tauringana and Afrifa, 2013)., effective management of their working capital is vital to SMEs than to larger firm (Afrifa and Padachi, 2016). Yet, there is still less attention paid to this issue academically, particularly in Malaysia (Wasiuzzaman, 2015). It is interesting to know how the working capital policy influences the profitability of SMEs in the Malaysia, as an emerging market, given that over the years Malaysian firms showed unsatisfactory working capital performance (Wasiuzzaman, 2015). Additionally, with the diverse economic conditions and culture, it is likely that investors in Malaysian have a different view and approach on the level of working capital compared to investors in the other countries. Therefore, the objective of this study is to investigate the link between WCM policy and financial performance of 103 Malaysian SMEs firm during the period 2008 to 2013. In this study, greater emphasised is given on the asset management as it is one of the important components of working capital. The result of the study supports the current use of a conservative WCM strategy.

This study is pertinent for a number of reasons. First, it contributes to the existing WCM literature. While at the international level, there is considerable literature on WCM (Wasiuzzaman, 2015), this study adds to the limited literature on WCM issues in Malaysia, which mainly focuses on SMEs. Second, this study helps the management of SMEs to prioritise their scarce resources to manage working capital component that are relatively significant for their profitability. Third, the finding is of important for regulators and policy makers in motivating and encouraging relevant parties to pay more attention on working capital.

The paper proceeds as follows. Section 2 reviews previous studies on issues related to working capital and performance. Then the research methodology employed in this study is introduced and explained in Section 3. Next the discussion of the results is presented in Section 4, followed by a conclusion and suggestions for future research in the last section.

LITERATURE REVIEW

Tradeoff theory delineates that firms need to keep a balance between liquidity and profitability (Abuzayed, 2012). The firm's WCM has a significant influence on its liquidity and profitability, hence the health of businesses (Singh and Kumar, 2014). The adopted policy choose by the firms will influence investment level in working capital (Afrifa, 2016; Garcia-Teruel and Martinez-Solano, 2007). Normally, a firm may adopts either an aggressive or conservative working capital policy by reducing or increasing the level of investment in working capital (Afza and Nazir, 2011; Nazir and Afza, 2009; Tauringana and Afrifa, 2013). The firm's ability in optimising the required amount of working capital in which could boost

*Corresponding author's email: NorEdi@uniten.edu.my

firm's profitability and creating more market value with minimization in risks was considered as an optimal policy (Nobanee, Abdullatif and AlHajjar, 2011).

The working capital policy provides a guideline for the firm to manage the current assets and current liabilities with the main objective of reducing default risk (Afza and Nazir, 2007). The conservative approach is a strategy that keeps a large amount of capital in current assets as compared to the fixed-assets, thus it will increase investment in working capital (Garcia-Teruel and Martinez-Solano, 2007). This strategy will help the firms to reduce the risk of short-term cash shortage, nevertheless for long-term it might affect the profitability. Additionally, expanding investment in working capital may result in opportunity cost of cash tied-up in the unproductive assets such as inventory and accounts receivable (Tauringana and Afrifa, 2013). Thus, by adopting this strategy, the firms obtain a lower expected profitability resulting in a lower risk (Nwankwo and Osho, 2010).

In contrast, firms adopting the aggressive approach will gain higher profitability that result from higher risk and lower working capital (Nwankwo and Osho, 2010). This approach would positively affect the profitability of the firm, by reducing the proportion of its total assets in the form of net current assets (Garcia-Teruel and Martinez-Solano, 2007). Nonetheless, the aggressiveness of WCM resulted in capital being minimised in current assets over fixed-asset might increase the liquidity risk. Additionally, firms that used the long-term capital to finance all of its fixed assets and used short-term credit to finance parts of its current assets could lead to risk of default and bankruptcy risk (Weinraub and Visscher, 1998).

Weinraub and Visscher (1998) is considered as one of the pioneers discussing the aggressive and conservative WCM policies. From their observation, the author found that each industry had a different WCM policies. Many researchers have established a relationship between WCM and profitability (Sing and Kumar, 2014). In the preceding studies, the findings on the WCM - profitability nexus are mixed. Some supported conservative strategy of WCM, arguing that the higher the working capital, the higher the profitability of the firm. The association between cash conversion cycle (CCC), a component used as a measure of working capital skill; and profitability was found to be positive and significant by researchers including Abuzayed (2012), Afrifa and Padachi (2016), Gill, Biger and Mathur (2010) and Samiloglu and Demirgunes (2008). The authors concluded that the higher the CCC, the higher the profitability of the firm.

Using data from 52 firms listed in Amman Stock Exchange for the period from 2000 to 2008, Abuzayed (2012) found a significant positive relationship between CCC and profitability. The result showed profitable firms less motivated in managing their working capital. Afrifa and Padachi (2016) made use of a sample of 160 Alternative Investment Market (AIM)-listed SMEs for the period from 2005 to 2010 to accentuate the relationship between WCM and profitability. The authors recorded that the profitability increase as CCC level increase; nevertheless it then starts to decline if it rises beyond a certain level. Similarly, Afrifa (2016) reported a strong concave relationship between net working capital and performance of 6,926 SMEs in the UK. However, when the manager is considering cash flow in working capital, the performance of the firms is improving.

In line with the aggressive strategy of WCM, Al-Shubiri (2011), Garcia-Teruel and Martinez-Solano (2007), Lyngstadaas and Berg (2016), Nobanee et al. (2011) and Pais and Gama, (2015) reported a negative relationship between WCM elements and profitability. In the Spain context,

*Corresponding author's email: NorEdi@uniten.edu.my

Garcia-Teruel and Martinez-Solano (2007) tested the effects of WCM of 8,872 Spanish SMEs on the firm profitability. The finding demonstrates that managers can create value by reducing their inventories and the number of days for which their accounts are outstanding.
A study conducted by Al-Shubiri (2011) for a sample of Amman Industrial Firms listed in Amman Stock Exchange for the period of 2004 to 2007 also indicates that firm's obtained a negative returns if the firm's follow aggressive working capital policy. The author argued that funds committed to working capital can be seen as hidden sources that could be utilized to improve profitability of the firm.

In a recent study, Pais and Gama (2015) examined the effects of WCM on the profitability of 6,063 Portuguese SMEs firms from the year 2002 to 2009. While, Lyngstadaas and Berg (2016) assessed the relationship between profitability of 21,075 SMEs and the inventory, accounts payable, accounts receivables and CCC variables for the period 2010 to 2013 in Norway. In Japan, Nobanee et al. (2011) examined the relation between CCC and firm profitability of 34,771 non-financial firms listed on the Tokyo Stock Exchange from 1990-2004 All of these studies concluded that by shortening the CCC firm's profitability could be improved. By shortening the length of CCC, external financing is unneeded, thus increasing profitability due to fewer borrowing cost and interest expense.

The studies for Malaysian sample were done by Zariyawati, Annuar, Taufiq and Rahim, (2009), Mohamad (2013) and Wasiuzzaman (2015). Zariyawati et al. (2009) examined the relationship between the length of the CCC and profitability of 148 companies in six different economic sectors over 1996-2006 periods. Their analysis provides a strong negative significant relationship between CCC and firm profitability. Mohamad (2013) examined the relationship between return factors and the firm's WCM policy of 370 companies listed in Bursa Malaysia. The finding indicated a significant positive relationship between both variables. In addition, the finding showed a negative/(positive) relationship between the degrees of aggressiveness/(conservativeness) of the firm's financing policy with the performance measurements. In a recent study, Wasiuzzaman (2015) examined the relationship between working capital efficiency and firm value of 192 firms spanning a period of ten years (1999-2008). The finding reveals that there is a negative impact of net operating working capital on firm value, implies that investors prefer firms which follow a more restrictive working capital policy. Nevertheless, these studies do not focus the effect of WCM and financial performance of SMEs. Based on the previous researches presented in this section, the following hypothesis is developed.

H1: There is a significance relationship between working capital policy and SME's financial performance

RESEARCH METHODOLOGY

Sample and Data

To accomplish the aforementioned research objectives, this study focuses SMEs listed in SME Corporation Malaysia for 6 years basis ranging from year 2008 to 2013. A sample of 250 annual reports of the selected SMEs was obtained from the Suruhanjaya Syarikat Malaysia. However, only 103 companies were valid for the analysis after deducting the companies with

*Corresponding author's email: NorEdi@uniten.edu.my

missing information and also companies that reported a loss for more than four consecutive years.

Variables

Financial performance is dependent variable, with Return on Total Assets (ROTA) as a proxy for profitability. Financial performance was considered as an indicator of a firm's financial health over a given period of time. The SME is categorized based by a low fixed asset that relied more towards accounts payable in funding its gross working capital (Padachi, 2006). Thus, in capturing its financial performance, ROTA could serve as an appropriate indicator in capturing the return towards total assets in which also cover for SME's total liability.

Following Weinraub and Visscher (1998), Nazir and Afza (2009) and Mohamad (2013), this study utilised investment policy (IP) as a proxy for working capital policy. The IP represent the level of investment in current assets versus fixed assets, which act as independent variable for this study. The conservative policy is used in explaining an increase in the level of investment in the current assets. Conversely, an aggressive policy was represents by investing fewer amounts in current assets. The following ratio was used to measure the degree of aggressiveness, in which the lower ratio indicates a relatively conservative policy:

$$\text{Investment Policy (IP)} = \text{Total Current Assets} / \text{Total Assets} \quad (1)$$

Others independent variables selected for the study are account payable turnover (APTO), account receivable turnover (ARTO) and cash (lgCASH) to represent the important component of working capital. ARTO was used to indicate the number of times per year the SMEs collect its average accounts receivable. It implies the SME's efficiency in using its asset in generating the revenue. The account receivable brings a positive impact towards an increase in working capital since it signify cash in terms of current assets components that should be received by the firms within a relatively short time period, ranging from a few days to years. The ratio can be calculated by dividing the credit sales with average account receivable. The higher the ratio indicates the SME's efficiency in the collection of receivable.

Account payable denotes the credit terms given by the suppliers or trade credit due to the supply of goods and services rendered to the firms (Okinyi, 2014). Account payable is also considered as a significant source of short-term financing available to the firms with minimum cost. The effective management of account payable might serve as indicator of SMEs overall operational effectiveness (Kallikkat, 2013). The higher the account payable might indicating problems in resolving payment to suppliers (Deloitte, 2015) while lower payables indicating firm's had waste the benefits of investment for the available cash by early settlement of the suppliers invoices (Kallikkat, 2013). The account payable turnover (APTO) measures the speeds of SME abilities in paying its suppliers. The ratio can be calculated by dividing the total purchases made from suppliers, or cost of sales with average account payable. The lower the ratio indicates inefficiency in payment policy thus could signify deteriorating in financial conditions.

lgCASH is used as proxy for cash. It is measured based on the log value of cash and cash equivalent hold by SMEs. Pastor and Gama (2013) believe SMEs hold large amounts of cash and cash equivalents for a various reasons. The higher level of cash hold might reduce the

*Corresponding author's email: NorEdi@uniten.edu.my

possibility of financial distress; representative towards firm's optimal investment policy; and potential contribution towards effective used of cost of external financing resources (Ferreira and Vilela, 2004).

Estimation

The relationship between the SME's financial performance and WCM policy is estimated using the following regression equations:

$$ROTA = \alpha + \beta_1 IP_1 + \beta_2 ARTO_2 + \beta_3 APTO_3 + \beta_4 lgCASH + e \quad (2)$$

where ROTA is return on total assets; IP, investment policy; ARTO, accounts receivable turnover; APTO, account payable turnover; lgCASH, log value of cash and cash equivalent and ε, residual error (unobservable factors that vary over time and affect ROTAi,t). in order to analyse the data, this study applied panel data methodology. A F-test was used to decide if fixed effect models are preferred over the pooled OLS, with the null hypothesis that all fixed effect intercepts are zero.

RESULTS AND DISCUSSION

Result and Analysis

The objective of this study is to investigate the link between WCM policy and financial performance of 103 Malaysian SMEs firm during the period 2008 to 2013. The following sections discuss the relevant analysis to achieve this objective.

Correlation Analysis

Table 1 presents the Pearson correlation matrix for the variables. Results indicated that there is no multicollinearity problems as the correlations were relatively low. According to Gujarati (2004), multicollinearity problems exist when the correlations value exceeded 0.80. The correlations results for ROTA indicated a positive significant coefficient with IP (0.258) and

Table 1. Correlations Analysis

	ROTA	IP	ARTO	APTO	lgCASH
ROTA	1	.258**	-.043	.035	.095*
IP	.258**	1	-.009	-.019	.085*
ARTO	-.043	-.009	1	.141**	-.088*
APTO	.035	-.019	.141**	1	-.026
lgCASH	.095*	.085*	-.088*	-.026	1

**. Correlation is significant at the 0.01 level (2-tailed).
*. Correlation is significant at the 0.05 level (2-tailed

lgCASH (0.095) at 1% and 5% significant level respectively. This result of the study is in line with the study conducted by Mohamad (2013). The magnitude of the Pearson correlation coefficient determines the strength of the correlation but there is no rule in assigning strength of association to particular values. Yet, some general guidelines are provided by Cohen, Cohen, West and Aiken (2003), indicating a coefficient with $r < 0.3$ does indicate a small coefficient. Although results of the estimated correlation coefficient indicate negative and positively

*Corresponding author's email: NorEdi@uniten.edu.my

correlated within the selected variable, however it still considers low; therefore, it is not large enough to cause any concern in the regression model.

Multiple Regressions

Table 2 contains result of estimation involving ROTA. To quantify the severity of multicollinearity for this study, the variance inflation factor (VIF) test was performed. The VIF results designate a very low level of multicollinearity since the VIF value for all variable tested is less than 10. VIF is simply the reciprocal of Tolerance (i.e., 1 divided by Tolerance). The results will indicates a sign of collinearity problem if the Tolerance value is less than 0.1; which is a VIF of greater than 10.

Based on Table 2, the Collinearity Statistics indicates that all the Tolerance values are greater than 0.1 (the lowest is 0.973). Thus results depict no problem of collinearity in this particular data set. Even though there is no formal VIF value, however, values of VIF that exceed 10 are often regarded as indicating multicollinearity (e.g., Hair, Anderson, Tatham and Black, 1995). In addition, there was independence of residuals, as assessed by a Durbin-Watson statistic 2.023. The Durbin-Watson statistic can range from 0 to 4, but a value of approximately 2 indicates that there is no correlation between residuals. Based on the results, the value is very close to 2, so it can be accepted that there is independence of errors (residuals).

Results of the regression analysis in Table 2 signify support the research hypothesis. The results for ROTA depicted a significant positive coefficient with IP (+6.461) at 1 % significant level. The positive coefficients of ROTA and IP signified a negative relationship between the degrees of aggressiveness in firm's investment policy with the performance measurements. Additionally, an increase in IP indicated a conservative's management of investment policy which emphasize more investment in liquid assets (Weinraub and Visscher, 1998).

The results support the implementation of conservative investment policy among SME's in Malaysia. The result is consistent with Nazir and Afza (2009) and Mohamad (2013) which also indicating a positive relationships between ROTA and conservative investment policy. Surprisingly, the result is similar with the study conducted by Mohamed (2013) for listed firms in Bursa Malaysia. The consistent result signifies that Malaysian SMEs have a working capital policy similar with its larger counterpart.

Table 2. Regression Results of ROTA and Working Capital Policy

	B	Std. Error	t	Sig.	Collinearity Statistics	
Variable /Indicator					**Tolerance**	**VIF**
IP	6.017	.931	6.461	.000**	.992	1.008
ARTO	-.014	.013	-1.042	.298	.973	1.028
APTO	.030	.025	1.220	.223	.980	1.021
lgCASH	1.197	.656	1.824	.069*	.985	1.015
R		0.274				
R Square		0.075				
F (p-value)		12.447(0. 000)				
df		(4,612)				
D-Watson		2.023				

** Correlation is significant at the 0.01 level (2-tailed)
*Correlation is significant at the 0.10 level (2-tailed).

*Corresponding author's email: NorEdi@uniten.edu.my

In addition, lgCASH (+1.824) also indicated a positive significant relationship with ROTA at 10% significant level. The finding implied that cash is an important component of working capital management for SMEs. The result supports the Kinz's theory in which firms hold a certain amount of cash for precautionary, transactional and speculative purposes (Pastor and Gama, 2013). Also, consistent with Abel (2008), cash is a vital component for every business in boosting firm's survival and wealth.

Overall, the regression results support hypotheses 1. As depicted in Table 2, the F statistics is substantiated at the 1% significant level for ROTA (12.447). Therefore, the results implied that the null hypotheses can be rejected at 1% level of significant. Hence the estimated regressions for hypotheses 1 is efficient for predictions, and the hypotheses can be accepted implying that there are an association between investment policy with SME's financial performance in Malaysia.

CONCLUSION

This paper seeks to investigate the influences of WCM policy towards SME's financial performance from the Malaysian perspective. This study tested ROTA model with the working capital policy for 103 selected SME listed with SME Development Corporation of Malaysia for the period of 2008 to 2013. Based on the findings of the study, it can be concluded that there is a significant relationship between firm's WCM policies and SME's financial performance.

This study provides significant contribution towards the literature on working capital policy for SME by identifying the conservative and aggressive working capital management from Malaysian perspective. However, the major limitation of the study is that it focused on a small SMEs sample size over a period of 2008-2013 and not extending it further and therefore generalization of the findings is restricted. In addition, the study do not considered moderating or mediating influence on the relationships between WCM and financial performance. For future study, it is suggested to use more samples of SMEs, testing with others variables as a proxy for SME's financial performance for an extended period; and also considering the moderating or mediating variables that can mitigate the inconclusive finding on the relationship between WCM and financial performance.

REFERENCES

Abuzayed, B. (2012). Working capital management and firms' performance in emerging markets: the case of Jordan. *International Journal of Managerial Finance*, *8*(2), 155–179. http://doi.org/10.1108/17439131211216620

Afrifa, G. A. (2016). Net working capital, cash flow and performance of UK SMEs. *Review of Accounting and Finance*, *15*(1), 21–44. http://doi.org/10.1108/RAF-02-2015-0031

Afrifa, G. A., & Padachi, K. (2016). Working capital level influence on SME profitability. *Journal of Small Business and Enterprise Development*, *23*(1), 44–63. http://doi.org/10.1108/JSBED-01-2014-0014

*Corresponding author's email: NorEdi@uniten.edu.my

Afza, T., & Nazir, M. S. (2007). Is It Better To Be Aggressive or Conservative in Managing Working Capital? *Journal of Quality and Technology Management*, *3*(2), 1–15. Retrieved from http://www.ciitlahore.edu.pk/Papers/Abstracts/146-8588087935136570808.pdf

Afza, T., & Nazir, M. S. (2011). Working Capital Management Efficiency of Cement Sector of Pakistan. *Journal of Economics and Behavioral Studies*, *2*(5), 223–235. Retrieved from http://www.researchgate.net/publication/233842286_Working_Capital_Management_Efficiency_of_Cement_Sector_of_Pakistan

Al-Shubiri, F. N. (2011). Analysis of the relationhsip between working capital policy and operating risk: An emperical study on Jordian Industrial companies. *Hebron University Research Journal*, *6*(1), 287–306. http://doi.org/10.1017/CBO9781107415324.004

Altman, E. I. (1968). Financial ratios, descriminant analysis and the prediction of corporate bankruptacy. *The Journal of Finance*, *XXIII*(4), 589–609.

Cohen, J., Cohen, P., West, S., & Aiken, L. (2003). Applied Multiple Regression / Correlation Analysis for the Behavioral Sciences. *2nd Ed Hillsdale NJ Lawrence Erlbaum Associates*, *Third Edit*, 703 S. http://doi.org/10.2307/2064799

Deloitte. (2015). *Strategies for optimizing your accounts payable*. Retrieved from https://www2.deloitte.com/content/dam/Deloitte/ca/Documents/finance/ca-en-FA-strategies-for-optimizing-your-accounts-payable.pdf

Ferreira, M. A., & Vilela, A. S. (2004). Why do firms hold cash? Evidence from EMU Countries. *European Financial Management*, *10*(2), 295–319. http://doi.org/10.1111/j.1354-7798.2004.00251.x

Garcia-Teruel, P. J., & Martinez-Solano, P. (2007). Effects of Working Capital Management on SME Profitability. *International Journal of Managerial Finance*, *3*(2), 164–177. http://doi.org/10.2139/ssrn.894865

Gill, A., Biger, N., & Mathur, N. (2010). The relationship between working capital management and profitability : Evidence from the United States. *Business and Economics Journal*, *2010*(1), 1–9. http://doi.org/10.18052/www.scipress.com/ILSHS.20.14

Gujarati, D. N. (2004). *Basic Econometrics*. *New York*. http://doi.org/10.1126/science.1186874

Hair, J. F., Anderson, R. E., Tatham, R. L., & Black, W. C. (1995). *Multivariate Data Analysis: With Readings*. *Statistics* (Vol. 151). http://doi.org/10.2307/2983017

Hashim, Ff. (2015). SMEs' impediments and developments in the internationalization process: Malaysian experiences Fariza. *World Journal of Entrepreneurship, Management and Sustainable Development*, *11*(2), 100–119. http://doi.org/10.1108/WJEMSD-05-2014-0012

Kallikkat, S. (2013). Effectively Managing Accounts Payable. Available at: https://clarity.sutherlandglobal.com/blog/accounting-minute/effectively-managing-accounts-payable. [Retrieved on 5 July 2016].

*Corresponding author's email: NorEdi@uniten.edu.my

Lyngstadaas, H., & Berg, T. (2016). Working capital management: evidence from Norway. *International Journal of Managerial Finance*, *12*(3), 295–313. http://doi.org/10.1108/IJMF-01-2016-0012

Mohamad, N. E. A. (2013). An analysis of working capital management with reference on listed companies in Bursa Malaysia. *American Journal of Economics*, *3*(6), 352–357. http://doi.org/10.5923/j.economics.20130306.16

Nazir, M. S., & Afza, T. (2009). Impact of Aggressive Working Capital Management Policy on Firms' Profitability. *IUP Journal of Applied Finance*, *15*(8), 19–30. http://doi.org/10.5897/AJBM11.326

Nobanee, H., Abdullatif, M., & AlHajjar, M. (2011). Cash conversion cycle and firm's performance of Japanese firms. *Asian Review of Accounting*, *19*(2), 147–156. http://doi.org/10.1108/13217341111181078

Nwankwo, O., & Osho, G. S. (2010). An empirical analysis of corporate survival and growth: Evidence from efficient working capital management. *International Journal of Scholarly Academic Intelectual Diversity*, *12*(1), 1–13. Retrieved from http://www.nationalforum.com/Electronic Journal Volumes/Nwankwo, Odi - An Empirical Analysis of Corporate Survival and Growth IJSAID V12 N1 2010.pdf

Okinyi, B. (2014). Effect of working capital management on financial performance of five star hotels in Nairobi country. *Interdisciplinary Journal of Contemporary Research in Business*, 2(1), 75–79.

Padachi, K. (2006). Trends in working capital management and its impact on firms' performance: An analysis of Mauritian small manufacturing firms. *International Review of Business Research Papers*, *2*(2), 45–58.

Pais, M. A., & Gama, P. M. (2015). Working capital management and SMEs profitability : Portuguese evidence. *International Journal of Managerial Finance*, *11*(3), 341–358. http://doi.org/10.1108/IJMF-11-2014-0170

Park, C. (1951). Working Capital and The Operating Cycle. *The Accounting Review*, *26*(3), 299–307.

Pastor, C. C., & Gama, P. M. (2013). Determinant factors of cash holdings : Evidence from Portuguese SMEs. *International Journal of Business and Management*, *8*(1), 104–113. http://doi.org/10.5539/ijbmv8n1p104

Ramiah, V., Zhao, Y., & Moosa, I. (2014). Working capital management during the global financial crisis: the Australian experience. *Qualitative Research in Financial Markets*, *6*(3), 332–351. http://doi.org/10.1108/QRFM-09-2012-0026

Samiloglu, F., & Demirgunes, K. (2008). The effect of working capital management on firm profitabilitu: evidence from Turkey. *The International Journal of Applied Eonomics and Finance*, *2*(1), 44–50.

*Corresponding author's email: NorEdi@uniten.edu.my

SME Annual Report. (2013). released by the Secretariat of the National SME Development Council on September 24, 2013. Available at http://www.google.com/url?sa=t&rct=j&q=&esrc=s&frm1&source=web&cd=1&ved=0CCUQFjAA&url=http%3A%2F%2Fwww.smecorp.gov.my%2Fvn2%2Fnode%2F717&ei=8M1AU9yKHM_OrQeFlIC4Cg&usg=AFQjCNHtd242V6O098JZm2e7g0Oc9 ghHKw; retrieved on May,18 2016

Singh, H. P., & Kumar, S. (2014). Working capital management: a literature review and research agenda. *Qualitative Research in Financial Markets*, *6*(2), 173–197. http://doi.org/10.1108/QRFM-04-2013-0010

Tauringana, V., & Afrifa, G. A. (2013). The relative importance of working capital management and its components to SMEs' profitability. *Journal of Small Business and Enterprise Development*, *20*(3), 453–469. http://doi.org/10.1108/jsbed-12-2011-0029

Wasiuzzaman, S. (2015). Working capital and firm value in an emerging market. *International Journal of Managerial Finance*, *11*(1), 60–79. http://doi.org/10.1108/IJMF-01-2013-0016

Weinraub, J. H., & Visscher, S. (1998). Industry practice relating to aggressive conservative working capital policies. *Journal of Financial and Strategic Decisions*, *11*(2), 11–18.

Yadav, R.A. (1986). Working capital management - A parametric approach, *The Chartered Accountant*, 535-552.

Zariyawati, M. A., Annuar, M. N., Taufiq, H., & Rahim, A. S. A. (2009). Working capital management and corporate performance: Case of Malaysia. *Journal of Modern Accounting and Auditing*, *5*(11), 47–54.

*Corresponding author's email: NorEdi@uniten.edu.my

THE EFFECTS OF ENVIRONMENTAL, SOCIAL AND GOVERNANCE ON THE CORPORATE PERFORMANCE OF MALAYSIAN GOVERNMENT-LINKED COMPANIES

Qian Long Kweh*, Bakhtiar Alrazi, Yee Chuann Chan, Wan Mohammad Taufik Wan Abdullah
Department of Accounting, Universiti Tenaga Nasional, Pahang, Malaysia

ABSTRACT

This study examines the impacts of ESG on the corporate performance government-linked companies (GLCs) in Malaysia. For the period 2006-2012, ESG disclosure data were extracted from the Sustainalytics ESG performance reports, while financial data were obtained from the Bloomberg database. Data development analysis (DEA) was used to estimate efficiency in the first stage; a regression analysis was performed to test the relationship between ESG and efficiency in the second stage. The empirical results of this study show that GLCs focused more on governance disclosures, followed by social and environmental aspects. Moreover, governance will improve firm efficiency, but social and environmental factors have no similar effect. In conclusion, this study provides insight on the limited literature on ESG and informs the relevant stakeholders on the important ESG components for financial and investment decisions.

Keywords: Environmental, Social and Governance (ESG), firm efficiency, data envelopment analysis, government-linked companies

INTRODUCTION

Environmental, Social and Governance (ESG) information is getting gradually more included into corporate communication (Arvidsson, 2010; Ihlen, 2008). ESG can also be known as 'extra-financial' information that help investors make investment decision by better assessments of risks and opportunities (Bassen & Kovacs, 2008). Prior studies (for example, Bachoo, Tan, & Wilson, 2013; Bebchuk, Cohen, & Wang, 2013; Clark, Feiner, & Viehs, 2015; Servaes & Tamayo, 2013; Wang, Lu, Kweh, & Lai, 2014) have extensively examined the effects of ESG on corporate performance; however, the results have been mixed and thus inconclusive.

The open debate could be due to the inconclusive findings and weaknesses in the measure of firm performance. To elaborate, single-dimensional measures of corporate performance used in the prior studies include return on assets (ROA) that are subject to interpretation and thus may not be sufficient for performance measurement (Feroz, Kim, & Raab, 2003). It is important to note that if a company has good sustainability management, its operating costs would be low and ultimately its efficiency will also improve in the long term (Bachoo et al., 2013).

*Corresponding author's email: qlkweh@uniten.edu.my

Consistent with the government's aspiration in the Economic Transformation Programme (ETP) to achieve its developed status, this study considers a contemporary measure of financial performance which is multidimensional. In contrast to single-dimensional measures of corporate performance, data envelopment analysis (DEA), a mathematical-programming method that incorporates multiple variables (Cooper, Seiford, & Tone, 2006), provides a comprehensive performance measure. Various attributes can be evaluated simultaneously to calculate an efficiency score for a decision-making unit (DMU). In other words, a holistic performance evaluation that provides aggregated activities information can be done via DEA (Homburg, 2001; Yeh, 1996). Taken together, this study examines the relationship between ESG and efficiency, which is considered in a holistic manner.

The paper proceeds as follows. The next section reviews prior studies, while the third section discusses methodology and data collection used in this study. The fourth section presents the empirical results. The final section concludes this study.

The Impacts of ESG on Corporate Performance in Malaysia

Under the fast-changing business environment, companies without SRI may face some problems such as decline in reputation, loss of customer loyalty and sales, and loss of intellectual capital. In other words, good ESG performance may create intangible values through minimizing unnecessary costs and solving potential conflicts with stakeholders. Ultimately, such companies will enjoy continuous support from stakeholders like consumers and investors who value ESG in business. This means that top management of a company has to consider ESG when making business decision. The growing number of ESG reports is an example whereby companies incorporate ESG into business practices (Lee & Saen, 2012).

The extant literature has examined the relationship between various aspects of ESG and firm performance. However, in this section, we only review studies that have been conducted in the Malaysian context. In order to highlight any gaps in which our study addresses, the focus of review is on (1) the scope of investigation, (2) sample used, (3) measurement of variables, and (4) the findings. While we only try to develop some trends here, a full list of studies reviewed is available in Table 1. The list is not exhaustive; however, we believe that we have covered a sufficient number of important journal articles.

*Corresponding author's email: qlkweh@uniten.edu.my

Table 1 Prior studies examining the association between ESG and FP in Malaysia (FP influences ESG)

No	Author(s)	Sample	ESG component	ESG measure	Firm performance	Findings
1	Amran et al. (2012)	100 PLCs (35 members of sustainability networks; 65 conveniently selected from several industries)	Environment	• Annual report and sustainability report disclosures 2009 • GHG disclosure (5 items, max: 12)	Lag return on assets	Positive and significant
2	Esa & Mohd Ghazali (2012)	27 GLCs	CSR	• Annual report disclosures 2005 & 2007 • Categories: human resource, value-added, environment, community, & product • 21 items (1,0 scoring)	Profit before tax over total assets	Positive but not significant
3	Othman et al. (2011)	117 PLCs from industrial products, properties, and plantation sectors (random)	ESG	• Annual report disclosures 2007 • Categories: citizenship (community & environment), workplace, and governance (marketplace) • 40 items (0-2 scale; max: 80)	Profit after tax over total assets (control variable)	Positive and significant
4	Sheikh Abu Bakar & Ameer (2011)	333 PLCs (6 sectors); only 131 reporting companies selected for further analysis	CSR	• Annual report disclosures 2007 • Readability score - using Flesh Reading Ease and Bullfighter composite index	• Earnings before tax over total assets • Tobin's q	Positive and significant
5	Wan Abd Rahman et al. (2011)	44 GLCs	CSR	• Annual report disclosures 2005-2006 • Categories: human resource, community, marketplace, & environment • 16 items (Number of sentences)	Net profit after tax over sales	Positive but not significant
6	Abd. Rahman et al. (2009)	Top 56 disclosing PLCs	Environment	• Annual report disclosures 2006 • Classified into detailed, superficial, and both	Net income over assets (high, medium, and low)	Not significant
7	Elijido-Ten (2009)	40 companies reported on the environment (79 firm-year observations)	Environment	• Annual report disclosures 2000-2001 • 19 items (0-3 scoring; number of sentences)	Average return on assets	Negative but not significant
8	Heaney (2009)	161 PLCs	Governance	• Year of analysis: 2000-2002 • Board size • Board independence	Book to market ratio	• Insignificant • Board size: Year 2000 (+)
9	Muhamad et al. (2009)	159 PLCs (random)	Governance	• Annual report disclosures 2006 • Disclosure index (1,0 scoring)	Net profit margin	Positive but not significant

*Corresponding author's email: qlkweh@uniten.edu.my

Table 1 Prior studies examining the association between ESG and FP in Malaysia (FP influences ESG)

No	Author(s)	Sample	ESG component	ESG measure	Firm performance	Findings
10	Othman et al. (2009)	Top 56 Shari'ah-approved PLCs	(Islamic) ESG	• Annual report disclosures 2004-2006 • Categories: Finance & investments, products and services, employees, society, environment, & corporate governance • 43 items (1,0 scoring)	Profit before tax	Positive but not significant
11	Shakir (2008)	81 PLCs from the properties sector	Governance	• Year of analysis: 1999-2005 • Block ownership (≥5%)	• Tobin's q • Market value	• Tobin's q (-) • Market value (+)
12	Amran et al. (2007)	Top 100 companies (PLCs)	Social	• Amount of cash donation for 2004 • 65 companies – collected from a previous study; 35 – from annual report/website	Profit before tax	Positive but not significant
13	Mohamad Ariff et al. (2007)	95 companies in the Corporate Governance rating	Governance	• Classified into top and bottom 50%	Net profit margin	Positive but not significant
14	Mohd Ghazali (2007)	87 Composite Index non-financial companies	CSR	• Annual report disclosures 2001 • 22 items (1,0 scoring)	Profit before tax over total assets	Positive but not significant
15	Haniffa & Cooke (2005)	139 non-financial PLCs (stratified)	CSR	• Annual report disclosures 1996 and 2002 • Categories: community, environmental, employee, product/ service, & value-added • 41 items (1,0 scoring; number of words)	Earnings after tax over total equity (control variable)	Positive and significant
16	Ahmad et al. (2003)	299 PLCs (randomly-selected)	Environment	• Annual report disclosures 1999 • Incidence (report/not)	Profit over assets	Positive but not significant

Overall, based on our review, the findings on the relationship between ESG (or its components) and firm performance have been inconclusive. However, the large majority found the influence to be insignificant (bin Abd. Rahman, binti Yusoff, & binti Wan Mohamed, 2009; Esa & Anum Mohd Ghazali, 2012; Hawani Wan Abd Rahman, Mohamed Zain, & Hanim Yaakop Yahaya Al-Haj, 2011). The same situation is said to happen for studies examining the impact of ESG on firm performance. For example, (Koe Hwee Nga, 2009) found that ISO 14000[1] adopters have significantly better ROE than the non-adopters. However, the same finding was not observed when using sales and market capitalisation as the proxies for financial performance. Additionally, Ponnu & Ramthandin (2008) found that companies with higher corporate governance rating have better ROE only when weighted rating score was used (but not when using the raw score) and the rating had no impact on the holding period return of the sample companies. This implies that different

[1] ISO 14000 is a series of environmental management standards developed and published by the International Organization for Standardization (ISO) for organizations. The ISO 14000 standards provide a guideline or framework for organizations that need to systematize and improve their environmental management efforts.

*Corresponding author's email: qlkweh@uniten.edu.my

DOI: 10.1051/shsconf/20173600022

measure could lead to different findings. Thus, the inconsistent findings offered by the previous studies warrant another separate research to be conducted.
Theories that have been applied to support this proportion "by engaging in ESG initiatives, companies can improve their financial performance" include good management theory (Fauzi, 2009), porter hypothesis (Wagner, Schaltegger, & Wehrmeyer, 2001), stakeholder theory (Saleh, Zulkifli, & Muhamad, 2011), and stewardship theory (Koe Hwee Nga, 2009). Therefore, we develop the following alternate hypothesis:
Hypothesis: Environmental, Social and Governance have impacts on corporate performance.

DESCRIPTION OF METHODOLOGY

Data Collection

The data for ESG is based on the Sustainalytics ESG Research data available in the Bloomberg made by companies in their annual reports, stand-alone reports, and websites. We choose secondary data for several reasons. Firstly, the data is easily accessible from databases. Secondly, we believe that companies would disclose information should they perceive the information is important for the stakeholders to make decisions. In this regard, we assume that disclosure, to a large extent, equates the actual practice. Thirdly, the use of secondary data is consistent with the previous studies in this area (Amran, Ooi, Nejati, Zulkafli, & Lim, 2012; Esa & Anum Mohd Ghazali, 2012; Hawani Wan Abd Rahman et al., 2011; Ponnu & Ramthandin, 2008; Saleh et al., 2011).

After removing observations with missing data, we have a final sample of 387 firm-year observations for analyses. Our sample companies are GLCs in Malaysia, covering the period from 2006-2012. This study does not rely on the list of GLCs published on The Putrajaya Committee on GLC High Performance (PCG) website as it is outdated at the time of this writing. We identify GLCs as Malaysian listed companies with 20% equity shares held by GLICs (Zin & Sulaiman, 2011).

Our study focus on the impacts of ESG on firm performance of Malaysian GLCs. Prior literature in Malaysia has largely focused on public listed companies in general. Esa & Anum Mohd Ghazali (2012) and Hawani Wan Abd Rahman et al. (2011) examined the association of ESG and firm performance among GLCs, however, the number of sample is rather small with 27 and 44 companies, respectively. Using GLCs as the focus is essential since we expect these companies to be more proactive in ESG initiatives than their counterparts, as a result of various Government initiatives on ESG.

Methodology

To provide answers to our research questions, we apply a two-stage approach. In the first (1) stage, we estimate efficiency performance of our sample companies using DEA. In the second (2) stage, this study performs regression analysis to examine the impact of ESG on efficiency performance.

*Corresponding author's email: qlkweh@uniten.edu.my

DOI: 10.1051/shsconf/20173600022

Performance Measure: Data envelopment analysis (DEA)

DEA, a widely used linear-programming-based composite tool, is developed by Charnes, Cooper, & Rhodes (1978) and extended by Banker, Charnes, & Cooper (1984). DEA, a mathematical technique comparing multiple inputs and outputs of decision-making units (DMUs) for measuring relative DMUs' efficiency, allows the identification of benchmarking. Instead of using merely uni-dimensional ratios and other individual financial variables, IC indicators such as human capital and structural capital can be accommodated so that possible interactions between them can be captured to derive efficiency scores using DEA. Moreover, DEA approach provides added information (Feroz et al., 2003).

Truncated Regression Model

Note that efficiency scores range from zero to one. As the truncated regression technique is able to offset the bias involved in estimating such parameters, this study adopts truncated regression proposed by Simar & Wilson (2007) to examine the impact of exogenous factors on operating efficiency and ESG score. In the regression analysis, we will run truncated regression for Equation (2). The following regression models are estimated:

$$EFF_{it} = \alpha_0 + \alpha_1 ENV_{it} + \alpha_2 SOC_{it} + \alpha_3 GOV_{it} + \alpha_4 FSIZE_{it} + \alpha_5 FLEV_{it} + \alpha_6 MTB_{it} + \alpha_7 OCF_{it} + \varepsilon \quad (2)$$

where

EFF_{it} = The efficiency score in year *t*.
ENV_{it} = The percentage of environmental-related disclosures of firm *i* at period *t*.
SOC_{it} = The percentage of social-related disclosures of firm *i* at period *t*.
GOV_{it} = The percentage of governance-related disclosure of firm *i* at period *t*.
$FSIZE_{it}$ = The natural logarithm of total assets of firm *i* at period *t*.
$FLEV_{it}$ = The ratio of the sum of short-term and long-term borrowings to total assets of firm *i* at period *t*.
MTB_{it} = The ratio of market value to book value of firm *i* at period *t*.
OCF_{it} = The ratio of operating cash flow to total assets of firm *i* at period *t*.

EMPIRICAL RESULTS

Table 2 presents the descriptive statistics of dependent variable and explanatory variables. The average of environmental score (ENV) in our sample is about 1.484. The mean of social score (SOC) is 2.863, and the mean of governance score (GOV) is about 34.606. Among the three pillar of ESG, governance seems to have the largest amount value, followed by social aspects. This suggests that the firms disclosed GOV information more than SOC and ENV. It is also noteworthy that the standard deviation of GOV and SOC are relatively high as compared to the ENV. Meanwhile, the mean firm size is approximately 9.185, while the mean of firm leverage is about 0.229.

*Corresponding author's email: qlkweh@uniten.edu.my

DOI: 10.1051/shsconf/20173600022

Table 2 Descriptive Statistics

Variable	Mean	Median	Maximum	Minimum	Std. Dev
EFF	0.541	0.472	1	0.020	0.336
ESG	54.475	30.165	100.000	9.211	40.668
ENV	1.484	0	49.612	0	5.554
SOC	2.863	0	60.938	0	9.166
GOV	34.606	48.214	73.214	0.103	21.829
FSIZE	9.185	9.227	13.369	4.633	1.667
FLEV	0.229	0.210	0.659	0	0.166
MTB	3.967	1.796	157.392	0.303	10.352
OCF	0.092	0.062	0.752	-0.117	0.116

Table 3 reports the Pearson correlation coefficients in a two-tailed setting. GOV is significantly and positively related to efficiency (EFF). This means that firms' performance is increased when the firms disclose the GOV information. Meanwhile, we only find a significantly positive correlation between governance (GOV) and ESG. From this study, we can see among the three pillars of ESG, firms are most likely to disclose GOV information because it can enhance the efficiency of the firm performance. Other correlation coefficients are all lower than 0.5. Besides, the maximum value of untabulated variance inflation factors (VIF) is about 2.0. These results suggest that there is no multicollinearity problem for multivariate analysis.

Table 3 Correlation Matrix

Probability	EFF	ESG	ENV	SOC	GOV	FSIZE	FLEV	MTB	OCF
EFF	1								
ESG	0.0216	1							
ENV	-0.0037	-0.1478***	1						
SOC	-0.0136	-0.2059***	0.8862***	1					
GOV	0.1185**	0.1925***	-0.0720	-0.0894*	1				
FSIZE	-0.0968*	0.0945*	0.0385	0.0857*	0.4354***	1			
FLEV	-0.2367***	-0.0974**	0.0352	0.0262	-0.0956**	0.1952***	1		
MTB	0.2341***	0.1399***	0.1992***	0.1121**	-0.0524	-0.1622***	0.0467	1	
OCF	0.3574***	0.07705	0.2297***	0.1236**	-0.1676***	-0.4277***	-0.1327**	0.6252***	1

*, ** and *** are statistical significances at the 10%, 5% and 1% levels, respectively.

Table 4 reports the outcomes of the regression analysis. The models are significant at the 0.01 significance level. The result shows that environment (ENV) has a negative relationship with efficiency (EFF) and the coefficient is significant. While previous studies that focus on environment issues find a positive results between environment and firm efficiency (Al-Najjar & Anfimiadou, 2012; Derwall, 2007; Sinkin, Wright, & Burnett, 2008), this study argues that the focus of GLCs could be the reason for the negative association.

*Corresponding author's email: qlkweh@uniten.edu.my

Table 4 Regression Results (Dependent variable:EFF)

	OLS		Tobit	
Variable	Coefficient	Prob.	Coefficient	Prob.
C	0.3323**	0.0194	0.2488***	0.0005
ENV	-0.0096*	0.0772	-0.0154*	0.0641
SOC	0.0030	0.4500	0.0053	0.3592
GOV	-0.0596	0.5218	-0.0457***	0.0003
FSIZE	0.0245*	0.0810	0.0363	0.5561
FLEV	-0.4153***	0.0027	-0.5392***	0.0004
MTB	0.0015	0.4792	0.0152	0.3959
OCF	1.0845***	0.0000	1.1873***	0.0000
Adjusted R-squared	0.1697			
F-statistic	12.2726***			
Log likelihood			-218.127	

* Statistical significance at the 10% level.
** Statistical significance at the 5% level.
*** Statistical significance at the 1% level.

Specifically, the role of ESG in GLCs can be different in the sense that government may lead and promote more spending on protecting the environment without considering the consequence of the high spending on ENV on its firm performance. The coefficient of social (SOC) is positive and insignificant; this finding is similar with Filbeck, Gorman, & Zhao (2009). For the governance aspect, Statman & Glushkov (2009) and Core, Guay, & Rusticus (2006) find that this aspect does not affect firm performance. Meanwhile, this study shows the coefficients of governance (GOV) are negative but only significant in the Tobin regression. These suggest that the SOC and GOV of ESG generally do not enhance firm performance. Firm size (FSIZE), market to book (MTB) and operating cash flow (OCF) are positively related to EFF but the coefficient of MTB is insignificant. However, firm leverage (FLEV) has a significantly negative association with efficiency.

DISCUSSION

This study may create greater awareness on ESG practices among companies in Malaysia. Through the use of GLCs as our sample, we may show that Malaysia government plays a leading role in leading the corporate sector to invest in ESG for sustainable long-term performance. The information obtained from this study provides insights to ponder on the effectiveness of the government initiative in launching the Government-linked Companies Transformation Programme (GLCTP) in May 2004. This study covers up to the final phase of the 10-year plan of GLCTP. By promoting investment in ESG, Malaysia will be able to attract more foreign direct investments or investors who value sustainability in business. This supports the government's first idea out of ten ideas contained in the 10th Malaysia Plan (2011-2015) which is internally driven, externally aware. This awareness will subsequently promote growth in private sector investment. This study is very important to many stakeholders in helping them improve the living standard and quality of life in many ways (from the perspective of environmental, social, and governance) as aspirated in the 10th Malaysia Plan through income and capacity building programmes, strengthening the social safety net and addressing the needs of the disadvantaged groups.

*Corresponding author's email: qlkweh@uniten.edu.my

CONCLUSION

We explore the impact of ESG on efficiency using GLCs in Malaysia from year 2006 to 2012. This study obtains data on the ESG disclosure level from the Sustainalytics ESG Research data available in the Bloomberg. The study finds that GLCs focus more on governance disclosures, followed by social and environmental disclosures. The regression analyses show that governance is positively associated with firm efficiency, but social and environmental factors have no similar effect. In summary, this study provides an insight on the limited literature on ESG and informs the relevant stakeholders on the important ESG components for financial and investment decisions.

There are limitations in this study that we need to acknowledge. First, the sample of companies comes from GLCs, leaving us unable to make general conclusions to all companies in Malaysia. It would be interesting to carry out a similar study that tests other companies in Malaysia to strengthen the results of this study. Second, the Sustainalytics ESG Research is used to test the ESG reporting, suggesting that future studies may utilise alternative ESG dataset. In other words, a comparison study can be performed in future research.

ACKNOWLEDGEMENTS

The researchers wish to thank the Ministry of Higher Education Malaysia (MOHE) for the research grant received to undertake this research (FRGS/1/2014/SS05/UNITEN/03/4).

REFERENCES

Al-Najjar, B., & Anfimiadou, A. (2012). Environmental policies and firm value. *Business Strategy and the Environment, 21*(1), 49-59.

Amran, A., Ooi, S. K., Nejati, M., Zulkafli, A. H., & Lim, B. A. (2012). Relationship of firm attributes, ownership structure and business network on climate change efforts: evidence from Malaysia. *International Journal of Sustainable Development & World Ecology, 19*(5), 406-414.

Arvidsson, S. (2010). Communication of corporate social responsibility: A study of the views of management teams in large companies. *Journal of Business Ethics, 96*(3), 339-354.

Bachoo, K., Tan, R., & Wilson, M. (2013). Firm value and the quality of sustainability reporting in Australia. *Australian Accounting Review, 23*(1), 67-87.

Banker, R. D., Charnes, A., & Cooper, W. W. (1984). Some models for the estimation of technical and scale inefficiencies in data envelopment analysis. *Management Science, 30*(9), 1078-1092.

*Corresponding author's email: qlkweh@uniten.edu.my

Bassen, A., & Kovacs, A. M. M. (2008). Environmental, social and governance key performance indicators from a capital market perspective. *Zeitschrift für Wirtschafts-und Unternehmensethik*(9/2), 182-192.

Bebchuk, L. A., Cohen, A., & Wang, C. C. (2013). Learning and the disappearing association between governance and returns. *Journal of financial economics, 108*(2), 323-348.

bin Abd. Rahman, S. A., binti Yusoff, R., & binti Wan Mohamed, W. N. (2009). Environmental disclosure and financial performance: An empirical study of Malaysia, Thailand and Singapore. *Social and Environmental Accountability Journal, 29*(2), 46-58.

Charnes, A., Cooper, W. W., & Rhodes, E. (1978). Measuring the efficiency of decision making units. *European Journal of Operational Research, 2*, 429-444.

Clark, G. L., Feiner, A., & Viehs, M. (2015). From the stockholder to the stakeholder: How sustainability can drive financial outperformance. *Available at SSRN 2508281.*

Cooper, W. W., Seiford, L. M., & Tone, K. (2006). *Introduction to data envelopment analysis and its uses: with DEA-solver software and references*: Springer Science & Business Media.

Core, J. E., Guay, W. R., & Rusticus, T. O. (2006). Does weak governance cause weak stock returns? An examination of firm operating performance and investors' expectations. *The Journal of Finance, 61*(2), 655-687.

Derwall, J. (2007). *The economic virtues of SRI and CSR.*

Esa, E., & Anum Mohd Ghazali, N. (2012). Corporate social responsibility and corporate governance in Malaysian government-linked companies. *Corporate Governance: The international journal of business in society, 12*(3), 292-305.

Fauzi, H. (2009). The Determinants of the Relationship of Corporate Social Performance and Financial Performance: A Conceptual Framework. *Issues in Social and Environmental Accounting, Forthcoming.*

Feroz, E. H., Kim, S., & Raab, R. L. (2003). Financial statement analysis: A data envelopment analysis approach. *Journal of the Operational Research Society, 54*(1), 48-58.

Filbeck, G., Gorman, R., & Zhao, X. (2009). The "Best Corporate Citizens": Are They Good for Their Shareholders? *Financial Review, 44*(2), 239-262.

Hawani Wan Abd Rahman, N., Mohamed Zain, M., & Hanim Yaakop Yahaya Al-Haj, N. (2011). CSR disclosures and its determinants: evidence from Malaysian government link companies. *Social Responsibility Journal, 7*(2), 181-201.

Homburg, C. (2001). Using data envelopment analysis to benchmark activities. *International Journal of Production Economics, 73*(1), 51-58.

*Corresponding author's email: qlkweh@uniten.edu.my

Ihlen, Ø. (2008). Mapping the environment for corporate social responsibility: Stakeholders, publics and the public sphere. *Corporate Communications: An International Journal, 13*(2), 135-146.

Koe Hwee Nga, J. (2009). The influence of ISO 14000 on firm performance. *Social Responsibility Journal, 5*(3), 408-422.

Lee, K.-H., & Saen, R. F. (2012). Measuring corporate sustainability management: A data envelopment analysis approach. *International Journal of Production Economics, 140*(1), 219-226.

Ponnu, C. H., & Ramthandin, S. (2008). Governance and performance: Publicly listed companies in Malaysia. *Journal of Business Systems, Governance and Ethics, 3*(1), 35-53.

Saleh, M., Zulkifli, N., & Muhamad, R. (2011). Looking for evidence of the relationship between corporate social responsibility and corporate financial performance in an emerging market. *Asia-Pacific Journal of Business Administration, 3*(2), 165-190.

Servaes, H., & Tamayo, A. (2013). The impact of corporate social responsibility on firm value: The role of customer awareness. *management science, 59*(5), 1045-1061.

Simar, L., & Wilson, P. W. (2007). Estimation and inference in two-stage, semi-parametric models of production processes. *Journal of Econometrics, 136*(1), 31-64.

Sinkin, C., Wright, C. J., & Burnett, R. D. (2008). Eco-efficiency and firm value. *Journal of accounting and public policy, 27*(2), 167-176.

Statman, M., & Glushkov, D. (2009). The wages of social responsibility. *Financial Analysts Journal, 65*(4), 33-46.

Wagner, M., Schaltegger, S., & Wehrmeyer, W. (2001). The relationship between the environmental and economic performance of firms. *Greener Management International, 2001*(34), 94-111.

Wang, W.-K., Lu, W.-M., Kweh, Q. L., & Lai, H.-W. (2014). Does corporate social responsibility influence the corporate performance of the US telecommunications industry? *Telecommunications Policy, 38*(7), 580-591.

Yeh, Q.-J. (1996). The application of data envelopment analysis in conjunction with financial ratios for bank performance evaluation. *Journal of the Operational Research Society*, 980-988.

Zin, N. M., & Sulaiman, S. (2011). *Government-linked Companies Blue Book (GLCs Blue Book) as a complement to Balanced Scorecard (BSC) in the Government-Linked Companies transformation program.* Paper presented at the 2010 International Conference on Business and Economics Research.

*Corresponding author's email: qlkweh@uniten.edu.my

FORMULATING A PERFORMANCE MEASUREMENT SYSTEM USING BALANCED SCORE CARD IN ISLAMIC PHILANTHROPY INSTITUTIONS

A'ieshah Abdullah Sani*, Amilia Saidin, Khair Syakira Bustamam, Siti Azrina Adanan & Khairunnisa Abd Samad
Faculty of Accountancy, Universiti Teknologi MARA, KM26, Jalan Lendu, 78000 Alor Gajah Melaka

ABSTRACT

The introduction of Balanced Score Card (BSC) as one of the management accounting tool to measure the performance of companies in both aspect of financial and non-financial position has been predominantly used since 1992 and had resulted positive energies for the profit making companies. Hence, the extension of the BSC is also required to help the social business organisation in delivering their multiple bottom lines in the most effective ways and provide organisations a mechanism to track their performance holistically through both quantitative and qualitative information. The growing of Islamic Philanthropy Institutions (IPI) especially in Malaysia with the objective to promote social development based on the charitable activities has been acknowledged by the society. Therefore, in order to ensure the sustainability of these institutions, a proper measurement system is required to ensure the final goal of the institutions is achievable and successful. Thus, the objective of this paper is to formulate the main indicators and variables that are suitable to measure the IPIs based on BSC system in four main perspectives which are financial perspective, stakeholder's perspective, internal process perspective and learning and growth perspective. The uniqueness of the establishment of IPIs would require different indicators and variables which are more appropriate as compared with the profit sector organisations. The outcome of this study could provide clear picture and guidelines in designing a proper performance measurement system to be implemented by current IPIs in Malaysia in order to maintain the sustainability and competitiveness of this institutions in future.

Keywords: Balanced Score Card, Islamic Philanthropy Institutions, Sustainability

INTRODUCTION

In recent years, the Islamic Philanthropy Institutions (IPIs) had been noticed for its existence in the current environment, especially in Malaysia. It is acknowledged by the society as a non-profit organization that promote social development based on charitable activities. Because of its ultimate goal for the society benefits, there is less concern on performance measurement especially on profit figures. However, with the increasing numbers of IPIs, people are questioning on the credibility and sustainability of this organizations in handling their resources as well as their activities. Alam (2010) in his study found that Muslims are less concerned about the use of their philanthropy, arguing that their responsibility ends with fulfillment of the obligation to give, and that charitable institution have to account for their use of funds. The rapid growth of IPIs with poor governance system would motivate the creation of proper system to be implemented in enhancing the quality of IPIs in Malaysia.

* Corresponding author's *email: aieshah124@bdrmelaka.uitm.edu.my , Contact no: 012-2546914*

On the basis of non-profit organization, it is more accurate to measure the performance of IPIs using non-financial measures instead of solely focusing on financial measures. Thus, this paper will focus on formulating a performance measurement that are suitable to apply for social enterprises, particularly IPIs using a well-known management tool, which is Balanced Scorecard (BSC). Since the introduction of BSC by Kaplan and Norton (1992), it had emerged throughout the years in order to keep applicable with the changing economy and business environment. The BSC allows the management of non-profit organization to innovatively evaluate on a practical basis mixing the financial and non-financial perspective which ultimately creating a report for decision making process (Hartnett & Matan, 2011). The four perspective measure identified are financial perspective, stakeholder's perspective, internal process perspective and learning and growth perspective. With the recent evolving BSC, it will further tied up the four perspective with the organization's strategic plan.

This conceptual paper will focus on existing literature of using BSC for performance measurement and reporting of social enterprise, particularly on IPIs and further propose a framework for IPIs in measuring their performance especially for sustainability purposes. The strength of BSC in strategy development will help the IPIs in meeting the organization's objective while delivering the best for the society.

LITERATURE REVIEW

ISLAMIC PHILANTHROHY INSTITUTIONS

The word of philanthropy is related with the noble work performed by an individual or a group of individual regardless either the contributions is being made to the other people, community or nation as a whole. In general, philanthropy can best be described as kind, pleasant, helpful, charity, voluntary, and aid work. Philanthropy can also be defined as contributions made by an individual or group in terms of time, money, labour, ideas and other things in a sincere manner without any compulsion and not expecting any rewards from the recipients (Muhammad Iqmal Hisham Kamaruddin & Nathasa Mazna Ramli, 2015). Philanthropy activities being taught by all religion and in Islam, these activities are considered as part of the human action in lives as it is supported by several Quranic verse on voluntary acts such as from verse 2:280 mentioned that: "And if someone is in hardship, then [let there be] postponement until [a time of] ease. But if you give [from your right as] charity, then it is better for you, if you only knew."

The growing of philanthropy organization in Malaysia is mushrooming from year to year in order to cater for the social development and fulfil different needs of the society. Basically, there are various types of non-profit organizations including Islamic philanthropic organizations in Malaysia and it can be classified into four main categories, which are (1) societies; (2) associations; (3) foundations; and (4) companies limited by guarantee. Based on annual statistic by the registry of society (ROS), the application for registration of non-profit organizations with the ROS has increased tremendously from year 2013 until august 2016. This comprised from 12,777 applications in year 2013, 30,445 in year 2014, 30,071 in year 2015 and 27,704 registrations until august 2016. All the non-profit organizations registered with the ROS has been classified into ten categories which are; (1) political organizations; (2) welfare organizations; (3) social organizations; (4) recreation organizations; (5) arts and cultural organizations; (6) human right organizations; (7) trading organizations; (8) security

* Corresponding author's *email: aieshah124@bdrmelaka.uitm.edu.my , Contact no: 012-2546914*

organizations; (9) religious organizations; (10) professions organizations; and (11) other organizations (Registrar of Society, 2016).

Nowadays, the magnitude of philanthropic giving in Muslim societies is quite high either through obligatory form of charity which is zakat or any other various form of voluntary acts such as waqf (perpetual voluntary charitable act), infak (voluntary gift for restricted purpose) and sadaqah (voluntary charity either in monetary or non-monetary form). Thus, the growing of Islamic philanthropic institutions (IPIs) particularly in Malaysia is significant in managing all these resources efficiently with the main goal to achieve the social development and solve the community problems effectively. Basically, the existence of IPIs in Malaysia is regulated under state government such as state zakat department, state Islamic religious councils (SIRCs), waqf institutions, state Islamic religious schools, baitulmal and others. The main state institutions is 14 IPIs which represent each state in Malaysia, however there is also some of the private institutions to cater for Islamic philanthropy activities such as Awqaf Holdings Berhad, Waqaf An-Nur Corporation Berhad and others that registered as non-profit organizations with ROS. Therefore, a need of proper measurement system is required to ensure that all the resources in IPIs are being utilized and managed properly and efficiently so that the ultimate goal of upliftment the economy of the poor and reduce social inequalities is achievable.

BALANCED SCORE CARD (BSC) MODEL IN IPIs

IPIs need to start measuring their performances in a systemic way, in order to support decision making and ensure accountability towards their stakeholders. In order to be as effective and successful than profit-making organizations, IPIs need to be highly innovative in creating reporting practices and measurement systems because they are not as developed and extensive than the measurement systems for financial reporting of profit making companies. However, because of hybridity nature of IPIs, performance measurement and reporting processes are very complex and challenging for them. Thus, this paper is aiming to propose a proper balanced score card model that could be utilized by IPIs in measuring their performance to deliver its multiple bottom lines goals. Balanced score card is one of the most common management tools that enable organizations to consider both hard and soft sides of business (London & Scorecard, n.d.).

The proposed framework of balanced score card that is suite with IPIs is based on the framework as adopted by non-profit organizations and social enterprises. According to Hartnett & Matan (2011), the application of the BSC able to transform their organization's strategy, set measurable goals and design a timetable for execution. BSC is one of the tool that able to measure and observe the cause and effect relationships between the organizations key objectives and an accurate report on leading and lagging initiatives. Therefore, in order to ensure the effectiveness of BSC application for IPIs, the measurement of BSC must be designed in a holistic form covering all four main elements of BSC which are financial perspective, stakeholder perspective, internal process perspective and learning and growth perspective.

* Corresponding author's *email: aieshah124@bdrmelaka.uitm.edu.my , Contact no: 012-2546914*

Financial perspective

This perspective is important in order to measure the financial sustainability of the IPIs. BSC could help IPIs to identify the key drivers or ways of achieving success in financial perspectives by identifying what is the key stakeholders want from the organisation and what processes it needs to put in place internally to deliver things. IPIs must be well understanding of the institutions' financial health and able to utilize it efficiently to achieve the strategic goal. Some of the common measurement in financial area includes the measurement of operating income, return on capital, and economic value added (R. S. Kaplan, 2009). However, IPIs or non-profit organizations are not focussing on profit oriented. The BSC model would be customized to ensure that IPIs have available timely data on funding sources where part of it to cover its cost of services and other overhead costs. The management of IPIs must be well informed about the financial health of the organization and be comfortable with the financial statements and budgets. These financial reports is significant in providing a solid basis for operations and build confidence with funders, donors, grantors and other sources of revenue (Hartnett & Matan, 2011).

Stakeholder perspective

This perspective is about the donor, volunteer or clientele experience, which is found by measuring satisfaction and retention as well as assessing the non-profit's market share in its position. Every non-profit should measure the attitude of its strongest and most loyal supporters to gain the most for the organization. Keeping donors and volunteers engaged and enthusiastic is required and through BSC, it able to identify the main key drivers and performance indicators which could provide an incredible advantage for any organization (Somers, 2005). IPIs need to widen its stakeholder perspective by considering more on both side which are the donors and the recipients as well as the IPIs impact towards enhancing the community development as a whole. Similarly, as for profit oriented companies, Kaplan and Norton (2001) highlighted the core of business strategy is based on customer value proposition which describes the unique mix of product, price, service, and image that a company offers. According to Martello, Fischer & Watson (2008), an organization must identify the right scope of customer attraction and its competitors in order to remain relevance in delivering to their customer's needs. The same principle also could be apply for IPIs whereby the satisfaction of recipients and donors is being measured appropriately to create value proposition.

Internal Process Perspective

This perspective involves measuring the cost, throughput and quality of the non-profit's key operational processes such as programs provided, services offered, and ability to address targeted recipient's needs. This internal focus gives a whole picture and a thorough understanding of how well the non-profit is running and can help them determine which programs and services are meeting the real needs of the community. Martello et al. (2008) indicate in improving internal processes, there should be a connection between the overall strategy and the improvements. There should also be a determination of how the strategy is to be measured. Typically, there are three main indicators fall under internal process which are cost related, time related and quality related (R. Kaplan, 2001). Hence, to measure the efficiency and effectiveness of the IPIs operation, it should include qualitative measurement in terms of process quality, cycle times and survey from public or recipients (Figge, Hahn,

* Corresponding author's *email: aieshah124@bdrmelaka.uitm.edu.my , Contact no: 012-2546914*

Schaltegger & Wagner, 2002). It is vital to assess the overall effect of each program in order to maintain the sustainability of IPIs and always relevance for the society.

Learning and Growth Perspective

Basically, this perspective looks at the non-profit's human capital which are its employees, volunteers and its board of directors in order to measure satisfaction, necessary skills, community connections, retention and adherence to the organization's mission. Since the staff and volunteers represent the organization's major resources, it is imperative that their performance is appropriately measured. Decisions on training and skill building can be based, in part, on their level of knowledge about the organization. The management of IPIs can also take into account the business skills needed to advance the mission, such as donor development, marketing and branding, leadership, communications and the use of technology to support every aspect of the organization (Martello et al., 2008). High performing boards, volunteers and staff are a prerequisite to the success of any non-profit (Zimmerman, 2009). The learning and growth perspective focuses on the creation of organizational value through employees and innovative practices. Nonfinancial metrics for this perspective may relate to employee turnover, employee cross-training and skill levels, and other product development indicators. Kaplan (2001) described that the learning and growth perspective involves a determination of employee capabilities and skills, technology, and a corporate climate needed to support a strategy. The human resources, technology and organizational climate must be aligned with the strategies within the other three perspectives giving the organization linkages among the four perspectives (Kaplan and Norton, 1996 b, 2001).

* Corresponding author's *email: aieshah124@bdrmelaka.uitm.edu.my , Contact no: 012-2546914*

PROPOSED BSC MODEL FOR IPIs.

The diagram below shows the proposed measurement variables using BSC for IPIs:

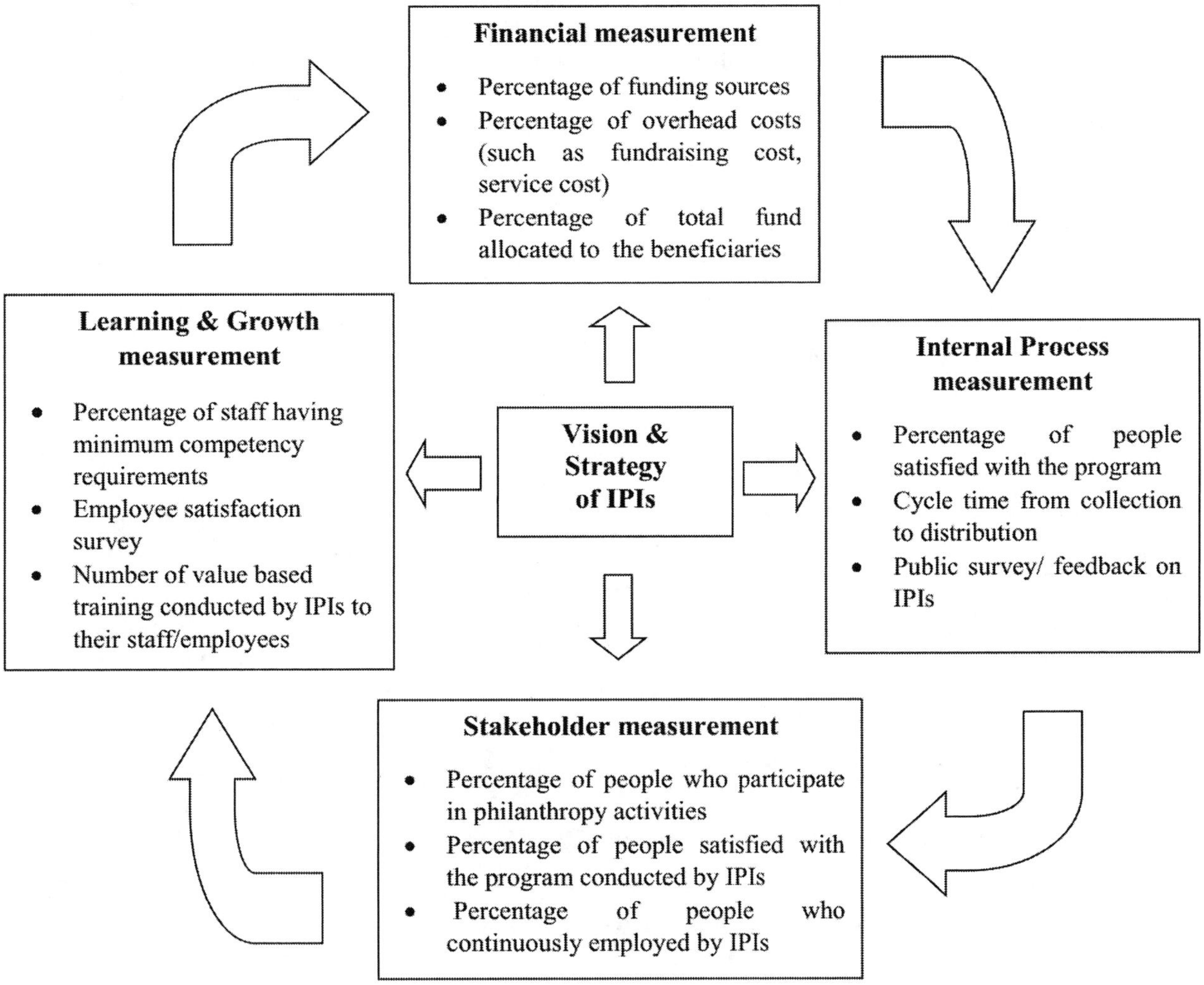

* Corresponding author's *email: aieshah124@bdrmelaka.uitm.edu.my , Contact no: 012-2546914*

CONCLUSION

In conclusion, the proposed BSC model is useful in order to track the performance of IPIs holistically through both quantitative and qualitative information. BSC is primarily an internal management tool which draw cause and effect relationship between organization's strategies, vision and mission with appropriate measurement (Hoque, 2014). However, this study only propose the basic framework of BSC that could be applicable for IPIs. The realization of this framework can be utilized using actual case study research for IPIs in Malaysia. This lead to the limitation of this study whereby further research should be done to test whether the proposed measurement is suitable and practical for IPIs. Having good measurement system for IPIs is significant in order to ensure the sustainability of the institutions as well as improve the governance of IPIs. Thus, this study could provide clear picture and guidelines in designing a proper performance measurement system to be implemented by current IPIs in Malaysia in order to maintain the sustainability and competitiveness of this institutions in future.

REFERENCES:

Alam, N. (2010). Islamic Venture Philanthropy: A Tool for Sustainable Community Development. *SSRN Electronic Journal*, (March 2010), 1–14. http://doi.org/10.2139/ssrn.1565859

Figge, F., Hahn, T., Schaltegger, S., & Wagner, M. (2002). The sustainability Balanced Scorecard – Theory and Application of a Tool for Value-Based Sustainability Management. *Greening of Industry Network Conference 2002, Gothenburg*, 1–32. http://doi.org/10.1002/bse.339

Hartnett, B., & Matan, R. (2011). The Balanced Scorecard: A Strategic Tool for the Nonprofit Sector, 15. Retrieved from http://sobel-cpa.com/sites/default/files/whitepaper.Jan2011 final.pdf

Hoque, Z. (2014). 20 years of studies on the balanced scorecard: Trends, accomplishments, gaps and opportunities for future research. *British Accounting Review*, *46*(1), 33–59. http://doi.org/10.1016/j.bar.2013.10.003

Kaplan, R. and Norton, D. (1992). The Balanced Scorecard – Measures That Drive Performance.‖ Harvard Business Review 70(1): 71-79.

Kaplan, R. (2001). Strategic Performance Measurement and Management in Nonprofit Organizations. *Nonprofit Management & Leadership*, *3*(Spring), 353 – 370. http://doi.org/10.1002/nml.11308

Kaplan, R. S. (2009). Conceptual Foundations of the Balanced Scorecard. *Handbooks of Management Accounting Research*, *3*, 1253–1269. http://doi.org/10.1016/S1751-3243(07)03003-9

London, S. E., & Scorecard, B. (n.d.). Social Enterprise Balanced Scorecard. *Chief Executive.*

Martello, M., Fischer, M. J., & Watson, J. G. (2008). Implementing a Balanced Scorecard in a Not-For-Profit Organization. *Journal of Business & Economics Research*, *6*(9), 67–80.

Muhammad Iqmal Hisham Kamaruddin, & Nathasa Mazna Ramli. (2015). Enhancing financial accountability of Islamic philanthropic organizations through financial

* Corresponding author's *email: aieshah124@bdrmelaka.uitm.edu.my , Contact no: 012-2546914*

disclosure. *Online Journal Research in Islamic Studies*, *Volume 2 N*(2015), 29–42. http://doi.org/10.15364/ris15-0201-03

Registry of Societies Malaysia (ROS) Website: http://www.ros.gov.my/index.php/en/semakan-pendaftaran, accessed on September 1, 2016.

Somers, A. B. (2005). Shaping the balanced scorecard for use in UK social enterprises. *Social Enterprise Journal*, *1*(1), 43–56. http://doi.org/10.1108/17508610580000706

Zimmerman, J. (2009). Using a Balanced Scorecard in a Nonprofit Organization. *Nonprofit World*, *27*(3), 10–12. http://doi.org/10.1108/13683040110397248

* Corresponding author's *email: aieshah124@bdrmelaka.uitm.edu.my , Contact no: 012-2546914*

RETIREMENT PLANNING: YOUNG PROFESSIONALS IN PRIVATE SECTOR

Ainol Sarina Ahmad Zazili*, Mohammad Firdaus Bin Ghazali**, Norlinda Tendot Binti Abu Bakar**,Mastura Binti Ayob[2],Irwani Hazlina Binti Abd Samad**
Faculty of Accountancy, Universiti Teknologi MARA Melaka, Faculty of Business & ManagementUniversiti Teknologi MARA Melaka***

ABSTRACT

The purpose of this study is to determine the factors influencing retirement planning among young professionals in private sector. There are three factors identified in this research which includes financial literacy, job satisfaction and savings behavior. Data used for this study are primary and secondary data such as from journal articles, periodicals and textbooks. A questionnaire is distributed and administered to extract data from the respondents consist of executives, non-executives and managers around Klang Valley, aged between 20 - 34 years old. The data is analyzed using frequency analysis, reliability test and Pearson correlation in order to obtain a clear findings and results. The findings show that financial literacy, job satisfaction and savings behavior has a positive association towards retirement planning. Furthermore, it is shown that financial literacy and saving behavior have a significant relationship with retirement planning. It is hope that this study will inform and encourage the young professionals to save and invest for the retirement.

Keywords: Retirement Planning, Young Professionals, Financial Literacy, Job Satisfaction and Saving Behavior

INTRODUCTION

This research paper discusses retirement planning among young professionals in private sector and factors that can be an encouragement for them to plan for the retirement. The concern over retirement planning is not only an interest of employees but also government. One of the retirement planning introduced by the Malaysian government is Employee Provident Fund (EPF) pension where every employee has been required to make a mandatory contribution to the Employee Provident Fund (EPF) pension scheme. The EPF is a social security institution formed according to the Laws of Malaysia, Employees provident Fund Act 1991 (Act 452) which provides retirement benefits for members. The members of EPF are from private and non-pensionable public sector employees. Currently, the employees' monthly statutory contribution rate reduced from 11% to 8% for members below age 60 and 5.5% to 4% for those age 60 and above, effective March 2016 until December 2017 salary.

Even though workings individuals in Malaysian private sectors have savings with EPF pension scheme, it is not guarantee that their fund is sufficient for retirement because it takes many years to accumulate financial fund to live comfortably during retirement. Some employees are concern about his EPF and some might not. Surprisingly, study has found that younger generation of workings individuals in private sectors perceived that retirement planning is a burden for them because it involves long-term planning (Moorthy, et al., 2012). If this issue is not addressed appropriately, employees' especially young professionals in private sectors may face the problems in their retirement life. Therefore, it is very crucial to identify the factors that lead the employees to plan for their retirement. Some of the factors discussed by prior literatures are financial literacy (Potts, 2013, Moorthy et al, 2012 & Lusardi & Mitchell, 2011), job satisfactions (Schultz & Wang 2010) and savings behavior (Stawski, Hershey & Jacobs-Lawson, 2007). There is a limited prior study done in

*Corresponding author's email: ainol1776@bdrmelaka.uitm.edu.my

Malaysia toward the association of retirement planning with financial literacy, job satisfaction and saving behavior especially among young professional working in private sector. Therefore, this research is done and questionnaires are distributed to 100 respondents of young professionals in private sectors around Klang Vallley area. The main purpose of this study is to determine the association between retirement planning with financial literacy, job satisfaction and savings behavior. Other than that, this research is intended to present the demographic profiles of young professionals in private sector towards the retirement planning.

LITERATURE REVIEW

This section provides literature review from various researchers on the association of retirement planning with financial literacy, job satisfaction and savings behavior.

Lusardi & Mitchell (2007) reviewed international evidence on financial literacy and found that financial illiteracy is common in many develop countries such as Australia, Japan, Korea as well as develop countries in Europe. A few studies have shown that retirement planning and financial literacy are interrelated. Individuals' inability to make self-beneficial financial decisions relating to consumer financing can have negative implications on the entire economy (Klein, Mandell & Schmid, 2009). Therefore, every individual is responsible for their own financial decisions in order to diversify its planning for better future in term of being the purchase and financing of a home and preparing for retirement (Hung, Parker & Yoong, 2009). There is evidence in a study done by Klein, Mandell & Schmid (2007) found that financial behavior is positively affected by financial literacy but it will be less certain for long term period if financial education on financial behavior is not practiced by young generation. Thus, lack of literacy and financial sophistication is critical because they have important consequences for lifetime wellbeing (Lusardi & Mitchell, 2011).

Research on the relationship between job satisfaction and retirement is quite limited. Study done by Clark, Mavromaras & Wei (2014) has found significant relationship between job satisfaction and retirement planning. Further, they explained the retirement decision of those who are currently active in the labor market and the job satisfaction on retirement, while significant for both males and females, is slightly stronger for females. Also, job satisfaction was found to be one of the factors that encourage plan for retirement. This kind of relationship perhaps will enhance individual's work related attitudes from time-to-time (Dendinger, Gary & Jacobson, 2005). This is supported by a study done by DeVaney & Chen (2003) which describes the attitude toward the job is related to their relationships with superior, top management and co-worker because it is an essential element of job satisfaction in order to have more of a sense of control to their retirement planning.

Another study found that there is a link between saving behavior with retirement planning. Lusardi (2003) states many individuals perform poor savings behavior as they assume that planning for retirement is a difficult task until they will experience that it is too late to make decision for retirement. Gedela & Prakasa (2012) even suggested that the tendency to save should be raised by appropriate incentives and policies since better retirement planning can be put in place with the higher domestic savings. For instance the strategy used to motivate employee involvement in retirement plan known as automatic enrollment (or negative election), has been proved to amplify enrollment in U.S. defined contribution plans (Benartzi & Thaler, 2007). Additionally, every employees is responsible for their own savings decisions in order to accumulate sufficient resources and have adequate retirement fund at the desired age and they must know to decide when to start saving, how much to save and how to invest their money to generate income during retirement period (Clark, d'Ambrosio, McDermed & Sawant, 2003). If most of potential retirees having lack of information about saving or investing for retirement, they will fail to plan ahead and take on financial risks without realizing it (Mbazigwe, 2013).

Based on a review of relevant literatures discussed above, there are two types of variables identified in this research. First variable is retirement planning which is the dependent variable. The second variable

*Corresponding author's email: ainol1776@bdrmelaka.uitm.edu.my

is independent variables such as financial literacy, job satisfaction and saving behavior. Figure 1 illustrates the theoretical framework of this research.

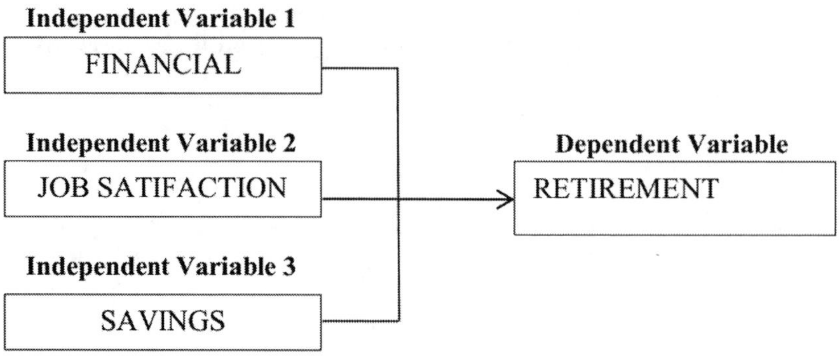

Figure 1: Theoretical Framework

RESEARCH METHODOLOGY

This part explained the method used for the collection of data, presentation and analysis of data. The questionnaires has been used to collect the information needed from young professional in private sectors around Klang-Valley who aged between 20 until 34 years old, consists of executive, non-executive and manager. According to the United Nations, Educational, Scientific and Cultural Organization (UNESCO) 'youth' is defined as those persons between the ages of 15 and 24 years. The ages between 15 to 24 years is not suitable to be used in this study because it only includes group of young teenagers who are still studying in secondary high schools or universities and not many of them get employed in Malaysian private and public sectors. In 1997, The National Youth Development Policy of Malaysia defines 'youth' as people between 15 to 40 years old. The young persons in his or her 20s or early 30s who is employed as a manager, or administrators in the organization would be considered a young professional while someone occupying the same position but had reached the age of 40s would be labeled as a professional, but not necessarily classified as young. Meanwhile, young professionals, as chosen in this study focus only on executive, non-executive and managers aged between 20 through 34 who are working in private sectors. They are selected because it is felt that these group are very likely to have valuable insight regarding what factors would be the most important when planning for retirement. The distribution of the questionnaires is generally to any private sectors employees without focus on any specific gender or position. The sample size in this research paper is limited to 100 respondents since the research paper must be finished in a short period. However, this sample size is considered appropriate for this research. As mentioned by Roscoe (1975) the rules of thumb for determining the sample size, the sample size should be larger than 30 and less than 500. The data is analyzed using IBM SPSS Statistic Data Editor to conduct frequency analysis and reliability analysis, mean analysis, Pearson correlation analysis, and regression analysis.

FINDINGS AND ANALYSIS

Frequency Analysis (Demographic Profiles)

Frequency analysis is used to measure the percentage of the respondents in term of their gender, age, race, religion, education level, marital status, employment status, position and monthly salary. Table 1 displays the demographic characteristics analysis among the young professional in private sector. Majority of the respondents are female (61.0%) and male only at 39.0 percent. Respondents aged between 24 to 27 years old contribute 44.0 % from the overall respondents that answering the questionnaire, followed by 25.0% from aged 28 to 31 years old, 17.0% from aged 32 to 34 years old

*Corresponding author's email: ainol1776@bdrmelaka.uitm.edu.my

and only 14.0% from aged 20 to 23 years old. It's clearly shown that the majority of the respondents are Malay (92.0%), followed by Chinese, Indian and other race with percentage contribution are 5.0%, 2.0% and 1.0% respectively.

Table 1 shows the educational background of respondents at different levels of education which are SPM/STPM, professional certificate, diploma, degree and master level. Majority of the respondents possess a degree certificate with percentage of 55.0%. 23 respondents holding diploma certificate, SPM/STPM (16.0%), Professional Certificate (4.0%) and the least are possessed by Master holder which is only 2.0%. Majority of the respondents are single with contribution percentage are 59.0% of overall respondents, while 41.0% are married.

Table 1: Demographic Characteristics Analysis

Item	**Frequency**	**Percentage (%)**
Gender		
Male	39	39.0
Female	61	61.0
Total	**100**	**100.0**
Age		
20 – 23	14	14.0
24 – 27	44	44.0
28 – 31	25	25.0
32 - 34	17	17.0
Total	**100**	**100.0**
Race		
Malay	92	92.0
Chinese	5	5.0
Indian	2	2.0
Others	1	1.0
Total	**100**	**100.0**
Education Level		
SPM/STPM	16	16.0
Professional Certificate	4	4.0
Diploma	23	23.0
Degree	55	55.0
Master	2	2.0
Total	**100**	**100.0**
Marital Status		
Single	59	59.0
Married	41	41.0
Total	**100**	**100.0**
Employment Status		
Permanent	68	68.0
Contract	14	14.0
Temporary	18	18.0
Total	**100**	**100.0**
Monthly Income		
Less than 2,000	39	39.0
2,001 – 3,000	31	31.0
3,001 – 4,000	16	16.0
4,001 – 5,000	6	6.0
5,001 – 6,000	3	3.0
6,001 – 7,000	2	2.0
7,001 – 8,000	1	1.0
8,001 – 9,000	2	2.0
Total	**100**	**100.0**
Position		
Non-executive	61	61.0%
Executive	33	33.0%
Manager	6	6.0%
Total	**100**	**100.0%**

*Corresponding author's email: ainol1776@bdrmelaka.uitm.edu.my

DOI: 10.1051/shsconf/20173600024

Next is employment status. The lowest percentage is contract workers (14.0%), 18.0% are temporary workers while 68.0% are permanent workers. A few number of respondent tend to have monthly income range between RM8,001 to RM9,000 (only 2.0%), between RM7,001 to RM8,000 (1.0%) followed with the highest percentage of monthly income range are from less than RM2,000 income that contribute 39.0% from total 100 respondents. Table 1, shows the position of respondents at their company. The majority of respondents are non-executives that contribute of 61.0% of them followed by the executive position that represents 33.0%. The least percentage is manager that only represents 6% of them.

Reliability Analysis

Table 2 below shows the reliability analysis used of the research that consists of independent variables and dependent variable. The reliability test is used to test the association between the groups of independent variables which are financial literacy, job satisfaction and savings behavior with the dependent variable i.e. retirement planning.

Table 2: Reliability Analysis

	Number of Items	Cronbach's Alpha
Independent Variables (IV)		
Financial Literacy	10	0.749
Job Satisfaction	10	0.795
Saving Behaviour	10	0.797
Dependent Variable		
Retirement Planning	10	0.832

According to the Table 2 above, the reliability for retirement planning is 0.832 which resulted in a good category. While the Cronbach's Alpha for financial literacy, job satisfaction and saving behavior are 0.749, 0.795 and 0.797 respectively.

Mean Analysis

Based on Table 3, the number of observation in this research is 100 respondents. The score from the Likert Scale are - "1" represented strongly disagree, "2" represented disagree, "3" represented agree and "4" represented strongly agree. Overall mean for the retirement planning is 3.0260 which clarify most of the respondents is agreed that they have a well planned for their retirement. While, mean 3.0040 for financial literacy, 3.0380 for job satisfactions and 3.1590 for saving behavior. The mean score for all the three independent variables shows that all the respondents are agree with the statements in the questionnaire.

Table 3: Mean Analysis

	Number of Observation (N)	Mean	Std. Deviation
Independent Variables (IV)			
Financial literacy	100	3.0040	.37977
Job satisfaction	100	3.0380	.35127
Saving behavior	100	3.1590	.36985
Dependent Variable			
Retirement planning	100	3.0260	.38261

Pearson Correlation Analysis

The Pearson correlation analyzes the direction, strength and the relationship between independent variable and dependent variable. It can be shown as in Table 4 below that there is positive moderate relationship and significant between independent variables (financial literacy, job satisfaction and saving behavior) and dependent variable (retirement planning).

*Corresponding author's email: ainol1776@bdrmelaka.uitm.edu.my

Table 4: Pearson Correlation Analysis

	Pearson Correlation	Sig. (2-tailed)
	Retirement Planning (DV)	
Financial literacy	.487**	.000
Job satisfaction	.401**	.000
Saving behavior	.477**	.000

** Correlation is significant at 0.01 level (2-tailed)

The correlation between financial literacy and retirement planning is 0.487 which have moderate relationship and it is significant. Besides that, job satisfaction and saving behavior are correlated with 0.401 and 0.477 with retirement planning respectively.

Regression Analysis

A multiple regression is used to shows the most factors that influence the retirement planning. It can be shown in Table 5 that there is strong statistical significance with $p<0.001$ with R^2 of 0.311. It means that only 31.1 percent changes in dependent variable are because of independent variable.

Table 5: Model Summary

Model	R	R Square	Adjusted R Square	Std. Error of the Estimate
1	.558[a]	.311	.289	.32255

Table 6: Coefficients

Model		Unstandardized Coefficients		Standardize Cooefficients	t	Sig.
		B	Std. Error	Beta		
1	(Constant)	.913	.332		2.753	.007
	Financial Literacy	.283	.108	.281	2.626	.010
	Job Satisfaction	.135	.114	.124	1.191	.237
	Saving Behavior	.270	.110	.261	2.451	.016

The results shown in Table 6 indicate financial literacy and saving behavior are significantly positive associated with retirement planning. Increases a unit of financial literacy will increase the retirement planning by 0.283 unit. As for saving behavior factor, as 1 unit increases, it will increase 0.270 unit retirement planning. This situation implies that the respondents acknowledged that saving behavior and financial literacy have an association with retirement planning.

CONCLUSION

Based on the findings above, it shows that there is positive and significant correlation between saving behavior, and financial literacy with retirement planning. Although there is significant association between financial literacy and retirement planning, it seems that the relationship is moderate. This finding is supported by Lusardi & Mitchell (2011) that reveals that there is positively association between financial literacy towards retirement planning. Financial literacy also can be good association for financial decision making and strongly correlated to the planning for retirement (Lusardi & Mitchell, 2007). Huston (2010) also believes that, there are relationships between financial literacy with the retirement planning management.

Furthermore, although there is positive association, but it is not significant with retirement planning. Research findings by Delpachitra & Beal (2002) also reveals that there is strong correlation between probabilities of having a suitable job with the retirement planning by age of an individual.

*Corresponding author's email: ainol1776@bdrmelaka.uitm.edu.my

Saving behavior also has been accepted as one of factors that influencing retirement planning. It shows that there is a positive significant association but still the association is moderate. It is supported by Ntalianis & Wise (2011) that states the saving behavior of individuals can be positively associated towards retirement planning with advance exposure to financial programs.

As for the recommendations, to attract more young professionals to save and invest for their future retirement planning, financial institutions should develop more attractive marketing strategy from time to time. With the attractive and in-depth explanations on important of investment and insurance, it will attract the young professionals to save. Encouragements from both government and private sectors are needed for the young professionals to plan for their retirement.

REFERENCES

Benartzi, S., & Thaler, R. H. (2007). Heiristics and Biases in Retirement Savings Behavior. *Journal of Economic Perspectives* , 81-104.

Clark, A. E., Mavromaras, K., & Wei, Z. (2014). Happy to Stay: Job Satisfaction and Retirement. NILS working paper series *Working Papers No: 211/2014*

Clark, R. L., d'Ambrosio, M. B., McDermed, A. A., & Sawant, K. (2003). Financial education and retirement savings. *Sustainable Community Development: What Works, What Doesn't and Why* (pp. 1-43). Washington DC: Federal Reserve System.

Dendinger, V., Gary, A., & Jacobson, J. (2005) Reasons for Working and Their Relationship to Retirement Attitudes, Job Satisfaction and Occupational Self-Efficacy of Bridge Employees. *The International Journal of Aging and Human Development*, 24.

DeVaney, S. A., & Chen, Z. S. (2003, May 28). Job Satisfaction of Recent Graduates in Financial Services. *Compensation and Working Conditions* Retrieved March 16, 2017, from https://stats.bls.gov
Lusardi, A. (2003). Planning and Saving for Retirement. *Research Gate*, 3-5.

Delpachitra, S., & Beal, D. (2002). Factors Influencing Planning for Retirement. *Economic Papers* , *21* (3), 1-13.

Gedela, R. & Prakasa, D.S. (2012). Determinants of Saving Behaviour in Rural and Tribal Households. *International Journal of Research in Social Science, 2 (3),* 109.

Huston S. J. (2010) Measuring Financial Literacy, Journal of Consumer Affairs. Volume 44, Issues 2 Summer 2010, page 296-316

Hung, A. A., Parker, A. M., & Yoong, J. K. (2009). Defining and Measuring Financial Literacy. *Rand Labor and Population*, 3.

Klein, Mandell, L., & Schmid, L. (2009). The impact of Financial Literacy Education on Subsequent Financial Behavior. *Financial Counseling and Planning Education*, 16-17.

Klein, Mandell, L., & Schmid, L. (2007). Motivation and Financial Literacy. *Financial Services Review*, 107.

Lusardi, A., & Mitchell, O. S. (2011). Financial Literacy Around the World: An Overview. *NBER Working Paper Series*, 5.

*Corresponding author's email: ainol1776@bdrmelaka.uitm.edu.my

Lusardi, A. & Mitchell, O.S. (2011). Financial Literacy and Planning: Implications for Retirement Wellbeing. *NBER Working Paper Series*, 2-3.

Lusardi., & Mitchell , O.S. (2007). Financial Literacy and Retirement Preparedness: Evidence and Implications for Financial Education. *Business Economics*, 39.

Lusardi, A. (2003). Planning and Saving for Retirement. *Research Gate*, 3-5.

Mbazigwe, C. (2013, October 07). *The meaning of Financial Literacy and its Importance*. Retrieved April 09, 2015, from futurpreneur Canada: http://www.futurpreneur.ca/en/2013/meaning-financial-literacy/

Moorthy, K.M., Chelliah, D.T., Sien, C.S., Leong, L.C., Kai, N.Z., Rhu. W.C., et al. (2012). A Study on the Retirement Planning Behaviour of Working Individuals in Malaysia. *International Journal of Academic Research in Economics and Management Sciences*, 1(2), 55

Ntalianis, M., & Wise, V. (2011). The Role of Financial Education in Retirement Planning. *Australasian Accounting, Business and Finance Journal , 5* (2), 23-37.

Potts (2013). Retirement Plans and saving Decisions: Western Kentucky University, United States. Retrieved from http://digitalcommons.wku.edu/cgi/viewcontent.cgi?article=1440&context=stu_hon_theses

Roscoe, J.T. (1975). *Fundamental Research Statistics for the Behavioural Sciences*, 2nd edition. New York: Holt Rinehart & Winston.

Shultz, K.S., & Wang, M. (2010). Employee Retirement: A Review and Recommendations for Future Investigation. *Journal of Management*, 36(1), 175.

Stawski, R., Hershey, D., & Jacobs-Lawson, J.M. (2007). Goal Clarity and Financial Planning Activities as Determinants of Retirement Savings Contributions. *The International Journal of Aging and Human Development*, 17.

*Corresponding author's email: ainol1776@bdrmelaka.uitm.edu.my

CAPITAL STRUCTURE INFLUENCE ON CONSTRUCTION FIRM PERFORMANCE

Nur Faezah Jayiddin*, Anita Jamil**, Saiyidi Mat Roni***
Center for Diploma Programme, Multimedia University, Malaysia, Faculty of Accountancy, Universiti Teknologi MARA, Melaka**, Accounting Research Institute, Universiti Teknologi MARA, Malaysia. ****

ABSTRACT

The interconnectedness between capital structure and firm performance is a topic of high interest among scholars and management alike. The scholars tend to unveil the why segment of the relationship, while the management looks into the how side to promote capital structure policy which can optimise the firm performance. While many studies have looked into this relationship across multiple industries and spanning across decades of data, the current study trains its lens on Malaysian public listed company companies which operate in the construction sector, and with data window between 2010 to 2014. This specific sector was chosen for their high gearing which renders firms to relatively high insolvency exposure emanating from interest rate fluctuations. The five-year timeframe was selected to isolate potential data contaminations streaming from global financial crisis which winds down in 2009. Financial data of the company were extracted from Bloomberg Terminal based on a pre-prepared list of Bloomberg tickers. A total of 225 observations were recorded in this study. Using Tobin's Q as a proxy for firm performance, this study finds a mixed result where short term debts ratio indicates a significant negative effect, while long term debt ratio presents a non-significant influence. Explanations on this output are therefore discussed in this paper.

Keywords: Capital structure, firm performance, panel data, debt ratio.

INTRODUCTION

Capital structure has been a popular research topic among the financial scholars. As such various studies were conducted to examine the effect of capital structure decision and firm performance (Adami, Gough, Muradoğlu, & Sivaprasad, 2013; Collins, Filibus, & Clement, 2012; David & Olorunfemi, 2010; Dawar, 2014; Hasan, Ahsan, Rahaman, & Alam, 2014; John, 2013; Khan, 2012). There are numerous research conducted to examine the relationship between capital structure and firm performance in developing countries such as Nigeria (Akeem, K, Kiyanjui, & Kayode, 2014; Collins et al., 2012; David & Olorunfemi, 2010), India (Chisti, Ali, & Sangmi, 2013; Dawar, 2014), Bangladesh (Hasan et al., 2014) and Jordan (Zeitun & Tian, 2007). As for example research conducted by Zeitun and Tian (2007) in Jordan which aim to investigate and examine the effect of capital structure on corporate performance using a panel data sample representing of 167 Jordanian companies during 1989-2003. Based on the finding, they found that firm's capital structure had a significantly negative impact on the firm's performance measurement for both accounting and market's measures in Jordan which is one of developing country.

One study worth mentioning is research done by Salim and Yadav (2012) who investigated the relationship between capital structure and firm performance of Malaysian listed companies

*Corresponding author's email: rashdanita@melaka.uitm.edu.my

from year 1995 to 2011. The researchers measured capital structure based on short-term debt ratio (STDR), long-term debt ratio (LTDR), total debt ratio (TDR) and growth (% of total asset). While the firm performance are proxy by both accounting measure and market measure which is ROA, ROE, Tobin's Q and earning per share (EPS). One control variable is use in the research study which is the firm size. The researcher focused on six industries in Malaysia namely construction industry, consumer industry, industrial product industry, plantation industry, property industry and trading and services industry. The researcher found that there is negative relationship between short-term debt (STD), long-term debts (LTD) and total debts (TD) with return on asset (ROA), return on equity (ROE) and earnings per share (EPS). Other than that, the study also found that there is significant positive relationship between short-term debts (STD) and long-term debts (LTD) to Tobin's Q.

There are two main issue to be highlighted in research done by Salim and Yadav (2012). The first issue is related to the findings which cover a long economic period from 1995 to 2011 and the second issue is related to the various industries covered in the study. Research conducted by Salim and Yadav (2012) shows various results regarding the relationship between capital structure and firm performance since they utilize various independent variable and dependent variable and tested on six industries in Malaysia. While the results provide meaningful insights, it also leads to a question whether the trends pointed in their study (and others) still relevant within a smaller context – industry specific. Therefore, this study attempts to uncover if the contentions are still consistent in post-Asian financial crisis for construction industry. The construction industry is chosen because the industry is capital intensive leading to relatively more susceptible to economic shocks.

LITERATURE REVIEW

There are differences between financial structure and capital structure as financial structure is related to the various ways a firm can use to raise its funds while capital structure represents the proportionate relationship between the long-term debt and equity capital (Pandey, 1999). Capital structure plays an important role in a firm especially when the firm wants to fulfil the need of their stakeholders through payment of dividend, payment of debt, payment of salaries on time and other financial obligation (Opoku & Adu, 2012). Capital structure is normally made of equity capital, preference capital and long-term loan (debt) capital. Debt capital such as long-term bonds is used by the firm to finance its investment decision by paying back the principal and interest in return. The combinations of different capital types will have differing impacts on firm performance. This is because each capital type has its own cost and incentives for the firms, thus defining or influencing the way the firms operate.

The result of a firm performance is the output where the stakeholders hope for a positive outcome. One of the debatable issues in finance is regarding the idea or concept of performance due to its multi-dimensional meanings (Zeitun & Tian, 2007). Study related to firms performance originates from organization theory and strategic management (Murphy, Trailer, & Hill, 1996). Performance measure can be divided into two categories which are financial performance or organizational performance. Profit maximization, maximizing profit on assets and maximizing shareholders' value are examples of measurement for financial performance. On the other hand, operational performance such as growth in sales and growth in market share referring to a wide definition of performance since they focus on determinants that contribute to financial performance (Hofer & Sandberg, 1987).

*Corresponding author's email: rashdanita@melaka.uitm.edu.my

DOI: 10.1051/shsconf/20173600025

In this study, the firm's performance was measured using Return on Asset (ROA). ROA is widely known as the most useful measure to determine the firms' performance (Long & Ravenscraft, 1984; Reece & Cool, 1978). ROA is viewed as a proxy to measure the ability of management to employ the asset within their control efficiently regardless of the source of financing (Ebrati, Emadi, Balasang, & Safari, 2013). Among the financial ratio, ROA is found to be the most popular and useful financial ratio. ROA can be derived by dividing net income plus interest expenses with total assets (Ebrati et al., 2013; Khan, 2012). The ratio was introduced by DuPont company back in 1919 where they used it as the top of their ratio triangle system (Jewell & Mankin, 2011). Since then, ROA has been used in the industry until now. Gibson (1987) conducted a survey toward Chartered Financial Analyst to investigate the importance of financial ratio. Based on the finding by Gibson (1987), all four different version of ROA was selected by at least 90% of the respondent as the main measure of profitability.

METHODOLOGY

This study aims to examine the relationship between the capital structure and firm performance in Malaysian construction public listed company (PLC) during the stable economic growth from 2010 to 2014. This study focus construction industry only as compared to study done by Salim and Yadav (2012) which include various industry in the research study. The range of year from 2010 to 2014 were selected because the Malaysia's quarterly Real Gross Domestic Product (RGDP) rate from 2010 to 2014 showed a positive economic growth (*Bank Negara Malaysia's Quarterly Bulletin*, n.d.).

A series of financial data were collected from a total of 45 construction companies listed in Bursa Malaysia as at August 2015. All companies listed in Bursa Malaysia under construction industry were selected. This means the entire population of construction companies are included in the study. Financial data of the company were extracted from Bloomberg Terminal. Prior of extracting data from the Bloomberg Terminal, a list of Bloomberg ticker were prepared to identify the company uniquely in Bloomberg to avoid any error in data extraction. A total of 225 observations were recorded in this study.

R-Statistic with an interface of R Studio version 0.99.486 were used to analyse the date. The primary package for this panel data analysis was PLM version 1.5-12. PLM package are useful to make an estimation of linear panel models straightforward and it provides allows to estimate various models and to make a robust inferences (Croissant & Millo, 2008). In order to capture the relationship between the capital structure and firm performance, this study formulates the following panel regression model:

$$y_{it} = \beta_0 + \beta_1 X_{1it} + \beta_2 X_{2it} + \beta_3 X_{3it} + \beta_4 X_{4it} + \beta_5 X_{5it} + \varepsilon_i + u_{it}$$

Where y_{it} is alternately Return on Assets (ROA) for firm, as a measure of performance of i in year t.

X_1 = Short-term debt (STDR)
X_2 = Long-term debt (LTDR)
X_3 = Total debt (TDR)
X_4 = Growth (GROW)

*Corresponding author's email: rashdanita@melaka.uitm.edu.my

X_5 = Firm's Size (SIZE)

ε_i = Unobserved or heterogeneity effect; and

u_{it} = residual error of firm i in year t

RESULTS AND DISCUSSIONS

Descriptive Statistic

In this study, descriptive analysis that covered minimum amount, 1st quarter amount, median amount, mean amount, 3rd quarter amount and maximum amount of the data set were presented. **Error! Reference source not found.**below shows the descriptive statistic for independent variable, dependent variable and control variable used in the study.

Table 1. Descriptive statistics

	STDR	LTDR	TDR	ROA	GROW	TA
N	45	45	45	45	45	45
Minimum	0.00000	-0.0005865	0.1590	-27.260	-0.89674	11.72
1st Quarter	0.02325	0.0123410	0.4441	1.030	0.00000	321.12
Median	0.09380	0.1553566	0.5657	3.620	0.08197	596.62
Mean	0.10427	0.1787671	0.5976	6.034	0.25045	1317.42
3rd Quarter	0.13487	0.2724188	0.6630	5.500	0.19432	1269.85
Maximum	1.07204	0.7580640	10.3183	576.100	16.69333	19730.69

Based on the descriptive statistic, the mean ($\overline{X}$) or average for return on asset (ROA) for construction industry from 2010 to 2014 is 6.0%. The mean for both ROA shows that the performance of construction firms in Malaysia from 2010 to 2014 is positive. Other than that, from the mean of ROA, it is obvious that construction firms in Malaysia have a slightly higher accounting performance compared to the overall market.

The minimum amount of ROA is -27.26% and the maximum amount for ROA is 576.10%. The interquartile range which is the difference between the 1st quartile and the 3rd quartile for ROA is 4.47%. This indicates that some of the construction firms are highly profitable as the maximum ratio stands at 576%.

As for short-term debt ratio (STDR), the minimum amount starts at 0% and maximum amount is 1.07% with a mean scored of 0.10%. The interquartile range for STDR is 0.11%. On the other hand, the long-term debt ratio (LTDR) recorded a minimum of -0.00% and a maximum amount of 0.76% with a mean of 0.17%. The interquartile range for LTDR is 0.26%. The mean score for LTDR is slightly higher compared to STDR by 0.07%. This shows that constructions firms in Malaysia utilized more long-term debt compared to short-term debt in the positive economic growth. It is interesting to note that construction companies started to increase their long-term debt to finance its business during positive economic growth in this study which cover from 2010 to 2014.

Capital structure and return on asset

Table 2. Capital structure effects on ROA

Coefficients:	Estimate	Std. Error	t-value	Pr(>|t|)
STDR	-19.278	32.755	-0.5886	0.556907
LTDR	-1.6402	25.339	-0.0647	0.948461
TDR	-13.6891	4.2432	-3.2267	0.001495**

*Corresponding author's email: rashdanita@melaka.uitm.edu.my

GROW	1.8707	2.3763	0.7873	0.432197
SIZE	-0.00084895	0.00497902	-0.1708	0.864573
R-Squared	0.069105			
F- Statistic	2.59821			
p-value	0.027019			

Note: ***, **, * Coefficients are significant at the 1%, 10% and 5% levels
The results are based on *fixed effect model* after considering outputs from Haussman Test.
ROA = net profit divided by total assets; STDR = short-term debt to total assets; LTDR = long term debt divided by long term debt plus equity; TDR = total asset minus total equity divided by total asset; GROW= annual growth rate of total asset; SIZE= total asset

Table 2 presented the relationship between capital structure and construction firms' performance in Malaysia based on ROA. F-test is used in this study to test whether the variation in the independent variable explained a significant portion of the variation in the dependent variable in the overall model. The *F*-statistic value of 2.59821 ($p < 0.05$), show that the independent variables are jointly statistically significant in the fixed estimates in explaining the dependent variables which is ROA. R^2 statistics value of 0.069 implies that the independent variables jointly account for about 6.9% variation on firm performance in the fixed effect model for ROA.

Further, only TDR has a significant negative relationship with ROA (TDR: $\beta = -13.6891$, $p < 0.01$) which means that one unit change in total debt will reduce ROA by -13.6891 unit. Other independent variables such as STDR, LTDR, GROW and SIZE show a negative relationship. However, the relationships are not statistically significant.

The result shows that STDR has no significant relationship to the ROA in construction firms in Malaysia (STDR: $\beta = -19.278$, $p>0.05$). The insignificant relationship between the STDR and accounting based measure which is ROA indicates that STDR has no significant impact on returns of construction firms in Malaysia. This result is consistent with the study by Ebrati, Emadi, Balasang, and Safari (2013), Hamid, Abdullah, and, Kamaruzzaman (2015), Hasan, Ahsan, Rahaman, and Alam (2014), Khan (2012), Olokoyo (2013), and Taani (2013). Table **2** also shows that the relationship between LTDR and ROA in construction firms in Malaysia is not statistically significant (LTDR: $\beta = -1.6402$, $p > 0.05$).

This result is align with the findings by previous researcher such as (Akeem et al., 2014; Ebaid, 2009; Ebrati et al., 2013; Gill, Biger & Mathur, 2011; and John, 2013). In United States, Gill et al. (2011) analyse the relationship between the capital structure and firm performance among the American service and manufacturing firms from 2005-2007 and they also found the insignificant relationship between long-term debt to accounting performance in service industry. They claim that American firms that operated in service industry recorded a higher gearing ratio. When firms utilize a higher gearing ratio, the profitability for the firms will reduce subsequently and the tax benefit also will start to disappear. Firms that operated in construction industry also may have a higher gearing ratio.

This is also evidence by the findings by Hung, Albert, and Eddie (2002) who found that gearing ratio for construction firms are higher compared to developer firms in Hong Kong. In Hong Kong, the gearing ratio are higher not due to debt but due to the cost of equity. However, in Malaysia, Baharuddin, Khamis, Mahmood, and Dollah (2011) found that construction firms in Malaysia own a higher asset tangibility and as the firms expand to a bigger capacity, construction firms will employ more asset. Thus, the demand for debt financing will increase.

*Corresponding author's email: rashdanita@melaka.uitm.edu.my

Table 2 also shows that there is significant negative relationship between TDR and ROA in construction firms in Malaysia (TDR: β = -13.6891, $p < 0.01$). A similar result were found by previous researcher in capital structure study such as Akeem et al. (2014), Azhagaiah (2007), Ebaid (2009), and Wahba (2013). Based on previous study in Malaysia, Hamid et al. (2015) found significant negative relationship between total debt and profitability for both family firms and non-family firms within firms listed in Bursa Malaysia for the year 2009-2011. This means firms will endure lower profitability as the amount of debt they employed increase. The researcher posit that profitable firms are depending on equity as their main financing choice (Shubita & Alswalhah, 2012).

Hamid et al. (2015) agree that the negative correlation between the total debts and the accounting performance support the pecking order theory introduce by (Myers, 1984). The concept of pecking order theory is related to the method firms choose their finance resources to support their operation which is firms prefer to utilize their internal funds obtained from retained earnings as first option for financing sources.

CONCLUSION

The contradiction between the general findings by Salim and Yadav (2012) and the findings by this study might happen due to several reasons. First is because the study done by Salim and Yadav (2012) ignored the macroeconomic indicator such as Gross Domestic Product (GDP) growth as the study selected a large data span which covers 17 years starting from 1995 to 2011. Along the duration of 1995 to 2011, there are two financial crises that happen in the global setting which affected Malaysia economy which is Asian Financial Crisis (AFC) 1997-1998 and Global Financial Crisis (GFC) that happen in 2007. This is important as Jong, Kabir, and, Nguyen (2008) found that the macroeconomic indicator such as real GDP growth give a significant impact towards capital structure.

Due to that, it can be note that study by Salim and Yadav (2012) has ignored the economic growth which might lead to contamination of data. The exclusion of the impact of the financial crisis in the study done by Salim and Yadav (2012) can cause bias toward their result findings. While as for this current study, the data chosen in this study were collected during the positive economic growth in Malaysia which is from 2010 to 2014 which covers 5 years span and are free from any financial crisis.

Secondly, the general findings by Salim and Yadav (2012) were obtained based on the different regression result of six different industry including construction industry. This may affect the validity of the findings as different industries sector have different capital structure and have different impact toward the firm performance (Bradley, Jarrell, & Kim, 1984; and Degryse, de Goeij, & Kappert, 2012). Thus, the general findings by Salim and Yadav (2012) cannot be used to reflect the impact of short-term debt to ROA for construction industry alone. While as for this study, it focuses specifically for construction industry.

Thirdly, Salim and Yadav (2012) used Pooled Ordinary Least Square (POLS) method to analyse their data which is generally known that the method are unable to control the individual or time-specific effect or in another term is called unobserved effect (Hasan et al., 2014; Manawaduge, Zoysa, & Chowdhury, 2011; and Seetanah, Seetah, Appadu, & Padachi, 2014). Thus, this will allow heterogeneity bias to occur that arise due to the omitting a time constant variable (Wooldridge, 2015). While for this study, panel data method are employed to control

*Corresponding author's email: rashdanita@melaka.uitm.edu.my

the individual-specific effects that usually unobservable which may be correlated with other explanatory variables included in the regression model (Hausman & Taylor, 1981).

The extent to which the POLS bias the result lies beyond the scope of the current study. However, as the result shows, the contradicting results between the current study and study done by Salim and Yadav (2012) provide a strong indication that heterogeneity in sample affects the generalizability of the results of the latter. While Salim and Yadav (2012) provides the general idea of the effect of capital structure on firm's performance, the current study look deeper into industry specific relationships of the variables. Given this contradictions and similarities, future policies on capital structure compositions should be made with precaution. We believe industry-specific and economic growth should be accounted for.

REFERENCES:

Abubakar, S. Y. (2002). Migrant Labour in Malaysia : Impact and Implications of the Asian Financial Crisis. *Crisis*, (January), 1–44.

Akeem, L. B., K, E. T., Kiyanjui, M. W., & Kayode, M. (2014). Effects of Capital Structure on Firm ' s Performance : Empirical Study of Manufacturing Companies in Nigeria, *3*(4), 39–57.

Azhagaiah, R. (2007). The Impact of Capital Structure on Profitability, *9*, 371–392.

Baharuddin, N. S., Khamis, Z., Mahmood, W. M. W., & Dollah, H. (2011). Determinants of capital structure for listed construction companies in Malaysia. *Journal of Applied ...*, *1*(2), 115–132. Retrieved from http://scienpress.com/upload/JAFB/Vol 1_2_6.pdf

Bradley, M., Jarrell, G. a, & Kim, E. H. (1984). On the Existence of an Optimal Capital Structure: Theory and Evidence. *Journal of Finance*, *39*(3), 857. http://doi.org/10.2307/2327950

Chisti, K. A., Ali, K., & Sangmi, M. (2013). Impact of Capital Structure on Profitability of Listed Companies (Evidence From India). *The USV Annals of Economics and Public Administration*, *13*(1), 183–191.

Collins, O. S., Filibus, I. E., & Clement, A. A. (2012). Corporate Capital Structure and Corporate Market Value: Empirical Evidence from Nigeria. *International Journal of Economics and Finance*, *4*(12), 193–202. http://doi.org/10.5539/ijef.v4n12p193

David, D. F., & Olorunfemi, S. (2010). Capital Structure and Corporate Performance in Nigeria Petroleum Industry : Panel Data Analysis. *Journal of Mathematics and Statistics 6*, *6*(2), 168–173.

Dawar, V. (2014). Agency theory, capital structure and firm performance: some Indian evidence. *Managerial Finance*, *40*(12), 1190–1206. http://doi.org/10.1108/MF-10-2013-0275

Degryse, H., de Goeij, P., & Kappert, P. (2012). The impact of firm and industry characteristics on small firms' capital structure. *Small Business Economics*, *38*(4), 431–447. http://doi.org/10.1007/s11187-010-9281-8

Ebaid, I. E.-S. (2009). The impact of capital-structure choice on firm performance: empirical

*Corresponding author's email: rashdanita@melaka.uitm.edu.my

evidence from Egypt. *The Journal of Risk Finance*, *10*(5), 477–487. http://doi.org/10.1108/15265940911001385

Ebrati, M. R., Emadi, F., Balasang, R. S., & Safari, G. (2013). The Impact of Capital Structure on Firm Performance : Evidence from Tehran Stock Exchange. *Australian Journal of Basic and Applied Scienc*, *7*(4), 1–8.

Gibson, C. (1987). How Chartered Financial Analysts View Financial Ratios. *Financial Analysts Journal*, *43*(3), 74–76.

Gill, A., Biger, N., & Mathur, N. (2011). The Effect of Capital Structure on Profitability: Evidence from the United States. *International Journal of Management*, *28*(4), 3–15.

Hamid, M. A., Abdullah, A., & Kamaruzzaman, N. A. (2015). Capital Structure and Profitability in Family and Non-Family Firms : Malaysian evidence. *Procedia Economics and Finance*, *31*(15), 44–55. http://doi.org/10.1016/S2212-5671(15)01130-2

Hasan, M. B., Ahsan, a. F. M. M., Rahaman, M. A., & Alam, M. N. (2014). Influence of Capital Structure on Firm Performance: Evidence from Bangladesh. *International Journal of Business and Management*, *9*(5), 184–194. http://doi.org/10.5539/ijbm.v9n5p184

Hausman, J., & Taylor, W. (1981). Panel data and unobservable individual effects. *Econometrica: Journal of the Econometric Society*, *49*(6), 1377–1398. http://doi.org/10.2307/1911406

Hofer, C. W., & Sandberg, W. R. (1987). Improving New Venture Performance: Some Guidelines for Success. *American Journal of Small Business*, *12*(1983), 11.

Hung, C. Y., Albert, C. P. C., & Eddie, H. C. M. (2002). Capital structure and profitability of the property and construction sectors in Hong Kong. *Journal of Property Investment & Finance*, *20*(6), 434–453. http://doi.org/10.1108/14635780210446469

Jewell, J. J., & Mankin, J. a. (2011). What Is Your Roa ? an Investigation of the Many Formulas for Calculating Return on Assets. *Academy of Educational Leadership Journal*, *15*(1980), 79–91.

John, A. O. (2013). Effect Of Capital Structure On Firm Performance : Evidence From Nigerian Manufacturing Industry. *International Journal of Innovative Research and Studies*, *2*(9), 469–480. Retrieved from http://www.ijirs.com/vol2_issue-9/40.pdf

Khan, A. G. (2012). The relationship of capital structure decisions with firm performance: A study of the engineering sector of Pakistan. *International Journal of Accounting and Financial Reporting*, *2*(1), 245–262. http://doi.org/10.5296/ijafr.v2i1.1825

Long, W. F., & Ravenscraft, D. J. (1984). The Misuse of Accounting Rates of Return: Comment. *The American Economic Review*, *74*(3), 494–500.

Manawaduge, A., Zoysa, A. De, & Chowdhury, K. (2011). Capital Structure and Firm Performance in Emerging Economies : an Empirical Analysis of Sri Lankan Firms. *Corporate Ownership & Control*, *8*(4), 253–263.

*Corresponding author's email: rashdanita@melaka.uitm.edu.my

Murphy, G. B., Trailer, J. W., & Hill, R. C. (1996). Measuring performance in entrepreneurship research. *Journal of Business Research*, *36*(1), 15–23. http://doi.org/10.1016/0148-2963(95)00159-X

Olokoyo, F. O. (2013). Capital Structure and Corporate Performance of Nigerian Quoted Firms : A Panel Data Approach. *African Development Review*, *25*(3), 358–369.

Opoku, E. F., & Adu, J. K. (2012). The Impact of Capital Structure and Profitability of Listed Banks on the Ghana Stock Exchange. *Social and Basic Sciences Research Review*, *1*(2), 74–91.

Pandey, I. . (1999). *Financial Management* (8th ed.). New Delhi: Vikas Publishing House Pvt. Ltd.

Reece, J. S., & Cool, W. R. (1978). Measuring Investment Center Performance. *Harvard Business Review*, *56*(3), 28.

Salim, M., & Yadav, R. (2012). Capital Structure and Firm Performance: Evidence from Malaysian Listed Companies. *Procedia - Social and Behavioral Sciences*, *65*(ICIBSoS), 156–166. http://doi.org/10.1016/j.sbspro.2012.11.105

Seetanah, B., Seetah, K., Appadu, K., & Padachi, K. (2014). Capital structure and firm performance : evidence from an emerging economy. *The Business {&} Management Review*, *4*(4).

Taani, K. (2013). The Relationship between Capital Structure and Firm Performance: Evidence from Jordan. *Global Advanced Research Journal*, *2*(11), 542–546.

Wahba, H. (2013). Capital structure, managerial ownership and firm performance: evidence from Egypt. *Journal of Management & Governance*, *18*(4), 1041–1061. http://doi.org/10.1007/s10997-013-9271-8

Wooldridge, J. (2015). *Introductory Econometrics: A Modern Approach* (6th ed.). United States: Cengage Learning.

Zeitun, R., & Tian, G. G. (2007). Capital structure and corporate performance : evidence from Jordan. *Australasian Accounting Business and Finance Journal*, *1*(4), 40–61

*Corresponding author's email: rashdanita@melaka.uitm.edu.my

HUMAN CAPITAL EFFICIENCY AND FIRM PERFORMANCE: AN EMPIRICAL STUDY ON MALAYSIAN TECHNOLOGY INDUSTRY

Azlina Rahim*[1], Ruhaya Atan[2] and Amrizah Kamaluddin[2]
Faculty of Accountancy, Universiti Teknologi MARA Melaka[1]
Faculty of Accountancy, Universiti Teknologi MARA, Puncak Alam, Selangor[2]

ABSTRACT

This study examined the relationship between human capital efficiency and firm's performance in Malaysian technology industry. Using accounting data, this study reviewed annual reports of all technology companies listed under Main Market and Ace Market of Bursa Malaysia in year 2009. The study applied Value Added Intellectual Coefficient (VAIC^{TM}) methodology developed by Ante Pulic to measure human capital efficiency. The results showed that there was not much difference in terms of human capital efficiency between the Main Market and Ace Market. Results from correlation analysis indicate that human capital efficiency has significant and positive relation with firm's performance. Findings from this study may be useful to companies' managers to make better decision pertaining to the proper deployment of their strategic asset, namely human capital.

Keywords: human capital efficiency, VAIC^{TM} , technology industry

INTRODUCTION

Human capital is the components that emerged from the concept of intellectual capital (Bontis et al., 2000; Tayles et al., 2007). Human capital is the most important asset that exists within a firm. It represents the human factor in an organisation where by combination of intelligence, skills, knowledge, aptitudes and expertise that gives the organisation its distinctive character which those traits contributing to production and profitability, thus improve organizational performance (Bontis et al., 2000 Tayles et al., 2007; Gazor et al., 2013). Additionally, Yusuf (2013) argued that the ability of a corporate organization to successfully implement business strategies solely depends on efficient use of intangibles asset, particularly human capital.

Efficiency in using resources plays an important role in determining the strength of the organization. Measuring human capital performance has become an essential issue for companies in today's business world since it may help them to get the right perspective on human capital. A proper performance measurement tool could provide the firms with the necessary information for creating an action plan in order to improve human capital contribution to the organizational success.
The Value Added Intellectual Coefficient (VAICTM) has been introduced by Ante Pulic which can be used to measure the efficiency of intellectual capital within a company. The introduction of these monetary intellectual capital measurement methods provides new opportunities for companies and their stakeholders. This is because it provides a concrete basis for comparing the intellectual capital of different companies (Kujansivu and Lonnqvist, 2005).

*Corresponding author's email: azlina@bdrmelaka.uitm.edu.my

Even though the term intellectual capital is widely used in recent times by the research community in the developed countries, there have been very few studies that have used emerging economies for evaluating the implications of intellectual capital for the industries. Therefore, it becomes necessary to understand whether this resource is being efficiently utilized by the industries in the emerging economy to their advantage in creating value over time (Kamath, 2007).

The aim of this study is to provide an empirical view of the present state of human capital (HC) in Malaysian technology companies for year 2009. This paper investigated the relationship between human capital efficiency and firm's performance.

There is not much literature on human capital efficiency especially in Malaysia. Most of the studies focusing on the intellectual capital performance which majority of the studies in Malaysia focus on the finance sector. This paper differs from other Malaysian studies since it uses a sample of specific industry i.e. technology industry. Therefore, it contributes in the selection of sample size. Furthermore, the study measures the utilization of human capital efficiency in both Main Market and Ace Market of Bursa Malaysia. Thus, the results from this study may benefits the government and accounting professional bodies for future regulatory impact especially in drawing up future guidelines and policies pertaining to human capital performance. Additionally, it may facilitate the accountants and managers of the companies for better allocation of resources in those organizations.

The remainder of this article is structured in the following manner. The next section reviews the empirical background of the issue addressed in this study, followed by data and methodology. Then, the results of the study. The final section presents the conclusion and implications of the study.

LITERATURE REVIEW

Definition of human capital

There have been many attempts made by the researchers to define human capital. Different authors give different interpretation of this concept. Even though the definitions differ from one to another, many definitions have offered general view of the human capital. The underlying concepts in these definitions include the notions that human capital is emphasis on the skills and knowledge of employees rather than on the physical assets of a company Muhammad (2009). Human capital definitions by various authors summarized in Table 1.

Table 1: Summary of HC definitions

Authors	Definition of human capital
Edvinsson and Malone (1997)	The knowledge, expertise, innovative ideas and the capability of employees to solve problems in the organization
Becker, Huselid and Ulrich (2002)	The productive efforts of an organization's workforce
Fincham and Roslender (2003)	It is the only property which generates value
Chen, Zhu and Zie (2004)	No value can be generate without human capital
Ting and Lean (2009)	Human capital include innovation, capacity, creativity, know-how and previous experience, teamwork capacity, employee flexibility, tolerance for ambiguity, motivation, satisfaction, learning capacity, loyalty, formal training and education.
Baron (2011)	It comprises of knowledge, skills, abilities to develop and innovativeness possessed by the employees of an organisation

*Corresponding author's email: azlina@bdrmelaka.uitm.edu.my

Micah, Ofurum and Ihendinihu (2012)	The energies, skills, talents and knowledge of people which are, or which potentially can be applied to the production of goods or rendering useful services

The concept of Human Capital Efficiency (HCE)

Value Added Intellectual Coefficient (VAIC™) developed by Ante Pulic is an analytical tool for measuring the efficiency of intellectual capital within a company. It was designed to enable management, shareholders and other relevant stakeholders to efficiently monitor and evaluate the efficiency of firm's total resources and each major resource component. The model gives a new insight into how value creation efficiency in the company is measured and monitored using accounting based figures.

One of the VAIC™ components is human capital efficiency (HCE). HCE measures the value added by the human resources of an organization (Kwarbai and Akinpelu, 2016). Many literatures have discussed the advantages of using the VAIC™ methodology usefully (Chan, 2009; Chen et al., 2005; Firer and Williams, 2003; Goh, 2005; Mavridis, 2004; Tseng and Goo, 2005; William, 2000). The availability of these studies adds further credibility to the usage of this methodology.

According to these researchers, VAIC™ produces quantifiable, objective and quantitative measurements without the requirement of any subjective grading and awarding of scores or scales. It aids further computation and statistical analysis of a large sample size that may run into thousands of data items collected over a period of time.

Furthermore, it uses relatively simple and straightforward procedures in the computation of the necessary indexes and coefficients, which may be simple to understand, especially for management and business people who are accustomed to traditional accounting information. These coefficients enable the management to visualize the value creation efficiency of resources in the company. The higher the coefficient, the better management utilizes the company's value creation potential.

Even though there are several measurement methodologies, the most suitable method to measure the intellectual capital efficiency and relate it to the value of the organization is VAIC™ created by Pulic and classified under ROA group method (Jurczak, 2008). The introduction of these monetary intellectual capital measurement methods provides new opportunities for companies and their stakeholders. This is because it provides a concrete basis for comparing the intellectual capital of different companies (Kujansivu and Lonnqvist, 2007).

Human Capital Efficiency and Firm Performance

Previous studies have shown that there is a significant relationship between human capital efficiency and organizational performance (Goh, 2005; Makki et al, 2008; Ting and Lean, 2009; Ghosh and Mondal, 2009). According to Plink and Barning (2010) human capital positively affects organizational performance because it can generate significant value for companies and provide them with sustainable competitive advantage.

Basically, these study focus on the intellectual capital as a whole. Riahi-Belkaoui (2003) investigated the effect of intellectual capital on multinational firms in the US. The sample of this research included 84 firms. The results revealed that there is a significant positive relationship between intellectual capital and firm performance Firer and Williams (2003)

*Corresponding author's email: azlina@bdrmelaka.uitm.edu.my

DOI: 10.1051/shsconf/20173600026

studied the relationship between the efficiency of value added and corporate performance. The results of their study indicated that the relationship between the efficiency of intellectual capital and corporate performance is limited and mixed. Mavridis (2004) applied the same method to analyze the data from Japanese banks and confirmed the existence of significant performance differences among the various groups.

Tseng and Goo (2005) studied the relationship between intellectual capital and the firm market value and financial performance in Taiwan. The data were extracted from 500 Taiwanese firms. The results of this research showed that there is a significant positive relationship between intellectual capital and firms' market value.

Another study on intellectual capital performance was conducted by Shiu (2006) where she examined 80 annual reports from Taiwan listed technologies firms. Based on the 2003 annual reports ,results showed that the VAIC has a significant positive correlation to profitability and market valuation.

Kamath (2007) also adapted the same model for measuring the value-based performance of the Indian banking sector for a period of five years from 2000 to 2004. The study confirms the existence of vast differences in the performance of Indian banks. There is an evident bias in favour of the performance of foreign banks compared with domestic banks.

Ghosh and Mondal (2009) analyzed the relationship between intellectual capital and the performance of pharmaceutical and software companies in India. Based on 80 companies, they concluded that there is a significant positive relationship between intellectual capital and Profitability. On the other hand, there is no significant relationship with productivity.

Zéghal and Maaloul (2010) measured intellectual capital and its relationship with financial performance and market value of commercial and industrial companies in UK. The study used data from 37 multinational companies in Serbia from 2006 to 2008. The results of this study reveal that there is a significant positive relationship between intellectual capital and the financial and economic performance of these companies.

Only few study that focus on the human capital. Yusuf (2013) investigated the relationship between human capital efficiency and financial performance of Nigerian banks. The study concluded that efficient utilization of human capital does not have any significant impact on the return of equity of banks.

Parham and Heling (2015) studied the efficiency of human capital and its impact on the financial performance of Dutch production companies. The study revealed that there is positive relationship between human capital efficiency and all three corporate performance measures' namely return on asset, return on equity and employee productivity. The result shows a strongly significant relationship between human capital efficiency and employee productivity.

In Malaysia, one of the earliest study on intellectual capital performance was conducted by Goh (2005), where he measured the intellectual capital performance of commercial banks for the period 2001 to 2003. The results found that, all banks have relatively higher human capital efficiency than structural and capital efficiencies. As expected, foreign banks are intellectually more efficient than the domestic banks. The results revealed that there were significant differences between rankings of bank according to efficiency and traditional accounting measures.Hazlina and Zubaidah (2008) examined the intellectual capital and performance for

*Corresponding author's email: azlina@bdrmelaka.uitm.edu.my

companies listed in the Bursa Malaysia Main Board for the year 2005-2006.They found significant positive relationship to firms' profitability. However, there is no significant relationship between intellectual capital value and firms' market valuation for companies listed in Main and Second Boards.

Nik Maheran and Md Khairu (2009) investigated the intellectual capital efficiency and its performance in Malaysian financial sector. The data were taken from 18 companies under financial sector for the year 2007.It was found that intellectual capital has significant and positive relationships with company's performance measured by profitability and Return on Assets.

Further, Ting and Lean (2009) studied the intellectual capital performance and its relationship with financial performance of financial institutions in Malaysia for the period 1997 to 2007. The study revealed that intellectual capital performance and Return on Assets are positively correlated among the finance sector.

The recent study in Malaysia was conducted by Tze et al. (2011) where they review the annual reports of Food and Beverage (F&B) companies from 2008 to 2010. Those companies selected were under consumer sector listed in Bursa Malaysia. The findings indicated that the beverage companies have greater VAIC compared to food companies over the 3 years period.

As one of the emerging market in Asia, there is a practical need for Malaysia to determine the human capital efficiency and understand the impact of human capital efficiency on firms' performance. Thus, the firms will get to know the utilization of their resources, so that they can decide on their future resource allocation decisions.

RESEARCH METHODOLOGY

Sample Selection and Data Collection

The study examined the annual reports of all technology companies listed under Main Market and Ace Market of Bursa Malaysia in year 2009. The technology industry was selected because; these companies rely on intellectual capital in the operation of their business as compared to other traditional sector companies. The data from this study were extracted from the company's annual reports. Bontis (2003) mentioned that annual reports are considered as an important source of company information by external users such as stakeholders.

All the annual reports were downloaded from the Bursa Malaysia website. Based on Bursa Malaysia web site, there are a total of 93 companies listed under Technology industry. However, due to some missing data and negative value added, some companies were eliminated. Therefore, the final data consists of 19 companies from the Main market and 36 were from Ace market. The details were tabulated in Table 2.

Table 2: Descriptive analysis of the sample

	Original data		Final data	
	Frequency	**Percentage**	**Frequency**	**Percentage**
Main market	22	23.7	19	34.5
Ace market	71	76.3	36	65.5
Total	**93**	**100**	**55**	**100**

*Corresponding author's email: azlina@bdrmelaka.uitm.edu.my

The VAIC™ model

The VAIC™ model was selected to measure the intellectual capital efficiency and its components namely human capital efficiency, capital employed efficiency and structural capital efficiency. Human capital efficiency indicates the value added by human capital which is the main focus in this study. This method has been widely used by researchers (Chen et al., 2005; Kamath, 2007; Nazari & Herremans, 2007; Chan, 2009; Ghosh & Mondal, 2012). Therefore, the selection for this model in Malaysian context is justifiable. In fact, VAIC™ provides a unique measurement that can be used for comparative analyses across various companies, time periods and industries, both internationally and locally (Ranjith, 2007).

FINDINGS AND DISCUSSION

Descriptive studies

Table 3 presents the mean, standard deviation, minimum and maximum values of all variables for the year 2009. The mean of HCE is 2.64 with a range from 1.01 to 10.08. On average, the Malaysian technology companies created RM 2.64 for every one ringgit employed. This result is consistent with Kamath (2008) and Firer and William (2003).

Table 3: Descriptive statistics for selected variables

Variable	*Mean*	*Std Deviation*	*Minimum*	*Maximum*
HCE	2.6411	2.04851	1.01	10.08
ROA	.0741	.06459	.00	.22

Based on the mean and std deviation, there is not much difference in the mean and std deviation of human capital efficiency between Main Market and Ace Market. Both markets were utilizing their human capital well. The detail statistics for each market was tabulated in Table 4.

Table 4: Comparison of mean between Main market and Ace market

	Mean		Std. Deviation	
	Main market	Ace market	Main market	Ace market
HCE	2.7426	2.5875	2.12574	2.03517

This result has been supported statistically by conducting the independent samples t-test. The test was conducted to compare means between Main market and Ace market. Based on the test of equality of variances (Levene's test), it shows that the variances of the variable that is HCE for the two groups are equal (p-value = 0.995 > 0.05). Therefore, it indicates that there was no significant difference in HCE for Main market and Ace market. The result was tabulated in Table 5.

Table 5: Independent sample test

		HCE Equal variances assumed	HCE Equal variances not assumed
Levene's Test for Equality of Variances	*F*	.000	
	Sig.	.995	
t-test for Equality of *Means*	*t*	.303	.302
	df	53	36.145
	Sig. (2-tailed)	.763	.765

*Corresponding author's email: azlina@bdrmelaka.uitm.edu.my

Mean Differences	.19298	.19298
Std. Error Difference	.63601	.63980
95% Confidence Interval of the difference		
Lower	-1.08270	-1.10442
Upper	1.46866	1.49038

Correlation Analysis

The secondary data was checked for normality and the data was not normally distributed. Therefore, the non-parametric tests Spearman's rank-order correlation was applied to test the relationship between the variables. The findings indicate that HCE ($r=0.539$, $p< 0.01$) was significantly and positively correlated to ROA.

The correlation implies that human capital efficiency was significant factor in the technology industry; companies with higher human capital would be more profitable than others.Table 6 presents the results of the Spearman's correlations.

Table 6: Correlations (2-tailed)

	HCE
ROA	0.539*

Note: * indicates correlation is significant at the 0.01 level in the two-tailed test

CONCLUSIONS AND IMPLICATIONS

The purpose of this study is to provide an empirical view of the present state of human capital in Malaysian technology companies for year 2009. This study also measures the human capital efficiency and firm's performance in Malaysian technology industry. The data were extracted the annual reports of all technology companies listed under Main Market and Ace Market of Bursa Malaysia in year 2009.

The study reveals that the mean of HCE is 2.64 with a range from 1.01 to 10.08. However, there is not much difference in the mean and std deviation of human capital efficiency between Main Market and Ace market in the technolgy industry. Spearman's rank-order correlation was applied to test the relationship between the variables. The empirical findings revealed that there is a significant positive relationship between HCE and ROA.

Findings from this study are subject to some limitations that provide initiatives for future research. The study was limited to one year sample. Therefore, the avenues for further research will include more data from more sample companies covering longer periods of time. In addition, it could be useful to further investigate the relationship of human capital and other firms' characteristics such as market value, return on investment and others.

In spite of some existing limitations, this study contributes to the human capital literature in several ways. First, human capital in a specific industry. Second, this study provides the empirical understanding on the human capital efficiency for both Malaysian markets; Main Market and Ace Market. Further, the study discuss in detail the human capital efficiency that been published in Malaysian annual reports.

*Corresponding author's email: azlina@bdrmelaka.uitm.edu.my

The results of this study may assist the companies to be more aware of other aspects of corporate reporting particulary on the human capital disclosure. In addition, it may benefit the regulators and standard setting bodies for future regulatory impact especially in setting framework for Malaysian companies.

REFERENCES

Abdul Latif, S. and Fauziah, S. (2007), "Intellectual capital management in Malaysian public listed companies', International Review of Business Resaerch paper, Vol 3, No 1, pp. 266- 278.

Becker, B., Huselid, M.A. & Ulrich, D 2002 , Six key principles for measuring human capital performance in your organization, working paper

Baron, A. (2011). Measuring human capital. Strategic HR Review, 10(2), 30–35. doi:10.1108/14754391111108338

Bontis, N. William, C. and Richardson, S., (2000), "Intellectual capital and business performance in Malaysian industries", Journal of Intellectual Capital, Vol 1, No 1, pp 85-100.

Chen, J, Zhu, Z & Xie, H 2004 'Measuring intellectual capital: a new model and empirical study', Journal of Intellectual Capital, Vol.5, No.1, pp. 195-212

Chan, K. H. (2009), "Impact of intellectual capital on organizational performance", The Learning Organization, Vol. 16, pp. 4-21.

Chen, M.-C., Cheng, S.-J., Hwang, Y. (2005), "An empirical investigation of the relationship between intellectual capital and firms' market value and financial performance", Journal of Intellectual Capital, Vol. 6 No.2, pp.159-76.

Damanpour F. (1991), "Organizational innovation: a meta-analysis of effects of determinants and moderators", Academy of Management Journal, Vol 34, No 3, pp 550–90.

Edvinsson, L. (1997), "Measuring intellectual capital at Skandia", Long Range Planning, Vol. 30 No.3, pp.266-373.

Edvinsson, L., Malone, M.S. (1997), Intellectual Capital: Realizing Your Company's True Value by Finding Its Hidden Brainpower, Harper Business, New York, NY.

Fincham, R, Roslender, R (2003), 'Intellectual capital as management fashion: a review and critique', European Accounting Review, Vol.12,No.4, pp.781-795.

Firer, S and Stainbank, L. (2003), Testing the relationship between intellectual capital and a company's performance, Meditari Accountancy Research Vol. 11, pp 25–44

Firer, S. and Williams, S.M. (2003), "Intellectual capital and traditional measures of corporate performance", Journal of Intellectual Capital, Vol. 4 No.3, 348-360.

*Corresponding author's email: azlina@bdrmelaka.uitm.edu.my

Goh, P.C. (2005), "Intellectual capital performance of commercial banks in Malaysia", Journal of Intellectual Capital, Vol. 6 No.3, pp.385-96.

Guthrie J. and Petty, R. (2000), "Intellectual Capital Literature Review: Measurement, reporting and management", The Journal of Intellectual Capital, Vol 1, pp155 – 176.

Ghosh, S. and Mondal A. (2009), "Indian software and pharmaceutical sector IC and financial performance", Journal of Intellectual Capital, Vol. 6 No.3, pp.369-88.

Hazlina, H. and Zubaidah Z.A (2008), "Relationship between intellectual capital and firms' performance: evidence from public listed companies in Malaysia", Proceeding International Accounting Business Conference in Johor Bahru, Malaysia

Jurczak, J. (2008), "Intellectual Capital Measurement methods", Economics and Organization of Enterprise, Vol. 1, No.1, pp 37- 45

Kamath, B. G. (2007), "Intellectual capital performance of Indian banking sector", Journal of Intellectual Capital, Vol. 8 No. 1, pp96-123.

Kamath, B. G. (2008), "Intellectual capital and corporate performance in Indian pharmaceutical industry", Journal of Intellectual Capital, Vol. 9 No. 4, pp 684-704..

Kaplan, R.S. and Norton, D.P. (2004), Strategy Maps: Converting Intangible Assets into Tangible Outcomes, Harvard Business School Press, Boston, MA.

Kujansivu, P. and Lonnqvist, A. (2005), "How do investments in intellectual capital create profits?", Frontiers of E-business Research, pp. 304-318.

Kujansivu, P. and Lonnqvist, A. (2007)." Investigating the value and efficiency of intellectual capital". Journal of Intellectual Capital, Vol. 8, pp. 272-287.

Lin, C. Y.-Y., & Chen, M. Y.-C. (2007), „Does Innovation Lead to Performance? An Empirical Study of SMEs in Taiwan", Management Research News, Vol 30, No 2, pp 115-132.

Lev, B. (2001), Intangibles: Management, Measurement, and Reporting, Brookings Institution Press, Washington, DC.

Luthy, D.H. (1998), "Intellectual capital and its measurement", Proceedings of the Asian Pacific Interdisciplinary Research in Accounting Conference (APIRA) Utah State University, Logan, Utah USA.

Mavridis, D.. (2004), "The intellectual capital performance of the Japanese banking sector", Journal of Intellectual Capital, Vol. 5 No.1, pp.92-115.

Mavridis, D. and Kyrmizoglou, P. (2005), "Intellectual capital performance drivers in the Greek banking sector", Management Research News, Patington, Vol. 28 No.5, pp.43-62.

*Corresponding author's email: azlina@bdrmelaka.uitm.edu.my

Micah L., Ofurum C., Ihendinihu J (2012), "Firms Financial Performance and Human Resource Accounting Disclosure in Nigeria", International Journal of Business and Management, Vol 7, No 14, pp 67-75

Montequin, V.R., Fernandez, F.O., Cabal,V.A., and Gutierrez, N.R. (2006), "An integrated framework for intellectual capital measurement and knowledge management implementation in small and medium-sized enterprises", Journal of Information Science, Vol 32,pp 525-538.

Nik Maheran N. M. and Md Khairu I. (2009) ."Intellectual Capital Efficiency and Firm's Performance: Study on Malaysian Financial Sectors", International Journal of Economics and Finance, Vol 1 No. 1, pp 206-212

Oke, A., Burke, G. and Myers, A. (2004). "Innovation types and their impact on performance in UK SMEs", The 11th proceeding. Cranfield School of Management, Cranfield University

Parham S. and Heling G. (2015), "The relationship between human capital efficiency and financial performance of Dutch production companies", Research Journal of Finance and Accounting; Vol. 6, No 8, pp. 188-201.

Pulic, A. (1998), "Measuring the performance of intellectual potential in knowledge economy", available at: www.vaic-on.net (accessed 13 July 2009).

Pulic, A. (2001), "Value creation efficiency analysis of Croatian banks 1996-2000", available online at www.vaic-on.net (accessed 28 August 2009).

Pulic, A. (2004), "Intellectual capital – does it create or destroy value?", Measuring Business Excellence, Vol. 8 No.1, pp.62-8.

Ranjith, B. A. (2007),"The Impact of Intellectual Capital on Investors' Capital Gains on Shares: An Empirical Investigation of Thai Banking finance& Insurance Sector", International Management Review; Vol. 3, No 2, pp. 14-25.

Riahi-Belkaoui, A. (2003), "Intellectual capital and firm performance of US multinational firms. A study of the resource-based and stakeholder views", Journal of Intellectual Capital, Vol 4,No 2, pp. 215-226.

Shiu, H. J. (2006), "The application of the value added intellectual coefficient to measure corporate performance: evidence from technological firms", International Journal of Management, Vol. 23 No. 2, pp. 356-365.

Schoenecker, T. and L. Swanson. (2002), "Indicators of firm technological capability: Validity and performance implications", IEEE Transactions on Engineering Management 49 (February): pp 36-44.

Sveiby, K.-E. (1997), The New Organizational Wealth: Managing and Measuring Knowledge-Based Assets, Berrett-Koehler Publishers Inc., San Francisco, CA .

Tan, H.P., David, H. and Hancock, P (2007). "Intellectual capital and financial returns of companies", Journal of Intellectual Capital, Vol 8, No. 1 , pp 76-95.

*Corresponding author's email: azlina@bdrmelaka.uitm.edu.my

Ting, W.K. and Lean H. H. (2009), " Intellectual capital performance of financial institutions in Malaysia", Journal of Intellectual Capital, Vol 10 No 4, pp 588- 599.

Tseng, C. and Goo, J. (2005), "Intellectual capital and corporate value in an emerging economy: empirical study of Taiwanese manufacturers", R&D Management, Vo. 35 No. 2, pp. 187-201.

Wang, W., Chang, Ch. (2005), "Intellectual capital and performance in causal models. Evidence from the information technology industry in Taiwan", Journal of Intellectual Capital, Vol 6., No. 2, pp. 222-236.

Yusuf I. (2013), "The relationship between human capital efficiency and financial performance: an empirical examination of quoted Nigerian banks", Research Journal of Finance and Accounting; Vol. 4, No 4, pp. 148-154.

*Corresponding author's email: azlina@bdrmelaka.uitm.edu.my

GOODS AND SERVICES TAX (GST): CHALLENGES FACED BY BUSINESS OPERATORS IN MALAYSIA

Loo Ern Chen*, Mohd Shukry Bin Md Taib[1]
[*]*Faculty of Accountancy Universiti Teknologi MARA Melaka,* [1]*Account Division, Accountant General's Department of Malaysia*

ABSTRACT

The objective of this study is to identify the factors affecting business operators' awareness on the implementation of Goods and Services Tax (GST) and the perceived issues and problems they may likely encounter with the implementation of GST in Malaysia. Questionnaires were distributed to 302 registered members (business operators) of three Chambers of Commerce in Melaka. The findings of this study show that more than 70% of the respondents are aware of the GST mechanism and issues. The majority of the respondents also agreed that to comply with GST legislations, business operators need to have sufficient knowledge on appropriate and adequate documentation thus leading to increased compliance costs.

Keywords: Goods and Services Tax, awareness, compliance costs

INTRODUCTION

Goods and Services Tax (GST), also known as Value Added Tax (VAT) is a form of expenditure or consumption tax. Prior to the implementation of GST, among the expenditure taxes in force were the Sales Tax[1] and Service Tax.[2] Sales Tax, a single stage tax was imposed on goods manufactured in or imported into Malaysia. Service Tax, also a single stage tax was imposed on taxable services provided by taxable persons.

In the 2005 Budget Speech, it was announced that GST would be introduced, effective from 1 January 2007, but in 2006, the Government deferred GST implementation indefinitely. In the 2010 Budget Speech, the Minister of Finance announced the readiness of the Government to implement GST, and finally the Goods and Services Tax Act 2014 came into force effective from 1 April 2015,[3] replacing the Sales and Service Tax (SST). The initial announcement and subsequent deferment on the implementation of GST was partly to give more time to business operators to change their business processes, to develop software and to provide training to their personnel pertaining to compliance with the GST legislations,[4] and mechanism.[5]

---

[1] Sales Tax was imposed under the Sales Tax Act 1972.
[2] Service Tax was imposed under the Service Tax Act 1975.
[3] GST is imposed under the Goods and Services Tax Act 2014.
[4] GST legislations comprise of the Goods and Services Tax Act 2014, Goods and Services Tax Regulations 2014, the various gazetted GST Orders (all of which have legal effects).

*Corresponding author's email: looern@yahoo.com

GST is a multi-stage tax, imposed on the value added to goods or services in the production and distribution chain, and the tax is effectively on the increase in the value or price of goods or services as they pass through that chain.

[5] Generally, the operations of GST mechanism may be guided by the multitude of GST guidelines and the numerous DG Decisions.

Generally, under the GST legislations,[6] upon achieving the specific threshold[7] from making taxable supplies,[8] a business operator is liable to be registered as a taxable person (aka GST registrant). Since GST is to be collected at various stages of the production and supply chain, a GST registrant is required to charge GST (output tax)[9] on taxable supplies made to its customers. Generally, a taxable supply is taxed at either standard rate[10]or zero rate.[11] For a taxable person, any GST paid is referred to as input tax.[12] A taxable supply will go through a few processes or supply chain before it reaches the final consumer. Suppliers (who are taxable persons) along the supply chain will pay GST when they acquire input and will collect GST from customers on their output at the time of making supply. Suppliers (who are taxable persons) can set off their input tax[13] against output tax and remit the net GST to Royal Malaysian Customs Department (RMCD).

Under the GST regime, GST registrants will need to manage documentation processes and keep all relevant records to account for input and output tax, which may be subject to audit by RMCD. GST registrants are required to keep business and accounting records related to GST for a period of seven years. These records can be kept either in Bahasa Melayu or English. The following are the list of records that should be kept by GST registrants:

(a) All records of goods or services supplied by or to a taxable person including tax invoices, statements of sales, receipts, credit notes, debit notes and export declaration forms;
(b) All records of importations of goods;
(c) Physical books of account, financial statements and paper based source documents including computer printouts of business and accounting records;
(d) Electronic records; and
(e) Details of accounting system, including charts, codes of accounts, instruction manuals, system and programmes documentation and specification, etc.

GST registrants are required to file their GST returns to RMCD not later than the last day of the following month after the end of a taxable period. Taxable period of a GST registrant depends on its taxable turnover. Those with annual gross taxable turnover of less than RM5 million are required to submit their tax returns quarterly. A monthly submission of tax return is applicable to those whose gross annual taxable turnover is more than RM5 million.

---

[6] Section 20, Goods and Services Tax Act 2014.
[7] The threshold is gross turnover of at least RM500,000 over a period of 12 months (or any period within less than 12 month) in relation to the supply of taxable goods or services. The RM500,000 threshold is only applicable to turnover of taxable supplies (i.e. standard rated supply and zero rated supply), but exclude turnover of exempt supplies.
[8] A 'taxable supply' means a supply of goods or services which are standard rated (i.e. 6%) and zero-rated, but excludes exempt supply (Section 2(1), Goods and Services Tax Act 2014).
[9] 'Output tax' means tax on any taxable supply of goods or services made by a taxable person in the course or furtherance of its business in Malaysia (Section 2(1), Goods and Services Tax Act 2014).
[10] The standard rate of 6% is gazetted under the Goods & Services Tax (Rate of Tax) Order 2014.
[11] The types of goods and services subjected to zero rate are gazetted under Goods & Services Tax (Zero-Rated Supply) Order 2014.
[12] 'Input tax' means tax on any supply of goods or services (paid) to a taxable person; and tax paid or to be paid by a taxable person on the importation of goods, and the goods and services are used or to be used for the purposes of any business carried on or to be carried on by the taxable person [Section 2(1), Goods and Services Tax Act 2014].
[13] Claim of input tax credit is only allowable for making of 'taxable supply', which means that input tax in relation to exempt supply is not an allowable input tax credit.

Challenges faced by business operators

Business operators, as intermediate parties need to be aware of, and be ready to accept that GST would impact on their business operations. Among the impacts is the GST charged on the acquisition of taxable supplies necessary for the operations of their businesses. Business operators would need to identify the GST paid that is allowable as input tax credit, as non-allowable input tax credit would lead to increase in cost of doing business.

Business operators could be facing challenges in identifying and distinguishing between taxable supplies (i.e. standard rated supply and zero rated supply) where input tax credit is allowable. The challenges is also in understanding the differences in making zero rated supplies and exempt supplies, [14] as both do not attract any output tax, although input tax credit is allowable for the former, but not for the latter.

Awareness towards GST not only implies readiness of business operators for the implementation of GST, but also their potential acceptance and support for the GST regime. Business operators may not accept or support a change in the tax system unless they could recognise and are aware of the implications on their businesses. Feedback from business operators and the public is of utmost important in order to examine their readiness for the implementation of GST. By taking cognition of feedback from stakeholders, the relevant government agencies can then design appropriate plan and strategy for the implementation of GST.

With the implementation of GST, business operators need to change their accounting system and business processes and must be ready to institute proper accounting record keeping, train their personnel and have proper software to ensure successful documentation and recording for the purpose of GST compliance. Thus, there is possibility that business operators might face issues and problems pertaining to the compliance with GST legislations.

One importance aspect of the implementation of GST is whether business operators, who are GST registrants, as agents to collect tax understand GST issues and the mechanism as applied in Malaysia. To ensure a successful implementation of GST, business operators need to know when they are required to register, to determine the input and output taxes, and when to remit the net GST to the RMCD. They also need to be aware of the penalties for non-compliance.

It is also important to identify business operators' level of awareness, and the issues and problems that business operators are likely to encounter with the implementation of GST. These could be useful inputs for the RMCD to plan necessary programmes, trainings and dissemination of appropriate GST information to enhance awareness, knowledge, support and acceptance of the GST regime.

[14] The types of goods and services exempted from GST are gazetted under Goods & Services Tax (Exempt Supply) Order 2014.

Objective of Study

The objective of this study is to identify the factors affecting business operators' awareness on the implementation of GST and the perceived issues and problems they may likely encounter with the implementation of GST.

This paper consists of six parts. Following the introduction in Part 1, Part 2 provides brief discussion on prior studies on awareness, issues and problems with regards to the implementation of GST. The conceptual framework and hypothesis development are presented in Part 3. Part 4 discusses on the research methodology for this study which includes research method, research sampling, research instrument and data collection. Part 5 presents the findings and analysis of this study. Lastly, Part 6 elaborates on the conclusion of findings, contributions and limitation of study.

LITERATURE REVIEW

Awareness of GST among business operators is important to ensure that they collect the appropriate taxes from their customers and comply with all other requirement of GST. Prior studies (Choong & Lai, 2006; Saira, Zariyawati & May, 2010; Samsuddin, et. al., 2014) showed that awareness of GST implementation is related to education background, levels of education, types of employment and initiatives of tax authority in conducting GST seminars, talks or dialogues.

Palil, Zaini and Mamat (2013) examined consumers' readiness, perception and acceptance of GST in Malaysia and found that the majority (73.2%) of the respondents were aware about the implementation of GST while 53.6% believed that GST would have significant impact of their livelihood. Saira Saira, K., Zariyawati, M.A., & May et al. (2010) found that 95% of the respondents were aware that GST is a major source of government revenue, although around 50% did not understand the operations of the GST system. This study also revealed that non-accounting students did not understand what GST is all about. This implies that tax knowledge and education could probably have bearings on the awareness of the implementation of GST.

Another study found that 77.2% of the respondents were aware about the implementation of GST but most of them were not ready for GST implementation (Abdullah, Idrus & Mehat, 2013). 52.5% of the respondents also perceived GST to be not suitable in Malaysia and 47.5% were aware that GST might affect their spending behaviour. In this study, respondents' level of education showed positive relationship with their level of understanding towards GST. Besides, taxpayers' ethnicity is also found to be associated with their level of awareness towards tax (Jaidi, Noordin & Kassim, 2013). A study investigated the level of awareness among educators found that 93.8% of the respondents were aware of the GST system, although 60.2% indicated that the level of information provided by the relevant authorities was moderately useful (Samsuddin, et. al., 2014).

Public education on the GST implementation appears to be the right direction that tax authority needs to conduct widespread GST education campaigns, form call centres to disseminate information on GST implementation to the public and business operators nationwide (Choong & Lai, 2006; Chan, 2009). These programmes include seminars, field visits, hotlines, dialogue sessions, brochures, media campaign and direct mailers. In order to reduce the burden of small businesses, tax education such as free seminars on tax compliance, use of technology, electronic lodgement and others should be organised by the relevant authority (Pope, 2001).

In Singapore, a broad taxpayer education programme was launched to educate the public on GST (Jenkins & Khadka, 1998). The Inland Revenue Authority of Singapore (IRAS) conducted dialogue session to get input from businesses on their business process and designed the GST system that would not be unreasonably complex. The trade associations also collaborated with the IRAS to educate business owners by providing logistic support while the IRAS provided technical support. The education programmes explained the rational of GST implementation and enhanced public awareness on the GST system. Most of the small and medium enterprises (SMEs) owners' awareness of tax laws and procedures are the results from training, seminars and visitations by revenue officers (Lubua, 2014).

A proper channel of communication to share and to deliver information and knowledge and to cultivate positive attitude towards GST system is essential for enhancing the level of acceptance on the implementation of GST (Zhou, Tam & Heng-Contaxis, 2013). For instance, in the United Kingdom, Chinese taxpayers did not apply for tax credit due to lack of awareness and access to information on tax credits (HM Revenue & Customs, 2010). Therefore, public awareness and good education campaign could ensure the acceptance and smooth introduction of a new tax system (Yew, 2009) and also to raise sufficient revenue to finance governments' expenditures (Abdullah, 2012).

The implementation of GST will require business operators to keep appropriate and adequate accounting records (Alebel, 2011). Proper documentation and accounting records are essential in the event of tax audit and investigation. The lack of appropriate records could lead to SMEs' failure to comply with tax legislations (Abdul-Jabbar, 1996). Thus, in order to comply with the GST legislations, SMEs might need to hire new personnel to handle all documentation related to GST.

SMEs operators also encountered difficulties in complying with various and complicated tax legislations and regulations due to their limited resources and inadequate expertise (Pope & Abdul-Jabbar, 2008). Thus, the implementation of GST would be another added tax compliance burden. SMEs faced a relatively higher compliance cost upon GST implementation, which is regressive for smaller organizations (Pope, 2001; Palil et al, 2013; Grinberg, 2006). GST compliance cost among businesses would be high, particularly for businesses that are involved with various taxable and exempted supplies (Palil & Ibrahim, 2012).

Compliance costs can be divided into two categories; namely GST start-up (or implementation) cost and annual recurring (or on-going) cost (Brown & Marsden, 2002). GST implementation cost comprised of additional expenditure on computer upgrading, staff cost to adjust the pricing of products, training and IT consultation cost, cost for extra bookkeeping requirements and also time costs spent by business owners on GST related activities, such as spending more time on paperwork and on collecting taxes. Study on GST compliance costs in the service sector in Queensland showed that the mean start-up cost is around $5,610[15] per organization while the mean for recurrent cost is approximately $5,600 (Breen, Bergin-Seers & Sims, 2002).

On-going GST cost is a substantial burden in relation to GST compliance. Delayed payments by firms' debtors created cash flows problems when the businesses need to pay their GST commitments. A study on 868 small businesses by Rametse (2010) found that the preparation for GST implementation was a heavy management and financial burden for each business, with an average of $7,888 start-up cost being incurred on equipment, professional fees and stationery. In addition, in terms of time costs, 131 hours were spent, costing to $2,882 to seek professional accounting and information technology advice, and 58 hours (about $1,276 in monetary term) in preparing GST.

In Australia, total GST implementation costs were almost double ($4.5 billion) as compare to the first-year GST operating costs ($2.5 billion), and that high GST implementation costs have negative consequences on economy in various ways (Tran-Nam, 2000). Therefore, in order to reduce compliance costs a simplified method of computing such tax should be introduced (Chan, 2009).

In Malaysia, before the implementation of GST, the expected GST compliance cost for SMEs was RM28,456 (Palil, et. al., 2013). The annual compliance cost consist of internal cost of RM16,772 (58.87%), external cost of RM7,388 (26.01%) and additional cost of RM4,296 (15.42%). Expected cost could increase due to external sourcing of services for GST tax computation and tax planning. On an average, an additional cost of RM6,336 was expected to be incurred on external sourcing of services, and that most of the companies expected an increase in internal cost for tax computation and tax planning after the implementation of GST.

Relative to sales or assets value, SMEs faced higher compliance costs as compared to large businesses (Weichenrieder, 2007; Slemrod & Venkatesh, 2002). In general, compliance cost vary significantly, depending on business size (Breen, Bergin-Seers & Sims, 2002). In Nigerian, an average of ₦67,995 was spent on tax compliance cost for SMEs who were engaged in VAT transactions (Eragbhe & Modugu, 2014). It is also noted that on average, VAT compliance cost burden was higher than compliance cost for other types of taxes, such as income tax, withholding tax, and customs and excise duties.

CONCEPTUAL FRAMEWORK AND HYPOTHESIS DEVELOPMENT

This study has identified four factors namely, business experience, ethnicity of business operators, business size and business initiatives that might have association with business

[15] The $ monetary value refers to is the Australian dollar.

operators' awareness of, and the perceived issues and problems[16] on the implementation of GST they might encounter (see Figure 1).

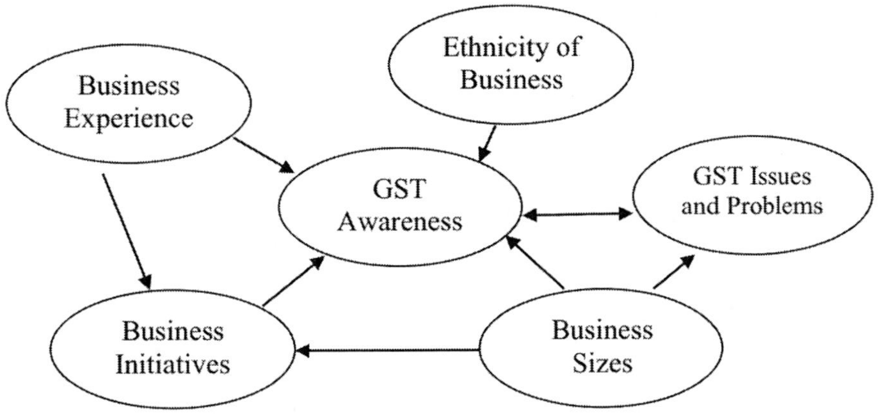

Figure 1: Conceptual Framework

Business experience refers to prior experience in dealing with SST. Although the mechanism and approaches between SST and GST are different, knowledge and experience with SST would be an added advantage to business operators. Previous experience with SST made an important difference on careful preparation of VAT returns (Casanegra de Jantscher, 1987), as business operators with experience in dealing SST would likely to be more aware of the mechanism, regulations and requirements of GST. They could be more knowledgeable in dealing with issue and problems related GST. In addition, business operators with longer period of operations would probably have better understanding of the business environment.

The implementation of GST requires businesses to keep appropriate accounting records (Alebel, 2011). However, SMEs have experienced difficulties in maintaining proper records for business and taxation purposes (Hanefah & Al-Mureshi, 1991). The lack of appropriate record keeping had led to SMEs' failure to comply with tax legislations (Abdul-Jabbar, 1996) and also being audited by tax authorities (Yong, 2011). Hence, business experience in tax audit might influence taxpayers' documentation practices and their tax compliance costs. Therefore, the following hypotheses are formulated:

H_1: There is a relationship between business experience and GST awareness.
H_2: There is a relationship between business experience and perceived issues and problems (in relation to record keeping and compliance cost)

Ethnicities are groups of community with a set of shared cultural values, beliefs and norms to solve their basic social problems (Basu & Altinay, 2002). In New Zealand, Maori and Pacific SMEs were found to have difficulties in tax payments as compared to Europeans and Asians (Yong, 2011). In Nigeria, different ethnic groups faced different social, technical and logistic problems when dealing with the VAT system (Oladipupo & Izedonmi, 2013).

[16] In this study, "GST issues and problems" refer to the perceived issues and problems (as per Table 6) that business operators are likely to face on the implementation of GST.

In the context of this study, ethnicity of business operators consists of three major groups: Malay, Chinese and Indian business operators who are members of Chambers of Commerce. It is noted that when the Government decided to implement GST, the Chinese Chambers of Commerce had organised various seminars, dialogues and relevant GST courses to create awareness among its members. Hence, members of the Chinese Chambers of Commerce might be more aware on the implementation of GST. It is expected that the ethnicity of business operators has association with their awareness of GST.

In terms of documentation and compliance costs, it is possible that different ethnic communities may have different documentation practices such as filing and records keeping and to a certain extent ethnicity of business operators may face different GST compliance cost. Thus, it is hypothesised that:

H_3: Ethnicity of business operators is associated with their awareness of GST.
H_4: There is an association between ethnicity of business operators and GST issues and problems (related to records and compliance cost).

Business operators' awareness of, and perceived issues and problems that they might face on the implementation of GST might also be influenced by business size. This study classified business size into two categories, namely businesses with threshold of less than RM500,000 and with threshold of more than RM500,000. Business operators achieving the threshold of more than RM500,000 are liable to be registered as GST taxable persons. These business operators are expected to source for relevant information on GST, and therefore be more aware of GST implementation. In contrast, business operators who have no legal obligation to register may not be so interested, thus would likely be less aware of GST implementation and of issues and problems related to GST implementation.

The implementation of GST could give rise to additional compliance burdens and compliance costs in relation to record keeping and could create major problems for businesses (Palil et al., 2013). SMEs faced a relatively higher cost of GST implementation (Pope, 2001) would incur relatively more costs on resources to manage their tax affairs (Palil et al, 2013). Generally, compliance cost vary significantly, depending on business size (Breen, Bergin-Seers, & Sims, 2002). As such, the hypotheses formulated are:

H_5: There is a relationship between business size and awareness of GST.
H_6: There is a relationship between business size and GST issues and problems (related to records and compliance cost)

Business initiatives refer to initiatives to seek for more information on GST. These include attending seminars, talks, trainings and purchase of publications related to GST implementation. Study found a substantial difference in knowledge between taxpayers who had attended and those who did not attend any tax seminars, talks or trainings (Palil, 2010). If business operators take more of such initiatives, they are expected to be more aware and have better understanding of GST, hence, able to identify any possible problems in relation to GST that may arise.

It is noted that the main driving factor of tax compliance costs is due to insufficient knowledge of business owners (European Commission, 2007). SME operators lacking in tax knowledge faced more challenges with income tax and VAT (Maseko, 2014). Insufficient technical knowledge on tax matters likely to be even more critical among the self-employed who have to deal with GST affairs (Saad, 2014).

Thus, it is envisaged that business operators who engaged in business initiatives to enhance their tax knowledge are more aware of GST and able to deal with any major issues and problems in relation to GST compliance, such as costs of compliance and documentation. Thus, the hypotheses are formulated as follows:

H_7: There is a relationship between business initiatives and awareness of GST.
H_8: There is a relationship between business initiatives and GST issues and problems (related to records and compliance cost)

RESEARCH METHODOLOGY

To address the objective of this study, a quantitative research using survey is conducted. Quantitative research is generally used to measure respondents' behaviour, knowledge, attitudes or opinion, and also used to answer questions such as how long, how much, how often, who, what and when (Cooper & Schindler, 2014). Survey is chosen to gather primary data in order to describe a population too huge to observe directly.

Research Sampling

Hair, Money, Samouel and Page (2007) defined population as a group of knowledgeable people. Target population is defined in terms of elements, geographical boundaries and time (Sekaran & Bougie, 2009) which allows researchers to make some forms of interpretations (Cooper & Schindler, 2014). The population of this study consists of registered members of three Chambers of Commerce in Melaka, namely the Malay Chambers of Commerce of Melaka, Melaka Chinese Chamber of Commerce and Industry and Melaka Indian Chamber of Commerce and Industry.

Stratified random sampling was used to select respondents from the three Chambers of Commerce in Melaka. The advantage of stratified random sampling is the possibility to compare information among groups (Sekaran & Bougie, 2009). It is because stratified random sampling divides the population into homogenous subgroups that have differences and non-overlapping which enables researchers to make comparison between different ethnicity of the business operators in relation to awareness and perceive issues and problem of GST implementation.

Table 1: Registered Members of Chambers of Commerce, Sample Size and Response Rate among Respondents

Chambers of Commerce	Number of Registered Members *	Sample Size (50%)	Response Rate **	
			No	%
Malay	144	72	64	88.9
Chinese	399	200	90	45.0
Indian	60	30	30	100
Total	478	302	184	

* Source: Registers of the respective Chambers of Commerce in Melaka.
**The response rate is based on the number of questionnaires distributed to and returned by the members of the respective particular Chambers of Commerce. Collectively, for all the three Chambers of Commerce, the response rate is 60.93%.

50% of members from each Chamber of Commerce in Melaka were selected, which formed a total sample size of 302 for this study (see Table 1).

Research Instrument

The questionnaire consists of four sections. Section 1 elicits data on the nature of business, ethnicity of business operators, the business annual gross turnover and GST registration status. This section also elicits data on whether the business was licensed under SST, and whether business operators had attended any seminars or training related to GST.

Section 2 elicits responses relating to the business operators' level of awareness on the implementation of GST in Malaysia. Section 3 comprised of 13 statements, aimed at identifying perceived issues and problems likely to be faced by the business operators when complying with the GST legislations. Respondents were requested to response to the statements based on a five-point Likert-scale (of which 1=strongly disagree, 2=disagree, 3=neither agree not disagree, 4=agree and 5=strongly agree).

Section 4 of the questionnaire provided an open-ended question allowing respondents to provide further comments and general opinion on the implementation of GST in Malaysia.
In order to check the validity and clarity of statements listed in the questionnaire, a pilot test was conducted where 10 sets of questionnaire were distributed to selected academics and business operators in Melaka. Subsequently, amendments were made to ensure clarity of wordings used in the questionnaires.

Data Collection

Questionnaires were distributed to the 302 selected samples (See Table 1). Only 60.93% of usable sets of questionnaire were returned. Sandford (1995) stated that response rate of survey study would probably be around 30%. Therefore, the response rate for this study was reasonably high and adequate for analysis.

FINDINGS AND ANALYSIS

Profile of the respondents

With regard to ethnicity of business operators, 34.8% are Malays, 48.9% are Chinese and 16.3% are Indians. In terms of industry sectors, 37.5% are from the service sector, followed by trading (27.7%), manufacturing (10.9%) and transport and storage (5.4%), while the "other" category (such as hotel, hire purchase, food and beverage, bookstore, insurance and motorcycle businesses) making up of the remaining 18.5%.

Table 2 below shows the cross tabulation results of business with annual gross turnover and registration for GST. Compulsory registration is only applicable to those with annual gross turnover exceeded RM500,000. Of the total respondents, 37% had registered (i.e. "compulsory registration"), while 14% who ought to be registered have yet to do so. Of the total respondents, 6% opted for voluntary registration, citing the rationale as "to comply with the law,[17] increase profitability of their business, to be eligible for input tax credit and to enjoy any benefit as may be provided by tax authority". On the other hand, the reasons given by those who did not opt for voluntary registration are "having to incur additional costs of doing business, did not have staff to handle GST, increased billing charges on customers, GST is very complicated, waste of money and time to learn about GST and GST does not bring benefits to their business, they do not agree with GST implementation, running a small business with not enough time to manage GST matters".

Table 2: Respondent's Business Turnover and Registration for GST

Annual Gross Turnover	GST Registration				Total Number	%
	Yes		No			
	No.	%	No.	%		
Less than RM500,000	11	6	80	43	91	49
More than RM500,000	68	37	25	14	93	51
Total	79	43	105	57	184	100

Of the total respondents, only 36.4% were licensed SST taxable person. Only slightly more than half (52.5%) of the respondents have attended seminars, dialogues, training or conference related to GST. The reasons cited by the remaining 47.5% for not attending any courses on GST are that "they did not get any invitation, headquarters are in-charge for GST training, GST has yet to be implemented[18] and not willing to incur extra cost to attend GST courses".

Reliability and Validity Test

Reliability and validity tests were conducted on the statements listed in the questionnaire. Reliability is a test on how consistent a measuring instrument measures whatever concept it is measuring (Sekaran & Bougie 2009; Hair et al., 2007). Cronbach's alpha reliability coefficient indicates how well the items in a set are positively associated to one another.

[17] The word 'voluntary" implies that voluntary registration is in fact not mandatory under the GST legislation.
[18] The questionnaires were administered about 6 months prior to the coming effect of GST on 1 April 2015.

Internal consistency is considered reliable when Cronbach's alpha is closer to 1 (Sekaran & Bougie, 2009).

As presented in Table 3, the Cronbach's Alpha values of more than 0.8 for the variables (awareness and issues and problems) suggest that the instrument used is valid and of a high degree of reliability.

Table 3: Reliability Test

Variables	Cronbach's Alpha	Cronbach's Alpha Based on Standardized Items	No of Items
Awareness	0.925	0.927	28
Issues and Problems	0.902	0.909	13

Validity is defined as evidence that the instrument, technique or process used to measure a concept does indeed measure the intended concept (Sekaran & Bougie, 2009). Validity is also related to the term accuracy (Hair et al., 2007). If the Kaiser-Meyer-Olkin (KMO) value is greater than 0.5, the variables used are acceptable and valid (Kaiser, 1974). The value of KMO (0.787) presented in Table 4 shows that the variables for awareness, issues and problems used in the instrument are acceptable.

Table 4: Validity Test

Kaiser-Meyer-Olkin (KMO) Measure of Sampling Adequacy		.787
	Approx. Chi-Square	8620.745
Bartlett's Test of Sphericity	df	1711
	Sig.	.000

 DOI: 10.1051/shsconf/20173600027

GST Awareness among Business Operators

The majority (>70%) of the respondents are aware of the GST mechanism. The detail findings of awareness of GST mechanism are presented in Table 5.

Table 5: GST Awareness among Business Operators

Awareness of GST Mechanism	Yes		No	
	Number	%	Number	%
GST will be charged at 6% when implemented on 1st April 2015.	177	96.2	7	3.8
GST is a multi-stage tax.	134	72.8	50	27.7
The 3 categories of GST supply are:				
i. standard rated supply	167	90.8	17	9.2
ii. zero-rated supply	148	80.4	36	19.6
iii. exempt supply	143	77.7	41	22.3
Under the GST regime, any business with an annual threshold of more than RM500,000 is required to register with the Royal Malaysia Customs Department (RMCD).	150	81.5	34	18.5
Any business with an annual threshold of RM500,000 and below may apply for voluntary registration with the RMCD.	133	72.3	51	27.7
Generally any business that is registered for GST is eligible to claim input tax.	122	66.3	62	33.7
The following goods … are not subject to GST:				
i. sugar	165	89.7	19	10.3
ii. flour	162	88	22	12
iii. fish	144	78.3	40	21.7
iv. frozen food	112	60.9	72	39.1
The following services … are exempted from GST:				
i. public transportation	143	77.7	41	22.2
ii. banking	144	78.3	40	21.7
iii. education	156	84.8	28	15.2
When the output tax is less than input tax, business operator is entitled to refund from Royal Malaysia Customs Department.	134	72.8	50	27.2
Business and accounting records related to GST must be kept for a period of 7 years.	143	77.7	41	22.2
The Royal Malaysia Customs Department is an entity responsible to collect GST from the registrants of GST in Malaysia.	161	87.5	23	12.5
Businesses are required to:				
i. submit GST returns not later than the last day of the following month after the taxable period	128	69.6	56	30.4
ii. remit GST not later than the last day of the following month after the taxable period of GST return	130	70.7	54	29.3
Penalty may be imposed on any business registered under GST for following offences :				
i. deficiency on the net tax payable	137	74.5	47	25.5
ii. no GST return is made	134	72.8	50	27.2
iii. GST return is submitted without payment	132	71.7	52	28.3
iv. GST return is submitted with insufficient payment	129	70.1	55	29.9
For refund purposes, payment will be made within				
i. 14 working days if it was submitted online	148	80.4	36	19.6
ii. 28 working days for manual submission.	133	72.3	51	27.7
Businesses registered under GST are required to issue tax invoice within 21 days when the supply is deemed to have taken place.	136	73.9	48	26.1

Perceived Issues and Problems Faced by Business Operators

More than two-thirds of the respondents agreed with the perceived issues and problems that they might face upon the implementation of GST. The details of the findings on such perceived issues and problems are presented in Table 6.

Table 6: Issues and Problems faced by Business Operators on GST Implementation

Statements	SD No. (%)	D No. (%)	N No. (%)	A No. (%)	SA No. (%)	Mean	STD
With the implementation of GST:							
•Your business will face problems to keep all relevant business records to comply with GST legislations.	5 (2.7)	26 (14.2)	24 (13)	105 (57.1)	24 (13)	3.64	0.97
•All documents to be kept for 7 years will lead to storage problem for your business.	7 (3.8)	22 (12)	34 (18.5)	106 (57.5)	15 (8.2)	3.54	0.94
•The compliance costs to implement GST in your business will be higher.	10 (5.4)	18 (9.9)	30 (16.3)	95 (51.6)	31 (16.8)	3.65	1.05
•Your business will need to hire new personnel to manage GST matters.	10 (5.4)	29 (15.8)	41 (22.3)	79 (42.9)	25 (13.6)	3.43	1.08
•Your existing staff will have to attend training courses in order to learn to manage GST matters.	8 (4.3)	23 (12.5)	20 (10.9)	114 (62)	19 (10.3)	3.61	0.98
•Your business will face problems in setting the prices of goods and services provided to customers.	10 (5.4)	21 (11.5)	33 (17.9)	96 (52.2)	24 (13)	3.56	1.03
•Your business still does not know how GST will affect your business.	23 (12.5)	41 (22.3)	26 (14.1)	75 (40.8)	19 (10.3)	3.14	1.24
Your business will have problems to comply with GST such as:							
• keep proper accounting records	6 (3.3)	20 (9.9)	41 (22.3)	106 (57.6)	11 (6)	3.52	0.89
• filing GST returns	6 (3.3)	18 (9.8)	31 (16.8)	112 (60.9)	17 (9.2)	3.63	0.90
• remit GST	6 (3.3)	21 (11.4)	40 (21.7)	110 (59.8)	7 (3.8)	3.49	0.87
• claim input tax credit	4 (2.2)	16 (8.7)	44 (23.9)	108 (58.7)	12 (6.5)	3.59	0.83
The information and explanation on GST provided by RMCD is insufficient to deal with GST implementation.	10 (5.4)	26 (14.2)	50 (27.2)	74 (40.2)	24 (13)	3.41	1.06
GST will increase compliance cost in running your business.	8 (4.3)	22 (12)	25 (13.6)	109 (59.2)	20 (10.9)	3.60	0.98

Note: SD = Strongly Disagree; D = Disagree; N = Neither Agree nor Disagree; A = Agree; SA = Strongly Agree, STD = Standard Deviation

Kruskal-Wallis Test was used to test whether there is any significant difference among three different groups of ethnicity namely Malay, Chinese and Indian in terms of awareness, issues and problems on the implementation of GST. At 0.05 significance levels, there is no statistically significant difference among the three groups of ethnicity with regards to awareness (χ^2=1.306, p=0.520) and issues and problems (χ^2=4.437, p=0.109) on the implementation of GST.

Correlation Analysis

Correlation analysis is used to measure strength association between two or more variables. Correlation results presented in Table 7 shows that business experience (BE) in dealing with the SST has no significant relationship with level of awareness (AW, r=.130) and perceived issues and problem (IP, r=.007) likely to be faced by business operators on the implementation of GST. Therefore, H_1 and H_2 are rejected.

Ethnicity of business operators (EBO) shows no significant association with the level of awareness (AW, r=.016) and perceived issues and problems (IP, r=.102) likely to be faced by business operators on the implementation of GST. Thus, H_3 and H_4 are rejected. This result contradicts with the findings of Jaidi et al. (2013) which revealed that taxpayer's awareness levels were associated with their ethnicity. Similarly, this result also contradicts with the findings of Yong and Hulst (2014) who indicated that cultural value of ethnic groups are associated with their tax compliance practices and perceptions.

At 0.01 significance levels, business size (BS) shows significant relationship (r=.224) with the level of awareness (AW) on the implementation of GST, while at 0.05 significance level, business size (BS) shows significant relationship (r=.184) with perceived issues and problems (IP) faced by business operators on the implementation of GST. The results indicate that business operators who have an annual turnover of more than RM500,000 are more aware of the GST mechanism. This result implies that business operators who are liable to be registered would take initiatives to acquire knowledge and identify issues related to GST. Thus, it can be concluded that H_5 and H_6 are accepted.

Table 7: Spearman Correlations

		BE	EBO	BS	BI	AW	IP
Business Experience	**BE**	1	-.071	.161*	.102	.130	.007
Ethnicity of Business Operators	**EBO**		1	-.073	.084	-.016	.102
Business Size	**BS**			1	.337**	.224**	.184*
Business Initiatives	**BI**				1	.283**	.323**
Awareness	**AW**					1	.346**
Issues & Problems	**IP**						1

* Correlation is significant at the 0.05 level (2-tailed)
** Correlation is significant at the 0.01 level (2-tailed)

At 0.01 level of significance, business initiatives (BI) shows significant relationship with awareness (AW, r=.283), issues and problems (IP, r=.323) towards the implementation of GST among business operators. This shows that initiatives taken by business operators to attend courses related to GST do have influence on their level of awareness of GST and on perceived issues and problems that may arise when GST is implemented. Therefore, H_7 and H_8 are accepted. This result is consistent with the study of Saira, Zariyawati and May (2010) that lack of knowledge and information on GST could lead to low level of awareness regarding the implementation of GST.

It is noted that at the significance level of 0.01, awareness (AW) of the implementation of GST and perceived issues and problems (IP) likely to be experienced are positively correlated (r=.346).

CONCLUSION, LIMITATION AND FUTURE RESEARCH

More than 70% of the respondents are aware of the GST mechanism, which implies that businesses operators do have adequate knowledge on GST in order to face the challenges arising from the implementation of GST. This finding is consistent with that of previous research by Palil, et. al. (2013) but contradicts the findings by Saira, Zariyawati and May (2010).

Most of the business operators agreed with the perceived issues and problems in keeping relevant business records, storage of document, hiring of new personnel and training for existing staff in managing GST matters would lead to higher compliance costs, which is consistent with the finding by Alebel (2011). The respondents also agreed that they would face problems in re-setting prices of goods and services, filing GST returns, remitting GST due, as well as in claiming input tax refunds.

Business size and business initiatives are found to have positive relationship with awareness of GST, and perceived issues and problems they are likely to face. This study found that business operators with annual gross turnover of more than RM500,000 are more likely to have higher level of awareness pertaining to GST and are more aware of the perceived issues and problems that they may encounter.

Contribution of Study

This study provided useful evidence pertaining to business operators' level of awareness on, and concerns regarding the perceived issues and problems associated with the implementation of GST. Besides, the findings would be useful inputs for the relevant tax authorities, policy makers and other stakeholders to formulate plans and strategies to address the issues and problems that are likely to be encountered by business operators and to enhance the level of compliance.

Limitation of Study

It is acknowledged that the findings of this study cannot be generalised to the whole population of Malaysia business operators as the samples are restricted to the registered members of the three Chambers of Commerce in Melaka. Besides, this study only examined the association of four factors (business experience, ethnicity of business operators, business size and business initiatives), level of awareness on GST, and the perceived issues and problems likely to be encountered by business operators.

Recommendation for Future Research

For future research, the target respondents could be extended to those in other States in Malaysia. A comparative study among business operators in different States would be useful to examine whether there is any significant difference between business operators in the different States. Future study could also include other possible factors that may be associated with the level of awareness as well as issues and problems likely to be encountered in relation to compliance with GST legislations.

REFERENCES:

Abdul-Jabbar, H. (1996). The implementation of scheduler monthly tax deduction: The case of small and medium industries in Kedah Darul Aman. *Master Thesis*, Universiti Utara Malaysia.

Abdullah, H., Idrus, A.A.M., & Mehat, N. (2013). Kefahaman dan kesediaan pengguna terhadap pelaksanaan Cukai Barang dan Perkhidmatan (GST): Satu tinjauan awal. *Prosiding Perkem*, *2*(8), 688–702.

Abdullah, K. (2012). Revenue mobilization measures and current tax issues. *IMF-Japan High Level Tax Conference for Asian and Pacific Countries*, 1-14.

Alebel, A.M.S. (2011). Goods and Services Tax: problems and effects of implementation. *Business E-Bulletin*, *1*(1), 55–64.

Basu, A., & Altinay, E. (2002). The interaction between culture and entrepreneurship in london's immigrant businesses. *International Small Business Journal, 20*(4), 371-393.

Breen, J., Bergin-Seers, S., & Sims, R. (2002). The impact of the introduction of the GST on small business in Australia. *Asian Review of Accounting*, *10*(1), 89–104.

Brown, K., & Marsden, S. (2002). The impact of the gst on the community services sector in Queensland. In *The Fifth International Conference on Tax Administration*, 1-15.

Casanegra de Jantscher, Milka. 1987. *Problems in administering a value-added tax in developing countries: an overview*. Development Research Department discussion paper no. DRD 246. Washington, DC: World Bank

Chan, T. (2009). Goods and Services Tax in Malaysia. *Accountants Today*, 33.

Choong, K.F., & Lai, M.L. (2006). Towards Goods and Services Tax in Malaysia: A Preliminary Study. *Global Business & Economics Anthology*, *1*, 75–86.

Cooper, D.R., & Schindler, P.S. (2014). *Business Research Methods* (12th Edition). Boston: Mc-Graw-Hill.

Eragbhe, E., & Modugu, K.P. (2014). Tax compliance costs of small and medium scale enterprises in Nigeria. *International Journal of Accounting and Taxation*, *2*(1), 63–87.

European Commission. (2007). Simplified tax compliance procedures for SMEs: Final Report of the Expert Group. *European Commission: Enterprise and Industry Directorate-General.*

Government of Malaysia. (2014). *Law of Malaysia, Goods and Services Tax Act 2014*. Percetakan Nasional Malaysia Berhad.

Government of Malaysia. (2014). *Federal Government Gazette, Goods and Services Tax Regulations 2014*. Attorney General Chamber.

Grinberg, I. (2006). Implementing a progressive consumption tax: Advantages of adopting the VAT Credit-Method System. *National Tax Journal*, *LIX*(4), 929–954.

Hair, J.F.J., Money, A.H., Samouel, P. & Page, M. (2007) *Research Methods for Business.* United Kingdom: John Wiley & Sons, Inc.

Hanefah, M., & Al-Mureshi, A.L. (1991). Accounting in small and medium enterprises. *Akauntan Nasional.*

HM Revenue & Customs (2010), Engagement with tax credits among Chinese and Indian people. *A study conducted for HM Revenue and Customs. Her Majesty's Research Report 117.*

Jaidi, J., Noordin, R., & Kassim, A.W.M. (2013). Individual taxpayers' perception towards self-assessment system: A case of Sabah. *Journal of the Asian Academy of Applied Business, 2*(1), 56-65.

Jenkins, G.P., & Khadka, R. (1998). Value Added Tax policy and implementation in Singapore. *International Bureau of Fiscal Documentation*, *9*(2), 35–47.

Lubua, E.W. (2014). Influencing tax compliance in SMEs through the use of ICTs. *International Journal of Learning, Teaching and Education Research*, *2*(1), 80-90.

Maseko, N. (2014). The impact of personal tax knowledge and compliance costs on tax compliance behaviour of SMEs in Zimbabwe. *Elite Research Journal of Accounting and Business Management*, *2*(3), 26-37.

Oladipupo, A.O., & Izedonmi, F.P. (2013). Public perceptions and attitude towards Value Added Tax (VAT) in Nigeria. *Scientific Research, 5*, 126-135.

Palil, M.R. (2010). Tax knowledge and tax compliance determinants in self assessment system in Malaysia. Degree of Doctor of Philosophy Thesis, University of Birmingham.

Palil, M.R., & Ibrahim, M.A. (2012). The impacts of Goods And Services Tax (GST) on middle income earners in Malaysia. *World Review of Business Research*, *1*(3), 192–206.

Palil, M.R., Ramli, R., Mustapha, A.F., Syuhada, N., & Hassan, A. (2013). Elements of compliance costs: Lesson from Malaysian companies towards Goods and Services Tax (GST). *Asian Social Science*, *9*(11), 135–147.

Palil, M.R., Zain, N.H.M., & Mamat, N.H. (2013). Consumer Readiness, Perceptions and Acceptance of Goods and Services Tax (GST): Middle and Lower Income Earners Perspective. *Proceeding-Kuala Lumpur International Business, Economics and Law Conference*, *1*, 353–360.

Pope, J. (2001). Estimating and alleviating the Goods and Services Tax compliance cost burden upon small business. *Revenue Law Journal*, *11*(1), 1–18.

Pope, J., & Abdul-Jabbar, H. (2008). Small and medium-sized enterprises and tax compliance burden in Malaysia: Issues and challenges for tax administration. *Small Enterprise Research, 16*(1), 1-14.

Rametse, N. (2010). An International perspective on small business implementation costs of a new tax and managerial benefits received. Transforming black townships into economic powerhouses. *Proceedings of the 2010 Soweto International Conference on Entrepreneurship & Development, South Africa, 27-28 January*, 1–24.

Saad, N. (2014). Tax knowledge, tax complexity and tax compliance:taxpayers' view. *Procedia - Social and Behavioural Sciences, 109*, 1069-1075.

Saira, K., Zariyawati, M.A., & May, L.Y. (2010). An exploratory study of Goods and Services Tax awareness in Malaysia, 265–276. Retrieved from repo.uum.edu.my/3178/1/S14.pdf

Samsuddin, A., Meor Ruslan, M.I., Abd Halim, A., Zahari, N.F., & Mohamad Fazi, N.F. (2014). Educators' awareness and acceptance towards Goods and Services Tax (GST) implementation in Malaysia: A study in Bandar Muadzam Shah, Pahang. *International Journal of Business, Economics and Law*, *4*(1), 1–10.

Sandford, C. (1995). *Tax compliance costs measurement and policy.* (p.379). Fiscal Publications, Bath.

Sekaran, U., & Bougie, R. (2009). *Research methods for business a skill building approach* (Fifth Edition). UK: John Wiley & Sons Ltd.

Slemrod, J., & Venkatesh, V. (2002). The income tax compliance cost of large and mid-size businesses: A Report to the IRS LMSB Division, Office of Tax Policy Research, University of Michigan Business School.

Tran-Nam, B. (2000). The Implementation Costs of the GST in Australia: Concepts, Preliminary Estimates and Implications. *Journal of Australian Taxation*, 331–343.

Weichenrieder, A.J. (2007). Survey on the taxation of small and medium-sized enterprise: Draft Report on Response to the Questionnaire. Retrieved from http://www.oecd.org/tax/tax-policy/39597756.pdf

Yew, T.E. (2009). When should malaysia implement GST? *New Straits Times*, p.8.

Yong. S. (2011). Tax compliance and small and medium enterprise operators: An intra-cultural study in New Zealand. Doctoral Thesis, Auckland University of Technology, New Zealand.

Yong, S., & Hulst, G. (2014). Multiculturalism and tax compliance of business entrepreneurs. In conference proceeding. *Accounting and Finance Association of Australia and New Zealand.* Auckland, New Zealand.

Zhou, L.Z., Tam, J., & Heng-Contaxis, J. (2013). The introduction of Goods and Services Tax in Malaysia: A policy analysis. *Centre for Public Paper Series Asian Strategy & Leadership Institute*, 1-29.

AN OVERVIEW OF GOVERNANCE AND ACCOUNTABILITY OF TAHFIZ INSTITUTIONS IN MALAYSIAN: RELIGIOUS COUNCILS PERSPECTIVE

Hamidah Bani*[1], Mohd Yassir Jaaffar[1], Maheran Katan[2], Abd Halim Mohd Noor[3]
Faculty of Accountancy, Universiti Teknologi MARA Melaka[1], Faculty of Business & Management, Universiti Teknologi MARA Melaka[2], Center for Islamic Philanthropy and Sicial Finance[3]

ABSTRACT

Tahfiz education is one of the popular streams of education in Malaysia and well accepted by Muslim community. Originally the education is being offered in "pondok", a traditional religious school. Developing through times, it is now also being offered in tahfiz centers as well as in modern tahfiz institutions. These entities specialize in producing tahfiz graduates who can memorize and recite the whole Al-Quran and more than 91% of them are privately owned. The applicability of good governance and accountability measures is indeed highly desirable in these Islamic entities as it deals with the responsibility in fulfilling sacred religious obligation. Though the number of tahfiz education providers (TEP) has grown up tremendously from only 58 in 1999 to 278 in 2011, literatures on governance and accountability of TEP are still lacking as the areas are rarely examined directly. The lacking may lead to unfavorable inferences that TEP are behaving in the manners contrary to the interest of the stakeholders. An ideally accountable governance of tahfiz centers will gain trust from parents and various stakeholders and will indicate the achievement in fulfilling one of the objective of Maqasid Shariah to preserve and promote the religious faith through preserving the memorization and knowledge of Al Quran (the primary source of Islamic Law) by the huffaz. The objective of this research is to investigate external governance mechanism and the accountability of Tahfiz centers as viewed by religious regulatory bodies. A qualitative research approach is undertaken by interviewing representatives from selected State Islamic Religious Councils, State Islamic Foundation and National Religious Departments in Malaysia. Using thematic analysis, findings of this study revealed that there are two-tier governance infrastructures of tahfiz institutions in Malaysia. However, the enforcement of the governance structure is still at infancy level. The accountability frameworks of TEP management divulged in four dimensions; ultimate upwards, upward, parallel and downwards in organizing, teaching and learning, and compliance activities and legal requirements. The accountability can be measured using five accountability mechanisms namely report and disclosure statements, performance assessment and evaluations, participation, self-regulation, and social audit.

Keywords: Tahfiz, Accountability, Governance Mechanism, Islamic

*Corresponding author's email: snorhafiza@melaka.uitm.edu.my

INTRODUCTION

An Overview of Governance and Accountability of Tahfiz Institutions in Malaysian: Religious Councils Perspective

Tahfiz schools have become avenues for Islamic education focusing on teaching students to memorize and recite the whole Al-Quran. The recorded number of centers offering the stream is mushrooming since the years 2000 with majority are privately owned centers (Bani, Katan, Mohd Noor & Abdul Fatah, 2014). The Tahfiz school stream is also supported by The Ministry of Education through its commitment in implementing the Ulul Albab Tahfiz Model (TMUA) pioneer project in national religious schools (SMKA) and fully residential schools (Ministry of Education, 2016). Apparently there are 577 tahfiz centers in Malaysia; twenty (20) are state government schools, ten (10) tahfiz colleges and 547 are privately owned (JAKIM, 2016). This data provides evidences on parents' inclination in sending their children to tahfiz schools operated by either Ministry of Education (MOE), states or private entities.

The privately owned centers may be set up by an individual, public or private entities. As such they have different governance practices according to the founders' norms. Besides differences in the organizational structure, the centers may also have different curriculum. In general, there are two distinct stream; solely tahfiz education of fully Quranic teaching, and an integration of Quranic teaching and academics. A privately owned center may adopt any curriculum or teaching methods (Hashim, 2015). Due to these differences, there is a dire need of an efficient and standardize corporate governance that would result in systematic education accountability system. The education accountability has to comply with regulations, adhere to professional norms and results driven (Anderson, 2005). With the aim of understanding their complex structure, this study provides an overview of the external governance mechanism and accountability of these tahfiz centers from the regulatory bodies' perspective. Data were collected through several interviews with representatives from selected State Islamic Religious Councils, State Islamic Foundation and National Religious Department in Malaysia.

The Development of Tahfiz Education in Malaysia

Tahfiz education were initially conducted in "pondok", a traditional religious school. Then in 1966, tahfiz was institutionalized by the government when Tunku Abdul Rahman Al-Haj Prince, first Prime Minister of Malaysia, announced the aspiration to set up an Institute for the Study of the Quran in Malaysia. Started on March 1, 1966, Ma'ahad Tahfiz Quran wal Qiraat was set up and tahfiz classes were carried out at National Mosque Lecture Hall with eight pioneering students selected from various states in Malaysia. Efforts to establish Ma'ahad Tahfiz were made to preserve and promote the religious faith among Muslims. Since then tahfiz centers are opened in every states of Malaysia. In addition, tahfiz centers are also set up by private entities. The main objective of establishing tahfiz center is to produce tahfiz graduate who are valuable human capital needed for the development of the nation (Azmil, Abd. Halim, & Misnan,2013).

In terms of the employability of tahfiz graduates on 16 April 1983, the certificate of eligibility produced by Darul Quran was recognized by the Public Service as equivalent to diploma from MARA Institute of Technology (now known as UiTM). This recognition enables Darul Quran graduates to be accepted as civil servants. Darul Quran graduates are also recognized by local and overseas universities where they can further their studies in any field of their choice. As such, tahfiz graduates graduates are equipped with both religious knowledge and professional expertise.

*Corresponding author's email: snorhafiza@melaka.uitm.edu.my

Corporate Governance

Corporate Governance (CG) is a process of managing, directing, controlling, regulating, be in charge, planning and conducting affairs of an institution. In Arabic term *Howkomat al Sharikat* (حوكمةالشركات) literally means the governance of companies (Adwam, World Bank Group 2013). In Al Quran it is termed as *Tadbir* (تَدْبِي) (Al Burey, 1985). It refers to rational act of examining the consequence (s) of an affair so that a creditable result may be worked out and thus obtained.

The CG of any Islamic institution needs to escalate the stakeholders' relations in the systems and processes. Managers are required to act in a responsible manner in all areas of their activities they are accounted for. Efficient governance will undertake responsible activities, policies or programs to discharge accountability to various stakeholders so as to maximize the institution performance and the delivery outcome. On top of that it is to ensure that the operations conform to requirement of the law, regulations, standards and community expectation on accountability, transparency, justice and Theoretically, this resembles the application of stakeholder theory (Iqbal and Mirakhor, 2004; Hasan, 2008; Hasan and Harahap, 2010; Bani, 2014). The theory specifies that decision makers utilize resources efficiently, to manage them in the best interest of their principal stakeholders and to maintain good relationship between different stakeholders (Grais and Pellegrini, 2006). Indirectly, CG also covers decision-making activities; who make the decision on CG, who will benefit from CG and what resources to use to materialize CG (Lewis, 2005).

Philosophically, efficient governance primarily needs to ensure efficient relationship with the ultimate stakeholder that is the resource owner (Allah) through compliance of *Shari'ah*. The *Tawhidic* paradigm (Unity of God) conceptually provides illustrations of the link between the spirit of *Tawhid* with all management functions (planning, organization, leading, and controlling) and ensures harmony between (*Al-Akhirah* - the Hereafter) and the worldly (*Al Duniya*). The mission and objectives will be achieved through continuous enforcement in promoting goodness (*amr bil ma"aruf*) and prevention of evil (*wannaha anil munkar*). Based on this paradigm, harmonized management approach, in accordance to the requirements of the religion (*ad-din*), is an important ingredient to achieve the objective of *Shari'ah* (*Maqasid Shari'ah)* of any Islamic institutions successfully. Commercially, an effective governance of tahfiz centres must be able to achieve its economic objective financial stability that will protect the economic interest of the shareholders and in return increase the value of the centre. Socially tahfiz centres must be efficient in protecting the interest of affected parties or entities (stakeholders) such as employees, the general community and environment surrounding the tahfiz center.

Governance Framework of Tahfiz Schools in Malaysia

Malaysia implements Public Sector Governance of school system since independence. The landscape of education system changed as a result of smart partnership which allows stakeholder active participation in the school system as a result of new initiatives introduced in 2006. The Education Development Master Plan – EDMP 2006-2010 (Ministry of Education, 2006) and National Key Result Areas (NKRA-one of the national development plan) implementation allows the existence of various types of schools, subjects and specializations. More private schools enter into the education sector, more private religious schools seek registration and vocational subjects in schools are given more emphasis. As a result this initiative provides more educational choices to parents and students. In line with this, statistics shows an increase in number of students attending religious schools offering tahfiz education (JAKIM, 2016)

*Corresponding author's email: snorhafiza@melaka.uitm.edu.my

Tahfiz schools dealing with religious education activities falls under the category of not for profit sector. The framework of governance for Tahfiz schools is based on public sector governance best practices. Tahfiz schools operating under the supervision of MOE and state government have less autonomy and generally react to government initiatives enforced. Privately owned tahfiz schools have high autonomy as they lead the centers on their own and update the government on their policy and may receive many state initiatives. Effective governance of tahfiz schools is very important as effective and accountable governance will gain trust from parents and various stakeholders and achieve of the objective of Maqasid Shariah to preserve and promote the religious faith through preserving the memorization and knowledge of Al Quran (the primary source of Islamic Law) by the huffaz (Bani, Katan, Mohd Noor and Jaafar, 2016).

In Malaysia, tahfiz education is being offered in schools under MOE, schools under state supervision and private schools. Tahfiz schools may be named differently by the operators such as Mahaad, Tahfiz Center, Ulul Albab School and others. Tahfiz schools under MOE in general are under of MOE's governing authority. Private religious schools or centers offering the teaching of Al Quran together with national academic curriculum are also subject to MOE and religious council governing authority. Whilst, private tahfiz institutions or centers focusing on the teaching of Al Quran only and not registered with MOE are under the governance of the State Religious authority. As such there are two external governance mechanisms for tahfiz schools; MOE and State Religious Council (refer to Figure 1)

The external control mechanisms generally monitor the legal compliance and management adequacy of tahfiz schools or centers. The authority issues regulatory guidelines and suggest guidelines for best practices which work as boundaries to the operations. An efficient and good governance of tahfiz centers are expected to report their compliance status to external stakeholders. Though there are two external governance mechanisms directly regulating and controlling tahfiz schools or centers, this study limits the discussion on the mechanism from the state religious council perspectives.

*Corresponding author's email: snorhafiza@melaka.uitm.edu.my

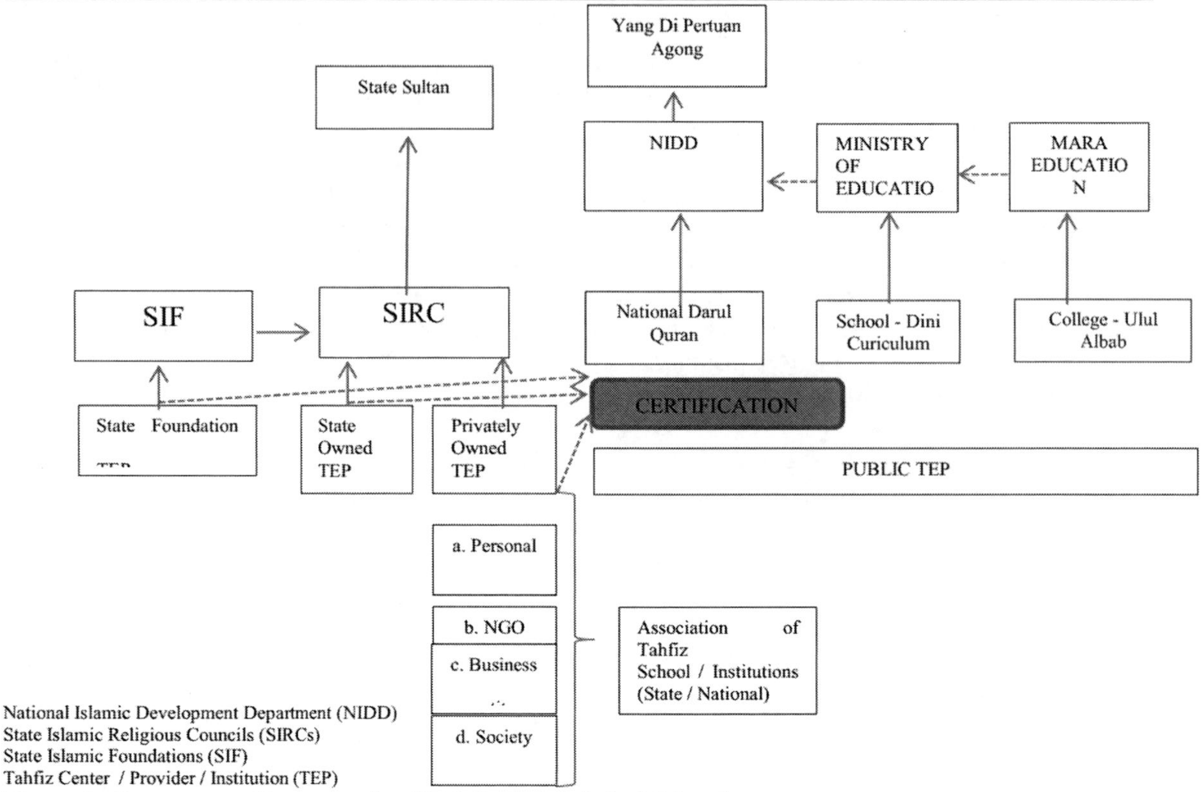

Figure 1: Governance Framework of Tahfiz Schools in Malaysia

Regulatory Framework Of Tahfiz School Religious Council Perspective

Private Tahfiz schools are categorized as Islamic religious school in Malaysia. As such these schools are subject to state regulatory enactment of Islamic religion according to the state where they are registered. The state is responsible in creating a regulatory framework as a mechanism to ensure that the objective of tahfiz schools is aligned with the goal of the sector and the regulation set able to influence the behavior of the players in the sector. Besides state enactment of legislation for the administration of Islamic religion, tahfiz schools are also enacted to State Enactment Controlling Islamic Religious Schools. These enactment acts as external governance mechanism in regulating tahfiz institution. Table 1.0 highlights regulatory framework of Tahfiz Schools in three different states. Relevant sections in the State enactments are being referred for discussion.

*Corresponding author's email: snorhafiza@melaka.uitm.edu.my

Table 1: State Enactment applicable for Tahfiz institutions

State No	State Enactment and Control	Content of Enactment
State 1	State Enactment Enactment No 7 2002 Section 108	Seksyen 108. Religious school. Subject to the Control of Islamic Religious Schools Enactment 2002 [Enactment No. 3 of 2002] the Majlis shall have power to register Islamic Religious schools in the State
	Enactment Controlling Islamic Religious Schools No 3 of 2002 and Registration rules 2003 (Part i – iiv)	i. Introduction ii. Registrar/ Deputy Registrar iii. Registration of religious schools iv. Board of Management v. Inspectors of Islamic Religious Schools vi. Inspection of Religious Schools by Registrar vii. Miscellaneous
State 2	State Enactment Section 120	Subject to the control of Islamic Religious School Enactment 1988 Enactment No 2 the Majlis shall have power to register Islamic Religious School in the State
	Enactment Controlling Islamic Religious Schools Enactment 2 1988 (Part i– vi)	i. Introduction ii. Registrar/ Deputy Registrar iii. Registration of religious schools iv. Board of Management v. Inspectors of Islamic Religious Schools vi. Miscellaneous
State 3	State enactment Section 96(1) Section 96(2)	Section 96(1) The Majlis shall have power to registerand control Islamic Religious School and Islamic education institution in the State and any related matters to it. Section 96(2) The Majlis can exempt any Islamic religious school or institutions from registration and monitoring
	No specific enactment on controlling the Islamic religious school or institutions	

Table 2: The content of States' enactment controlling state religious schools in State 1 and State 2 .

Part i.	As Preliminary part this section provides interpretations of term and terminology used in the enactment.
Part ii.	Explains procedures in appointing the registrars etc. and the power and duty of registrar
Part iii.	Explain procedures in registering the religious school, application to establish religious school, registration and revocation of registration, procedure in appealing for registration, prohibition to advertise religious school and prohibition in change of address of premise of religious school.
Part iv.	Explaining the instruments in establishing board management and its membership, qualification of the members of the board, board duties and responsibilities, and power of board
Part v.	Explain the appointment and right of the inspector to inspect school and the right of Majlis to authorize any person to inspect school
Part vi.	(Only in State 1) Explain the provision which allow registrar to inspect school and the power of the registrar on inspection, procedure in inspection, power to close unregistered religious school, power to investigate and power to enter school without warrant.

Part iiv (State 1) and Part vi (State 2)

The last part in the enactment on miscellaneous items provide the power of Advisory Board to give assistance to Majlis, the right of Majlis to give exemption for all provisions of the enactments except for registration, conviction for obstruction to registrar etc in carrying out their duties and responsibilities, the liability of the director, manager, secretary or officers for offence made by the corporate body, general penalty of an offence, jurisdiction to try offences, Director's written consent in the institution of prosecution, protection against suits and legal proceedings, the position of officers as public servant, documentation and regulation for the Islamic school operations.

*Corresponding author's email: snorhafiza@melaka.uitm.edu.my

In different to State 1 and State 2, State 3 did not have specific enactment in controlling Islamic religious schools. As such, in the State 3 Tahfiz centers administration uses general law as stated in section 96 (1) and 96(2). In that, the state enactment may exempt the Islamic religious school from registration and monitored by the Majlis. State 1 and State 2 have similarities in their Enactment Controlling Islamic Religious Schools except for an additional section elaborating the inspection of Religious Schools by Registrar in State 1. Finding from these comparisons may deduce that the legislations and regulations for tahfiz institution in Malaysia are available in the state enactment but they have yet to be standardize and enforced nationally. Standardized regulations is needed so that evaluation on the compliance to legislations, rules and regulations and assessment of the sector goals achievement is align to religious and society expectations.

Governance Players in Tahfiz School and Their Accountability

Several governance players in Islamic school are highlighted in the enactment. Each player has their own accountability in managing, controlling and monitoring Islamic school operation. Applying the general principles in the enactment for Religious schools the key players in tahfiz schools are Yang Dipertuan Agong, Directors and officers in National Religious Department, Director and officers in Islamic studies at Ministry of Education (MOE), Registrar comprised of Directors and officers appointed at State religious department, board of management also known as board of trustee, school inspectors and teachers. These players have various accountabilities (acknowledgement of responsibilities) for policies, decisions and actions within the ambit of their role in the tahfiz schools. Based on the findings from interviews and review of enactments controlling religious schools in selected states, this study suggests the uniform key players in tahfiz school governance and their accountability as in Table 2.0.

Table 3: Key Players and Governance Accountability of Tahfiz Schools

	Players	Structure	Accountability
1	Yang Di Pertuan Agong	Legislative power for authorization and delegation	a) Notify in the gazette the effective date of the enactment b) Approved the appointment of registrar and deputy registrars c) Make regulation (with the advice of Majlis) to ensure the provision of the enactment are carried out into effect
2	National Religious Department	Director, Assistant Director and officers at National and State department	a) Undertake national or state level monitoring activities b) Distributes the allocated fund - Zakat - Ministry of Education (Private Education sector funds)
3	Ministry of Education	National Education Department – Islamic Studies	Undertake national level monitoring for Tahfiz schools under MOE
4	Registrar at State religious department (Majlis)	Registrar Deputy Registrar Officer	a) ensure tahfiz schools are properly administered b) Ensure that the state of affairs of Tahfiz schools are informed to Majlis c) Keep register of all registered tahfiz school (name, address, members of the board, school curriculum and information deemed necessary by registrar d) Direct , control make delegation to assistant registrar, revoke delegation e) Inspect tahfiz schools f) Inspect unregistered premises g) Seize any materials to furnish

*Corresponding author's email: snorhafiza@melaka.uitm.edu.my

			evidence of offence against enactment h) Close unregistered Tahfiz Schools i) Investigate into commission of any offence under enactment j) To enter premises without warrant on reasonable ground
5	Board of Management	(Excluding state government religious school) Appointed by each private tahfiz school At least 7 persons School Principal as secretary	Trustee for the school a) Acquire / purchase assets b) To borrow from time to time c) To make arrangement with the state or federal government Responsibility a) Maintain the school b) Comply with the direction of registrar c) Maintain book of record - Teacher and staff - Students, age and sex - Financial sources of the school - Curriculum of the school - Other information required by registrar d) Administer school fund e) Submit report to registrar f) Comply curriculum made by Majlis
6	School Inspector	Inspector appointed by Majlis	a) make school inspection - Teaching the permitted curriculum - Textbook used - Teachers employed - Compliance on enactment b) conduct investigation on any matters as may be authorized by the Registrar c) make inspection of school account d) report irregular matters to registrar e) perform other duties as directed by Registrar
7	Teacher	Employed by MOE or tahfiz school	a) Qualified and trained teacher b) Jointly liable on the offence of the Tahfiz school c) Teaching and learning of Tahfiz school students d) Adhere to teaching curriculum set by SIRC
8	Association of Private Tahfiz Schools / Institutions	State Association and National Association	To protect the interest of Private Tahfiz schools in Malaysia (especially the registered member)

In summary, governance of Tahfiz institution has two tier governance infrastructure; state and tahfiz school infrastructure. At state level, the state management will issue rules and regulations binding to all Tahfiz schools, advice the board management and monitor the compliance. Tahfiz governance players who are involved in these activities are Yang Di Pertuan Agong, Directors, Deputy Directors and officers at National Islamic Religious Department; and Registrar, Deputy Registrar and officers at State National Islamic Religious Department ; who

*Corresponding author's email: snorhafiza@melaka.uitm.edu.my

are accountable to undertake various responsibility as specified in Table 2.0 above. At Tahfiz School level, board management and teachers are trusted to have standard procedures in ensuring the compliance to the state rules and regulations. In attesting that the enactment, rules and regulation issued by the state authority are complied, an audit work will be undertaken by the school inspector. As such the inspector represents the third governance mechanism i.e the independent auditors to tahfiz schools. The detail accountabilities of this player are specified in the above Table 3.0.

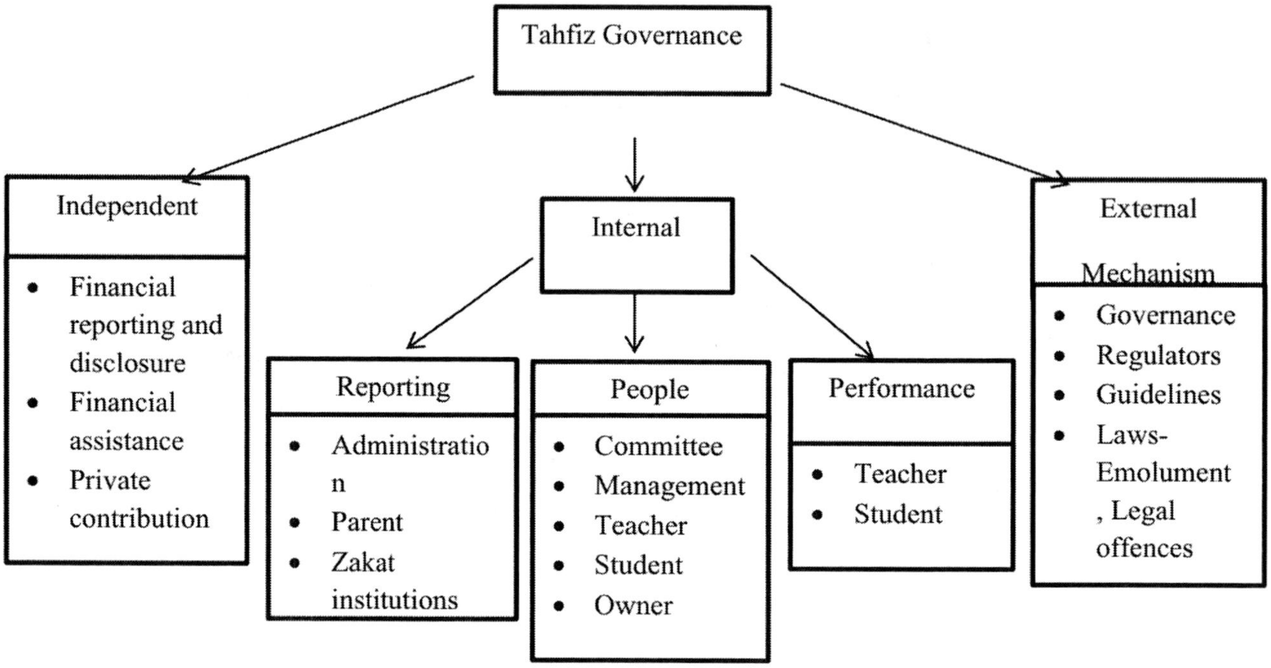

Tahfiz Governance Mechanism

As a summary to accountability issue in Tahfiz education, key Tahfiz schools governance players are accountable to the achievement of Maqasid Shariah in the aim to protect Ad-din particularly through producing tahfiz graduates with the capability of preserving the memory on the main source of Shariah (Al Quran) and the teaching. Based on this objective, accountability mechanism by Ebrahim (2010) is broadening by the inclusion of ultimate accountability. Ultimate accountability is the accountability of an individual man to perfect himself / herself to Allah derived from the concept of khalifah (vicegerence) in which man become a trustee of Allah resources.

The management of TEP is accountable to other stakeholders such as nature and Universe, donors, regulators, human resources, students, parents and society. According to Llyod (2005) these can be further categorized as upwards accountability, downwards accountability and parallel accountability. The upward accountability refers to accountability / responsibilities to bodies that give them financial aid and legal base such as donors, foundations and governments. In the context of Tahfiz schools, upward accountability is the accountability to the aid giver such as Ministry of Education, Zakat institutions, and donors who provide assistance to the TEP's operations both in financial and non-financial form as well as State and National Religious regulators bodies who perform oversight function. The horizontal accountability refers to the accountability between Tahfiz schools and the employees, peers or the management of TEP who directly involved in the Tahfiz schools day to day operations. The

*Corresponding author's email: snorhafiza@melaka.uitm.edu.my

downwards accountability referred to collective accountability of Tahfiz schools to their beneficiaries whom they provide their services in perfecting the quality of tahfiz graduates, such as fulfilling the parents aspirations and serving the surrounding communities.

The summary of the proposed graphical accountability line suitable for tahfiz schools dealing with ultimate upwards, upward, parallel and downwards is captured in figure 3 below Bani, Katan,Jaafar, Mohd Noor,(2015) . Kementerian Pengajian Tinggi

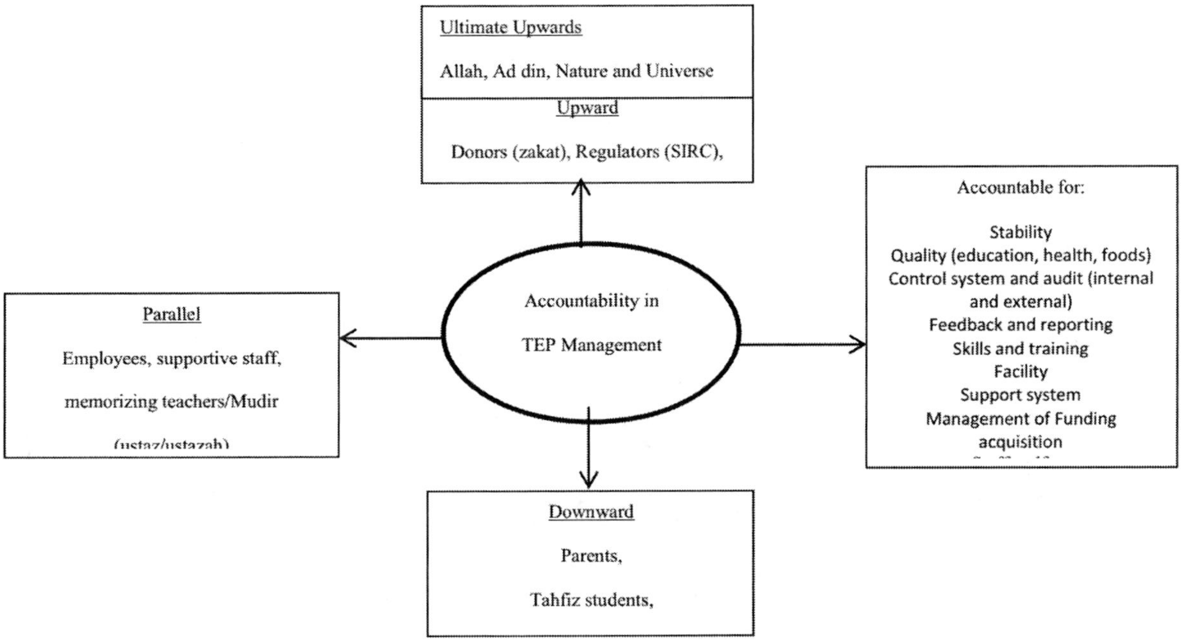

Figure 3: Accountability in Tahfiz Education Providers (TEP) Management

DIRECTION OF FUTURE RESEARCH

Tahfiz school governance mechanism research is still at infancy level. While main stream research on tahfiz education revolves around teaching and learning method (Ashraf, Ab Halim Tamuri, Kamarulzaman & Nurul Hudaa,2013; Azmil, Ab Tamuri (2012), memorization technique (Abdul Hafiz & Hasimah, 2004;) and rules and students (Norlizah, Fathiyah, Ahmad Fauzi, Lukman & Wan Marzuki , 2015) high expectation has been placed by the Muslim community on Tahfiz performance. Hence, future research may be directed to evaluate the performance of Tahfiz Schools, the compliance to enactment, ethical principles of Private Tahfiz school management and accountability of Private Tahfiz Education Providers as well as fund management. It is expected that the best practice of managing private tahfizs can be unveiled and knowledge will help other private tahfiz schools calrify their role systematically to uplift their performance.

*Corresponding author's email: snorhafiza@melaka.uitm.edu.my

DOI: 10.1051/shsconf/20173600028

REFERENCES

Abdullah, A. H., Salamon, H., Suratman, A. S., Noor, S. S. M., Jasmi, K. A., & Samat, A. B. (2005). Sistem Pembelajaran Dan Kaedah Hafazan Al-Qur ' Yang Efektif : Satu Kajian Ustaz Abdul Hafiz Bin Haji Abdullah Ustaz Kamarul Azmi Bin Jasmi.

Azmil, H., Abd. Halim, T., & Misnan, J. (2013). Latar Belakang Guru Tahfiz Dan Amalan Kaedah. *Journal Of Islamic Education*, *1*(1), 28–39.

Administration of the religion of Islam State of ……2003, Percetakan Nasional Malaysia Berhad, Kuala Lumpur

Bani, H., Katan, M., Noor, A. H. M., & Fatah, M. M. A. (2014). Applying Stakeholder Approach In Developing Accountability Indicators, (September) iCARE Proceeding

Bani, Katan,Jaafar, Mohd Noor,(2015),Tahfiz Education Provider Accountability Model . (FRGS/1/2015/SS01/UITM/02/9), Kementerian Pengajian Tinggi Malaysia

Ebrahim, A. (2003). Accountability In Practice : Mechanisms for NGOs, *31*(5), 813–829.

Enactment Controlling Islamic Religious Schools No 3 of 2002 and Registration rules 2003 Control of Islamic Religious Schools , Percetakan Nasional Malaysia Berhad, Kuala Lumpur

Guskey, T. R. (2007). Leadership in the age of accountability. *Educational Horizons*, 29–34.

Henard, F & Mitterle, A. (2012), Governance and quality guidelines in Higher Education. A Review of Governance Arrangements and Quality Assurance Guidelines, Organization of for Economic Corporation and Development (OECD)

Hashim, A. (2015). Correlation between Strategy of Tahfiz Learning Styles and Students Performance in Al-Qur'an Memorization (Hifz). *Mediterranean Journal of Social Sciences*, *6*(2), 85–92.

Hassan, N. C., Fakhruddin, F. M., Fauzi, A., Ayub, M., & Mutalib, L. A. (2015). Tahfiz Schools Entry Requirement and Characteristics of Tahfiz Students, *1*(3), 234–241.

Hanushek, E. A., & Raymond, M. E. (2004). The Effect Of School Accountability Systems On The Level And Distribution Of Student Achievement, (May 2004), 406–415.

Jabatan Kemajuan Islam Malaysia (2016), Data Masjid, Surau, Kafa dan Sekolah Agama Tahun 2016 , Bahagian Kemajuan Islam

Kementerian Pengajian Tinggi, (2015) Tahfiz Education Provider Accountability Model. KPT, FRGS/1/2015/SS01/UITM/02/9)

Jensen, C., & Meckling, H. (1976). Theory Of The Firm : Managerial Behavior , Agency Costs And Ownership Structure I *3*, 305–360.

Kai, J. (2009). A Critical Analysis of Accountability in Higher Education. *Chinese Education*

*Corresponding author's email: snorhafiza@melaka.uitm.edu.my

& Society, *42*(2), 39–51.

Lloyd, R. (2005). The Role of NGO Self-Regulation in Increasing Stakeholder Accountability, (210180), 1–15.

Lessing, J.,Morrison, D.& Nicolae (2012)"Educational institutions, corporate governance and not for profit" Corporate Governance e-Journal, 1-22,ISSN 1836 -1110

Ministry of Education (2016). Surat Pekeliling Iktisas KPM Bil 1 2016 Perlaksanaan Tahfiz Model Ulul Albab di Sekolah Menengah Kementerian Pendidikan Malaysia. http://www.moe.gov.my/v/pekeliling-ikhtisas-view?id=1094

Nakpodia, E. D., & Okiemute, A. R. (2011). Teacher's accountability in Nigerian education system: Perceptions of teachers and administrators in Delta State. *International NGO Journal, 6(7)*, 152-158. doi: 10.5897/ngoj11.009

Norlizah , Fathiyah , Lukman & Wan Marzuki (2015), Tahfiz Schools Entry Requirement and Characteristics of Tahfiz Students, International E-Journal of Advances in Education, Vol.1, Issue 3, December

Samah, R. A. (2016, January). More Maahad Tahfiz schools should be built around the country:Zahid. *New Strait Times*. Retrieved from http://www.nst.com.my/news/2016

Shuaib, F. S. (2008). Strengthening Administrative Institutions Of Islamic Law In Malaysia :, *16*, 443–464.

Siraj, S. A. (2012). An Empirical Investigation Into The Accounting , Accountability And Effectiveness Of Waqf Management In The State Islamic Religious Councils (Sircs).

Urbanski, A. (1998). Teacher Professionalism and Teacher Accountability: Toward a more Genuine Teaching Profession. *Educational Policy, 12*(4), 449-457. doi: 10.1177/0895904898012004006

*Corresponding author's email: snorhafiza@melaka.uitm.edu.my

THE DETERMINANTS OF CORPORATE SOCIAL RESPONSIBILITY DISCLOSURE: THE CASE OF MALAYSIAN GOVERNMENT-LINKED COMPANIES IN MALAYSIA

Assoc. Prof. Dr Hanim Norza Baba
Faculty of Accountancy, Universiti Teknologi MARA Melaka

ABSTRACT

The implementation of Government-Linked Company (GLC) Transformation Program 2005/06 by government is one effort to promote Corporate Social Responsibility (CSR) disclosures among its government-linked companies (GLCs). CSR issues are being stressed in the Silver Book included in the GLC Transformation Manual under the GLC Transformation Program 2005/06. It is questionable as to whether the introduction of the Silver Book really reflects goods prospects for government-linked companies to disclose their CSR, and whether there are any other factors that will influence the GLCs in Malaysia to disclose their CSR. Thus, the objective of this study is to examine whether the introduction of the Silver Book affect the CSR disclosure among Malaysian GLCs as well as to examine the determinants of CSR disclosure, focusing on the profitability, board size and board independence. Multiple linear regression analysis is being used to examine the relationship between all the independent variables and dependent variable. Findings show that there is an increasing trend in CSR disclosure among Malaysian GLCs from year 2011 until 2015. Two variables i.e. board size and board independence has been found to have a significant positive relationship with the CSR disclosure. This study gives implications to various parties such as Malaysian Government, Bursa Malaysia, Security Commission and other relevant parties in to improve CSR awareness, practices, disclosures and quality in GLCs.

Keywords: GLC, CSR disclosure, board size, board independence

INTRODUCTION

Corporate social responsibility (CSR) disclosure is not new in Malaysia. The degree of awareness and engagement in CSR is not at an ignorance stage even though we are not a member of Organization for Economic Co-operation and Development (OECD) and only recognized as a developing country (Lo & Yap, 2011). Research by Mohd Ghazali (2007) shows that there is an increase in CSR disclosures in Malaysian companies' annual reports as compared to the previous study done by Andrew, Gul, Guthrie, and Teoh (1989) and Baba & Mohd Hanefah (2001). Change in public concern and awareness on environmental issues make companies in Malaysia provide environmental disclosures within the annual reports (Nik Ahmad, & Sulaiman, 2004; Baba, 2004). According to Abd Rahman, Mohamed Zain and Yaakop Yahaya Al-Haj (2011), the traditional role of company's corporate reporting currently is being extended to social and environmental information, not only to financial

*Corresponding author's email: hanorza@melaka.uitm.edu.my

information. Businesses' roles today also include an element of CSR and accountability and not only restricted to profit making purposes (Mohd Ghazali, 2007).

In Malaysia, Bursa Malaysia has made it mandatory for all listed companies to disclose CSR in their annual reports. Disclosing the social and environmental accounting practices would give better image and reputation to the companies as well as to increase the chances to secure contracts from the the government as it is seen as being socially responsible. CSR helps improve financial performance, enhance brand image and increase the ability to attract and retain the best workplace, contributing to the market value of the company (Said, Hj Zainuddin, & Haron, 2009).

A number of CSR disclosure researches in Malaysia concerned with the determinants, level, theme and the form of disclosures (Baba & Mohd Hanefah, 2001; Mohd Ghazali, 2007; Saleh, 2009; Abd Rahman et al., 2011; Lassaad & Khamoussi, 2012); the current state, nature, quality and quantity of disclosures (Jaffar, Jamaludin, & Che Abdul Raman, 2007; Ribeiro & Aibar-Guzman 2010; Sibao & Huaer, 2009; Lo & Yap, 2011); to examine the determinants of CSR disclosures among public listed companies (PLCs) in Malaysia (Baba & Mohd Hanefah, 2001; Said et al., 2009; Al Arussi, Selamat, & Hanefah, 2009).

Issues concerning social and environmental matters have become more important in Malaysia nowadays. Companies should be aware of this responsibility and take action to ensure that they fulfill their obligations, role and status to the public (Baba, 2004). Companies will seek legitimacy if they are aware of the changes in public perception and awareness of the business to be socially responsible (Haider, 2010). In order to protect the companies self-interest, the disclosures would show their positive contribution thereby improving the companies' image and reputation as well as legitimize their activities and managing the society.

The government also emphasizes that its controlled companies, GLCs, to be actively participate in the social and environmental responsibility, despite the growing pressures on the PLCs to be socially and environmentally responsible (Abd Rahman et al., 2011). The implementation of GLC Transformation Program 2005/06 by government is one effort to promote the CSR disclosures among GLCs. CSR issues are being stressed in the Silver Book, one of the initiatives included in the GLC Transformation Manual under the GLC Transformation Program 2005/06 (Esa & Mohd Ghazali, 2012), and Khazanah Nasional Berhad, a key Government-Linked Investment Companies (GLIC), was tasked with the responsibility of monitoring Malaysian GLCs on the implementation of the CSR (Ching & Wad, 2009). It is questionable, however, as to whether the introduction of the Silver Book really reflects goods prospects for GLCs to disclose their CSR, and whether there are any other factors that will influence the GLCs in Malaysia to disclose their CSR.

LITERATURE REVIEW

According to Siregar and Bachtiar (2010), board size has a positive and non-linear (quadratic and concave) relationship with CSR. This indicated larger board will be able to exercise better monitoring, but too large board will make the monitoring process ineffective. Firm size was found to have a positive effect on CSR, indicated that larger firms have more resources to devote to social activities and a larger asset base over which to spread the costs

[Type here]

of social responsibility. In addition, larger firms also face more pressure to disclose their social activities for various groups in society.

A study by Mulyadi and Anwar (2012) that investigated the relationship between corporate governance and CSR disclosure of listed corporation in Indonesia Stock Exchange, focused on institutional ownership and independence of board toward the CSR disclosure. They developed CSR disclosure index based on Global Reporting Initiatives (GRI) indicators as a measurement of CSR disclosure. The study found that board independence and institutional ownership did not have influence to corporate CSR disclosure, while profitability was influential to corporate CSR disclosure at 1% significant.

A study by Esa and Mohd Ghazali (2012) in Malaysia investigated the change in the level of CSR disclosure among Malaysian GLCs for the years 2005 and 2011. The study also investigated whether corporate governance attributes influence CSR disclosure in corporate annual reports of Malaysian GLCs. Results from paired-sample t-tests showed that there was an increase in the extent of CSR disclosure at the 1 percent level, and multiple regression analysis revealed that board size was positively associated and statistically significant with the extent of CSR disclosure at the 1 percent level. The results implied that level of CSR disclosures increase after the introduction of the Silver Book in 2006, and GLCs with a larger board size disclosed significantly more CSR information than others in the sense that boards with more members with diverse experience and backgrounds exposed more healthier and livelier discussion on corporate social activities and hence improved investment in those activities.

Profitability

Profitability is the ability of company to earn profit or gain that would sustain the company's long-term and short-term growth (Lucyanda & Siagian, 2012). One of the factors enabling management to disclose CSR activity in high level of flexibility is profitability (Mulyadi & Anwar, 2012). According to Abd Rahman et al. (2011), high gap between the minimum and maximum values on profitability ratio indicating that certain companies are enjoying a good financial performance while some are not and their study found that profitability do not have any relationship with the level of CSR disclosure among Malaysian GLCs. Esa and Mohd Ghazali (2012) also found that there is no relationship between profitability and CSR disclosure. Pressure to disclose CSR is more influenced by ''public'' rather than ''economic'' pressure (Esa & Mohd Ghazali, 2012). Said et al. (2009) on the other hand found that there is a significant positive relationship between profitability and the extent of CSR disclosure in Malaysian non-financial companies. This may due to the reason that profitable firms may have larger profit base to reach the costs of social responsibility and have more resources to allocate to social activities. More profitable firms would support more extensive social responsibility programs than less profitable firms because they have supplementary funds to do so (Luethge & Han, 2012). Furthermore, management of profitable firms has more flexibility and there would be lack of restrictions to engage in and to disclose the social activities (Siregar & Bachtiar, 2010).

Board Size

According to MCCG (revised 2011), every board should examine its size, with a view to determine the impact of the number upon its effectiveness and the board should include a

[Type here]

balance of executive directors and non-executive directors (including independent non-executives) such that no individual or small group of individuals can dominate the board's decision making. As stated in the Green Book of the GLC Transformation Manual, board structure and board composition is the foundation of board effectiveness. To ensure that the board can effectively discharge its roles and responsibilities, there must be a sufficient number of directors held on the board. The size also should not become too large and unwieldy, which could then compromise board dynamics and the accountability of individual directors.

Vafeas (2000) proposes that a smaller board size can enhance the quality of financial reporting and, hence give better quality of information, while a larger board is expected to be less effective, as the monitoring responsibility will be dispersed among many directors. This may be because of the possibility that small board has better discussion of financial reporting numbers among their members as compared to a large board (Mohamad, Abdul Rashid, & Shawtari, 2012). According to Siregar and Bachtiar (2010), larger board will be able to exercise better monitoring, but too large board will make the monitoring process ineffective. Esa and Mohd Ghazali (2012) found that board size positively associated and statistically significant with the extent of CSR disclosure among Malaysian GLCs. It shows that GLCs with a larger board size disclosed significantly more CSR information than others. They suggests that boards with more members have diverse experience and backgrounds so that they are more exposed to healthier and livelier discussion on corporate social activities and hence make more investment in those activities. On the contrary, Said et al. (2009) found that there is no relationship between board size and CSR disclosure.

Board Independence

MCCG (2012) strengthening the role of board by reinforcing the board independence. According to the MCCG (2012), the board should undertake an assessment of its independent directors annually and there must be majority of independence directors sit on the board. The Green Book in GLC Transformation Manual further provides that there should be no more than two Executive Directors and at least one-third of board is independence because a board with a balanced composition will ensure that no individual or small group of individuals will dominate decision-making.

Independent directors contribute to a good reporting system and when the investors observe that a firm has independent directors they believe that there is a monitoring function that is independent from the management to ensure such good reporting system (Siagian & Tresnaningsih, 2011). Conversely, Esa and Mohd Ghazali (2012) argue that independent directors are a cost-efficient substitute for information disclosure. They found that larger proportion of independence directors disclose less CSR. Eng and Mak (2003) also found the same result where the higher the proportion of independence directors, the lower the information disclosed. Said et al. (2009) in their study does not found any relationship between board independence and the disclosure of CSR.

Firm Size

Firm size is considered as a control variable because it has been found to influence the CSR disclosures. According to Aldama, Amar and Trostianki (2009), size of the company is strongly associated with the CSR structuring and CSR strategies in terms of its employees

[Type here]

and revenues. Larger company has bigger chance to find a structured area and tend to declare more sustainable development-related policies. Since large companies have the more resources to collect, analyze and report data, they tend to disclose more information than small size companies (Al-Shubiri et al., 2012). Larger companies are supposed to disclose more CSR information in order to expose or portray their corporate citizenship, in so doing legitimizing their existence (Mohd Ghazali, 2011). The firm size for Malaysian GLCs is a significant item to measure the disclosure of CSR information (Abd Rahman et al., 2011). Nevertheless, study by Bayoud, Kavanagh, & Slaughter (2012) found that levels of CSR disclosure do not seem to be affected by company size in Libyan companies.

Legitimacy Theory

Legitimacy is a generalized perception or assumption that the actions of an entity are desirable, proper, or appropriate within some socially constructed system of norms, values, beliefs, and definitions (Suchman, 1995, p. 574). Legitimacy theory is used as a communication mechanism to inform or manipulate the perceptions of the firm's actions. Islam and Deegan (2008) state that legitimacy theory asserts that organisations attempt to ensure that their activities are perceived by outside parties as being legitimate in order to ensure that they are perceived as operating within the bounds and norms of their respective societies. Besides that, legitimacy theory posits that businesses are bound by the social contract in which the firms agree to perform various socially desired actions in return for approval of its objectives and other rewards, and this ultimately guarantees its continued existence. Thus, firms use communication or accounting to defend or maintain their legitimacy in the eyes of the society.

According to Islam and Deegan (2010), there is a 'social contract' between an organization and the society (societies) in which it operates, and there will be some negative implications for the ongoing survival of the organization if there is any breach of the social contract. In a sense management seeks congruence between the outside perceptions of its own organization's social values and what is deemed by society to be appropriate social conduct, management seeks to maintain or establish its legitimacy (Deegan & Gordon, 1996). Observer groups such as audiences or relevant publics evaluate the organization based on their perceptions and assumptions regarding congruence between their values and organizational values and if the relevant audiences are not aware of the divergence, that organization may materially diverge from expectations and still maintain its legitimacy (Mobus, 2005). In addition, O'Sullivan and O'Dwyer (2009) emphasize that accountability and moral legitimacy pay attention to the societal values and expectations from an organizational perspective. In a nutshell, company should provide CSR information by reporting the CSR activities to the society and behave according to the legitimacy theory in order to remain competitive and survive in the market, as well as to make itself accountable to the public and society.

HYPOTHESES DEVELOPMENT

Singhvi and Desai (1971) argue that when the company achieves a high margin of profit and there is a high rate of return, the managerial groups are motivated to disclose more information in order to show off good reputation to the consumers, shareholders, investors and other stakeholders. Contrary, if the company suffers losses or there is a low rate of return, the managerial groups may disclose less information in order to cover the reasons for such losses or declining profits (Al Arussi et al., 2009). One of the factors enabling management to disclose CSR activity in high level of flexibility is profitability (Mulyadi & Anwar, 2012).

However, literature shows that the results on the association between profitability and CSR disclosure were mixed. Mohamed Zain and Janggu (2006), Jaffar et al. (2007) and Said et al. (2009) found that there is a positive association between profitability and CSR disclosure, while Abdul Rahman et al. (2015) and Esa and Mohd Ghazali (2012) found that there is a no association or relationship between profitability and CSR disclosure. They found that profitability does not influence CSR disclosure made by companies. Due to the mixed findings, the first hypothesis will be as follows:

H_1: There is a significant relationship between profitability and CSR disclosure among GLCs in Malaysia.

One of the most important elements of corporate governance mechanism is a board of directors who has the role to oversee and ensure that agents have properly managed the business (Said et al., 2009). MCCG (revised 2011) does not state the exact number on how many directors should be on the board to ensure the effectiveness but it suggests that there should be a balance between executive and non-executive directors on the board. The Green Book in the GLC Transformation Manual also does not state on the appropriate number, however, in order to ensure board can effectively discharge its roles and responsibilities, there must be a sufficient number of directors held on the board, where it should not become too large and unwieldy, which could then compromise board dynamics and the accountability of individual directors. Research regarding the investigation of the relationship between board size and CSR disclosure is rarely conducted (Esa & Mohd Ghazali, 2012). Only few researches have focus on this factor (board size). Siregar and Bachtiar (2010) and Esa and Mohd Ghazali (2012) in their study found that there is a positive relationship between board size and CSR disclosure. Their findings show that larger board size has higher influence on CSR disclosure, as larger board size exercises better monitoring and controlling function. Said et al. (2009) on contrary do not found any relationship between board size and CSR. The next hypothesis developed is:

H_2: There is a significant relationship between board size and CSR disclosure among GLCs in Malaysia.

In corporate governance context, when arriving at board decisions, independent directors are supposed to perform a monitoring role ensuring that shareholders' interests are taken into consideration (Esa & Mohd Ghazali, 2012). The Green Book in GLC Transformation Manual as well as MCCG 2012 encourages companies in Malaysia to have strong independence on board. There should be at least one-third independent directors sit on the board to monitor and provide independent view and judgment. Literature regarding the relationship between board independence and CSR disclosure are found to have a mixed result. Eng and Mak (2003) and Esa and Mohd Ghazali (2012) found a negative relationship between board independence and CSR disclosure whereas Donnelly and Mulcany (2008) found a negative relationship. Said et al. (2009) found no relationship between board independence and CSR disclosure. The mixed findings lead to the following hypothesis:

[Type here]

DOI: 10.1051/shsconf/20173600029

H3: There is a significant relationship between board independence and CSR disclosure among GLCs in Malaysia.

Table A: The operationalization of variables

Variables	Abbreviation	Operationalization	Reference
Dependent Variable			
Corporate Social Responsibility Disclosures	CSRD	Number of words from all four themes i.e. human resources, community, marketplace and environment stated in annual reports	Ng (1985)
Independent Variables			
Profitability	PROFIT	Profit before tax over total assets	Esa and Mohd Ghazali (2012)
Board Size	BSIZE	Total number of directors sit on the board	Said et al. (2009)
Board Independence	IND	Percentage of independent directors to total number of directors sit on the board	Esa and Mohd Ghazali (2012)
Control Variable			
Firm Size	FSIZE	Log10 of total assets	Abd Raman et al. (2015)

RESEARCH METHODOLOGY

Research design for this study is based on quantitative approach and longitudinal study of which investigate the relationship between CSR disclosures with corporate governance characteristics and firm characteristics. Annual reports for five years starting from year 2011 until 2015 of 16 GLC listed on Bursa Malaysia are taken from the Bursa Malaysia website (www.bursamalaysia.com). This study does not rely on the list of GLCs published on The Putrajaya Committee on GLC High Performance (PCG) website as it is outdated at the time of this writing. The author identify GLCs as Malaysian listed companies with 20% equity shares held by GLICs (Zin & Sulaiman, 2011). The findings of this study was analyzed using Statistical Package for Social Science (SPSS) version 22.0. Descriptive statistics will be used to identify the frequency distribution, central tendency (mean, median and mode) and dispersion (maximum, minimum and standard deviation) of all variables. Multiple regression analysis will be used to examine the relationship between the dependent variable i.e. CSR disclosures and the independent variables i.e. firm characteristics and corporate governance characteristics. Multiple regression analysis is used to find the result of all hypotheses developed in this study. The regression model of this study is as follows:

$$CSRD = \beta_0 + \beta_1 PROFIT + \beta_2 BSIZE + \beta_3 IND + \beta_4 FSIZE + \varepsilon$$

[Type here]

Where:

CSRD	= total CSR disclosures
βo	= intercept coefficient, when all other independent variables are zero
PROFIT	= profit before tax over total assets
BSIZE	= total number of directors sit on the board
IND	= percentage of independent directors to total number of directors sit on the board
FSIZE	= log10 of total assets
ε	= standard normal, randomly distributed error term

FINDINGS AND DISCUSSION

Descriptive statistics is executed to describe and summarize the characteristics of the sample. Tables presented below show the descriptive results for all variables utilized in this study based on the corporate annual report from year 2015 until 2015. Tables below report the minimum, maximum, mean and standard deviation.

Table B: Descriptive Statistics of firm's characteristics, corporate governance characteristics and control variable

Variables	*N*	*Minimum*	*Maximum*	*Mean*	*Std. Deviation*
2015					
PROFIT	16	0.006	17.115	1.18419	4.250660
BSIZE	16	6	12	9.19	1.515
IND	16	0.300	0.625	0.47638	0.097712
FSIZE	16	8.36	10.94	9.5412	0.73186
2012					
PROFIT	16	0.004	14.229	0.93869	3.544552
BSIZE	16	6	12	9.19	1.601
IND	16	0.286	0.778	0.46588	0.124154
FSIZE	16	8.34	10.80	9.6152	0.65384
2013					
PROFIT	16	0.004	1.608	0.14481	0.393101
BSIZE	16	6	13	9.38	1.857
IND	16	0.300	0.778	0.49500	0.115590
FSIZE	16	8.38	11.11	9.7018	0.72616
2014					
PROFIT	16	0.009	19.270	1.26763	4.800925
BSIZE	16	6	15	9.69	2.301

[Type here]

IND	16	0.273	0.778	0.51038	0.124858
FSIZE	16	8.40	10.84	9.6857	0.62399
2015					
PROFIT	16	0.004	15.533	1.11094	3.856695
BSIZE	16	6	12	9.25	1.880
IND	16	0.333	0.778	0.50306	0.134819
FSIZE	16	8.47	10.82	9.7273	0.61960

Note: PROFIT = profitability measured by profit before tax over total assets; BSIZE = board size measured by total number of directors sit on the board; IND = board independence measured by percentage of independent directors to total directors sit on the board; FSIZE = firm size measured by log10 of total assets.

Table B presents the descriptive statistics for all independence variables i.e. firm characteristics and corporate governance characteristics as well as the control variable used in this study. The results show that the worst profitability that Malaysian GLCs obtain is in year 2013 because the minimum profitability is 0.004% and the maximum profitability is only 1.608%, while the optimum profitability that Malaysian GLCs obtain is in year 2014 where the minimum profitability is 0.009% and the maximum profitability is 19.270%. The results also show that mean or average profitability for Malaysian GLCs reduce from year 2011 to year 2012 and 2013, but increase back in year 2014. In year 2015, the average profitability decrease by 0.157% from the previous year.

The table also explains descriptive statistics including the minimum, maximum, mean and standard deviation for corporate governance characteristics in Malaysian GLCs. In term of board size, the results show that over the five years period studied, the minimum number of directors sit on Malaysian GLCs board is 6 while the maximum number of directors sit on Malaysian GLCs board is 15. In average, there are 9 directors sit on the board of Malaysian GLCs.

In term of board independence, the results indicate that not all Malaysian GLCs follow the guideline in the Green Book where there should be one-third of independent directors sit on the board to ensure the independence on the board. In average, the results show that there is no slight different in the percentage of independent directors sit on the board for the five consecutive years.

Finally, firm size as a control variable, the results show that there is slightly similar in term of size of the GLCs in Malaysia for the five consecutive years, in which the biggest size is 11.11 comes from year 2013. In average, these indicate that Malaysian GLCs are almost similar in size when they are measured in term of total assets held by the companies.

[Type here]

DOI: 10.1051/shsconf/20173600029

Table C: Descriptive Statistics of corporate social responsibility disclosures by general themes

	N	*Sum*	*Minimum*	*Maximum*	*Mean*	*Std. Deviation*
2011						
Human Resource	16	4062	0	873	253.88	283.257
Community	16	4789	117	2970	1093.13	907.319
Marketplace	16	2024	0	780	126.50	231.454
Environment	16	3608	0	879	225.50	248.012
CSRD	16	14483	245	4178	1699.00	1298.358
2012						
Human Resource	16	3986	0	1703	535.88	578.776
Community	16	6476	180	3145	880.75	758.435
Marketplace	16	2351	0	1265	226.00	411.919
Environment	16	5340	0	975	333.75	298.599
CSRD	16	18153	180	5823	1976.38	1634.691
2013						
Human Resource	16	5958	0	1386	459.00	392.190
Community	16	6951	172	2020	805.38	612.862
Marketplace	16	2604	0	754	162.75	193.735
Environment	16	3808	0	1934	471.06	575.318
CSRD	16	19321	172	5201	1898.19	1305.388
2014						
Human Resource	16	4881	0	5108	910.19	1231.643
Community	16	3087	279	9919	1859.56	2347.310
Marketplace	16	3556	0	2741	393.56	670.988
Environment	16	5414	0	2278	689.00	659.766
CSRD	16	16938	352	20046	3852.31	4629.711
2015						
Human Resource	16	5636	0	3622	785.44	839.297
Community	16	4564	248	6052	1737.31	1921.979
Marketplace	16	3364	0	2212	348.50	557.659
Environment	16	4614	0	1805	653.06	548.301
CSRD	16	18178	344	13129	3524.31	3164.919

Table C presents the frequency analysis for CSR disclosures by four general themes i.e. human resources, community, marketplace and environment as well as the total CSR disclosures from year 2011 until year 2015. It shows that the highest CSR disclosure in year 2011 is under community theme, while the lowest disclosure is under marketplace theme. The highest CSR disclosure in year 2012 is under community theme, while the lowest

 DOI: 10.1051/shsconf/20173600029

disclosure is under marketplace theme, same with the previous year. In year 2013, the highest CSR disclosure is under community theme, while the lowest disclosure is under marketplace theme, same also with the previous years. In year 2014, the highest CSR disclosure is under environment theme, but in average is under community theme. The lowest disclosure is again found to be under marketplace theme, same with the previous years. Lastly, in year 2015, the highest CSR disclosure is under human resources theme, but in average is under community theme. The lowest disclosure is still under marketplace theme, same with the previous years. Overall, the result indicates that the introduction of the Silver Book really reflect a CSR disclosure among Malaysian GLCs.

Correlation

Correlation measures the strength, direction and significant relationship between independent and dependent variables. According to Pallant (2011), multicollinearity may be a problem when the correlation exceeded 0.7. Hence, to verify for multicollinearity problem, Table D below shows the result of the correlation matrix.

Table D: Spearman's Correlations Coefficient Matrix

	CSRD	PROFIT	BSIZE	IND	FSIZE
CSRD	1.000	.187	.440**	.196	.372**
PROFIT		1.000	.217	.157	-.194
BSIZE			1.000	.197	.255*
IND				1.000	.093
FSIZE					1.000

Note: PROFIT = profitability measured by profit before tax over total assets; BSIZE = board size measured by total number of directors sit on the board; IND = board independence measured by percentage of independent directors to total directors sit on the board; FSIZE = firm size measured by log10 of total assets.

As shown in Table D, CSRD is significantly positive correlated with BSIZE at 0.440 and FSIZE at 0.372. Therefore, it can be concluded that multicollinearity is not a serious problem, and the use of parametric statistical analysis such as regression analysis for all variables is considered appropriate for this study.

Regression Analysis

Table E below summarizes the regression analysis results for the data of the relationship between the dependent variable i.e. corporate social responsibility disclosure (CSRD) and the independent variables i.e. profitability (PROFIT), board size (BSIZE), and board independence (IND).

[Type here]

DOI: 10.1051/shsconf/20173600029

Table E: Multiple Linear Regression Analysis

	Variables	Coefficients	Significant (T-Stat)	P-value
Dependent Variable	CSRD			
Independent Variables	(Constant)	1.709	0.66	0.511
	PROFITABILITY	3.250	1.992	0.055
	BSIZE	0.097	1.604	0.043*
	IND	1.326	1.466	0.037*
Control Variable	FSIZE	0.108	0.503	0.116
	N	80		
	R^2	0.179		
	Adjusted R^2	0.169		
	F-value	17.06		
	P-value	0		

*significant at 0.05

Note: PROFIT = profitability measured by profit before tax over total assets; BSIZE = board size measured by total number of directors sit on the board; IND = board independence measured by percentage of independent directors to total directors sit on the board; FSIZE = firm size measured by log10 of total assets.

Table E shows that F-value for the data is statistically not significant at the 1 percent level where the p-value is 17.060 (more than 0.05). It indicates that the independent variables in this study are not significant to measure the dependent variable. The R^2 is 0.179 and the adjusted R^2 is 0.169. It can be concluded that 17.9% variance in corporate social responsibility disclosure (CSRD) is explained by all the independent variables i.e. PROFIT, BSIZE, IND and another 82.1% variance in CSRD might be explained by the other factors or variables that have not been used in this study. Referring to the P-value, there are only two variables that make a statistically significant contribution (less than 0.05) toward CSRD which are BSIZE and IND, where the highest coefficient value is IND ($p = 0.037$). Both BSIZE (Beta = 0.097) and IND (Beta = 1.326) were found significantly positive correlated with the corporate social responsibility disclosure. This suggests that bigger board size has various background, skill and experience to promote CSR, thus lead to higher level of CSR disclosure, and higher proportion of independent directors sit on the board will have stronger monitoring control, thus lead to higher level of CSR disclosure among Malaysian GLCs.

Based on the results tabulated in Table E above, the regression model for corporate social responsibility disclosure in relation to firm' characteristics and corporate governance characteristics can be expressed as follows:

$$CSRD = 1.709 - 3.250(PROFIT) + 0.097(BSIZE) + 1.326(IND) + 0.108(FSIZE) + \varepsilon$$

[Type here]

This model suggests that CSR disclosure among GLCs in Malaysia could be significantly enhanced by increasing the number of directors sit on the board as well as increasing the involvement of independent directors sit on the board.

Relationship between profitability and CSR disclosure

The above result shows that there is no significant relationship between profitability and CSR disclosure. Therefore, hypothesis H1 is rejected. This result is consistent with Abdul Rahman et al. (2015) and Said et al. (2009). It indicates that profitability does not influence the CSR disclosure made by Malaysian GLCs.

Relationship between board size and CSR disclosure

The regression result shows that board size has significant positive relationship with the CSR disclosure. It indicates that large number of directors sit on the board provide better disclosure of CSR. Whereas, small number of directors sit on the board lead to weaker disclosure of CSR in Malaysian GLCs annual reports. Hence, hypothesis H4 is supported. This result is parallel with the results of previous studies conducted by the other researchers (Siregar & Bachtiar, 2010; Esa & Mohd Ghazali, 2012). This result indicates that larger board size is more effective and having various background, skills and knowledge enhance the disclosure of CSR activities in the company's annual reports.

Relationship between board independence and CSR disclosure

Based on the table above, there is a significant positive relationship between board independence and CSR disclosure. It shows that higher proportion of independence directors over total number of directors sit on the board provide greater CSR disclosure. On contrary, lower proportion of independence directors over total number of directors sit on the board lead to the smaller amount of CSR disclosed by Malaysian GLCs in their annual reports. Therefore, hypothesis H5 is supported. This result is consistent with Eng and Mak (2003) and Esa and Mohd Ghazali (2012). It indicates that higher percentage of independent directors sit on the board would enhance the monitoring and controlling function so that companies would provide sufficient disclosure on the CSR information to public and all stakeholders.

CONCLUSION

Malaysian government has made it mandatory for all public listed companies (PLCs) in Malaysia to report and disclose their CSR. In order to protect interest of all stakeholders, there is a need for a company to report their social responsibility. Reporting CSR will enhance public confidence and give better image and reputation for the companies. Malaysian GLCs, whose majority shares are owned by government also, need to report and disclose their CSR to obtain such benefits. As what has been presented in the descriptive analysis, the trend of CSR disclosure in Malaysian GLCs is increasing over the years, indicating that Malaysian GLCs are become more and more concern about their social responsibility in term of human resources, community, marketplace and environmental matters, which the most focus is on the community matters such as education, disaster relief, charity, sports and cultures. The increasing trend of disclosure also proves that the

[Type here]

introduction of the Silver Book really reflect a better CSR disclosure among Malaysian GLCs. Also, CSR disclosure made by Malaysian GLCs is influenced by certain corporate governance characteristics i.e. board size and board independence. Firm characteristics on the other hand do not influence the CSR disclosure made by Malaysian GLCs.

Accordingly, higher number of directors sits on the board increase the level of CSR disclosure due to the excessively skills and experience to improve CSR awareness in the companies. Higher percentages of independence directors sit on the board also increase the level of CSR disclosure because monitoring and controlling function are enhanced through the higher level of independence. In addition, although different firm or company has different characteristics, there is nothing to do with the level of CSR disclosure.

REFERENCES:

Abd Rahman, N.H., Mohamed Zain, M., & Yaakop Yahaya Al-Haj, N.H. (2011). CSR disclosures and its determinants: evidence from Malaysian government link companies. *Social Responsibility Journal*, 2(2), 181-201.

Al Arussi, A.S., Selamat, M.H., & Hanefah, M.M. (2009). Determinants of financial and environmental disclosures through the internet by Malaysian companies. *Asian Review of Accounting* , 17(1), 59-76.

Aldama, L.R., Amar, P.A., & Trostianki, D.W. (2009). Embedding corporate responsibility through effective organizational structures. *Corporate Governance*, 9(4), 506-516.

Al-Shubiri, F.N., Al-abedallat, A.Z., & Orabi, M.M. (2012). Financial and non financial determinants of corporate social responsibility. *Asian Ecomonic and Financial Review*, 2(8), 1001-1012.

Andrew, B.H., Gul, F.A., Guthrie, J.E., & Teoh, H.Y. (1989). A note on corporate social disclosure practices in developing countries: the case of Malaysia and Singapore. *British Accounting Review*, 21, 371-376.

Baba, H.N. & Mohd Hanefah, H.M. (2001). Corporate social disclosure practices in Malaysia. Presented at the 2nd Asean Academic Accounting Association Conference, Shangrila Hotel, Penang.

Baba, H.N. (2004). Corporate social responsiveness and environmental performance of small-medium enterprises: An Empirical Study. Unpublished PhD thesis.

Bayoud, N.S., Kavanagh, M., & Slaughter, G. (2012). Factors influencing levels of corporate social responsibility disclosure by Libyan firms: a mixed study . *International Journal of Economic and Finance*, 4(4), 13-29.

Bursa Malaysia (2006), CSR Framework for Malaysian PLCs, retrieved 3 April 2012, from http://ablemen.com/sustainability/lock/frameworks_introduction.php?page=frameworks&index=1

Ching, S.C. & Wad, P. (2009). Corporate social responsibility and offshore outsourcing: electrical and electronics firms in malaysia. *International Journal of Institutions and Economies*, 1, 253-282.

Corporate Governance Blueprint (2011), retrieved 3 April 2013 from http://www.sc.com.my/main.asp?pageid=1087&menuid=&newsid=&linkid=&type=

Deegan, C. & Gordon, B. (1996). A Study on the Environmental Disclosure Practices of Australian Corporations. *Accounting and Business Research* , 26(3), 187-199.

Donnelly, R. & Mulcany, M. (2008). Board structure, ownership and voluntary disclosure in Ireland. *Corporate Governance: An International Review* , 16(5), 416-429.

[Type here]

Eng, L.L., & Mak, Y.T. (2003). Corporate governance and voluntary disclosure. *Journal of Accounting and Public Policy* , 22, 325-345.

Esa, E. & Mohd Ghazali, N.A. (2012). Corporate social responsibility and corporate governance in Malaysian government-linked companies. *Corporate Governance*, 12(3), 292-305.

Islam, M.A., & Deegan, C. (2008). Motivations for an organisation within a developing countryto report social responsibility information; Evidence from Bangladesh. *Accounting, Auditing & Accountability Journal*, 21(6), 850-874.

Islam, M.A., & Deegan, C. (2010). Media pressures and corporate disclosure of social responsibility performance information: a study of two global clothing and sports retail companies. *Accounting and Business Research*, 40(2), 131-148.

Jaffar, R., Jamaludin, S., & Che Abdul Rahman, M.R. (2007). Determinant factors affecting quality of reporting in annual reports of Malaysian companies. *Malaysian Accounting Review* , 6(2), 19-39.

Lassaad, B.M., & Khamoussi, H. (2012). Determinanst of Communication about Corporate Social Responsibility: Case of French Companies. *International Journal of Contemporary Business Studies*, 3(5), 49-60.

Lo, B.C.W. & Yap, K.L. (2011). Are Malaysian Companies Ready For Corporate Social Responsibility? *Labuan e-Journal of Muamalat and Society*, 5, 11-15.

Lucyanda, J. & Siagian, L.G. (2012). The influence of Company Characteristics Toward Corporate Social Responsibility Disclosure . *The 2012 International Conference on Business and Management*, 601-619.

Luethge, D. & Han, H.G. (2012). Assessing corporate social and financial performance in China. *Social Responsibility Journal*, 8(3), 389-403.

MCCG (2012). Malaysian Code on Corporate Governance, retrieved 20 April 2013 from http://www.sc.com.my/eng/html/cg/cg2012.pdf

MCCG (revised 2007). Malaysian Code on Corporate Governance, retrieved 20 April 2013 from http://www.sc.com.my/eng/html/cg/cg2007.pdf

Mobus, J.L. (2005). Mandatory environmental disclosures in a legitimacy theory context. *Accounting, Auditing & Accountability Journal*, 18(4), 492-517.

Mohamad, M.H., Abdul Rashid, H.M., & Shawtari, F.A. (2012). Corporate governance and earnings management in Malaysian government linked companies; The impact of GLCs' transformation policy. *Asian Review of Accounting*, 20(3), 241-258.

Mohamed Zain, M. & Janggu, T. (2006). Corporate social disclosure (CSD) of construction companies in Malaysia. *Malaysian Accounting Review*, 5(1), 85-144.

Mohd Ghazali, N.A. (2007). Ownership structure and corporate social responsibility disclosure: some Malaysian evidence. *Corporate Governance*, 7(3), 251-266.

Mulyadi, M.S. & Anwar, Y. (2012). Influence of corporate governance and profitability to corporate CSR disclosure. *International Journal of Arts and Commerce*, 1(7), 29-35.

Nik Ahmad, N.N. & Sulaiman, M. (2004). Environmental disclosures in Malaysian annual reports: A legitimacy theory perspectives. *International Journal of Cost Management*, 14(1), 44-58.

O'Sullivan, N. & O'Dwyer, B. (2009). Stakeholder perspectives on a financial sector legitimation process: The case of NGOs and the Equator Principles. *Accounting, Auditing & Accountability Journal*, 22(4), 553-587.

Pallant, J. (2011). SPSS Survival Manual (4th edition): A step by step guide to data analysis using SPSS for Windows. *Open University Press* .

[Type here]

Putrajaya Committee (2006). GLC Transformation Manual, retrieved 3 April 2013 from http://www.pcg.gov.my/trans_manual.asp

Ribeiro, V.P. & Aibar-Guzman, C. (2010). Determinants of environmental accounting practices in local entities: evidence from Portugal. *Social Responsibility Journal*, 6(3), 404-419.

Said, R., Hj Zainuddin, Y., & Haron, H. (2009). The relationship between corporate social responsibility disclosure and corporate governance characteristics in Malaysian public listed companies. *Social Responsibility Journal*, 5(2), 212-226.

Saleh, M. (2009). Corporate Social Responsibility Disclosure in an Emerging Market: A Longitudinal Analysis Approach. *International Business Research*, 2(1), 131-141.

Siagian, F.T., & Tresnaningsih, E. (2011). The impact of independence directors and independence audit committees on earnings quality reported by Indonesian firms. *Asian Review of Accounting*, 19(3), 192-207.

Sibao, S. & Huaer, C. (2009). Economic globalization and the construction of China's corporate social responsibility. *International Journal of Law and Management*, 51(3), 134-138.

Singhvi, S.S. & Desai, H.B. (1971). An empirical analysis of the quality of corporate financial disclosure. *The Accounting Review*, 46(1), 129-138.

Siregar, S.V. & Bachtiar, Y. (2010). Corporate social reporting: empirical evidence from Indonesia Stock Exchange. *International Journal of Islamic and Middle Eastern Finance and Management*, 3(3), 241-252.

Suchman, M.C. (1995). Managing Legitimacy: Strategic and Institutional Approaches. *Academy of Management Review*, 20(3), 571-610.

Vafeas, N. (2000). Board structure and the informativeness of earnings. *Journal of Accounting and Public Policy*, 19(2), 136-160.

[Type here]

DOI: 10.1051/shsconf/20173600030

ACCEPTANCE OF HIBAH AS AN ALTERNATIVE MECHANISM IN MUSLIMS ASSET MANAGEMENT

Khairiah Ahmad[1], Rozaiha Ab Majid, Zaharah Abdullah, Siti Fatimah Noor Minhad, Shafinar Ismail[2]
[1]*Faculty of Accountancy, Universiti Teknologi MARA, Melaka*
[2]*Faculty of Business Management, Universiti Teknologi MARA, Melaka*

ABSTRACT

This study is aimed to identify the acceptance of *hibah* by beneficiaries based on cases registered in Melaka. Using secondary data collected from the Melaka Syariah Court and Amanah Raya Berhad Melaka, a number of *hibah* cases were identified from 2013-2015 and highlighted for further discussion in the paper. Interviews, telephone and email correspondences were conducted to further understand the issues concerning *hibah* cases and their reasoning. The findings indicated that the highest number of registered on *hibah* cases was recorded in 2015, increasing from 2013. The data result shows that there are more than 20 cases recorded in Melaka except only 12 cases in 2013. Due to increasing number of cases regarding *hibah*, it shows that citizen of Melaka is still low in understanding the concept and implementation of hibah. Therefore, it is a need to have a standard guideline regarding *hibah* to avoid any dispute about this matter in the future. Besides that, the dissemination of knowledge about *hibah* should be done in order to increase the awareness about the acceptable procedure of *hibah* because it is considered as the best alternative of asset management and to avoid unclaimed Muslim asset.

Keywords: Hibah, dissatisfaction, beneficiaries, court cases

INTRODUCTION

1.1 Background

Over the last five years, Malaysia has been experiencing an increase in the number of court cases over the dispute in regards to legal beneficiaries and heirs over the transfer of their inheritance to another person through the implementation of *hibah*. The cases are brought to Syariah Court and are heard under *The Administration of Islamic Law (Federal Territories) Act 1993 (Act 505) Part IV Sec. 46 (1) (2) (b) (vi) - Syariah High Court's Jurisdiction*.

In Melaka alone, the Melaka Tengah district recorded a progressive increase on *hibah* cases from 2010 to 2012. As reported in Mohd Yusof and Ahmad (2013), there were only 10 cases in year 2010 and significantly increased in 2011 and 2012 to 24 and 32 cases respectively. The increasing number of court cases showed that the implementation of *hibah* has resulted in growing dissatisfactions among related parties especially among family members who are also the beneficiaries. Eventually, these feelings of dissatisfaction may lead to a more complex dispute among related parties.

*Corresponding author's email: khairiah@melaka.uitm.edu.my

Therefore, it is essential to understand the reasons to this issue. The dissatisfaction may be due to the lack of knowledge and poor awareness on *hibah* among parties involved. *Hibah* may have not been done according to a specific procedure. For that, this research is conducted to understand the acceptance of *hibah* among the legal beneficiaries. Additionally, the significant of this study also lies on the fact that most of the previous research had only focused on issues relating to *zakat, baitulmal* and *wakaf* instead of *hibah*.

1.2 *Hibah* as an alternative mechanism for asset management

According to the Islamic term, *hibah* refers to a contract that entails granting ownership of the property to someone else during his/her lifetime without any reprisal (Muda, 2008) outlined five requirements of *hibah* which include: i). an *akad* (contract) of granting property ownership to a specific party; ii). the property that can be given as *hibah* is a physical property and not debt; iii). *hibah* contract is made without imposing any reprisal; iv). *hibah* contract is in effect during the lifetime of the *hibah* provider; v). *hibah* is made voluntarily and the contract is made with no intention to receive reward or honour a particular person. A *hibah* provider must be the genuine owner of the item/s intended for *hibah*. *Hibah* recipient can be anyone as long as he/she has the capability to own property (Muda, 2008).

Hibah refers to a transfer of ownership of an asset from a donor (*wahib*) to a recipient (*mawhub lahu)* without a reward (Bank Negara, 2016). In other words, *hibah* donor makes proposal (offer) to transfer the property to the beneficiary, and in return the beneficiary accepts the proposal, with no consideration (Mujani et al, 2011;Buang, 1993).

Utusan Online, 20 January 2014 reported that RM52 billion outstanding values of the estate owned by majority of Muslims failed to be distributed to the person entitled in the country and by referring to the Chief Marketing Officer of Amanah Raya Berhad. In 2011, RM42 billions of assets were frozen due to the failure to trace the legal beneficiaries (Shafie et. al, 2014). Therefore, it is crucial for Muslims to manage their inheritance property using a more effective and efficient approach.

Currently, the distribution of wealth particularly in regards to inheritance property among Muslims is managed through the law of succession (*fara'id*). However, the administration of property through this method in many instances were found to be less effective due to the current lifestyle people are usually occupied with day-to-day commitments and their chaotic schedules (Abdul Rashid et.al, 2013). On top of that, the process of distributing property via *fara'id* can be time consuming, complicated and costly (Hamid Ghul et.al, 2015). To make it worst, the process may disrupt relationships among family members due to the disputes brought forward. (Shafie et.al, 2014)

In relation to that, *hibah* can be viewed as an adequate solution to overcome issues concerning the use of *fara'id* particularly on the aspect that property may be transferred to family members with the terms of the donor's decision and choice (Abdul Rashid et.al, 2013). In addition, *hibah* is also a transfer of property inter vivos and takes immediate effect after the contract (Buang, 2008). Thus, the implementation of *hibah* will not only ensure the distribution of property to the legal beneficiaries is in accordance with donor's wishes but also to reduce the number of unsolved and complex cases in the distribution of properties through the process of *fara'id* (Abdul Rashid et.al, 2013).

Fara'id is known to be one complicated and less effective approach particularly on the fact that the distribution of wealth to legal beneficiaries must go through several stages involving several jurisdictions which may not be fully understood by the beneficiaries (Abdul Rashid et.al, 2013). On the other hand, the process of *hibah* is a more convenient approach as compared to *fara'id* because the distribution of wealth is done in accordance to donor's wishes. Under *fara'id*, the legal beneficiaries have the right over the deceased property and the asset must be distributed to the legal beneficiaries accordingly as stated under the Islamic Law. Therefore, each deceased's asset will be shared (or owned) by many individuals. For example, a piece of land may be owned by several owners with each one of them owning a very small size of the big land. As a result, the land or property cannot be fully developed according to one particular beneficiary's wish.

Under law of succession *(fara'id)*, distribution of wealth must follow the portion as stated in Islamic law for example; the daughter will only get half portion of the son. In *hibah*, the owner of the property has the right to decide the amount and proportion of property to be distributed to the heirs according to their needs and economic condition (Abdul Rashid et al., 2013).

Under *fara'id* law, the property will be distributed to the legal beneficiaries only when the person is passed away. Different from *fara'id*, in *hibah* the property can be distributed to heirs while the donor is still alive. But if the donor worried about being neglected after all property being distributed to the heirs, the donor can choose to use *hibah amanah*, where the property will only be transferred after the death of the donor (Yaacob, 2016).

It can be seen that *hibah* have advantage over *fara'id*, however, the current situation in Malaysia, there is no specific *hibah* law enacted which means that the State Legislature and Parliament has never approved the special statue of *hibah* for muslims to be implemented in Syariah Courts. The current status of *Hibah* cases in the Syariah Courts in Malaysia usually involves the verification of *hibah,* dispute on the extent or rate of *hibah* and withdrawal of *hibah* (Mohd Yusof and Ahmad, 2013).

1.3 Acceptance of beneficiaries on *hibah*

Hibah is a practice that is favourable (*sunat*) in Islam especially to the immediate family. This practice based on the *al-Quran, al-Sunnah* and consensus. To form a *hibah* contract, the following conditions must be fulfilled. Among the condition are must have a *hibah* provider, *hibah* recipient, goods or property for *hibah* and lastly *ijab* (contract) and *qabul* (grant). When a completed *hibah* contract meets the terms and condition, and there has been a surrender and acceptance of goods, then the property rightfully belongs to the *hibah* recipient even without repayment.

According to the Hanafi sect, it is '*makruh*' to withdraw *hibah* that has been given away. However, in the opinion of Syafie, Hambali and Maliki sects, when the *hibah* accompanied with the delivery and acceptance of goods then the *hibah* cannot be withdrawn unless the hibah is made by immediate family (eg: father, mother, grandfather, grandmother) to the children for as long as the property is not related to other people. Furthermore, based on the Prophet's saying meaning:

> "One who gets back his give is like a dog which vomits and then swallows that vomit."
> (By Al-Bukhari and Muslim)

Hibah cases in the Syariah Court in Malaysia, usually involves the verification of *hibah*, dispute on the extent or rate of *hibah* and the withdrawal or revocation of *hibah*. In the case of Eshah binti Abdullah and five others vs Che Aminah binti Abdul Razak and two others *(2004)* XVIII (I) JH 47 where the arising issue was whether or not the acceptance had been clearly stipulated after giving away or sufficient by other evidences to prove that was an acceptance. In this case, the donor had decease and there was no withdrawal of *hibah*. The building in dispute still bore the name of the deceased although the gift had been witnessed by others. The Terengganu Syariah Appeal Court accepted the appeal made by the petitioners based on the accepted *qarinah* (evidence) from the receiver, where all the three *hibah* receivers declared that they have been receiving the rental/lease (tenancy) of the disputed building since the deceased was still alive.

According to Mujani et al., (2011), in the case of Eshah Abdul Rahman vs Azuhar vs Ismail *(2004)*, the plaintiff had his piece of land transferred to his adopted son (defendant). After filed up form 14A of the National Land Code affecting the transfer, the plaintiff changed his mind demanding the land to be returned. By then, the defendant had developed the land and had constructed several buildings thereon. The issue was whether the transfer here in constituted a *hibah* and if so whether the plaintiff could demand from the defendant for its return. It was held that the transfer here represented a duly constituted *hibah* given that the plaintiff and defendant had complied with all conditions of *hibah*. In regards to the issue of revocation of the said *hibah*, the court opined that when a value which is inseparable from the subject matter of *hibah* was added there to as for instance when a building was constructed by the beneficiary on the *hibah* land, the donor could no longer retract the gift and demand for its returns.

One of the major issues highlighted on *hibah* in this paper is the dissatisfaction of the beneficiary on the distribution of properties. Based on the above cases and many others, *hibah* can be challenged in a Syariah Court both verbally or documented by an individual or a lawyer. However, it can only be questioned in terms of legality in accordance with Islamic law or if there is any beneficiary who disagrees with the content of the *hibah* after the demise of the donor (Mohd Yusof and Ahmad, 2013). For that matter, many cases on *hibah* in Malaysia generally involves proposal to revoke a gift *(hibah)* earlier given to someone and questioning the validity of a gift awarded to the third party after the demise of the donor (Oseni, 2012).

The increase dispute cases among family members due to the practice of *hibah* may lead to crime. Increase of crime will have great impact on the quality of life of the people. It is important to take actions to help in reducing the cases due to *hibah*. Effort to decrease the dispute will assist the government to achieve its National Key Result Area in reducing crime.

OBJECTIVE OF STUDY

The objective of this study is to identify the acceptance of *hibah* by beneficiaries based on the cases registered at Melaka Syariah Court and *Hibah Trust* product applied with Amanah Raya Berhad Melaka.

METHODOLOGY

This study is conducted based on the data collected from Melaka Syariah Court and Amanah Raya Berhad Melaka. Interview instrument is selected for data collection. Discussion among the research member on location of interview, how the interaction with the subjects would take place and also arranged the setting of the session to ensure the appropriate level of quiet, privacy, and primer data exclusively will be gathered (Business Graduate School, 2010). Following an application letter for setting the appointment to both the Melaka Syariah Court and Amanah Raya Berhad Melaka, stating the purpose of the interview and our objective to collect the data on the number of *hibah* cases in these organisations for the intended study.

An interview session with an officer from the Melaka Syariah Court was conducted regarding the registered *hibah* cases and their reasoning. The interview was carried out by having a preliminary meeting among the research member prior to the actual interview. Typically, such meetings were conducted roughly a week before the interview. This preliminary meeting is an opportunity to establish trust with the research member, review ethical considerations and complete consent forms. During this initial meeting, the research question has been reviewed and the lists of questions to be prepared for interview session. By going about it in this way it can aid in getting a richer description during the interview without asking too many questions to the interviewee.

Subsequently, the information data was also provided by the organization through email and telephone conversation to furnish research objective. The document reviews with the assistance of the interviewee, in the library of Melaka Syariah Court were also conducted to search if there are acts for *hibah*. On the other hand, the information and data from Amanah Raya Berhad Melaka were gathered through interviews via telephone conversation with the office, Puan Noraizah.

FINDINGS

4.1 Melaka Syariah Court

Based on the data collected from the Melaka Syariah Court from 2013-2015, the highest record on *hibah* was in 2015. The data result shows that there are an increase no of cases registered in Melaka state where 12 cases in 2013 followed by 22 cases in 2014 and 31 cases in 2015.

Figure 1: Trend of cases on *Hibah* in Melaka Syariah Court

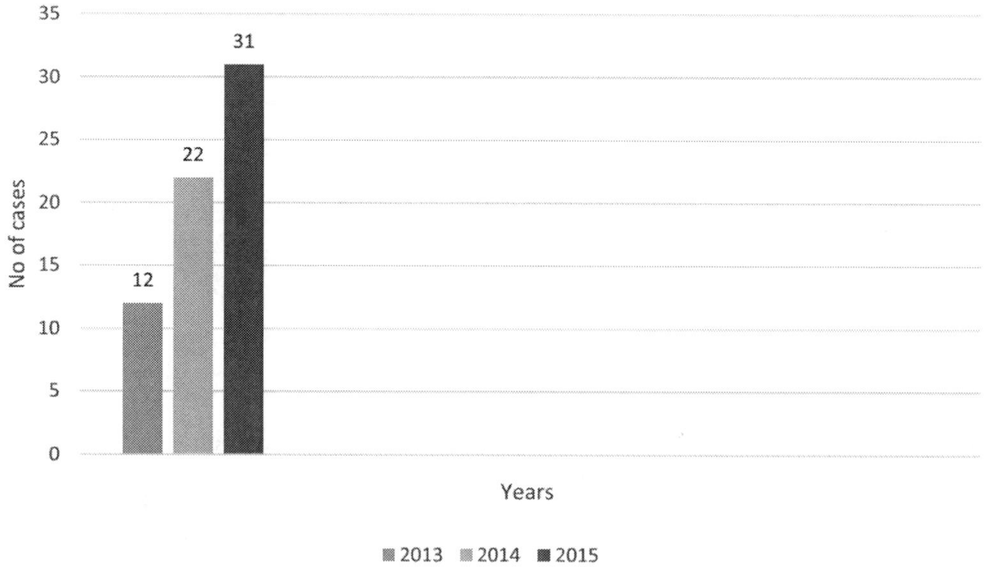

The existence of cases in Melaka Syariah Court has showed that there is still lack of acceptance or dissatisfaction by beneficiaries on *Hibah* which will lead them to file the case in Syariah Court. The increase number of court case shows that the giving of *hibah* has caused dissatisfaction among related parties especially family members who are the beneficiaries. Oseni (2012) mentioned that most cases in Syariah Court involving *hibah* and *wassiyyah* (will) - related involve family relations. This feeling of dissatisfaction may lead to dispute among related parties. Therefore, it is essential to understand the reasons to this problem. The dissatisfaction may be due to the lack of knowledge and awareness on *hibah*. *Hibah* may be done not according to any procedure.

Figure 2: No of cases registered in Melaka Syariah Court

Years	No. of cases
2013	12
2014	22
2015	31

Based on documentation observation session at Melaka Syariah Court, it is evidenced that no such of Act or Enactment on *hibah*, only the Enactment on *wassiyyah* (will). State government of Melaka should come out with a guideline on *hibah* so that potential donor can follow to avoid any unforeseeable problem in the future. According to Oseni (2012), in some cases the judge may refer the parties for out-of-court settlement which should begin with a psychotherapy procedure purposefully meant to remind of their family ties and the ephemeral nature of properties and worldly life in general.

4.2 Amanah Raya Berhad Melaka

Amanah Raya Berhad Melaka was only started operation in May 2013. Based from the data collected from Amanah Raya Berhad Melaka there are only 1 case in 2013 and 3 cases in 2014. There is no application for *Hibah* Trust product in 2015. According to Puan Noraizah (Officer of Amanah Raya Berhad Melaka), client prefers to choose *wassiyyah* (will) as compare to *Hibah* product. One of the major reasons is the difficulties in preparing the documentation for *hibah* as compare to will and the types of asset that can be given away (*Hibah*). Today, various institutions and agencies are involved in documenting *hibah* and there are also various *hibah* methods have emerged (Ahmad Yusof and Ahmad, 2013).

Figure 3: Number of registered *Hibah* Trust

Years	*Hibah* Trust
2013	1
2014	3
2015	0

Findings from these two organisations is consistent with Mohd Yusof and Ahmad (2013), where *hibah* are considered as new product and are recently introduced in Melaka (2013) upon realizing that it is a best method to reduce difficulties as well as to create awareness in property distribution among the Muslims in Melaka compare with will and inheritance.

LIMITATION

The present study has several limitations that may potentially influence the interpretation of the results. The data for this study were based on the *hibah* cases registered for the Melaka State only. This is because the Syariah Court is governing by the State Government and there is no standard *Hibah* Act/Enactment yet to be established by the Malaysian Government. Each state has their owned way of handling the matters regarding *hibah*. Thus, it is highly suggested that the further study focused on the implementation of *hibah* throughout Malaysia and from there we can make comparison regarding the implementation of *hibah* for each state in Malaysia.

CONCLUSION AND RECOMMENDATIONS

The proper disseminations of information and knowledge on *hibah* can allow Muslims in Malaysia to better understand their rights when facing a dispute. By having proper understanding on this matter, parties involve may avoid further dispute. This study revealed an increasing number of registered cases on *hibah* dispute in Melaka Syariah Court Melaka. The growing dispute cases among family members in lieu of the poor understanding and awareness on the practice of *hibah* have led to a substantial number of crime cases. This may eventually have an impact on the quality of life and their harmony. It is important to take actions to help in reducing the cases due to *hibah*. Effort to decrease the dispute will assist the government to achieve its National Key Result Area in reducing crime. Each state needs to establish a standard Muslim *Hibah* Act/Enactment as reference when matters about *hibah* arise. It will become a guideline for the Syariah Court to decide on the validity of a *hibah*. Besides that, the Fatwa Committee should issue a clear and strict fatwa regarding *hibah* that suits the current needs and parallel with the Islamic Law requirements. Furthermore, centralised monitoring and supervision should be implemented to any agencies that manage Muslim asset especially regarding the distribution of wealth through *hibah*. Therefore, further study should focus on determining the reasons behind these *hibah* cases particularly on the aspects of dissatisfactions among legal beneficiaries or heirs in the wealth distribution held through *hibah*. Subsequently, the focal point here is to develop a model or a guideline which can be applied in order to reduce dissatisfactions among legal beneficiaries when distributing wealth via *hibah*.

REFERENCES:

Abdul Rashid, R., Hassan, S.A. &Yaakub, N.I., (2013).*A Need For Legal Framework of Gift Inter Vivos (Hibah) In Malaysian Estate Planning*. International Journal of Business, Economics and Law, Vol 2. Issue 3 (June) ISSN 2289-1552

Bank Negara, (2016). *Hibah*. Guideline issued as reference on the Syariah rulings applicable to hibah.

Buang, A.H., (2008).*Appreciation of Syari'ah Principles in Property Management in Comtemporary Malaysian Society*. Shariah Journal, Vol16, Special Edition 555-566

Buang, S. (1993) Historic Judgement on article 121 of the federal constitution, Malaysian Law News.

Hamid Ghul, Z., Yahya, M.H. &Abdullah, A., (2015).*Wasiyyah (Islamic Will) Adoption and the Barriers in Islamic Inheritance Distribution among Malaysian Muslims*, International Journal of Humanities Social Sciences and Education, Vol. 2, Iss 1 pp. 1-11

Kamariah, M., (2004). *Halsbury's laws of Malaysia*. Malayan Law Journal, 23: 1-253.

Mohammad, M.T.S., (2015). *Theoretical and trustees' perspectives on the establishment of an Islamic social (Waqf) bank*, Humanomics, Vol. 31 Iss 1 pp. 37 – 73

Mohd Yusof, Y. &Ahmad, A., (2013).*Hibah As An Alternative Mechanism in Muslims Assets Management: A Study in Melaka Tengah*. South East Asia Journal of Contemporary Business, Economics and Law. Vol.3. Issue 3 (December) ISSN 2289-1560

Muda, M.Z., (2008), *Instruments of Hibah and Wills: Analysis of the regulations and applications in Malaysia*. Paper presented at the Hibah and Faraid National Convention 2008, Kuala Lumpur.

Mujani, W.K., Wan Hussain,W.M.H., Yaakub, N.I. &Abdul Rashid, R., (2011).*The concept of Law of Gift Inter Vivo Under Islamic Law and the Contracts Act, 1950*. International Business Management 5(6): 319-325, ISSN: 1993-5250.

Oseni, U.A., (2012).*Sharīʻah Court-annexed ADR: The Need for Effective Dispute Management in Waqf, Hibah and Wasiyyah Cases in Malaysia*. In: 14th Annual Conference of the Sharīʻah Legal Officers of Malaysia, 19th - 21st November 2012, One Lelang Hotel, Langkawi, Kedah. (Unpublished)

Sabirin, A., (2014), *Umat Islam rugi berbilion ringgit kerana harta pusaka.*Utusan online, 20 January 2014

Shafie, F., Wan Yusoff, W.Z. &Dawilah Al-Edrus, S.M., (2014).*Islamic Real Estate Management: Review on Issues and Challenges in Managing Inheritance Property in Malaysia.* International Real Estate Symposium 7th IRERS Kuala Lumpur

The Administration of Islamic Law (Federal Territories) Act 1993 (Act 505) Part IV Sec. 46 (1) (2) (b) (vi) - Syariah High Court's Jurisdiction.

Yaacob, O., (2006). *Pembentukan Trust Hibah Sebagai Alternative Perancangan Harta. In Siti Mashitoh Mahamood. Harta Amanah Orang Islam di Malaysia Perspektif Undang-undang dan Pentadbiran*. 174-175. Kuala Lumpur: Penerbit Universiti Malaya.

http://www.esyariah.gov.my/JurnalHukumSearch/htdocs/jurnalhukumsearch/JurnalHukumSearchDetailsPageBM.jsp?DetailsID=612,

http://www.trinitydc.edu/sps/files/2010/09/APA-6-BGS-Qualitative-Research-Paper-August-20142.pdf

CEO EMOLUMENTS DETERMINATION OF NON GOVERNMENTAL ORGANISATIONS (NGO) IN MELAKA: A CONCEPT PAPER

Masita Hassan*, Norlinda Tendot Binti Abu Bakar**, Afzan Sahilla Mohd Amir Hamzah**, Norfaezah Mohd Shahren**
Faculty of Accountancy, Universiti Teknologi MARA Melaka,Faculty of Business & Management , Universiti Teknologi MARA Melaka***

ABSTRACT

Non-governmental organisations (NGOs) are formed by mutual members in conducting activities to enhance social welfare of its members and public. The issue of governance of NGO arises when such social welfare fail to be achieved. Principal-agent theory explains that agent (elected CEO or key committees) is responsible to carry out NGO activities aligned with principal's interest (which described in NGO missions and goals). If CEO or key committees are being paid to govern NGO, emolument is primarily perceived as a signal to level of governance of NGO. This study objectively to determine factors that influence the decision of emoluments paid to elected CEO or key committees. Subsequently, this study is to analyze the level of governance of NGO. NGO registered in Melaka will be selected from Registrar of Societies (ROS) in Melaka. Therefore, data for this study will be collected using data from ROS Melaka. 594 registered NGO that located in Melaka will be contacted and interviewed to complete structured questionnaire. Data collected will be analyzed using SPSS software to determine the factors that influences emolument decision. Subsequently, data will be analyzed to determine the level of governance of NGOs in Malaysia, which related to emoluments decision making.

Keywords: NGO, CEO Emoluments, Organisation Categories, Board Composition Indicators And Board Individual Characteristics

INTRODUCTION

Non-governmental organisations (NGOs) are definitely important in filling the gap that government and private companies offering to public. NGOs are non-profit orientated organizations normally would not make any profit. Any access of funds generated from its activities, it will be ploughed back into the organisation. Source of funds (also known as revenues) for NGO are from donors and voluntary contribution to individual, other organisations or government agencies. Revenues and profit generated are then will be used to finance NGO activities including the payment of emoluments to its Board of directors (BOD) or its committee members.

It is common practice that emoluments are given to the Board of directors or its committee members. There is no legal statement in Malaysia that prevents such practice because emoluments given are as incentives and appreciation to the Board of directors (BOD) or its committee members in spending their time to manage NGO. An emolument is accepted as an alignment of "principle-agent relationship between managers and stakeholders" (Barros & Nunes, 2007). "Agent" would be the Chief Executive Officer (CEO) or managing director of the NGO and "principal" is NGO members or stakeholders. Large and medium sizes of NGOs need skilful managers and competent support staff to ensure its efficiency in management and strive to achieve the NGO's vision and mission. Nevertheless, it is a normal assumption of public that emoluments are the easiest way for villains to divert charitable funds into their own pockets (Abigail Barr & Owens, 2005)

*Corresponding author's email: hmasita@melaka.uitm.edu.my

In Malaysia, there are cases of irregularities on the financial management of NGO including disappearance of funds that amounting to millions of Ringgit. For example, the case of Malaysian Indian Education Development (MIED) involving missing of funds amounted to RM4 million (Anonymous, 2010, May 12). Members of the organisation and also the public will definitely doubt the ability of organisations' board in managing the organisation fund and governing the organisation towards its mission. Thus, it is essential to study Board of directors (BOD) or its committee members' emoluments in order to justify the management ability to run the activities. Hence, the study can conclude the level of good governance of the NGO. To fill up the gap, two objectives are set for this study. First, to document the determinant factors of the NGO's Board of Directors (BOD) or key committees' emoluments and the second objective is to analyse relationship of BOD or key committees' emoluments to the governance of the NGO.

The Society Act 1966 does not provide guideline on how to determine emoluments to their BOD or key committees. The absence of such guideline is giving out an opportunity to the NGO's BOD or key committees to commit an act of conflict of interest. The BOD or key committees are tended to follows their personal interest rather than the organisational goals. The example of Malaysian Islamic Economic Development Foundation's (YAPEIM) scandal (Anonymous, 2015, Nov 23) that claims the NGO has paid excessive allowances to its top management personnel arise doubt on ability of NGO determining the appropriate emoluments.

.

LITERATURE REVIEW

Emoluments and Management of NGO

An emolument is formally used to define money or another form of payment for the work done. Emoluments are inclusive salary, allowances, profit, salary, or fees from office or employment; compensation for services. Compensation to the BOD or committee members is a debatable issue which is reflecting the corporate governance. As founded by Alregab (2015), the higher the compensation that is paid to CEOs, the poorer the market performance expected and she confirms that most powerful CEOs may intervene in setting their compensation contracts, which contradicts corporate governance requirements.

Allowances for example, are inclusive food allowances and travel allowances remitted for BOD or committee members' expenses to carry out their duties or tasks. However, as explained in the problem statement, emoluments expenditure is the easiest way for dishonest BOD or committee members to divert NGO funds into their own pockets (Abigail Barr & Owens, 2005). Consequently, emolument is should be the first item to be audited in order to verify its integrity and good governance. It is fundamental to ensure good governance practiced in an organisation. However, it will be difficult to differentiate dishonest behaviour or incompetence staff or BOD or committee members if NGO has poor accounting record keeping. Logically, BOD or committee members should not be paid any emoluments, if there are appalling accounting records.

In view of profit oriented organisations, poor governance (such as fraud its own funds or resources) will lead to losses. In the end, the organisations will cease to exist because competitors will take over its customers. Subsequently, only profit orientated organisations that practice good governance will continue their business. Contradictory, inefficient NGO is rarely (or never) being punished. Nevertheless, whenever there is a bad governance of any NGO reported in the press or media, it will eventually affect incoming donors for NGO, even the NGO with very good governance.

Principal-agent relationship model has been widely use in evaluate management of profit-orientated organisation (Mayston, 1993). The model was developed by Jensen and Meckling (1976) presents the most influential framework that transforms the principal-agent relationship. As agency theory describes the conflict relationship between principal and an agent, where the agent (e.g. a manager) has to behave

*Corresponding author's email: hmasita@melaka.uitm.edu.my

on behalf of the principal (e.g. a shareholder). This model explains the mutual agreement between the principal and agent which is giving agent responsibility managing business assets (principally owned the business assets) in order to get profit. The agent (CEO or manager) elected by the principal (NGO members or BOD) to manage NGO assets towards achieving NGO missions and goals. The model will align the interest of principal and agents. Consequently, any discrepancy between the agent's and principal's interest will then create malfunction of NGO governance. In addition, if NGO agent (CEO or manager) is being paid to carry out their duties in managing NGO, their emoluments is the first signal of governance malfunction (Barros & Nunes, 2007). Jensen and Meckling (1976) add that, "If both parties to the relationship are utility maximizers, there is good reason to believe that the agent will not always act in the best interests of the principle".

Variables of Emolument Determinants

Barros and Nunes (2007), highlights that "board compositions, individual characteristics, organization type" and accounting indicators are positively related to CEO emoluments. Accounting indicator such as Return on Assets (ROA) is easily tested for profit orientated organization (Davila & Venkatachalam, 2004). However, NGO donors' collection is more appropriate to be used in this study. Another element that can be used as accounting performance indicators is number of NGO members. Members usually required to pay annual fees, thus could be considered as income to NGO in funding their activities including emoluments to its key committees. Ghosh and Sirmans (2005) have explored the impact of board characteristic on CEO compensation in Real Estate Investment Trusts (REITs) and found that CEO compensation is depends significantly on the usual economic measures of performance including firm size and return on assets. More important, CEO compensation is higher in REITs where the board is weak in monitoring due to large size, and older directors.

ROS in Malaysia categorized NGO into 14 categories. The categories are Religion, Youth, Mutual Beneficiary, Cultural, Occupation, Education, Commercial, Politics, Social & Recreational, Sport, Women, Community Welfare, general and others category. Profit orientated firms' characteristics are positively related to CEO compensations. Ashley & Yang (2004) reviewed that company stability influence compensation. Company size also does matter (Benito & Conyon, 1999). Therefore, it will be relevant to study whether NGO category influences the emolument payment. A study by Fiador (2013) on the other hand, found that the size of the NGOs is most positively influential factor in explaining an NGO's adoption of a governance framework. Another organizational characteristic which is age has received some attention in the governance literature. According to Loderer and Waelchli (2011), as organizations aged, their governance structure deteriorates because the complexity of expansion cause in weaker monitoring and ultimately poorer governance.

The compensation of a CEO receives also will depends, in part, on the types of programs and services offered by the charitable mission. Most of the charity taken place by the NGOs or NPOs. According to www.charitablenavigator.org, the median CEO pay at an educational institution is nearly $88,000 more than paid to the typical religious leader in the year 2013. Balsam and Harris (2015) found that the bonuses are positively associated with profitability, competition, firm size, available cash and the use of compensation consultants and committees, while negatively related to board oversight and charitable nature. The evidence consistent with the payment of bonuses incentivizing non-profit executives as, despite reduce contributions, future performance is positively associated with bonus compensation.

Board composition indicators that are studied by Barros & Nunes, (2007) include political alliance, duration served by the CEO and formal acquaintance of CEO with Board of Directors (BOD). Board composition is also known as a board structure and study by Roshayani et al. (2013) found that an optimum mix of board members matters in ensuring efficient resource strategy and consequently enhance the level of accountability in NPOs. Rose (2006), conducts a multivariate analysis using a sample of listed Danish firms found that the higher proportion of insider ownership increases the lawyer oriented, whereas a higher remuneration increases the business person oriented of the board. If the firm size increases, the international dimension of the board increases and if a firm's experiences less growth board structure becomes more traditional. Nevertheless, Kyereboah and Biekpe (2007) found a negative

*Corresponding author's email: hmasita@melaka.uitm.edu.my

relationship between firm age and the size of the board and the proportion of outside directors. The negative relationship with outside directors imply poor governance while the negative relationship between firm age and board size can be construed as ambiguous since the literature has not yet reached a consensus on which is best for good governance in terms of the larger or small board. However, a study by de Andres-Alonso et al. (2010) based on a sample of Spanish foundations, showed that the board size (board composition) and independence do not have a definitive effect on the entity's efficiency. Instead, the knowledge, diversity inside the boardroom and the active character of trustees has a positive influence on resource allocation.

Brickley, Horn, & Wedig (2010) also highlights board composition regarding emoluments decision making. This only can be established if BOD has the power to elect the CEO. A political alliance does not necessarily means a relationship with politicians. However, it includes the external support received by CEO in determining his/her emoluments (Elhagrasey, Harrison, & Buchholz, 1999). External support could be members of an NGO or high profile stakeholders. Nonetheless, some NGO's patron could influence management of the NGO because members of NGO highly valued patron's advices. Thus, it should be included in the study.

According to Hearn (2013), director's self-reward or compensation is measured through fixed base salary as well as total remuneration which provide a conservative estimate of the full private benefits of control of the directors. Hence, the study found that the larger board sizes have less effective governance mechanisms while enhanced expropriation of private benefits of control is closely associated with lower government promotion of private sector policies, media and analyst freedom and corruption control.

Individual characteristics are also influenced emolument payment to CEO. Public or NGO members' perceptions on CEO experiences, educations and capabilities could influence in determining CEO emoluments (Marchetti & Stefanelli, 2009). While study by Brick et al. (2006) found that relating future firm performance to the portions of a CEO and director compensation that are not related to firm characteristics, CEO characteristics or other governance variables. Logically, those who posses higher qualification are deemed perceived as more capable in managing the organisation compared to lower qualified personnel.

Moreover, Peni (2014) examines the association between the characteristics of the CEO and the Chairperson of the board and the firm performance and found that the age of the CEO or the Chair are mixed, while their experience and quality appear positively related to firm performance. Duc Vo and Thuy Phan (2013) also underlined that the elements of working experience of the board members and the compensation of board members have positive effects on the performance of the firms, while board size has a negative effect on the performance of the firms. In addition, the relationship between other board's characteristics such as educational level of board's members and firm's performance cannot be concluded from their study. Whilst Roberts and Zamora (2014) found that the board members who play both oversight and operational roles, is significantly impacted non-profit executive pay practices. This is consistent with Harris, Petrovits, and Yetman (2013), who profess that boards that actively oversee incentive compensation represent good governance. Contrary with Yetman and Yetman (2012), find that although boards can provide good governance, their effectiveness diminishes as board size increases. In fact, Aggarwal et al. (2012) found that the larger the board, the weaker the financials based pay for performance sensitivities.

RESEARCH DESIGN AND DATA COLLECTION

Researchers assumed that NGO registered in Melaka is more likely to fit our study. There are probability that NGO do not give any sort of payment to its key committees. In addition, there will be inactive NGOs but still registered with ROS and appeared in ROS listings. Researchers believe that inactive NGO should not be valid for this study. According to ROS statistics in December 2009, there are 1983

*Corresponding author's email: hmasita@melaka.uitm.edu.my

registered headquarter-NGO in Melaka. Therefore, researchers decided to determine the number of sample to be used in the study is based on rule of thumb principle as suggested by Newman (2003). Consequently, the study will be based on sample of 594 registered NGOs (30% x 1983).

Registered NGO are required to send its annual financial statements and report to ROS. Therefore, data such as board size, total revenue, funding details, and other financial data will be collected from ROS office based in Melaka. However, registered NGO could have delayed meeting ROS deadline in sending their report to ROS. Some details such on remuneration payment details and individual characteristics may not available at ROS office. Hence, second collection data will be conducted through telephone survey using structured questionnaires. Face to face interview will deemed as highly in cost to get such data. On the other hand, posted questionnaires are deemed to have low responses. Thus, telephone interview session is more appropriate to ensure NGOs responses and keeping cost of survey to minimum.

The questionnaire design will be based on the studies by Barros & Nunes (2007). The questionnaires will be divided into few sections including data on background of the NGO selected, board composition indicators and board individual characteristics.

The collected data from the questionnaires will be analyzed to determine whether emoluments paid to CEO or key committees are influenced by organisations categories, board composition indicators and board individual characteristics. Subsequently, data will be analyzed to establish the relationship between governance of NGO in terms of emoluments decision making.

CONCLUSION

The effectiveness operation of an NGO depends on how well the NGO is governed. Since the management primarily depends on CEO directives, it is assumed CEO is compensated according to his/her performance. The CEO individual missions should be aligned with NGO's mission. This study will provide results that vital in assisting NGO to improve its governance especially related to determining emoluments for its BOD and key committee members. Further results would highlight the necessity to develop code of governance of NGO in Malaysia.

References:

Abigail Barr M. F. & Owens T. (2005), The Governance of Nongovernmental organisations in Uganda, *World Development*, 33 (4), 657-679.

Aggarwal, R., Evans M.E., & Nanda, D. (2012). Nonprofit board: Size, performance and managerial incentives. *Journal of Accounting and Economics,* 53: 466-487.

Alregab, H. (2015). Does CEO pay enhance a firm's performance? An empirical investigation of UK listed companies. *Proceedings of the Eight Saudi Students Conference in the UK.* 145-164.

Anonymous, (2010, May 12), *Ex-CEO Denies Deceit Over RM4mil.* Retrieved July 05, 2010, from The Star Online: http://thestar.com.my/news/story.asp?file=/2010/5/12/courts/6237943&sec=courts.

Anonymous, (2015, Nov 23), *3,000 stop monthly contributions to YAPEIM fund*, retrieved Jan 05, 2016, from

Ashley, A. S., & Yang, S. S. (2004). Executive Compensation and Earnings Persistence. *Journal of Business Ethics, 50*, 369-382.

*Corresponding author's email: hmasita@melaka.uitm.edu.my

DOI: 10.1051/shsconf/20173600031

Balsam, S. & Harris, E. E. (2015). Nonprofit executive incentive pay.

Barros, C. P., & Nunes, F. (2007). Governance and CEO pay and performance in non-profit organisation. *International Journal of Social Economics*, 811-827.

Benito, A., & Conyon, M. J. (1999). The Governance of Directors' Pay: Evidence from UK Companies. *Journal of Management and Governance, 3*, 117-136.

Brick et al., (2006), *CEO Compensation, Director Compensation and Firm Performance: Evidence of Cronyism*, Journal of Corporate Finance, 12 (3), 403-423.

Brickley J. A., Horn R. L. & Wedig G. J. (2010). Board composition and nonprofit conduct: Evidence from hospitals. *Journal of Economic Behavior & Organization*, *76* (2), 196-208.

Charity Navigator. (n.d). Retrived on August 2016 from www.charitynavigator.org

Davila, A., & Venkatachalam, M. (2004). The Relevance of Non-financial Performance Measures for CEO Compensation: Evidence from the Airline Industry. *Review of Accounting Studies, 9*, 443-464.

De Andres-Alonso, P., Azofra-Palenzuela, V. & Romero-Merino, M.E. (2010). Beyond the disciplinary role of governance: how bards add value to Spanish foundations. *British Journal of Management,* Vol. 21, 100-114.

Duc Vo & Thuy Phan. (2013). Corporate governance and firm performance: Empirical evidence from Vietnam..

Elhagrasey, G. M., Harrison, J. R., & Buchholz, R. A. (1999). Power and Pay: The Politics of CEO Compensation. *Journal of Management and Governance, 2*, 309–332.

Fiador, V.O. (2013). Determinants of financial governance practices: evidence from NGOs in Ghana. *International Journal of Sociology and Social Policy,* Vol. 33 No. ½ . 114-130.

Ghosh, C. & Sirmans, C.F. (2005). On REIT CEO compensation: Does board structure matter? *The Journal of Real Estate Finance and Economics,* 30:4, 397-428.

Harris, E. C. Petrovits, M.H. Yetman. (2013). The effect of nonprofit governance on donations: Evidence from the revised Form 990. *Working Paper, Rutgers University Camden.*

Hearn, B (2013), *The Impact of Board Governance On Director Compensation In West African IPO Firms*, Research in International Business and Finance, Volume 28, May 2013, Pages 82–104

Jensen, M. C. & Meckling, W. H. (1976). Theory of the firm: Managerial behavior, agency costs and ownership structure. *Journal of Financial Economics,* 3(4): 305-360.

Kyereboah, C.A. & Biekpe, N. (2007). Corporate governance and financing choices of firms: a panel data analysis. *South African Journal of Economics*, Vol.74(4), 670-681.

Loderer, C.F. & Waelchli, U. (2011). Firm age and governance.

Marchetti, P., & Stefanelli, V. (2009). Does the compensation level of outside director depend on its personal profile? Some evidence from UK. *Journal of Management and Governance, 13*, 325-354.

*Corresponding author's email: hmasita@melaka.uitm.edu.my

Mayston, D. (1993). Principals, Agents and the Economics of Accountability in the New Public Sector. *Accounting, Auditing & Accountability Journal, 6* (3), 68-96.

Newman, W. L. (2003), *Social Research Method: Qualitative and Quantitative Approaches*, fifth edition, Allyn and Bacon Publication.

Peni, E. (2014). CEO compensation and Chairperson caracteristics and firm performance. *Journal of Management & Governance*, Vol. 18 (1), 185-205.

Roberts, A.A. & Zamora, V. (2014). Top corporate officers as board members: The impact on nonprofit executive compensation.

Rose, C. (2006). Board composition and corporate governance – A multivariate analysis of listed Danish firms. *Eur J Law Eco,* 21,113-127.

Roshayani Arshad, Noorbijan Abu Bakar & Normah Omar. (2013). Board composition and accountability of non-profit organizations. *The Journal of Applied Business Reserch.* Vol. 29 (4), 1021-1030.

Statistik Pertubuhan dan Laporan. (n.d.). Retrieved December 07, 2010, from Laman Web Rasmi Jabatan Pendaftaran dan Pertubuhan Malaysia: http://www.ros.gov.my/report.php.

Yetman, M.H. and R.J. Yetman. (2012). The effects of governance on the accuracy of charitable expenses reported by nonprofit organizations. *Contemporary Accounting Research,* 29, 738-767.

*Corresponding author's email: hmasita@melaka.uitm.edu.my

PARTIAL LEAST SQUARE APPROACH TO SECOND ORDER FACTOR IN BEHAVIOURAL STUDY OF ACCOUNTING INFORMATION SYSTEM

Mohd Tarmizi Ibrahim*, Saiyidi Mat Roni**

*Accounting Research Institute, Universiti Teknologi Mara, Malaysia**
*Faculty of Accountancy Universiti Teknologi MARA (Melaka) Malaysia***

ABSTRACT

Theories are developed to explain an observed phenomenon in an effort to understand why and how things happen. Theories thus, use latent variables to estimate conceptual parameters. The level of abstraction depends, partly on the complexity of the theoretical model explaining the phenomenon. The conjugation of directly-measured variables leads to a formation of a first-order factor. A combination of theoretical underpinnings supporting an existence of a higher-order components, and statistical evidence pointing to such presence adds advantage for the researchers to investigate a phenomenon both at an aggregated and disjointed dimensions. As partial least square (PLS) gains its tractions in theory development, behavioural accounting discipline in general should exploit the flexibility of PLS to work with the higher-order factors. However, technical guides are scarcely available. Therefore, this article presents a PLS approach to validate a higher-order factor on a statistical ground using accounting information system dataset.

Keywords: Second-order factor, partial least square, PLS

INTRODUCTION

Partial least square (PLS) which was pioneered by Wold (1985), lends its relevance in the theory building. This helps researchers to explain an observed phenomenon into components that can be understood, hence acted upon. Although theories are designed to simplify explanations as why and how things happen, the level of abstractions at which the components in a theoretical model are represented can render the model too complex to evaluate or measure with relatively precision.
In situations where variables are not directly measured, the researchers approximate the components that make up building blocks of a theory. The components, referred as latent variables (LVs) or factors are represented by a conjugations of related indicators. As shown in Figure 1, x_i and x_j are LVs which compose of their respective blocks of indicators $x_1, x_2, \ldots, x_6$. This level of abstraction known as first-order factor.

In some circumstances however, the researchers need understand the observed phenomenon at a higher level. For example, perceived behavioural control (PBC) in theory of planned behaviour by Ajzen (1991) is conceded as two LVs, namely *perceived self-efficacy (efficacy)* and *perceived controllability (control)*, at its lower level and a single factor (i.e., PBC) at a higher-order (see Ajzen, 2002). This higher-order component or also referred as a second-order factor is represented by x_a in Figure 1. At this level, x_i and x_j form the base values that are used to form x_a.

*Corresponding author's email: tarmizi644@bdrmelaka.uitm.edu.my

 DOI: 10.1051/shsconf/20173600032

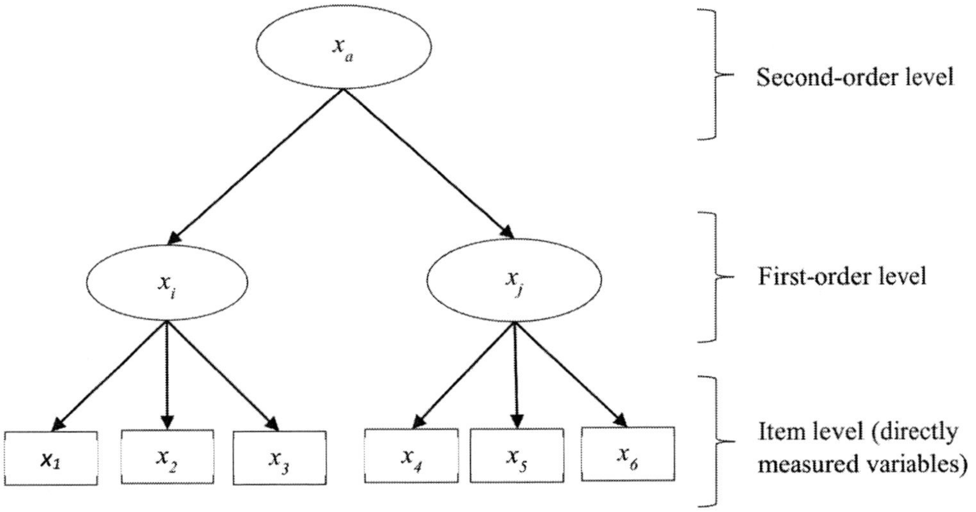

Figure 1.Reflective first- and second-order factors

As LVs can be measured at any level of abstraction (Hair, Black, Babin, & Anderson, 2010), a higher-order LV represents factors that apparently distinct, yet related, lower-order latent variables (Chen, Sousa, & West, 2005). The use of a higher-order in a structural equation modelling (SEM), particularly covariance-based SEM, results in more a parsimonious model.

For interpretative purposes, with sound theoretical backings and sufficient statistical parameters to support an aggregation of the first-order to higher-order factors, the use of second-order factor can provide a new dimension for the researchers to examine the phenomenon of interest. In this article, we use theory of planned behaviour to illustrate how PBC are populated into a single second-order factor which forms a basis for other statistical tests within the context of accounting information system.

Theory of Planned Behaviour

Theory of planned behaviour (TPB) is a widely-used theory used to explain behaviour in many disciplines. TPB consists of five components where the outcome variable is *actual behaviour* (behaviour) that can be predicted directly by *intention* and *perceived behaviour control* (PBC). Intention is further affected by *attitude*, *subjective norm*, and PBC. As mentioned earlier, PBC is a higher-order factor which is derived from a conjugation of efficacy and control. The relationship of these components is illustrated in Figure 2.

*Corresponding author's email: tarmizi644@bdrmelaka.uitm.edu.my

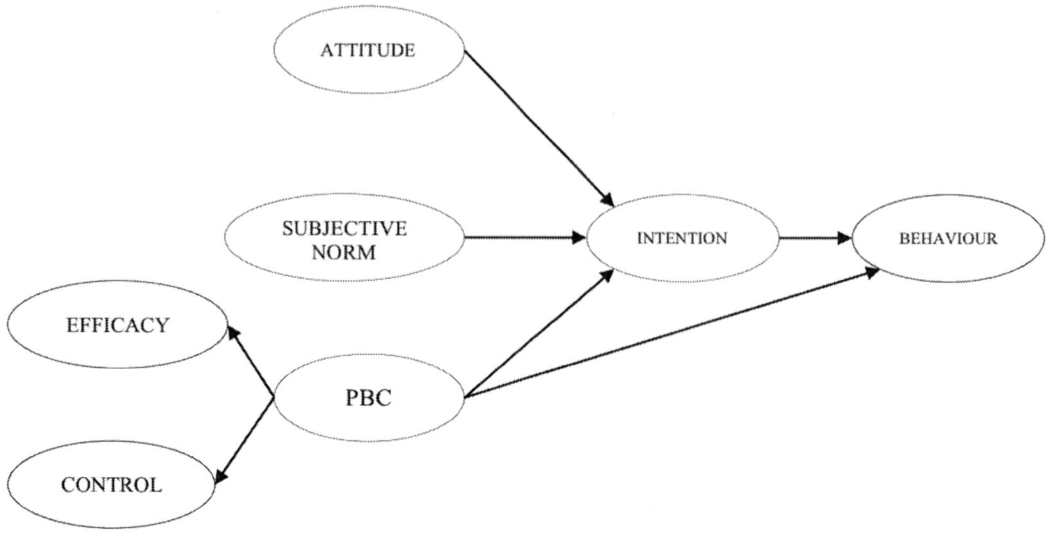

Figure 2. Theory of planned behaviour

The focus of this article is to illustrate how PBC can be unitarily designed at a higher-order level from two lower-order constructs. Therefore, discussion on the theoretical aspect of each component is not demonstrated.

METHODOLOGY

A total of 387 responses received from middle managers from regional industrial estate across Malaysia peninsular. This data are used to illustrate PLS approach to second-order factor. The survey instrument was designed to approximate the respondent's predispositions toward dysfunctional behaviour related to the use of firm's accounting information system. Three main components of theory of planned behaviour (intention, attitude, and subjective norm) were measured using items adapted from prior work. Perceived behaviour control on the other hand, was measured using its lower-order components, efficacy and control, based on five items. All items were put on a 7-point Likert scale. WarpPLS 5.0 was later used to build and test the model with the firs-order and second-order factors. The steps to validate the measurement model are detailed in the following section.

*Corresponding author's email: tarmizi644@bdrmelaka.uitm.edu.my

Table *1* summarises the number of items used to measure each TPB component.

WarpPLS 5.0 was later used to build and test the model with the firs-order and second-order factors. The steps to validate the measurement model are detailed in the following section.

*Corresponding author's email: tarmizi644@bdrmelaka.uitm.edu.my

Table 1. Item measure for construct

Construct/Factor	*N* items	References
Intention (INTENT)	5	Chatterjee (2008) Venkatesh, Morris, Gordon, and Davis (2003)
Attitude (ATT)	2	Chatterjee (2008) Venkatesh et al. (2003)
Subjective norm (SN)	3	Ajzen (1991) Chatterjee (2008) Venkatesh et al. (2003)
Self-efficacy (EFFICACY)	3	Thompson, Higgins, and Howell (1991) Venkatesh et al. (2003)
Controllability (CONTROL)	2	Ajzen (1991)

First-Order Component Validation

Theoretically, an item is said to have sufficient convergent validity when the item measures a latent variable (LV) or factor for which it is designed for. From statistical point of view, this usually means items having high loading (> .50) on their parent factor (Hair et al., 2010; Kline, 2010; Schumacker & Lomax, 2012) and crossload low on other factors give an indication of good convergent validity. The loadings should also be statistical significant as *p*-values are used as validation parameters in a confirmatory factor analysis. Additionally, average variance extracted (AVE) for each LV should exceed a minimum threshold of .50 (Hair et al., 2010; Urbach & Ahlemann, 2010), and a minimum composite reliability (CR) of the construct has to be at .70.

In order to assess discriminant validity of a construct, Fornell and Larcker (1981) suggest that, square-root of AVE of a construct has to be higher than the construct's correlations with other constructs. In addition, variance inflation factor (VIF), a measure of collinearity, is to remain below 5.0, although 3.30 is a normally practiced as a stringent threshold.

The result of the model with the first-order factor shows that the criteria for factor loadings, crossloadings, AVE, CR and multicollinearity VIF test were met. This is summarised in Table 2.

Table 2. Construct correlation matrix and average variance extracted (AVE) for first-order factor

	ATT	SN	EFFICACY	CONTROL	INTENT	VIF
ATT	(.975)					3.070
SN	.772	(.954)				3.291
EFFICACY	.642	.681	(.980)			3.595
CONTROL	.574	.636	.806	(.872)		2.996
INTENT	.770	.772	.711	.643	(.924)	3.433

Square-root of AVE is in brackets ( )on the diagonal
VIF = Variance inflation factor

Table 2 also shows EFFICACY and CONTROL correlation are relatively highly correlated (.81). This is unsurprising given that these LVs are the lower-order factors of PBC. Although they are highly correlated, variance inflation factor (VIF) indicates that these LVs are discernable into two separate constructs. This is consistent with the approach by Greene and D'Arcy (2010), and Djajadikerta, Mat Roni, and Trireksani (2015).

*Corresponding author's email: tarmizi644@bdrmelaka.uitm.edu.my

Second-Order Component Validation

The initial stage produces parameter estimates for the components and also computes their factor scores. The factor score for EFFICACY and CONTROL were then used as the scores to compute the higher-order PBC.

Hair, Black, Babin, Anderson, and Tatham (2006) suggest that a higher-order construct to be assessed in a similar manner as in the lower-order construct structure. Therefore, reliability, discriminant and convergent validity were checked again against acceptable threshold values.

Outputs from this stage show that EFFICACY and CONTROL had significant loading on PBC (loading = .95, $p < .001$). Further, AVE of PBC was .90. Table 3further supports discriminant validity of PBC with the square-root of AVE that was higher than the maximum shared variance among the constructs. Other parameter estimates for ATT, SN, and INTENT were all satisfactory.

Table 3. Construct correlation matrix and average variance extracted (AVE) for second-order factor

	ATT	SN	PBC	INTENT	VIF
ATT	(.975)				3.058
SN	.772	(.954)			3.291
PBC	.640	.693	(.950)		2.273
INTENT	.770	.772	.712	(.924)	3.406

CONCLUSION

The flexibility in partial least square (PLS) allows researchers to advance our knowledge to illuminate patterns in an apparently chaotic phenomenon. This flexibility also provides us lenses to look at a theory at different level of aggregation which gives us a broad understanding of an observed phenomenon.

As shown in the article, perceived behaviour control (PBC) from theory of planned behaviour (TPB) was theoretically vindicated as unitary at a higher-order of abstraction which was a result of a conjugation of two distinct latent variables. Using dataset in behavioural study in accounting information system (AIS), the theoretical argument was confirmed through statistical validations with PLS. When there is a sound theoretical underpinning supporting the existence of lower- and higher-order factors, researchers are encouraged to validate the connotation with statistical evidence. Having examining the models at both levels can enhance discussions and contribute to the pool of existing knowledge.

REFERENCES

Ajzen, I. (1991). The theory of planned behavior. *Organizational Behavior and Human Decision Processes, 50*(2), 179-211. doi:10.1016/0749-5978(91)90020-t

Ajzen, I. (2002). Perceived Behavioral Control, Self-Efficacy, Locus of Control, and the Theory of Planned Behavior1. *Journal of Applied Social Psychology, 32*(4), 665-683. doi:10.1111/j.1559-1816.2002.tb00236.x

*Corresponding author's email: tarmizi644@bdrmelaka.uitm.edu.my

Chatterjee, S. (2008). *Unethical behavior using information technology.* (Ph.D. 3370378), Washington State University, United States -- Washington. ABI/INFORM Complete; ProQuest Central; ProQuest Dissertations & Theses (PQDT) database.

Chen, F. F., Sousa, K. H., & West, S. G. (2005). Teacher's Corner: Testing Measurement Invariance of Second-Order Factor Models. *Structural Equation Modeling: A Multidisciplinary Journal, 12*(3), 471-492. doi:10.1207/s15328007sem1203_7

Djajadikerta, H. G., Mat Roni, S., & Trireksani, T. (2015). Dysfunctional information system behaviors are not all created the same: Challenges to the generalizability of security-based research. *Information & Management, 52*(8), 1012-1024. doi:http://dx.doi.org/10.1016/j.im.2015.07.008

Fornell, C., & Larcker, D. F. (1981). Evaluating structural equation models with unobservable variables and measurement error. *JMR, Journal of Marketing Research, 18*(1), 39.

Greene, G., & D'Arcy, J. P. (2010, 16-17 June 2010). *Assessing the impact of security culture and the employee-organisation relationship on IS security compliance.* Paper presented at the 5th Annual Symposium on Information Assurance 2010, New York.

Hair, J. F., Black, W. C., Babin, B. J., & Anderson, R. E. (2010). *Multivariate data analysis* (7 ed.). Upper Saddle River, NJ, USA: Prentice-Hall, Inc.

Hair, J. F., Black, W. C., Babin, B. J., Anderson, R. E., & Tatham, R. L. (2006). *Multivariate data analysis* (6 ed.). New Jersey: Pearson Prentice Hall.

Kline, R. B. (2010). *Principles and Practice of Structural Equation Modeling, Third Edition* (3 ed.). New York: Guilford Publications.

Schumacker, R. E., & Lomax, R. G. (2012). *A Beginner's Guide to Structural Equation Modeling : Third Edition* (3 ed.). Hoboken: Taylor and Francis.

Thompson, R. L., Higgins, C. A., & Howell, J. M. (1991). Personal computing: Toward a conceptual model of utilization. *MIS Quarterly, 15*(1), 125-143.

Urbach, N., & Ahlemann, F. (2010). Structural Equation Modeling in Information Systems Research Using Partial Least Squares. *Journal of Information Technology Theory and Application, 11*(2), 5-40.

Venkatesh, V., Morris, M. G., Gordon, B. D., & Davis, F. D. (2003). User Acceptance of Information Technology: Toward a Unified View. *MIS Quarterly, 27*(3), 425-478.

Wold, H. (1985). Partial least square. In S. Kotz & N. L. Johnson (Eds.), *Encyclopedia of statistical sciences* (Vol. 6, pp. 581-591). New York: Wiley.

*Corresponding author's email: tarmizi644@bdrmelaka.uitm.edu.my

AR-RAHNU: A REVIEW OF LITERATURE AND FUTURE RESEARCH

Nor Aishah Mohd Ali*, Yang Chik Johari, Muhammad Mukhlis Abdul Fatah
Faculty of Accountancy, Universiti Teknologi Mara Melaka

ABSTRACT

Academicians, practitioners, and other faith-based stakeholders have recognised Ar-Rahnu or Islamic pawn broking as one of micro financing tool to the Muslims. The increasing need for Ar-Rahnu services is influenced by the awareness of the Muslims on the availability of such service and the benefit of using it. In line with this development is academician is growing interest in Ar-Rahnu research, thus create opportunities for interested scholars to assess the development of research in this discipline. This paper purported to review published studies in this discipline since the inception of Ar-Rahnu concept to the current year to identify key trends or issues in Ar-Rahnu research and propose future research agenda. This review, which is the first of its kind, is hoped to be valuable to the present and future Ar-Rahnu scholars interested in the evolution of literature and identifying opportunities for future research.

Keywords*:* Ar-Rahnu, Islamic Pawn Broking, Shariah, Micro Financing Tool

INTRODUCTION

The academicians, practitioners, and concerned stakeholders have acknowledged Ar-Rahnu as an important component of Islamic micro-financing tool. Ar-Rahnu or better known as Islamic pawn broking can be defined as a method of providing short-term financing to a person by pawning his or her gold to the banks or any Islamic pawnshop as a security. It is one of the micro financing facilities that is available for medium and lower income group who are seeking financial assistance to meet two purposes; firstly, for precautionary purposes and secondly, to meet the individual or family's needs. The example of precautionary purpose is whenever a person encounters unexpected situation such as accident and death, which requires them to have cash (Amin and Chong 2011) whilst the needy are those under the lower income group or small businesses with limited capital or excluded from the mainstream financial system (Hisham et al. 2013).

Muslim users consider Ar-Rahnu as an alternative to conventional pawn-broking services, which offer lending money, or instant cash to the needy. Though Ar-Rahnu has existed since 1992 in Malaysia with the establishment of Muassasah Gadaian Islam Terengganu (MGIT) and there is substantial research on Islamic financing product, less attention has been paid to Ar-Rahnu study. The extremely few studies on Ar-Rahnu remain disintegrated, and there is limited of published empirical research on trend or development in Ar-Rahnu research (ARR). The lack of

* Corresponding author's email: aishah72@melaka.uitm.edu.my

such study limits our understanding of the current state, issues, and future direction of ARR. Hence, the main aim of this paper is to review development in ARR. To scrutinize this development, a review of research on Ar-Rahnu was performed. The paper uses a chronological review of the literature (Neely 2016; Taticchi et al. 2010) in the field of Ar-Rahnu and its related areas, as presented in the previous section. The review is drawn from an extensive examination of the breadth of Ar-Rahnu related literature without constraints as to a narrow time frame or limited journal list, although special attention was directed to leading journals in the publication of Islamic accounting, shariah, banking, economics, and finance related study.

METHODOLOGY

As the main objective of this paper is to review development in ARR, the following steps were followed. The first step was to choose a group of relevant journals likely to include a study in Ar-Rahnu and Islamic pawn-broking. Due to the scarcity of published journal articles, the search also included seminar and conference proceedings as well as masters and doctoral dissertations. This study primarily used the library and other academic archives. The second step was to identify the articles with main topics related to Ar-Rahnu. At this point, the paper focuses on articles discussion Ar-Rahnu from theoretical or practical perspectives written in two languages which are English and Malay languages. To perform out this identification, a bibliographic search was conducted using the following keywords: Ar-Rahnu, Islamic pawn broking.

Later, the paper extended the procedure in two ways. First, the study included in the search 'Ar-Rahnu and Islamic microfinance', since it was realised that there are articles looking at both Ar-Rahnu and the microfinance services. Second, the paper examined the bibliography of each of the articles identified by the initial search, to ensure that the possibility of overlooking a significant number of articles in the subject matter is minimised. Then, each article gathered was read and carefully analysed to ensure that it is indeed wholly or partly discussion on Ar-Rahnu. Hence, even though the title, keywords or abstract did not mention Ar-Rahnu specifically, the authors were able to identify the articles related to Ar-Rahnu as exhibited in Appendix 1. The analysis of the development on Ar-Rahnu is mainly looking at the number of articles, research themes, research methods (primarily their data collection methods and research instruments) and research context.

RESULTS AND DISCUSSION

Number of articles

A total of 50 papers were identified consisting of 20 published articles from 16 different journals, 18 international conference papers, five (5) dissertation and seven (7) working papers. The earliest paper included in the dataset was published in 1986 and the most recent in 2016. To maintain the quality of the study, articles published in the predatory journals and articles of information in nature presented in the Ar-Rahnu Regional Secretariat hosed by the Yayasan Pembangunan Ekonomi Islam Malaysia (YAPEIM) in 2002, 2004, 2007 and 2011 are not included in the study. Out of these 50 articles, 16 articles are conceptual in nature where researchers performed library archive. The rest of the article (34) consist of 23 quantitative and 11 qualitative empirical studies respectively. The graphical distribution of the number and types of articles are depicted in the below pie chart.

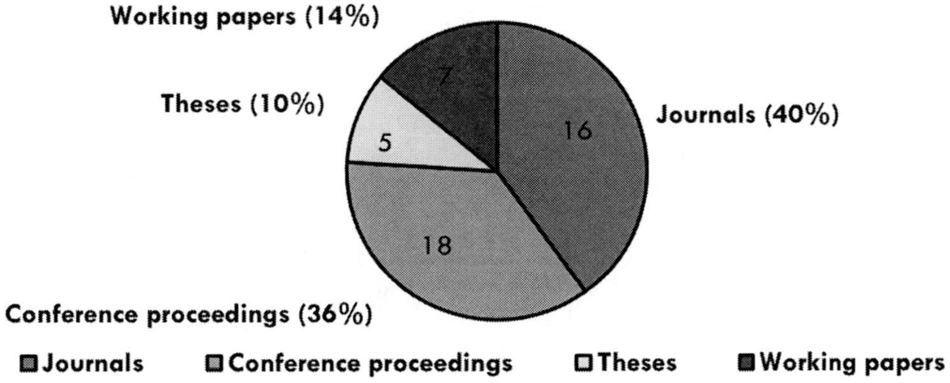

Figure 1: Distribution of articles

It is also evident from Appendix 1 that a few of the studies conducted are dominated by a limited number of researchers i.e. from Southeast Asia particularly Malaysia. Only one (1) study was performed outside the Southeast Asia, which is in India. The questions are why is that so? Why are there only a few contributors to this area? A possible answer is that, as alluded to earlier, many researchers focus on other Islamic financing study and ignore the significance of Ar-Rahnu as a new area to be explored. At that time, too, Ar-Rahnu was almost non-existent or not mandatory in many Muslim countries' jurisdiction.

Perhaps the surge in interest in Ar-Rahnu in the recent decade could be credited to the rise among the Muslim users of Ar-Rahnu services driven by better understanding of Ar-Rahnu concept by the Muslim consumers as well the increase in Ar-Rahnu providers, especially in Malaysia. Hence, it is not surprising that the dominant country context of most studies in Ar-Rahnu is Malaysia. This can be observed in the ensuing presentation of the findings by themes, methodology, and context. In terms of the number of articles published per year, the range was from three (3) to the year 1986 to 2000 to a maximum of 30 articles in the year 2011 and above, giving an average of 1.7 article per year. As presented in Table 1, overall there was an increase in the number of articles during the period examined, with an average of 0.2 articles per year in the period 1986-2000, 1.7 articles per year in the period 2001-2010 and 5 articles per year in the period 2011 - 2016.

 DOI: 10.1051/shsconf/20173600033

Table 1: Number of Ar-Rahnu articles in the period 1991-2016

Period	Number of articles	Average number of articles per year
1986 - 2000	3	0.2
2001 – 2010	17	1.7
2011 - 2016	30	5
Total	**50**	**1.7**

Results by themes

To review the trends in Ar-Rahnu since the inception of Ar-Rahnu scheme in the IFI in 1993, this paper divides the period into three: 1986-2000, 2001-2010 and 2011-2016. From the 50 articles, 16 articles are related to customer's related matters while other themes are micro finance and Ar-Rahnu, Ar-Rahnu efficiency, opportunities and challenges and general studies on Ar-Rahnu. Country cases account for most articles as depicted in Table 2. Figure 2, on the other hand, shows the concentration of the research theme graphically. It seems that customer related issue has become the favourite issue to be studied by most researchers.

Though the conventional pawn broking has been in existence for decades with act governing their operation, such as Pawn Broking Act 1972 (revised in 2004), Islamic pawn broking has only became widely used in the year 1990 onwards. In Malaysia, the Central Bank of Malaysia (CBM) have collaborated with YAPEIM and Bank Kerjasama Rakyat (Bank Rakyat) to offer Islamic pawn broking service in 1993 with its operation in six (6) branches. As for research on Ar-Rahnu, only in 1997, the first conceptual paper specifically on Ar-Rahnu was published in an international journal discussing the role of Ar-Rahnu as an instrument to micro enterprise credit (Ismail and Ahmad, 1997).

As shown in Table 2, the second period (2001-2010) extended the themes with more researchers exploring issues and challenges in providing Ar-Rahnu, efficiency, acceptance, and satisfaction of the Ar-Rahnu services among customers, comparative study between conventional pawn broking and Ar-Rahnu as well as other issues such as Ar-Rahnu model/framework and Ar-Rahnu keeping fee. This period also sees increases in the Ar-Rahnu providers in Malaysia. According to Abdul Razak (2008), Bank Rakyat, which officially started their operation in Ar-Rahnu back in 1993 with only six (six) branches, has expanded to 112 branches nationwide in 2007. This situation shows that the Malaysians acceptance towards Al-Rahnu has increased tremendously and more Malaysians started to choose Al-Rahnu as an alternative to conventional pawnshop. Furthermore, Muassasah Gadaian Terengganu (MGIT), the pioneer to Ar-Rahnu operation in 1992 has also experienced increase in customers, that is, from 12,732 customers in 1992 to 26,632 customers in 2005. Due to the increased number of customers, it is said that the value of pawned items has increased from RM9.85 million in 1992 to RM 28.9 million in the year 2005 (Abdul Razak, 2008). MGIT has also offered the best services to the Ar-Rahnu customers by charging 0% safekeeping fees, which is also in accordance with *Qardh Al-Hasan* model of financing. At the same period, Permodalan Kelantan Berhad also has been set up under Kelantan State Economic Development Corporation in March 1992 to help the needy people and only charged the minimal safekeeping fees to the customers (Abdul Razak, 2008). The findings by Abdul Razak (2008) is also supporting another acceptance study by Taher and Shafiai (2008) who remarked that the acceptance of the consumers in Malaysia keep increasing every year.

DOI: 10.1051/shsconf/20173600033

Table 2: Summary of previous studies by year and themes on Ar-Rahnu

Themes		Years		
		1991 - 2000	2001 - 2010	2011 - 2016
General issues		Wan Abdullah (1999)	Yaacob (2002); Ismail (2004); Abdul Razak (2008)	
Efficiency of Ar-Rahnu services			Maamor and Ismail (2010)	Samsudin (2012)
Microfinance and Ar-Rahnu		Mohamed (1986); Ismail and Ahmad (1997)	(Samsudin 2002)	Abdul Khir et al. (2012); Abdullah (2012); Mokhtar and Zambahari (2013)
Opportunities and Challenges of Ar-Rahnu operation			Bhatt and Sinnakkannu (2008)	Cheong and Sinnakkannu (2012); Hisham et al. (2013); Bahari et al. (2015)
Customer related matters	Acceptance of Ar-Rahnu among the consumers		Mohd. Taher and Shafiai (2008)	Amin (2011); Koe and Abdul Rahman (2012); Mat Noor et al. (2012); Mansor et al. (2014); Abdul Hamid, Abdul Rahman, and Abdul Halim (2014) ;Abdi (2015); Baharum, Maamor, and Othman (2015); Abdul Hamid et al. (2016)
	Profile of customers		Abdul Hamid and Aziz (2003)	Abdul Razak et al. (2013);
	Customer satisfaction		Sanusi and Johari (2008)	Wan Daud (2011); Othman, Hashim, and Abdullah (2012)
	Customer benefits			Mohd Azli, Palil, and Noor (2016)
Jurisdiction and Shari'ah governance of Ar-Rahnu				Mulazid (2012); Rasmin and Markom (2014); Shah and Yaacob (2016)
Study on difference between conventional and Islamic pawn broking			Mohamed Naim (2004)	Abdul Razak (2011); Yaacob (2014)
Other matters - Ar-Rahnu framework - Ar-Rahnu model			Khan and Nisar (2004); Ismail and Sanusi (2005); Ibrahim and Salleh (2006); Amin et al. (2007);Mohamed, Ismail, and Maamor (2007); Mohamad and Salleh (2008); Mohd.Taher et al. (2010)	Yaacob, Ahmad, and Ibrahim (2012); Othman, Hashim, and Abdullah (2013); Sharif et al. (2013); Ezahar (2014); Razak and Muhammad (2014);Abdul Razak et al.(2014)

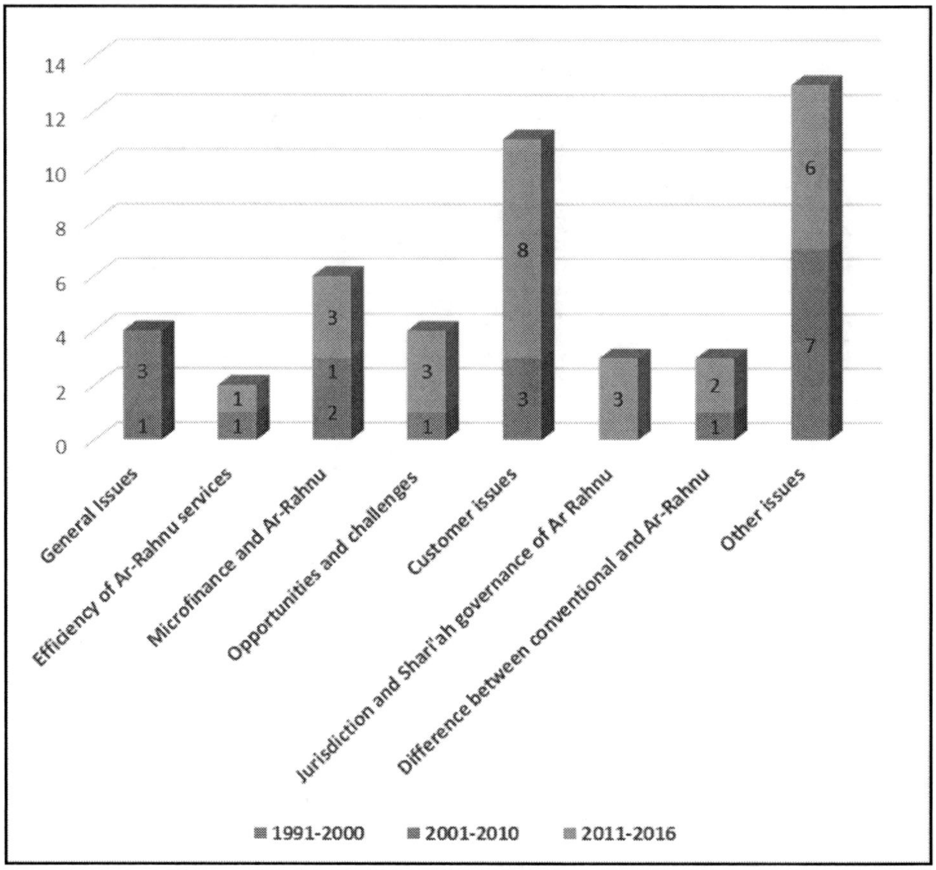

Figure 2: Distribution of articles by year and themes

On a different theme, some researchers of Ar-Rahnu have commented on the Ar-Rahnu model practised in Malaysia. Bhatt (2008) elaborated the four (4) concepts within the Ar-Rahnu operation, which are *Qaardhul Hasan* (benevolent loan), *Wadiah Yad Amanah* (trustworthiness), *Al-Ujrah* (safekeeping) and *Wadiah Yad Dhamanah* (guaranteed safekeeping). *Qardh Al-Hasan* is an interest free loan, which is granted to the applicant who wishes to pawn his valuable items such as gold. Under this concept, the borrower is required to pay the same amount that he has borrowed to redeem the valuable items at the maturity date of the agreed period. *Wadiah Yad Amanah,* on the other hand, requires the borrower to produce returnable collateral to ensure the repayment of the loan. If something happens to the collateral items, the lender has no responsibility to replace the missing valuable items. *Al-Ujrah*, on the other hand, is a concept where lenders can charge the safekeeping fees to the borrower to ensure the collateral items will be in good condition. The lending institutions take precautionary measure by providing security and insurance to ensure the collateral items will be retuned back to the borrower safely after the borrower has settled his debt. These safekeeping fees are called as *Al-Ujrah*. Lastly, the *Wadiah Yad Dhomanah* refers to the safekeeping with a guarantee. In this situation, the lender will be

DOI: 10.1051/shsconf/20173600033

responsible for replacing the collateral items if the items are missing or being stolen by somebody.

The differences between conventional and Islamic pawn shop also been highlighted by Mohamed Naim (2004) and Bhatt (2008). The customers who chose Ar-Rahnu to conventional pawnshop due to few reasons. Ar-Rahnu is argued to be based on shariah principles, which avoid any interest-based transaction. Besides, it is easier to apply with fast approval, and the cost is considered cheaper than conventional pawn broking. In addition, the storage fees are fixed, and the customers can enjoy longer financing terms with no penalties imposed on them. Finally, if the collateral items that failed to be redeemed will be sold to the third parties and the surplus of the sales will be returned to the borrower (Bhatt, 2008; Mohamed Naim, 2004).

The first study on Ar-Rahnu efficiency was conducted by Maamor & Ismail (2010) and expanded by Samsudin (2012). The study by Maamor & Ismail (2010) across several Ar-Rahnu operators, highlighted that Ar-Rahnu could be a medium for providing funds for the micro and small enterprise. In this case, the efficiency of Ar-Rahnu is being studied and the factor that effects the efficiency Ar-Rahnu also being monitored. Efficiency, to these researchers, was found to be affected by variables such as the firm size, financial self- sufficiency, subsidy, and a group of business. Samsudin (2012), on the other hand, studied the efficiency of Ar-Rahnu scheme to assist the small enterprise in four South-eastern countries which are Malaysia, Brunei, Thailand, and Indonesia revealed that Ar-Rahnu is a medium used by lower income groups as a micro financing tools to earn cash due to the financing limitation imposed by the financial institution. Normally, people tends to use Ar-Rahnu as a micro financing tools due to the reason that financial institution imposed strict regulations to the public who needs the bank's financing.

In the final period (2011 to 2016), recurring themes includes microfinance and Ar-Rahnu, opportunities and challenges, acceptance of Ar-Rahnu, customers' satisfaction, difference between conventional and Ar-Rahnu and customers' acceptance on Ar-Rahnu. These themes are complemented by a new theme area of study related to ARR, which is the jurisdiction, and shari'ah matters related to Ar-Rahnu. Based on the review, most of the researchers focused on the acceptance of Ar-Rahnu among the consumers as their research theme. This studies were undertaken by Amin (2011); Koe, Zalindah, and Rahman (2012); Noor et al. (2012); Mansor et al. (2014); Rasmin and Markom (2014); Abdul Hamid, Abdul Rahman, and Abdul Halim (2014); Abdi (2015); Baharum, Maamor, and Othman (2015) and Hamid et al. (2016) respectively.

The research performed by Amin (2011) focus on an empirical study on the perspective of Muslimah in using Ar-Rahnu in eastern Malaysia particularly in Labuan and Kota Kinabalu. The study found out that religious obligation, transaction cost, and information of Ar-Rahnu were instrumental in determining the use of Ar-Rahnu in Eastern Malaysia. Subsequently, Koe and Abdul Rahman (2012) emphasized on five factors that influence the customers' acceptance and intention to use Ar-Rahnu that is attitude, social influence, religious obligation, and attractive price. Mansor et al. (2014) have conducted the research specifically the demographic elements in capturing the customers' acceptance towards Ar-Rahnu offered by MAIDAM in Dungun, Terengganu. The most recent study on customer by Mohd Azli et al. (2016) discussed how the Ar-Rahnu services benefitted the customers. The empirical study which propose a relationship value model of Ar-Rahnu transaction found that the dimension of benefits in Ar-Rahnu in the form of quality service, gold investment, business capital, transparent, staff competency, and product feature have a positive and significant effect on long-term relationship between Ar-Rahnu customer and institutions.

On the other hand, some researchers focused on microfinance theme and Ar-Rahnu such as Abdul Khir et al. (2012); Abdullah (2012); Mokhtar and Zambahari (2013). According to Abdul Khir et al. (2012), Ar-Rahnu is essential in micro credit facility in competing for the conventional pawnshop but at the same time, the shariah issues should not be compromised. Mokhtar and Zambahari (2013), however, suggested that Ar-Rahnu could be the most suitable form of micro financing to cater the needs of individuals and small businesses, which has difficulties in obtaining the funding from financial institutions due to the restriction imposed.

Furthermore, Cheong and Sinnakkannu (2012); Hisham et al. (2013); Bahari et al. (2015) focused on opportunities and challenges faced by Ar-Rahnu operators in Malaysia. Ar-Rahnu definitely has opportunities in Malaysia due to cost efficiency and religion, but there are many challenges need to overcome before it can be one of the best medium to improve the quality of life. Mulazid (2012); Nur Hayati Rasmin and Markom (2014); Shah and Yaacob (2016) highlighted on jurisdiction and shariah related matters in Ar-Rahnu. In this situation, the researchers conduct the study to determine the sufficiency of the regulation made by the internal parties to protect both parties in case of dispute or the need for specific laws to govern this Ar-Rahnu practice in Malaysia.

In summary, Ar-Rahnu research began with general issues and Ar-Rahnu operation within the Islamic micro-financing tool and then focus on efficiency, opportunities, and challenges of conducting Ar-Rahnu as well as customer's related matters. In 2010 onwards, the dominant theme was the acceptance of Ar-Rahnu among the consumers. Throughout, researchers in the first and second phase have been discussing on various issues related to AR operation. However, learning from issues and challenges, more studies are needed exploring new themes to tackle issues pertaining to standardized regulation for Ar-Rahnu and shari'ah-compliant Ar-Rahnu towards achieving Maqasid As-shari'ah.

RESULTS BY METHODOLOGY AND CONTEXT

Early studies were merely discussion on the history of Ar-Rahnu services, and how it is being operated in Malaysia (as Malaysian scholars are dominating the articles published), in addition to the literature review discussion on the specific areas discussed. The first empirical study was only noted the second period specifically in 2003 by Abdul Hamid and Abdul Aziz, using a questionnaire to determine the profile of the pawn broking providers. This is followed by another study by Mohamed Naim (2004), who performed a qualitative study by way of content analysis, on the issue of the difference between the conventional pawn broking and Islamic pawn broking (IPB) or Ar-Rahnu. Then, most of the studies were conceptual using the same method of content analysis (Ismail & Sanusi, 2005; Ibrahim & Salleh, 2006; Bhatt & Sinnaknu, 2008 and Mohd Taher et al., 2010). Few other studies that used quantitative method i.e. questionnaires distribution was performed by Amin et al. (2007) and Maamor & Ismail (2010). In the third period (2011 to 2016), studies that are more empirical have been performed with a concentration on the topic of AR related to the customers. The studies were mostly using quantitative method through questionnaires survey (Abd Razak et al., 2013; Abd Hamid et al., 2014; Baharum et al., 2015, and Abd Hamid et al., 2016).

16 out of 27 qualitative studies performed are using content analysis where researchers wrote conceptual paper based on their library research on the verses from Al-Quran, hadith, the relevant acts, and prior literature (Ismail, 1997; Ismail and Sanusi, 2005); Ismail, 1997 and Hisham et al., 2013). Other non-content analysis qualitative studies used the interview as the medium for data gathering whilst all quantitative studies (22) were performed using questionnaires that ware distributed to the users of Ar-Rahnu services.

CONCLUSION, FUTURE RESEARCH, AND LIMITATIONS

In conclusion, prior studies tend to focus on the same issues of customers' acceptance and satisfaction as well as Ar-Rahnu and microfinance issue. Throughout the review, some theories and models that have been used by the researchers to support the study were service quality model by Wan Daud (2011), hierarchy of effect model (Yaacob, 2014) and theory of reasoned action (Abdi, 2015) which are all Master theses. Future studies may venture into comparative analysis between Ar-Rahnu and other informal finance as a micro-finance provider to the medium and lower income groups. Furthermore, comparative analysis between lending from Ar-Rahnu and the personal loans offered by banks can also been focused besides comparative study on the operation and performance of Ar-Rahnu among the Muslim countries as well as extending the previous studies on standardising the legislation between the Islamic pawn-broking service providers.

The review of studies presented here are not exhaustive but merely a preliminary attempt to initiate further research in the area so that Ar-Rahnu will not be part of forgotten market. The limitation of the study is that it utilised limited empirical studies. The substantial material may be available in Arabic or other languages which if analysed could have yielded further insight into the development of Ar-Rahnu. Notwithstanding this limitation, this study offers insight and practical help to scholars in the field and novice researcher on this topic by formulating a research agenda for subsequent study. Finally, this paper is invaluable to those Ar-Rahnu professionals who wish to gain a better understanding of the current situation.

REFERENCES

Abdi, Mohamed Farah. 2015. "Determinants of Customer's Intention to Use Islamic Pawn Broking (Ar-Rahnu) Scheme: A Case Study of (UPSB) Staffs in University Utara Malaysia." Master thesis, Universiti Utara Malaysia.

Abdullah, Rose. 2012. "The Roles of Micofinance in Brunei Darul Salam." In *Proceeding Konvensyen Ar-Rahnu Serantau 2012*, , 273–96.

Abdullah, Wan Abd Rahman Khudzri Wan. 1999. "Historical Development of Pawning Practice in Malaysia." *Jurnal Pembangunan Sosial* 1: 66–80.

Amin, Hanudin. 2011. "Modelling Ar-Rahnu Use in Eastern Malaysia: Perspectives of Muslimah." *Journal of Islamic Banking and Finance* 7(3): 63–76.

Azli, Rafidah Mohd, Mohd Rizal Palil, and Shifa Mohd Noor. 2016. "Effect of Ar-Rahnu Benefits on Long-Term Relationship Between Customer and Institutions." In *2nd International Conference on Islamic Perspective of Accounting, Finance, Economics and Management*, Yogjakarta.

Bahari, Nor Fadilah, Shafina Faisal, Wan Shadila Shahar, and Tajul Azli Shahadan. 2015. "The Development Of Islamic Pawnbroking In Malaysia , Application And Its Challenges." In *First International Conference on Economics and Banking (ICEB-15)*, , 230–38.

Baharum, Nadiah Nabilah, Selamah Maamor, and Azizah Othman. 2015. "Examining The Factors That Influence Customer ' S Intention To Use Ar-Rahnu At Post Office : A Case Study In Kedah." *Journal of Islamic Economics, Banking and Finance* 11(4): 86–99.

Bhatt, Payal, and Jothee Sinnakkannu. 2008. "International Islamic Finance Conference 2008 Peer Reviewed Paper Ar-Rahnu." In *6th International Islamic Finance Conference 2008*,.

Cheong, Calvin W. H., and Jothee Sinnakkannu. 2012. "Ar-Rahnu : Opportunities and Challenges in Malaysia." *SSRN* 2112809: 1–11.

Daud, Wan Noraini Wan. 2011. "Kepuasan Pelanggan Dan Kualiti Perkhidmatan Skim Ar-Rahnu." UUM.

Ezahar, Raudzatul Jannah. 2014. "A Study of Islamic Practice at Financial Institutions." UUM.

Fairooz, Mohamed, and Abdul Khir. 2012. *Critical Appraisal of the Rahn-Based Islamic Microcredit Facility*.

Hamid, Mohamad Abd, Ishak Abd Rahman, Ahmad Nafis, and Abd Halim. 2016. "Key Factors Influencing Customers to Use Ar-Rahnu (Islamic Pawn Shop) in Malaysia: Evidence from Bank Rakyat." *The Macrotheme Review* 5(1): 108–18.

Hamid, Mohamad Abdul, Ishak Abdul Rahman, and Ahmad Nafis Abdul Halim. 2014. "Factors Affecting the Acceptance on Ar-Rahnu (Islamic Based Pawn Broking): A Case Study of Islamic Banking in Malaysia." *The Macrotheme Review* 3(4): 22–35.

Hamid, Shaari Abd, and Azlina Abdul Aziz. 2003. "Development of Islamic Pawn-Broking Services: Differentiating Profiles of Their Respective Patrons." In *International Islamic Banking Conference 2003*, , 1–12.

Hisham, S, S Abdul Shukor, A B Ummi Salwa, and Kamaruzaman Jusoff. 2013. "The Concept and Challenges of Islamic Pawn Broking (Ar-Rahnu)." *Middle-East Journal of Scientific Research 13* 13: 98–102.

Ibrahim, Uzaimah, and Safinar Salleh. 2006. "The Objective of Al-Rahn and Their Achievement in Charge/Mortgage in Islamic Home Financing: An Analysis." In , 8–10.

Ismail, Abdul Ghafar. 2004. "Skim Ar-Rahnu: Ke Arah Pengukuhan Kerjasama Serantau Pengalaman Malaysia." In *Proceeding Konvensyen Ar-Rahnu Serantau 2004*, 69–86.

Ismail, Abdul Ghafar, and Nor Zakiah Ahmad. 1997. "Pawnshop as an Instrument of Microenterprise Credit in Malaysia." *International Journal of Social Economics* 24(11): 1343–52.

Ismail, Abdul Ghafar, and Nur Azura Sanusi. 2005. "A Framework for Regulating Pawnshops Why Do, What Areas and Syariah View." In *Malaysian Finance Association (MFA)'s 7th Annual Conference*, , 1–10.

Khan, Javed Ahmed, and Shariq Nisar. 2004. "Collateral (Al-Rahn) as Practiced by Muslim Funds of North India." *Journal of King Abdul Aziz University : Islamic Economics* 17(1): 17–34.

Koe, Wei-loon, Nor Zalindah, and Abdul Rahman. 2012. "The Use of Ar-Rahnu by Islamic Bank Customers in Malaysia." In *International Conference on Science, Technology and Social Science (ICTSS) 2012*,11–18.

Maamor, Selamah, and Abdul Ghafar Ismail. 2010. "The Ar-Rahnu Efficiency and Its Determinants." *Journal of Islamic Economics, Banking and Finance* 6(1): 105–26.

Mansor, Norudin, S A Bashir Ahmad, Semanat Abu Bakar, and Irwan Ismail. 2014. "A Demographic Analysis on Customer Acceptance towards Islamic Pawn Broking in Malaysia." *Asian Social Science* 10(2): 27–36.

Mohamad, Shamsiah, and Safinar Salleh. 2008. "Upah Simpan Barang Dalam Skim Ar-Rahnu : Satu Penilaian Semula." *Jurnal Fiqh* 5: 47–65.

Mohamed, Nafisah, Abdul Ghafar Ismail, and Selamah Maamor. 2007. *Competition and Outreach in Ar-Rahn Industry*.

Mohamed, Rugayah. 1986. "The Role of Pawnbrokers as Non-Institutional Creditors in Malaysia." *Kajian Ekonomi Malaysia* 22(2): 30–39.

Mokhtar, Imani, and Shah Rizal Zambahari. 2013. "Ar-Rahnu: A Short Term Financing Alternative." In *5th Islamic Economics System Conference (iECONS 2013)*, , 4–5.

Mulazid, Ade Sofyan. 2012. "Kedudukan Sistem Pegadaian Syariah Dalam Sistem Hukum Nasional Di Indonesia." *Innovatio* XI(2): 293–322.

Naim, Asmadi Mohamed. 2004. "Sistem Gadaian Islam." *Islamiyyat* 26(2): 39–57.

Neely, Andy. 2016. "The Evolution of Performance Measurement Research – Developments in the Last Decade and a Research Agenda for the Next." (June).

Noor, Norfaizah Mat, Maimum Abdullah, Noraina Ismail, Rositah Bakar, and Siti Sarah Mohd Yusni. 2012. "A Study of Customer Awareness toward Ar-Rahnu Scheme at Sg Buloh." In *AFBE 2012 Conference Paper (UNITEN)*, , 268–79.

Othman, Azizah, Norashidah Hashim, and Syahrina Abdullah. 2012. "Customer Satisfaction Levels of Service Quality at Ar-Rahnu, Kelantan." In *3rd International Conference on Business and Economic Research (3rd ICBER 2012) Proceeding*,.

———. 2013. "Perkembangan Ar-Rahnu Di Terengganu : Kajian Kes Terhadap Ar- Rahnu Majlis Agama Islam Dan Adat Melayu Terengganu (MAIDAM)." In *Prosiding PERKEM VIII*, , 951–59.

Rasmin, Nur Hayati, and Ruzian Markom. 2014. "An Overview on Implementation of Ar-Rahnu in Malaysia." In *Prosiding PERKEM IX*, , 561–70.

Rasmin, Nut Hayati, and Ruzian Markom. 2014. "Ar-Rahnu Governance Laws in Malaysia: An Analysis in Banks and Non-Banks Financial Institutions." In *International Conference of Global Islamic Studies 2014*, , 116–27.

Razak, Azila Abdul. 2008. "Malaysian Practice of Ar-Rahnu Scheme: Trends and Development." In *Malaysian Study of Islam*,.

———. 2011. "Economic and Religious Signi Cance of the Islamic and Conventional Pawnbroking in Malaysia : Behavioural and Perception Analysis."

Razak, Azila Abdul, and Fidlizan Muhammad. 2014. "Kemapanan Dan Daya Saing Pajak Gadai Islam Di Malaysia." In *Prosiding PERKEM IX*, , 551–60.

Razak, Azila Abdul, Fidlizan Muhammad, Mohd yahya Mohd Hussin, Suraini mohd Rhouse, and Emilda Hashim. 2013. "Peranan Institusi Pajak Gadai Islam Dalam Pembangunan Sosio-." In *Prosiding PERKEM VIII*, , 1352–63.

Samsudin, Abibullah. 2002. "Skim Ar-Rahnu: Satu Model Mikro Kredit Di Malaysia." In *Proceeding Konvensyen Ar-Rahnu Serantau 2002*, , 1–12.

———. 2012. "Kajian Keberkesanan Skim Ar-Rahnu Dalam Memantap Dan Memperkukuh Usahawan Kecil Di Negara2 Serantau." In *Proceeding Konvensyen Ar-Rahnu Serantau 2012*, , 353–84.

Sanusi, Nur Azura, and Mohamad Shukri Johari. 2008. "Permintaan Perkhidmatan Pajak Gadai: Perspektif Pengguna." *Malaysian Journal of Consumer and Family Economics*: 20–29.

Shah, A A, and S E Yaacob. 2016. "Critical Literature on Governance and Supervision of Ar-Rahnu in Malaysia)." *Journal of Advanced Review on Scientific Research* 18(1): 26–42.

Sharif, Dziauddin, Amir Shaharuddin, Nurul Aini Muhamed, Nasif Sidquee Pauzi, and

Mohamad Zaid Mohd. 2013. "The Improvement of Ar-Rahn (Islamic Pawn Broking) Enhanced Product in Islamic Banking System." *Asian Social Science* 9(2): 36–47.

Taher, Muhammad Saiful Islamic Mohd, and Muhammad Hakimi Mohd Shafiai. 2008. *Faktor Penerimaan Pelanggan Terhadap Skim Ar-Rahnu Di Kedai Ar-Rahnu, Bank Rakyat Alor Setar Kedah.*

Taher, Muhammad Saiful Islamic Mohd, Siti Aisyah Yusof, and Norridzwan Abidin. 2010. "Prinsip Al-Rahn Sebagai Asas Pelaksanaan Skim." *Voice of Academia* 5(1): 1–16.

Taticchi, Paolo, Paolo Taticchi, Flavio Tonelli, and Luca Cagnazzo. 2010. "Performance Measurement and Management : A Literature Review and a Research Agenda Performance Measurement and Management : A Literature Review and a Research Agenda." *Measuring Business Excellence* 4(1): 4–18.

Yaacob, Balqis. 2014. "Faktor-Faktor Yang Mempengaruhi Kesedaran Usahawan Kecil Muslim Terhadap Skim Ar-Rahnu."

Yaacob, Matahir. 2002. "Membangun Golongan Pendapatan Rendah Menerusi Pajak Gadai Islam - Mikro Kredit Serantau." In *Proceeding Konvensyen Ar-Rahnu Serantau 2002*, , 13–32.

Yaacob, Mohd Rafi, Ghazali Ahmad, and Mohamed Dahlan Ibrahim. 2012. "Developing a Customer's Oriented Ar-Rahnu (Islamic Pawn Broking) Model for Cooperatives in Microfinancing in Malaysia." In *Proceeding The 13th Malaysia Indonesia Conference on Economics, Management and Accounting (MIICEMA) 2012.*

Appendix 1: List of Ar-Rahnu related literature

No	Author(s)	Title	Year	Published/Unpublished
1	Mohamed	The Role of Pawnbrokers as Non-Institutional Creditors in Malaysia	1986	*Kajian Ekonomi Malaysia*
2	Ismail & Ahmad	Pawnshop as an Instrument of Microenterprise Credit in Malaysia	1997	*International Journal of Social Economics*
3	Wan Abdullah	Historical Development of Pawning Practice in Malaysia: From Pajak Gadai to Al Rahn	1999	*Jurnal Pembangunan Sosial*
4	Yaacob	Membangun Golongan Pendapatan Rendah menerusi pajak gadai Islam – Model Mikro Kredit Serantau (Developing Lower Income Group through Islamic Pawn broking – Regional Micro Credit Model)	2002	Proceeding of Konvensyen Ar-Rahnu Serantau 2002
5	Samsudin	Skim Ar-Rahnu: Satu Model Mikro Kredit di Malaysia (Ar-Rahnu Scheme: A Micro Credit Model in Malaysia)	2002	Proceeding of Konvensyen Ar-Rahnu Serantau 2002
6	Abd Hamid & Abd Aziz	Development of IPB services – Differentiating Profiles of their respective patrons	2003	Proceeding at Islamic Banking Conference
7	Mohamed Naim	Sistem Gadaian Islam (Islamic Pawn Broking System)	2004	*Islamiyyat*
8	Khan & Nisar	Collateral (Al-Rahnu) as Practiced by Muslim Funds of North India	2004	*Journal of King Abdul Aziz University: Islamic Economics*
9	Ismail	Skim Ar-Rahnu: Pengukuhan Kerjasama Serantau Pengalaman Malaysia (Pawn broking Scheme: Strengthening the Regional Malaysian Experience)	2004	Proceeding of Konvensyen Ar-Rahnu Serantau 2004
10	Ismail & Sanusi	A Framework for Regulating Pawnshops: Why do, what area and Syariah Review	2005	Malaysian Finance Association7th Annual Conferenc
11	Ibrahim & Salleh	The Objectives of Ar-Rahn and their Achievement in	2006	The International conference on Islamic

DOI: 10.1051/shsconf/20173600033

		Charge/Mortgage in Islamic Home Financing: An Analysis		Jurisprudence and the Challenges in the 21st century, IIUM
12	Amin, Dahlan & Supinah	An Ar-Rahnu Shop Acceptance Model (ARSAM)	2007	*Labuan e-J of Muamalat & Society*
13	Mohamed, Ismail & Maamor	Competition and Outreach on Ar-Rahnu Industry	2007	Working paper UKM EKONIS
14	Bhatt & Sinnakkannu	Ar-Rahnu (Islamic Pawning Broking) Opportunities and Challenges	2008	Proceedings of Malaysian 6th International Islamic Finance 2008
15	Abdul Razak	Malaysian Practice of al-Rahnu Scheme: Trends and Development	2008	Proceedings at Malaysian study of Islam UK
16	Sanusi & Johari	Demand of Pawn Shop Services: Consumers Perspective	2008	*Malaysian Journal of Consumer and Family Economics*
17	Mohd Taher & Mohd Shafiai	Customer Acceptance Factors on Ar-Rahnu Scheme at Kedai Ar-Rahnu Bank Rakyat, Alor Setar Kedah	2008	Proceeding of PERKEM III
18	Mohamad & Salleh	Upah Simpan Barang dalam Skim Ar-Rahnu: Satu Penilaian Semula (Safekeeping Fee in Ar-Rahnu Scheme: A Revaluation)	2008	*Jurnal Fiqh*
19	Mohd Taher, Yusof & Abidin	Prinsip Al-Rahn sebagai Asas Perlaksanaan Skim Pajak Gadai Islam di Malaysia (Al-Rahn Principle as the Basis to Islamic Pawn Broking Scheme in Malaysia)	2008	*Voice of Academia*
20	Maamor & Ismail	The Ar-Rahnu Efficiency and its Determinants	2010	*Journal of Islamic Economics, Banking and Finance*
21	Abdul Razak	Economic and Religious Significance of the Islamic and Conventional Pawn Broking in Malaysia: Behavioural and Perception Analysis	2011	Doctoral dissertation, University of Durham
22	Amin	Modelling Ar-Rahnu Use in Eastern Malaysia – Perspectives of Muslimah	2011	*Journal of Islamic Economics, Banking and Finance*

23	Wan Daud	Kepuasan Pelanggan dan Kualiti Perkhidmatan Skim Al-Rahnu Bank Rakyat (Customer Satisfaction and Service Quality of Bank Rakyat Al-Rahnu Scheme)	2011	Master dissertation, UUM
24	Cheong & Sinnakkannu	Ar-Rahnu: Opportunities and Challenges in Malaysia	2012	SSRN 2112809 electronic journal)
25	Othman, Hashim & Abdullah	Customer Satisfaction Levels of Service Quality at Ar-Rahnu, Kelantan	2012	Proceeding of Business and Economics Research (ICBER)
26	Yaacob, Ahmad & Ibrahim	Developing a Customers' Oriented Ar-Rahnu (Islamic Pawn Broking) Model for Cooperatives in Microfinancing in Malaysia	2012	Proceeding of 13th Malaysia, Indonesia conference on Economics, Management and Accounting
27	Abdul Khir, Badri & Hussain	Critical Appraisal of the Rahn-based Islamic Microcredit Policy	2012	ISRA Research paper
28	Mat Nor, Abdullah, Ismail, Bakar & Mohd Yusni	A Study of the Customer Awareness toward Ar-Rahnu scheme at Sg Buloh	2012	Proceeding of AFBE 2012 conference UNITEN
29	Mulazid	Kedudukan Sistem Pergadaian Syariah dalam Sistem Hukum Nasional di Indonesia (Shariah Pawn Broking System in the Indonesian National Legislation System)	2012	*Innovatio*
30	Koe & Abdul Rahman	The Use of Ar-Rahnu by Islamic Bank Customers in Malaysia	2012	Proceedings of the International Conference on Science, Technology and Social Sciences 2012
31	Abdullah	The Roles of Microfinance in Brunei Darul Salam	2012	Proceeding of Konvensyen Ar-Rahnu Serantau 2012
32	Samsudin	Kajian Keberkesanan Skim Ar-Rahnu dalam Memantap dan Memperkukuh Usahawan Kecil di Negara2 Serantau (The Study on the Effectiveness of Ar-Rahnu Scheme in Strengthening the Small Entrepreneurs in the Regional Countries)	2012	Proceeding of Konvensyen Ar-Rahnu Serantau 2012

33	Hisham, Abdul Shukor, Ahmad Bustamam & Jusoff	The Concept and Challenges of Islamic Pawn Broking (Ar-Rahnu)	2013	*Middle-East Journal of Scientific Research 13*
34	Mokhtar & Zambahari	Ar-Rahnu: A Short-term Financing Alternative	2013	Proceeding of 5th Islamic Economic System Conference (iECONS 2013)
35	Sharif, Shaharuddin, Muhamed, Pauzi & Mohd Zin	The Improvement of Ar-Rahnu (Islamic Pawn Broking) Enhanced Product in Islamic Banking System	2013	*Asian Social Science*
36	Abdul Razak, Muhammad, Mohd Hussin, Mohd Rhouse & Hashim	Peranan Institusi Pajak Gadai Islam dalam Pembangunan Sosio-ekonomi (Islamic Pawn Broking Function in Socio-economic Development)	2013	Proceeding of PERKEM VIII
37	Othman, Hashim and Abdullah	Perkembangan Ar-Rahnu di Terengganu: Kajian kes terhadap Ar-Rahnu Majlis Agama Islam dan Adat Melayu Terengganu (MAIDAM) (The Development of Ar-Rahnu in Terengganu: Case study of Islamic Council and Malay Custom Terengganu (MAIDAM))	2013	Proceeding of PERKEM VIII
38	Yaacob	Faktor-faktor yang Mempengaruhi Kesedaran Usahawan Kecil Muslim terhadap Skim Ar-Rahnu (Factors Influencing the Small Muslim Entrepreneurs on Ar-Rahnu)	2014	Master dissertation UUM
39	Mansor, Ahmad, Abu Bakar & Ismail	A Demographic Analysis on Customer Acceptance towards Islamic Pawn Broking in Malaysia	2014	*Asian Social Science*
40	Rasmin & Markom	Ar-Rahnu Governance laws in Malaysia: An Analysis in Banks and Non-banks Financial Institutions	2014	Proceeding of International Conference of Global Islamic Studies
41	Abdul Hamid, Abd Rahman & Abdul Halim	Factors Affecting the Acceptance on Ar-Rahnu (Islamic Pawn Broking) a Case Study of Islamic Banking in Malaysia	2014	*The Macrotheme Review*

DOI: 10.1051/shsconf/20173600033

42	Rasmin & Markom	An Overview on Implementation of Ar-Rahnu in Malaysia	2014	Proceeding of PERKEM IX
43	Ezahar	A Study of Islamic Practice in Al-Rahnu at Financial Institutions	2014	Master dissertation UUM
44	Abdul Razak, Muhammad, Mohd Hussin & Mahjom	Kemapanan dan Daya Saing Pajak Gadai Islam di Malaysia (Sustainability and Competitiveness of Islamic Pawn Broking in Malaysia)	2014	Proceeding of PERKEM IX
45	Abdi	Determinants of Customer's Intention to use Islamic Pawn Broking Scheme: A Case Study of (UPSB) staffs in UUM	2015	Master dissertation UUM
46	Baharum, Maamor & Othman	Examining the Factors that Influence Customer's Intention to use Ar-Rahnu at Post Office: A Case study in Kedah	2015	*Journal of Islamic Economics, Banking and Finance*
47	Bahari, Fisal, Shahar & Shahadan	The Development of Islamic Pawn Broking in Malaysia, Application and its Challenges	2015	Proceeding of First International Conference on Economics and Banking (ICEB-15)
48	Abd Hamid, Abd Rahman & Abd Halim	Key Factors Influencing Customers to use Ar-Rahnu (Islamic Pawn Broking) in Malaysia: Evidence from Bank Rakyat	2016	*The Macrotheme Review*
49	Shah & Yaacob	Critical Literature on Governance and Supervision of Ar-Rahnu in Malaysia	2016	*Journal of Advanced Review on Scientific Research*
50	Mohd Azli, Palil & Mohd Noor	Effect of Ar-Rahnu Benefits on Long-term relationship between Customer and Institutions	2016	Proceeding of 2nd International Conference on Islamic Perspective of Accounting, Finance, Economics, and Management (IPAFEM)

THE DETERMINANTS OF FTSE4 GOOD BURSA MALAYSIA

Nor 'Asyiqin Abu*, Prof. Madya Dr Hajjah Hanim Norza Baba, Siti Aisyah Basri, Zaleha Mahat
Faculty of Accountancy, Universiti Teknologi MARA Melaka

ABSTRACT

This study was conducted to identify the factors that influence the companies listed in Bursa Malaysia to be indexed as FTSE4Good Bursa Malaysia (F4GBM). The companies awarded as F4GBM are those with good sustainable practices. This study investigates the potential factors that explain the good sustainable practice (i.e. corporate size, financial performance, industry attachment and sharia-compliant). Analysis reveals that only corporate size which proxy by market capitalization has significant relationship with good sustainable practice.

Keywords: Sustainability, Firm characteristics, FTSE4Good, Corporate Social Responsibility (CSR)

INTRODUCTION

In 2014, Bursa Malaysia Bhd announced the launch of the Environmental, Social and Governance (ESG) Index. Known locally as the FTSE4Good Bursa Malaysia (F4GBM) Index, it requires Malaysian companies to innovate and adopt more sustainable practices in order to comply with this new mandate. Developed in collaboration with FTSE as part of the globally benchmarked FTSE4Good Index Series, the F4GBM Index is aligned with other leading global ESG frameworks such as the Global Reporting Initiative and the Carbon Disclosure Project (Bursa Malaysia, 2014).

According to this new mandate, all constituents of the F4GBM Index must meet internationally benchmarked criteria that measure things such as the effort put into environmental conservation, the impact of social responsibility initiatives on the community, as well as the practice of good governance through responsible and ethical decision-making.

The index also forces companies to look at value from a non-financial perspective such as a company's environmental and societal initiatives. The F4GBM Index is not only in line with the Prime Minister's vision in Budget 2014, but is also the latest step in Bursa Malaysia's sustainability roadmap (Bursa Malaysia, 2014).

As at December 2015, F4GBM Index with a total of 34 constituents, comprising public listed companies (PLCs) from across the small, medium and large market capitalisation segments. Index constituents are drawn from the top 200 shortlisted companies on the FTSE Bursa Malaysia EMAS Index and reviewed in June and December every year (Bursa Malaysia, 2015). FTSE Bursa Malaysia EMAS Index comprises the constituents of the FTSE Bursa Malaysia Top 100 Index and the FTSE Bursa Malaysia Small Cap Index. Being listed in F4GBM Index means the companies have been practicing good sustainability efforts and reporting. The question is, "Does the listed companies shared the same characteristics?" To

*Corresponding author's email: norasyiqin@melaka.uitm.edu.my

answer this question, this paper will identify the determinants which influence the listing in F4GBM Index.

LITERATURE REVIEW

The determinants of sustainability reporting can be categorised into two main categories; internal and external. The internal determinant encompasses corporate size, financial performance, social performance, environmental performance, and ownership structure. Conversely, external determinant covers corporate visibility, sector affiliation, country-of-origin, and legal requirements (Hahn and Kuhnen, 2013).

2.1 Internal Determinants of sustainability reporting

The most frequently investigated determinants are corporate size and financial performance. Corporate size usually measured by total assets, turnovers, sales, number of employees and market capitalization. From the previous literature, corporate size has a positive effect of sustainability since larger companies are more visible and therefore greater stakeholder scrutiny and pressure. In addition, higher profitability will increase the ability of a company to bear the costs of sustainability reporting (Kent and Monem, 2008). A high level of indebtedness, leverage and gearing can be assumed to decrease the ability and flexibility of a company to bear the cost of reporting or face the consequences of disclosing potentially damaging information (Kent and Monem, 2008).

Social and environmental performance can also measure by the number of fines for environmental wrongdoings. However, the results are inconsistent. According to Belal and Cooper (2001), companies with a lot of wrongdoings may be more actively engaged in reporting to mitigate legitimacy threats. While later study by Brammer and Pavellin (2006) find no significant effect between extend of reporting with social performance.

Ownership structure will also influence reporting practice. Haddock (2005) finds that listed companies disclose a higher level of sustainability information. While Amran and Haniffa (2011) indicate that state-owned and/or government shareholding is associated with a higher extend of sustainability reporting since the respective organizations are subject to more stringent reporting and scrutiny. Brammer and Pavelin (2006) indicate that the quality of reporting is negatively influenced by a concentrated ownership structure and Cormier and Magnan (2003) indicate a positive influence of foreign ownership on the level of sustainable reporting.

2.2 External Determinants of sustainable reporting

According to previous literatures, external determinants of sustainability reporting mainly covers aspects of corporate visibility, sector affiliation, country of origin and legal requirements (Hahn and Kuhnen, 2013). Corporate visibility means the exposure of the companies to media. The companies may start sustainability reporting or increase the depth of their disclosure in order to mitigate reputational risk (Haddock and Fraser, 2008). Instead of media exposure, corporate visibility is measured through supply chain position and branding issues.

Businesses to consumer (B2C) companies are more likely to engage in reporting activities rather than business to business (B2B) companies. It means that the extent of reporting depends

[Type here]

on a company's closeness to market. While company name and brand name confirm a positive influence on the adoption and extent of reporting (Haddock, 2005).

A company's sector or industry attachment is the most frequently addressed external determinant. Companies from industries with high social and environmental impacts may need to engage in sustainable reporting in order to respond to stakeholder pressure (Parsa&Kouhy, 2008)

A part from industry differences, reporting practices may also vary across countries and regions due to different cultural and social norms or governmental regulations (Sotorrio and Sanshez, 2010). The results indicate differences with respect to adoption, extent, volume and quality of reporting. The study also confirmed an increase in the sustainability reporting is due to tighten regulations.

From the previous literatures, this study has determined that corporate size, financial performance, sector or industry attachment have influence on sustainability reporting. Therefore, this study will use the same factors with an additional determinant 'sharia-compliant' to identify whether these factors influence F4GBM listing. Figure 2.1 illustrate the conceptual framework developed for this study.

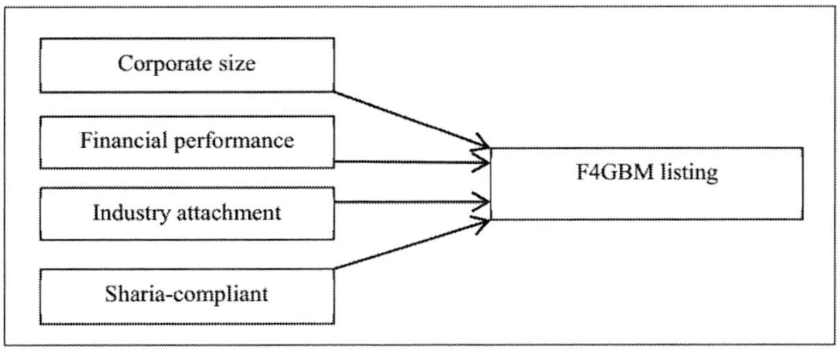

Figure 2.1: Framework of the determinants of FTSE4Good Bursa Malaysia

METHODOLOGY

The sample used comprises of the companies listed on the FTSE Bursa Malaysia Emas Index as at December 2015. Out of 254 being listed, only 34 entitle as constituents of F4GBM Index. F4GBM listing is considered as dependent variable. Score two (2) was given if the company is constituents of F4GBM Index and score one (1) if otherwise.

The independent variables and their respective measurement are depicted in Table 3.1.

Independent variables	Measurement
Corporate size	▪ Total assets ▪ Market capitalization
Financial performance	▪ Revenue ▪ Total debts

[Type here]

Industry attachment	Types of sector. FTSE Bursa Malaysia EMAS Index divided into 10 sectors
Sharia-compliant	Sharia-compliant =2 or non-Sharia-compliant=1

The data of independent variables was obtained from the latest annual report available in Thomson Reuters Datastream. This study has adopted the descriptive statistical analysis technique (SPSS) in analyzing the data.

RESULTS AND DISCUSSION

1. Descriptive Analysis

Descriptive analysis is conducted to generate information about the data collected. Figure 4.1 illustrates the number of company according to sectors. FTSE Bursa Malaysia EMAS Index mainly consists of ten (10) sectors. 32 percent of the constituents come from trade and service sectors, followed by 16 percent from industrial product sectors, almost 10% from consumers and also finance.

Figure 4.1: Number of samples according to sectors

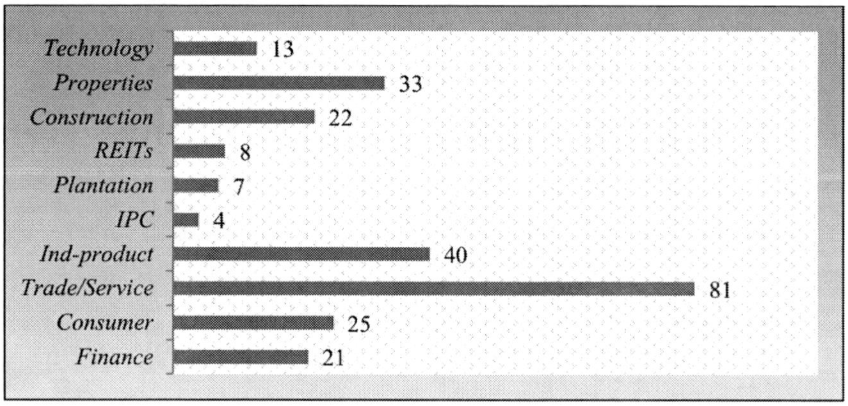

Figure 4.2: Number of F4GBM according to sectors

[Type here]

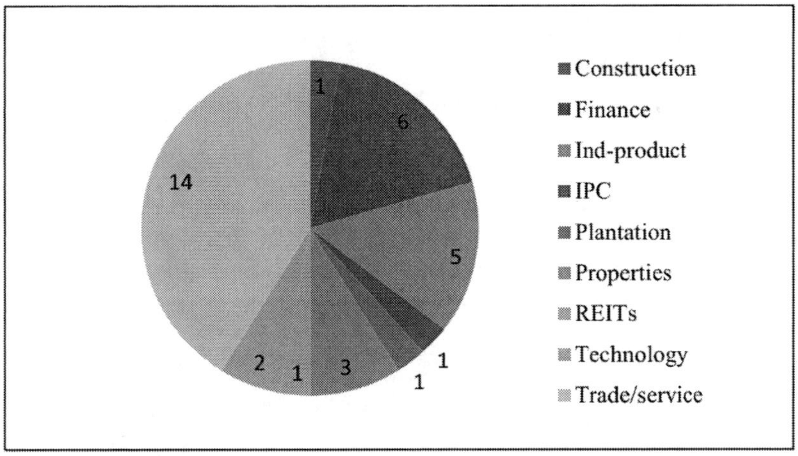

As shown in Figure 4.2, 74 percent of F4GBM constituents are comes from trade and service, finance and industrial products sector. These three sectors conquering F4GBM Index since they are dominate constituents in FTSE Bursa Malaysia EMAS Index.

In addition, trade and service become the majority constituent in F4GBM since this industry is very close to consumer. As mentioned by Haddock (2005), the extent of reporting depends on a company's closeness to market. The term 'close to consumer' is defined as companies that supply goods or services directly into consumer markets instead of supplying to another business entity. Haddock-Fraser and Fraser (2008) demonstrated that companies focused on supplying directly to the end-consumer were more likely to be pro-active in environmental reporting.

Figure 4.3 provides the number of sharia-compliant companies. 76 percent of constituents in FTSE Bursa Malaysia EMAS Index are sharia-compliant. Interestingly, almost 90 percent (30 out of 34) companies listed in F4GBM are sharia-compliant.

In order to be listed as sharia-compliant, the company needs to fulfil the following requirements:

- Companies are not permitted to be involved in any of the following core activities: banking or any other interest-related activity, such as lender and brokerages, but excluding Islamic financial institutions; alcohol; tobacco; gaming; arms manufacturing; life insurance; and, pork and non-halal production, packaging and processing or any other activity related to pork and non-halal food.

- Ratios of debt and debt service in combination that is acceptable and indicative of an appropriate use of leverage relative to their assets.

- Income from cash or near cash equivalents are appropriate levels of receivables to assets.

- Liquid assets to illiquid assets that not exceed the percentages permitted.

- Cash and cash equivalents to total assets that not exceed the percentage permitted. (Bursa Malaysia, 2016)

[Type here]

As illustrated, most companies prefer to be labelled as sharia-compliance since empirical evidence has proven that performance of sharia stocks tends to be better compared the conventional portfolio return. Investors are also increasingly getting positive about Sharia-compliant funds, which have shown to be resilient against the global recession. Due to their relatively low risk and dependability compared to conventional funds, the demand for sharia-compliant unit trust funds has grown significantly. So now, even for non-Muslim clients, fund managers recommend a mixed portfolio consisting of sharia-compliant and conventional funds for conservative investors looking to achieve their financial goals (Reddy and Fu, 2014).

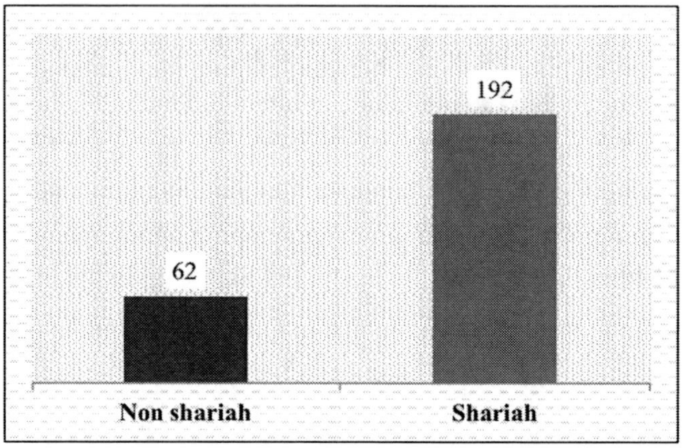

Figure 4.3: Number of samples according to sharia-compliant

1.1 *Corporate size and financial performance*

All proxies of corporate size and financial performance only explain 26 per cent of the variance (R Square) in F4GBM listing which highly significant, as indicated by the F-value of 21.386, $p<.05$ in the table below. An examination of the t-value indicates that market capitalization contribute to the prediction of good sustainable practice.

Model Summary[b]

Model	R	R Square	Adjusted R Square	Std. Error of the Estimate
1	.506[a]	.256	.244	.28949

a. Predictors: (Constant), TDEBT, MKTCAP, REVENUE, TOTALASSETS
b. Dependent Variable: FTSE4Good

ANOVA[a]

Model		Sum of Squares	df	Mean Square	F	Sig.
1	Regression	7.169	4	1.792	21.386	.000[b]
	Residual	20.783	248	.084		
	Total	27.953	252			

[Type here]

DOI: 10.1051/shsconf/20173600034

a. Dependent Variable: FTSE4Good
b. Predictors: (Constant), TDEBT, MKTCAP, REVENUE, TOTALASSETS

Coefficients[a]

Model		Unstandardized Coefficients		Standardized Coefficients		
		B	Std. Error	Beta	t	Sig.
1	(Constant)	1.056	.020		52.430	.000
	TOTALASSETS	1.434E-9	.000	.270	1.982	.049
	REVENUE	1.290E-8	.000	.223	2.139	.033
	MKTCAP	1.017E-8	.000	.399	4.157	.000
	TDEBT	-1.506E-8	.000	-.390	-2.532	.012

a. Dependent Variable: FTSE4Good

From the coefficients results, it shown that increasing total assets, revenue and market capitalization was associated with an increase likelihood of good sustainable practice but increasing in total debts was associated with reduction in the likelihood of good sustainable practice.

It is often argued that companies' adoption of sustainability strategies should grant competitive advantages over companies that do not adopt them. In addition, the information on sustainability practices that the company develops and discloses should facilitate the development of better systems of internal control, decision making, and cost saving. Thus, if the goods or services offered by the company possess elements that differentiate them from those of competitors, this will produce differences in sales and turnover (Adams and Zutshi, 2004).

Furthermore, some studies claim that sustainability practices favour stakeholders' legitimation, offering a positive image of the firm and improve its reputation (Adams, 2002). All these opinions support the findings of positive relationship between assets, revenue and market capitalization with good sustainable practice.

CONCLUSION

This paper explored the issue of whether or not corporate size, financial performance, industry attachment and sharia-compliant were factors influencing good sustainable practice for the FTSE Bursa Malaysia EMAS Index. It was established that a positive statistical association exists between corporate sizes which proxy by market capitalization with good sustainable practice. Interestingly, it was established that 90 percent companies listed in F4GBM are sharia-compliant. Close to consumer companies were more likely to be associated with good sustainable practice since 74 percent of companies listed in F4GBM are from trade and service industries.

The implications of this research are important to policy makers and government as well as investors in terms of identifying which companies are potentially leading the good sustainable practice and identifying why they are taking that action.

[Type here]

There are a number of limitations to this research study. By its nature it is an exploratory study considering general association patterns within a quantitative data set. As noted above, the good sustainable practice is being associated with close to consumer companies which are trade and service industries. However, further research would be needed to determine whether this perceived need by the companies is responding to an actual need of the consumer.

REFERENCES

Adams, C. (2002). Internal Organizational Factors Influencing Corporate Social and Ethical Reporting: Beyond Current Theorising, *Accounting, Auditing and Accountability Journal* 15(2), 223–250.

Adams, C. and Zutshi, A. (2004). Corporate Social Responsibility: Why Business Should Act Responsibly and Be Accountable, *Australian Accounting Review* 14(3), 31–39.

Amran, A. and Haniffa, R. (2011). Evidence in development of sustainability reporting: case of a developing country. *Business Strategy and the Environment*, 20 (2011), pp. 141–156.

Belal and Cooper, (2001). The absence of corporate social responsibility reporting in Bangladesh. *Critical Perspectives on Accounting*, 22 (2011), pp. 654-667.

Brammer and Pavelin, (2006).Voluntary environmental disclosures by large UK companies.*Journal of Business Finance & Accounting*, 33 (2006), pp. 1168-1188.

Bursa Malaysia,(2014). Bursa Malaysia Announces Launch of Environmental, Social and Governance Index. Retrieved 20 April 2015, from http://www.bursamalaysia.com/corporate/media-centre/media-releases/3081.

Bursa Malaysia,(2015). Bursa Malaysia Announces Constituent Additions to ESG Index. Retrieved 20 April 2015, from http://www.bursamalaysia.com/corporate/media-centre/media-releases/3853.

Bursa Malaysia,(2016).FTSE Bursa Malaysia Hijrah Shariah Index. Retrieved 11 August 2016, from http://www.bursamalaysia.com/market/islamic-markets/products/islamic-capital-market/shariah-compliant-listed-equities.

Cormier, D. and Magnan, M. (2003). Environmental reporting management: a continental European perspective. *Journal of Accounting and Public Policy*, 22 (2003), pp. 43–62.

Haddock, J. (2005). Consumer influence on internet-based corporate communication of environmental activities: the UK food sector. *British Food Journal*, 107 (2005), pp. 792–805.

[Type here]

Haddock, J. and Fraser, I. (2008). Assessing corporate environmental reporting motivations: differences between "close-to-market" and "business-to-business" companies. *Corporate Social Responsibility and Environmental Management*, 15 (2008), pp. 140–155.

Hahn and Kuhnen, (2013). Determinants of sustainability reporting: a review of results, trends, theory, and opportunities in an expanding field of research. *Journal of Cleaner Production*, 59 (2013), pp. 5–21.

Kent and Monem, (2008).What drives TBL reporting: good governance or threat to legitimacy.*Australian Accounting Review*, 18 (2008), pp. 297-309.

Parsa, S. and Kouhy, R. (2008). Social reporting by companies listed on the alternative investment market. *Journal of Business Ethics*, 79 (2008), pp. 345-360.

Reddy, K. and Fu, M. (2014). Does shariah compliant stocks perform better than the conventional stocks? A comparative study of stocks listed on the Australian Stock Exchange. *Asian Journal of Finance & Accounting*, 6 (2014), pp.155-170.

Sotorrio, L. and Sanchez, J.L.F. (2010). Corporate social reporting for different audiences: the case of multinational corporations in Spain. *Corporate Social Responsibility and Environmental Management*, 17 (2010), pp. 272-283.

Djajadikerta, H. G., Roni, S. M., & Trireksani, T. (2015). Dysfunctional information system behaviors are not all created the same: Challenges to the generalizability of security-based research. *Information & Management, 52*(8), 1012-1024. doi:http://dx.doi.org/10.1016/j.im.2015.07.008

MOTIVATING FACTORS FOR SUSTAINABLE ACCOUNTANT POTENTIALS IN MALAYSIA

Sharifah Norhafiza Syed Ibrahim*, Nur Farahah Mohd Pauzi, Adriana Shamsudin, Mohd Syazwan Karim, Khairiah Ahmad
Faculty of Accountancy, Universiti Teknologi MARA Melaka

ABSTRACT

Economic Transformation Plan (ETP) emphasizes that the country has a pool of resources to steer towards Malaysia's aspirations to become a developed and high-income nation. The ETP Roadmap highlights accountancy, an integral function in all businesses as part of the significant ETP driver. However, Malaysia still lacks the required number of qualified accountants by the year 2020. This challenges higher learning institutions to produce suitable accounting graduates. Despite taking an accounting subject in schools, not all these students later opt to enrol accounting in higher learning institutions as many factors influence their career choices. This study aims to explore these motivating factors. A quantitative approach was applied whereby primary data were collected through a questionnaire survey. Respondents were students from seven elite schools in Melaka. The study found eight motivating factors; family members in the accounting profession, public accountant as a career choice, students' interest in receiving relevant information, choosing private accountants as a career, parents, school counsellor and career talk. However, only four of these factors were significant to students' decisions to pursue professional accounting courses. In addition, parents' influence plays a greater role in motivating students to opt for professional accounting courses as compared to an accounting degree.

Keywords: Motivating Factors, Accounting Major, Professional Accounting Programs

INTRODUCTION

Accountants have a significant role to play in helping to drive Malaysian's economy ahead. They no longer merely responsible to keep the ledger books balanced but they also act as guardians of the 'checks and balances' to ensure the well-being of the company they work for. As Malaysia moves forward to be a developed country by the year 2020, the structure and size of the economy will change, influenced by regional and domestic developments. The private and public sectors will also become more competitive. Such a progression would demand the services of accountants across all sectors. Hence, the availability of talent both in number and with required competencies in various economic sectors would be highly critical.

Accountancy profession in Malaysia is regulated and developed by Malaysian Institute of Accountants (MIA). MIA membership is mandatory for those holding themselves out or practising as an accountant in the country. MIA's membership encompasses both legal and market recognition for accountants in Malaysia. Members of MIA are conferred with the designation of Chartered Accountant Malaysia. This designation is given to those who are professional in accountancy, business, and finance with recognised accountancy qualification and relevant working experiences. As a body for all of the accountants in Malaysia, MIA's responsibilities include education and quality assurance as well as enforcement in order to

*Corresponding author's email: snorhafiza@melaka.uitm.edu.my

maintain the credibility of the profession and the public interest. The functions of MIA comprise of determining the qualifications for membership admission; providing education and training in accountancy; regulating the practice of accountancy, and promoting the interests of the accountancy profession. At present, there are more than 32,000 MIA members working throughout the states of Malaysia in various industries. Nevertheless, the country is still short about half of the figure whereby Malaysia needs approximately 60,000 qualified accountants by the year 2020 which will be another four more years to go.

The reason for Malaysia being so far behind the required number of accountants is still a questionable mark for everybody. Is that because MIA does not play their role? According to a survey conducted by MIA from the discussion of Committee to Strengthen the Accountancy Profession (CSAP) together with major employers such as Chief Financial Officer (CFO) of commercial banks agreed that among the characteristics to be employed as professional accountants would include qualification as a professional accountants; must have broad skills, not limited to finance and business only; practical and market relevant knowledge; possess soft skill such as management, problem-solving ability, technology savvy; and strong written and spoken communication skills.

Principally, MIA does not have the ability to provide assurance on the baseline competency of its membership. The effectiveness of MIA to enforce compliance with its standards is being questioned. This is important for public confidence and the image of the accountancy profession at large. Employers do not consider MIA as a professional body and only recognise other professional accountancy qualifications such as Malaysian Institute of Certified Public Accountants (MICPA), Association of Certified Chartered Accountants (ACCA), Institute of Chartered Accountants in England and Wales (ICAEW) and Certified Practising Accountant (CPA) Australia which caused limited professional career opportunities for those who do not possess such professional qualifications (CSAP Report, 2014). Besides that, the number of Bumiputera students who possess professional accountancy qualifications is small. Therefore, it is essential for MIA to provide strong leadership in accountancy education and clear competency parameters for training purposes. Certainly, this would enable better alignment between market expectations, professional competencies, and professional accountancy education offered at tertiary levels.

Furthermore, some of our accountancy graduates are lacked in communication skill especially the Bumiputera students. Even though higher education institutions have taken necessary strides to enhance the quality of accountancy graduates, there are still other issues which needed to be addressed. Talent migration is another reason why Malaysia has a small number of professional accountants since it affects the supply of talent in adequate numbers for the needs of the economy. However, Malaysia has a comparative advantage in various industry segments. Therefore, proper planning and follow-up are required to preserve and leverage this advantage.

According to the report done by CSAP in 2014, several actions should be taken by Malaysia in order to sustain the accountancy profession in this country. The recommendations include the introduction of a new regulatory structure which will assist in developing the accountancy profession while promoting public interests through the formulation and enforcement of profession standards; ensuring adequate supply of professional accountants for the economy that fulfils market expectation in terms of knowledge, skills and values; improving the effectiveness and efficiency of the disciplinary structure for the enforcement of professional standards to ensure protection of public interest; ensuring financial support for education,

[Type here]

research and capacity building for the accountancy profession; creating a conducive and inclusive environment to attract students to pursue professional accountancy profession; fostering support for MIA members to enhance their skills and competency to match the market expectation; strengthening the Small and Medium Practices to render professional services for SMEs; strengthening the accountancy education as well as positioning Malaysia as the hub for accountancy education and training for this region; and enhancing the competitive edge of the profession by developing specialised competencies and skills, for example, in the area of Islamic Finance (CSAP Report, 2014).

Objective of the study

Approaching ETP portrays Malaysian government is really committed to becoming a high nation income and developed country. One of the detailed ETP roadmaps stated that the role of the qualified accountant is very crucial in executing ETP (Malaysian Institute of Accountant (MIA), 2015). In relation to this roadmap, Malaysian government aims to increase the qualified accountants to 60,000 by the year 2020. According to Wilder & Stocks (2004), recruiting quality students is a key success to produce professional accountants. The purpose of this study is to explore the motivating factors influencing secondary school students in pursuing the accounting courses. The key way to determine student's career choice is by gaining an understanding of the factors that affect student's career decision (Byrne, Willis, & Burke, 2012).

Besides, this study put an effort to determine whether there is a significant relationship between all the motivating factors and student's decision to pursue accounting courses. Lastly, this study attempts to suggest programs to inculcate student's interest in accounting courses particularly in professional programs such as ACCA, Chartered Institute of Management Accountants (CIMA), CPA Australia, MICPA and many more.

Problem Statement

ETP emphasizes that the country has a pool of resources which will steer towards Malaysia's aspirations to become a developed and high-income nation. As highlighted in the ETP Roadmap, accountancy is part of the significant driver and reflects as an integral function in all businesses. In line with the roadmap, MIA was given the big responsibility of developing a sustainable number of competent Chartered Accountants to realise this aspiration. However, based on MIA (2015), the report emphasizes that there is a decline of interest in choosing the accountancy as a preferable profession due to the mentality of long working hours and less glamorous profession as compared to any other professions. According to Sugahara and Boland (2006), the profession becomes unpopular among the brightest business students as they claim that accounting profession to have lower salary compared to other occupation like a consultation or computer-related position.

Over the latest decade, the number of students taking up accounting major has been declined considerably in various part of the world (Zyl & Villiers (2011), Sugahara & Boland (2006), Ahmed, Alam & Alam (1997)). According to Jackling & Calero (2006), in Australia, there was a drastic increment of students taking undergraduates degrees but the proportion of choosing accounting as a major study has decreased tremendously. Meanwhile, Sugahara, Hiramatsu & Boland (2009) discovered that the declining numbers of Certified Public Accountants Profession in Japan were mainly due to the requirement in their education

[Type here]

DOI: 10.1051/shsconf/20173600035

system of forcing the students to pass a very competitive examination based on technical knowledge and requiring mechanical memorization.

Based on MIA (2015), as at 30th June 2015, the number of professional accountants in Malaysia stands at 31,395 members. 53% from this number came from local graduates and another 43% were admitted as members via the professional membership route. In Malaysia, there are fifteen universities offering accounting programmes at undergraduate level such as University of Malaya, Universiti Kebangsaan Malaysia, Universiti Teknologi MARA, Universiti Utara Malaysia, Universiti Putra Malaysia, Universiti Islam Antarabangsa Malaysia, Universiti Sains Malaysia, Universiti Tenaga Nasional, Universiti Multimedia, Universiti Malaysia Terengganu, Universiti Malaysia Sabah, Universiti Industri Selangor, Universiti Sultan Zainal Abidin and Universiti Sains Islam Malaysia. The following table shows the admission of members to MIA from respective universities for the year 2015 and 2014.

Table 1: Admission of members to MIA by local universities

	Name of University	2015	2014
1	Bachelor of Accounting, University of Malaya	49	52
2	Bachelor of Accounting (Honours), University Kebangsaan Malaysia	44	52
3	Bachelor of Accounting (Honours), Universiti Teknologi MARA	255	219
4	Bachelor of Accounting (Honours), Universiti Utara Malaysia	159	14
5	Bachelor of Accounting (Honours), Universiti Putra Malaysia	61	-
6	Bachelor of Accounting (Honours), Universiti Islam Antarabangsa Malaysia	36	31
7	Bachelor of Accounting (Honours), Universiti Sains Malaysia	32	28
8	Bachelor of Accounting (Honours), Universiti Tenaga Nasional	26	20
9	Bachelor of Accounting (Honours), Universiti Multimedia	66	49
10	Bachelor of Accounting (Honours), (Information System) Universiti Utara Malaysia	28	-
11	Bachelor of Accounting (Honours), Universiti Malaysia Terengganu	14	14
12	Bachelor of Accounting (Honours), Universiti Malaysia Sabah	16	17
13	Bachelor of Accounting (Honours), Universiti Industri Selangor	16	13
14	Bachelor of Accounting (Honours), Universiti Sultan Zainal Abidin	1	-
15	Bachelor of Accounting (Honours), Universiti Sains Islam Malaysia	5	-

Source: MIA (2015), MIA (2014)

With rapid changes in business environment, it is insufficient for the accounting profession to supply the existing resources to serve the entire business requirement. Therefore, this study was conducted to search for the factors or reasons on why the accounting profession become less favourable among brightest secondary school accounting students.

PRIOR STUDIES

Several existing studies such as Ahmed & Alam (1997) and Byrne, Willis & Burke (2012) which have investigated which factors influence students' decision to choose accounting find rewards and job satisfaction are a few of the significant motivating factors. However, most of these studies conducted surveys among tertiary students. Nevertheless, students who aspire to become accountant are more likely have decided so as early as when they are in schools (Byrne, Willis and Burke (2012). Byrne, Willis & Burke (2012) also found that accounting teachers do not influence this decision. Hence, awareness programs such as career talks and seminars should provide as much information as possible so that school students are more inclined to pursue accounting studies (Byrne, Willis & Burke, 2012) and consequently, to work as professionally qualified accountants. This study reviewed prior research related to the five aspects of factors that motivate students' decisions to pursue an accounting degree

[Type here]

and professional accounting courses. These factors were categorised into five themes; awareness, family, interest, career choice and rewards. They were later constructed in the questionnaires.

Awareness

To date, Malaysia is still far behind the target of producing 60,000 qualified accountants by the year 2020. Concerns have been raised in the past few years about the shortage of qualified accountants by many sectors. Apparently, the shortage of qualified accountants could be due to less exposure been made to students as early as secondary school level. According to MIA (2014), none of the career and awareness talks has been conducted for secondary schools from July 2013 until Jun 2014. But, MIA (2015) showed an improvement with one session of the career and awareness talk has been conducted for secondary schools. Therefore, in order to attract more accounting graduates, necessary exposure effort should be conducted at early stage possible so as to attract potential students.

Nowadays, higher learning education has become a priority to students after they had completed their high school education. According to Blackwell & Pinder (2014), students' decision to attend college is encouraged by an inner drive to achieve a better way of life and not influenced by their family. In fact, students begin to make their career choices in their childhood and were fixed by the time they completed their school education (Smyth & Hannan, 2000; Jackman & Hollingworth, 2005). Therefore, it is essential to give exposure to students regarding the choices they would have to pursue their study in a higher learning institution so that they can make the right decision to choose the pathway in order to achieve their dream.

Apart from that, the decision made by the students either to pursue their education in accounting field or not would also depend on their perception towards the accounting course itself as well as their performance during high school level. It is believed that the students might be affected by the perceived difficulty of the accounting course (Allen, 2004) and ultimately discouraged them from pursuing studies in the accounting field. Furthermore, exposure to the accounting profession could adversely affect students' perception and career aspiration (Marriott & Marriott, 2003). Hence, students should be exposed to various sources to acquire information about future profession and career opportunities in order to make any considerations from the valid bases and be prepared for the future undertakings.

Family

Empirical evidence regarding the influence from different reference group such as parents, friends, teachers, and career advisors on students' choice of an accounting discipline is inconclusive. Results of the study done by Balckwell & Pinder (2013) revealed that the first-generation college students, unlike the third-generation college were not encouraged by family to attend college but their inner drive to attend college to achieve a better way of life for themselves led them being the first in their families to attend and graduate from college. First-generation students in this study referred to respondents' age between 53 to 72 years old meanwhile the third-generation students referred to respondents' age between 41 to 52 years old.

According to Dalci, Arasli, Tumer, & Baradani (2013), referents (parents, teachers, career advisors and peers) were significantly associated with Iranian student's decision to major in

[Type here]

accounting. The results also indicated that Iranian students were sensitive to referents' opinions in choosing an accounting major. This finding might be due to the fact that people were closely knitted to each other and they as a group must be considered before individuals within the Islamic society. Additionally, civil and external wars have increased the importance of hierarchical order in Iran. Thus, it was surprising that Iranian students attached profound importance on opinions of the referent group such as their parents. Moreover, the findings of Paolillo & Estes (1982), Mauldin, Crain & Mounce (2000) and Geiger & Ogilby (2000) revealed that in the USA instructors have a profound influence on students' decision to major in accounting. However, Gul, Andrew, Leong & Ismail (1989) found a contradict findings in Australia whereby parents did not play a significant role in students' choice of majors.

Interest

Being a qualified accountant requires some strong personal traits such as dedication, commitment, and passion for knowledge and self-enhancement. Thus, individuals with such traits are more likely to succeed in becoming a qualified accountant. Prior studies investigated whether positive attitude towards accounting influence students' decisions to major in accounting and to pursue accountant as a career. Using the theory of reasoned action (TRA) to assess intentions and behaviours in employment choice, levels of interest in accounting and attitudes towards the work of accountants between Australian and international students' differ significantly (Jackling & Calero, 2006). They further identified the positive impact of government policy on motivating international students towards accounting and the accounting profession. In addition, earlier exposure to accounting in schools as compared to non-accounting and non-business subjects may influence students' interest to pursue accounting majors (Sugahara, Hiramatsu, & Boland, 2009).

A study on commerce students in Australia to examine both the accounting education decision and the decision to pursue and remain in an accounting career by using a logistic regression model incorporating demographic and academic factors, as well as students' perceptions of the work of accountants, to predict intention to become an accountant indicated that the three significant factor influencing intention to work as an accountant were the perception of importance of generic skills, intrinsic interest in the discipline area, and course satisfaction (Jackling & Calero, 2006). However, a study on Iranian students did not find a genuine interest in accounting, perception of the accounting course, and perception of the accounting profession significant in influencing students' decision to study accounting (Dalci et al., 2013).

As the type of students changes over time, some factors such as intrinsic factors such as personal interest might not be significant motivating factors in the late 90s as found by Ahmed and Alam (1997) but were significant in the recent study such as found by Byrne, Willis and Burke (2012). Hence, new studies may reveal new significant motivating factors as well as some contradicting factors.

Career Goal

Jackling & Calero (2006) investigated on students' perception of accountants' work with their intention to become an accountant. There was a strong support when the student perceived generic skills such as the ability to work in a teamwork, soft skills and writing skills were important, they intend to pursue an accounting career compared to those students

[Type here]

who do not perceive these skills are important. The bad image of accountants that has been shown in television or film will not influence on career decisions, particularly of first-year university students. This is also supported by Sugahara, Hiramatsu & Boland (2009) who found that the decisions on career choice of Japanese students were not also be influenced by negative images of the certified Public Accountant profession. Meanwhile, Sugahara and Boland (2006) found that those students who were looking for necessary skills were more likely to choose an accounting profession as their chosen career path as they believed that the profession will have good future prospects, advancement opportunities, and powerful authority.

In addition, Sugahara and Boland (2006) also investigated the relationship between job market condition and students' career choice whereby job market condition refers to job availability, job security, job variety and adventure as well as flexibility in career option. The results discovered that job availability and job security was perceived to have a higher influence on students' career decision. The career is believed to be a stable profession in terms of employability. The is supported by Cilliers & Charl (2011) who revealed that students' career choice was influenced mainly by job security, job satisfaction, and potential future earnings. Moreover, Byrne, Willis & Burke (2012) also found that the students' career choice was mainly driven by job satisfaction, good working condition, and career aptitude. Furthermore, Dalci et al., (2013) investigated the factor of career option among Iranian students which revealed that the relationship between job market factors has a significant relation to career choice decision. This finding could be attributed to the war, high inflation rates, foreign embargos, and high unemployment rates in Iran since the students were more likely of looking forward to financial and job market issues when choosing a career under such conditions.

Additionally, Law (2010) discovered that flexibility of career options has the significant influence on the likelihood of accounting students to choose general accounting career rather than a non-accounting career. Non-accounting students from business, psychology, and education all agreed that the supply of jobs in accounting is better than any other field. Notably, students who have the intention to pursue careers in audit field perceived that they would have more client interaction, better future job opportunities, greater knowledge of the whole business process. In addition, students who aimed to pursue careers in tax field were looking forward to more stable daily routine, developing more specialised skill and building a more collaborative relationship with clients (Dalton, Buchheit & McMillan, 2014).

Rewards

Abraham Maslow is prominent with his theory of human motivation in 1943. Maslow proposed five basic needs in describing the stages of human growth arranging from lower to the higher stage for physiological, safety, love, esteem and self-actualization. Based on Maslow's theory, self-esteem needs can be met by rewarding or appreciating people for their efforts (Maslow, 1943). This sort of esteem is important to encourage people to develop themselves to become a better person. Moreover, self-esteem portrayed that all humans having a need to feel respected, accepted and valued by others. Some people incline to engage in a profession for the purpose of gaining recognition and subsequently, being recognized. Based on that theory, typically students may tend to pursue their study at the university level or professional level for the recognition and to get financial incentive from their employment.

[Type here]

Plenty of prior researchers revealed that high salary was one of the factors influencing student's career decision (Ahmed, Alam, & Alam, 1997; Felton, Dimnik, & Northey, 1995; Hermanson, Dana & Hermanson, Roger, 1995; Lowe & Simons, 1997; Tan & Laswad, 2006). According to Carpenter & Foster (1977), there was a three-dimensional model of factors influencing student's career decision, namely, intrinsic factor (intellectual interest, job satisfaction), extrinsic factor (availability of employment, remuneration) and interpersonal factor (influence of others such parent and siblings). However, the findings of majority researchers were inconsistent with the Carpenter & Foster's model in describing factors influencing student's career decision such as Felton, Buhr, & Northey, (1994); Lowe & Simons (1997) and Paolillo & Estes (1982).

The early study in exploring factors influencing student's career decision, to begin with, Paolillo & Estes (1982) who ranked potential earning earned by employment as one of the most important factors. Subsequently, many researchers found that financial reward or potential earnings have highly significant relationship in influencing student's decision in choosing a career in accounting (Cohen & Hanno, 1993; Felton et al., 1994; Gul, Andrew, Leong, & Ismail, 1989; Nelson, Vendrzyk, Quirin, & Allen, 2002). Overall, the findings from previous studies indicated that students were motivated by financial reward, high salary or potential earnings to choose accounting as their career.

RESEARCH DESIGN AND DATA COLLECTION

A quantitative approach was applied whereby primary data were collected through the questionnaire survey from which the numbers were used to indicate the motivation factors that might influence the students to pursue their education in accounting particularly professional accounting courses. According to Soni & Kodali (2012), the most highly preferred data collection method for survey research is questionnaire as compared to interview or combination of other types of data collection method as it could provide a much more valid measure (Ulmer & Wilson, 2003).

The questionnaire were constructed based on the motivating factors suggested by prior studies. It consisted of ten questions covering gender, school and two questions on decisions to major in accounting and to pursue professional accounting programs as well as six questions on the factors influencing these decisions. The six questions were related to awareness, family background, attitude, career plan, and rewards. Answers to these questions provide thirteen motivating factors as summarised in Table 2 below. Following Agarwala (2008), these factors were grouped into three types; intrinsic factors, extrinsic factors, and interpersonal factors. Intrinsic motivation reflects the natural human propensity to learn and assimilate whereas extrinsic motivation was argued to reflect external control (Ryan & Deci, 2000). Interpersonal factors were influenced by the perceptions, attitudes, and expectations of others such as parents and friends (Agarwala, 2008). Unlike existing studies which require the respondents to rate the factors using a Likert-scale, this study asks respondents to choose the factors that they think influence them either to pursue accounting major or to pursue any professional accounting qualification.

DOI: 10.1051/shsconf/20173600035

Table 2: Motivating Factors to Pursue Accounting and Professional Accounting Qualifications

Intrinsic factors	Extrinsic factors	Interpersonal factors
1. Enthusiasm or Eagerness for related information	1. Perceived interest of accounting students Perceived interest of accountants,	1. Exposure from career talks and seminars
2. Career choice – public accountant	2. Monetary rewards earned by accountants,	2. Friends' influence
3. Career choice – private accountant	3. Good image of accountants Skills and experience gained by accountants	3. School counsellor influence
4. Career choice - other		4. Parents' influence
		5. Family members are accounting graduates
		6. Family members work as accountants or/and auditors

As for this study, the data was collected from the program known as 'Kelestarian Program Perakaunan' (KPP) which was held on 21 March 2016 at Auditorium Hall, Universiti Teknologi MARA, Jasin Campus. This program was attended by a total of 233 form five students from seven elite schools around Melaka who will be taking an accounting subject for their 'Sijil Pelajaran Malaysia' (SPM) examination in November 2016. The schools involved were Maktab Rendah Sains MARA Alor Gajah, Maktab Rendah Sains MARA Tun Ghafar Baba, Maktab Rendah Sains MARA Terendak, Sekolah Menengah Kebangsaan Tinggi Perempuan Melaka, Sekolah Menengah Kebangsaan (A) Sharifah Rodziah, Sekolah Menengah Kebangsaan (A) Sultan Muhammad and Sekolah Menengah Kebangsaan (A) Tun Perak.

KPP program is planned to be implemented for five consecutive years. It started in the year 2015 and 2016 was the second year the program was conducted. The program intends to give exposure to students regarding the accounting field particularly to an emphasis on the professional accounting courses as their preliminary information as a guide in choosing their track to pursue studies in higher learning institution so that the ultimate aim to become a qualified accountant is achievable within stipulated time period.

KPP program has been run for one full day starting from 8am until 5pm. The program consisted of four slots, namely, briefing regarding 'UiTM Di Hatiku' by Deputy Rector of the Academic Affairs, briefing on the accounting courses offered by UiTM Melaka by the faculty coordinator, briefing on the professional accounting courses such as Certified Accounting Technician (CAT), Certificate of Business Accounting (CBA), Association of Certified Chartered Accountants (UK) (ACCA) and Chartered Institute of Management Accountants (CIMA) by senior lecturer, Centre of Professional Accountancy (IPAC) Education, Shah Alam and the final and awaiting slot was 'The Great Journey' forum whereby three alumni of UiTM professional accounting courses from various backgrounds have been invited to share their experiences and tips while studying the professional accounting courses with all the students.

RESULTS AND FINDINGS

Table 3 shows descriptive statistics of the data. The largest group was from SMKA Sharifah Rodziah where all students have decided to pursue accounting course.

[Type here]

DOI: 10.1051/shsconf/20173600035

Table 3: Group Size by Gender and Schools

School Name	Male	Female	Total
MRSM Tun Ghafar Baba	13	10	23
MRSM Terendak	11	23	34
SMKA Sh Rodziah	0	32	32
SMKA Sultan Mohamad	27	0	27
SMK Tinggi Perempuan Melaka	0	36	36
SMKA Tun Perak	8	17	25
TOTAL	59	118	177

Table 4 summarises the number of students who decided to major in accounting by gender and school. Approximately 8% of the respondents did not intend to study accounting in their tertiary education, 25% of them had not decided yet and 67% decided to major in accounting. As the majority of the participants and hence the respondents were female, the number of female students who decided to choose accounting course was about 70%, which was doubled the number of male students.

Table 4: Group Size and Decisions on Majoring Accounting by Schools

Decision		Male	Female	Total
No	MRSM Tun Ghafar Baba	1	2	3
	MRSM Terendak	2	0	2
	SMK Tinggi Perempuan Melaka	0	9	9
TOTAL		3	11	14
Not sure yet				
	MRSM Tun Ghafar Baba	4	2	6
	MRSM Terendak	2	4	6
	SMKA Sh Rodziah	0	5	5
	SMKA Sultan Mohamad	11	0	11
	SMK Tinggi Perempuan Melaka	0	9	9
	SMKA Tun Perak	3	4	7
TOTAL		20	24	44
Yes				
	MRSM Tun Ghafar Baba	8	6	14
	MRSM Terendak	7	19	26
	SMKA Sh Rodziah	0	27	27
	SMKA Sultan Mohamad	16	0	16
	SMK Tinggi Perempuan Melaka	0	18	18
	SMKA Tun Perak	5	13	18
TOTAL		36	83	119

Results for a decision on pursuing professional accounting qualification are summarised in Table 5. Interestingly, the number of respondents who intend to pursue professional accounting qualification is higher than those of pursuing accounting major which is approximately 84% of the respondents. Indirectly, this may indicate there are some students who have decided to directly enrol for professional accounting program after SPM. Both the number of male and female students who are interested in studying for professional accounting courses were greater compared to those of pursuing an accounting degree. This may also indicate that being professionally qualified accountants is more appealing to this group of students. Since the questionnaires were completed after the KPP program had ended, these results showed that the program was successfully influencing students to opt professionally qualified accountants as their future careers.

[Type here]

Table 5: Group Size and Decisions on Taking Professional Accounting Course by Schools

Decision		Male	Female	Total
No	MRSM Tun Ghafar Baba	1	2	3
	MRSM Terendak	0	1	1
	SMKA Sh Rodziah	0	1	1
	SMKA Sultan Mohamad	2	0	2
	SMK Tinggi Perempuan Melaka	0	8	8
	SMKA Tun Perak	1	0	0
TOTAL		4	12	15
Not sure yet				
	MRSM Tun Ghafar Baba	2	0	2
	MRSM Terendak	0	3	3
	SMKA Sh Rodziah	0	1	1
	SMKA Sultan Mohamad	2	0	2
	SMK Tinggi Perempuan Melaka	0	4	2
	SMKA Tun Perak	0	1	4
TOTAL		4	9	13
Yes				
	MRSM Tun Ghafar Baba	10	8	18
	MRSM Terendak	11	19	30
	SMKA Sh Rodziah	0	30	30
	SMKA Sultan Mohamad	23	0	23
	SMK Tinggi Perempuan Melaka	0	24	24
	SMKA Tun Perak	7	16	23
TOTAL		51	97	148

The following table shows which factors are significantly correlated to students' decisions to pursue accounting major and to pursue professional accounting qualifications. The data were both correlated by parametric and non-parametric correlations. Both results were consistent. The results shown were based on a non-parametric analysis which was slightly lower both in coefficients and statistical powers than those from parametric analyses.

Table 6: Factors with Significant Relation (Spearman's rho correlation)

Factors	Pursue Accounting Major	Pursue Professional Accounting Qualifications
1.Pursue Accounting Major		0.319 (p=0.000)***
2.Exposure from career talks and seminars	0.125 (p=0.098)*	
3.Parents' influence	0.142 (p=0.059)*	0.155 (p=0.040)*
4.School counsellor influence	0.142 (p=0.060)*	
5.Family members work as accountants or/and auditors	0.683 (p=0.000)***	0.379 (p=0.000)***
6.Eagerness for related information	0.206 (p=0.006)***	
7.Career choice – public accountant	0.248 (p=0.001)***	0.230 (p=0.002)***
8.Career choice – private accountant	0.146 (p=0.053)*	
9.Career choice – other	-0.329 (p=0.000)***	

All correlations were based on Spearman's rho for non-parametric data
*correlation is significant at the 0.10 level (2-tailed)
**correlation is significant at the 0.05 level (2-tailed)
***correlation is significant at the 0.01 level (2-tailed)

This study finds that more factors are significantly related to the decision to pursue professional accounting qualification than to decision to pursue accounting major. Eight of these factors were positively related to students' decisions to opt for an accounting degree and one factor, choosing other than accountant as a career choice was negatively related to students' decision to major in accounting. The most powerful motivating factor or influence was family members working as accountants and auditors. It is likely that these family members portray a positive image of accountants who have positively motivated these Gen Y

[Type here]

students to study accounting in universities. This factor is related to perceived job-related factors like good working condition and job satisfaction which are suggested by existing studies such as Byrne, Willis and Burke (2012) as the most powerful motivating factor towards choosing accountant as a career. The second strong factor was a public accountant as a career choice, followed by students' level of keenness or eagerness for receiving information related to accounting major, professional accounting courses and from the experience-sharing session with experienced accounting educator and young qualified accounting professionals. Students with a reasonable prior interest in accounting and accounting profession are very excited to gather relevant information for a program such as KPP. The result in that support career talks and similar programs as a significant factor is consistent with Sugahara et al., (2009). However, exposure for a career talk and seminars, parents and school councillors were statistically significant but at lower significant levels.

There were only three factors which were significantly related to students' decision to pursue professional accounting courses. These factors were family members in the accounting profession, public accountant as a career choice and parents' influence. However, these factors were statistically significant at a higher level. Nevertheless, students' decision to major in accounting was significantly related to their decisions to pursue professional accounting courses as 119 of them have decided to enrol in accounting degree first.

CONCLUSION AND RECOMMENDATIONS

This study attempts to identify what factors motivate students to major in accounting and to take professional accounting courses. Data was collected via a questionnaire which was distributed to each participant of the KPP one-day program. The study finds that majority of the respondents were female and the majority of them intends to choose accounting as their major and greater numbers of respondents are interested in pursuing professional accounting qualification relative to an accounting degree.

Thirteen factors were extracted from the questionnaire and based on statistical analysis, seven factors were positively influencing students' decisions to major in accounting. Ranked in accordance to the statistical power, these factors are family members in the accounting profession, public accountant as a career choice, students' interest in receiving relevant information, choosing private accountants as a career, parents, school counsellor and career talk. Expectedly, one factor, choosing another profession as a career will have opposite effect on students' decision. Students whose ambitions are not related to accounting will not opt to study accounting.

Despite greater numbers of students decide to pursue professional accounting courses, this study only finds four significant motivating factors. These factors are a decision to major in accounting, family members in the accounting profession, public accountant as a career choice and parents. Students are more motivated by parents to take professional accounting courses as compared to enrol for accounting degree which may highlight the need for family moral support to ease the perceived difficulty related to professional accounting courses.

Since most of the KPP program contents focused on professionally qualified accountants, a higher number of students chose to do professional accounting courses relative to accounting degree showed that the KPP program had to a certain degree made a positive impact on students' inclination towards being professionally qualified accountants. Therefore, this study

[Type here]

recommends other universities to organise similar programs to increase school students' awareness towards accounting courses and to choose qualified accountant as their desired career as well as to scout for quality potential accounting students and accounting professionals.

REFERENCES

Agarwala, T. (2008). Factors influencing career choice of management students in India. *Career Development International*, 13(4), 362–376.

Ahmed, K., Alam, K. F., & Alam, M. (1997). An empirical study of factors affecting accounting students' career choice in New Zealand. *Accounting Education*, 6(4), 325–335.

Allen, C. L. (2004). Business students' perception of the image of accounting. *Managerial Auditing Journal,* 19(2), 235-258.

Blackwell, E. & Pinder, P. J. (2013). What are the motivational factors of first-generation minority college students who overcome their family histories to pursue higher education?. STEM Education Research Institute (SERI), Indiana University, Purdue University.

Blackwell, E., & Pinder, P. J. (2014). What are the motivational factors of first-generation minority college students who overcome their family histories to pursue higher education? *College Student Journal*, *48*(1), 45–56.

Byrne, M., Willis, P., & Burke, J. (2012). Influences on school leavers' career decisions – Implications for the accounting profession. *The International Journal of Management Education*, 10(2), 101–111.

Carpenter, P. & Foster, B. (1977). The career decisions of student teachers. *Educational Research and Perspectives*, 4(1), 23–33.

Cohen, J. & Hanno, D. M. (1993). An analysis of underlying constructs affecting the choice of accounting as a major. *Issues in Accounting Education*, 8(2), 219–238.

Committee to Strengthen the Accountancy Profession (CSAP) (2014). Report on the strengthening of the accountancy profession in Malaysia.

Dalci, I., Arasli, H., Tumer, M., & Baradani, S. (2013). Factors that influence Iranian students' decision to choose accounting major. *Journal of Accounting in Emerging Economics*, 3(2), 145-163.

Dalton, D. W., Buchheit, S. & McMillan, J. J. (2014). Audit and tax career paths in public accounting: An analysis of student and professional perceptions. *Accounting Horizons,* 28(2), 213-231.

Felton, S., Buhr, N. & Northey, M. (1994). Factors influencing the business student's choice of a career in chartered accountancy. *Issues in Accounting Education*, 9(1), 131–141.

[Type here]

Felton, S., Dimnik, T. & Northey, M. (1995). A Theory of reasoned action model of the chartered accountant career choice. *Journal of Accounting Education*, 13(1), 1–19.

Geiger, M. A. & Ogilby, S. M. (2000). The first course in accounting students' perception and their effect on the decision to major in accounting. *Journal of Accounting Education,* 18(2), 63-78.

Gul, F. A., Andrew, B. H., Leong, S. C. & Ismail, Z. (1989). Factors influencing the choice of the discipline of study-accountancy, engineering, law, and medicine. *Accounting and Finance,* 29(2), 93-101.

Hermanson, Dana, R., & Hermanson, Roger, H. (1995) to confirm the name. Are America's top business students steering clear of accounting? *Ohio CPA Journal*, 54(2), 1–9.

Jackling, B. & Calero, C. (2006). Influences on undergraduate students' intentions to become qualified accountants: Evidence from Australia. *Accounting Education: An International Journal,* 15(4), 419-438.

Jackman, S., & Hollingworth, A. (2005). Factors influencing the career choice of accounting students: a New Zealand study. *NZ Journal of Applied Business Research,* 4(1), 69-84.

Law, P. K. (2010). A theory of reasoned action model of accounting students' career choice in public accounting practices in the post-Enron. *Journal of Applied Accounting,* 11(1), 58-73.

Lowe, D. R. & Simons, K. (1997). Factors influencing the choice of business major some additional evidence: A research note. *Accounting Education*, 6(1), 39–45.

Malaysian Institute of Accountant (MIA). (2015). *Malaysian Institute of Accountants Annual Report 2015.*

Malaysian Institute of Accountant (MIA). (2014). *Malaysian Institute of Accountants Annual Report 2014.*

Marriott, P., & Marriott, N. (2003). Are we turning them on? A longitudinal study of undergraduate accounting students' attitude towards accounting as a profession. *Accounting Education: An International Journal,* 12(2), 113-133.

Maslow, a. H. (1943). A theory of human motivation. *Psychological Review, 50*(13), 370–396.

Mauldin, S., Crain, J. L., & Mounce, P. H. (2000). The accounting principles instructors' influence on students' decision to major in accounting. *Journal of Education for Business*, 75(3), 142-148.

Nelson, I. T., Vendrzyk, V. P., Quirin, J. J. & Allen, R. D. (2002). No, the sky is not falling: Evidence of accounting student characteristics at FSA schools, 1995–2000. *Issues in Accounting Education*, 17(3), 269–287.

[Type here]

Paolillo, J. & Estes, R. (1982). An empirical analysis of career choice factors among accountants, attorneys, engineers, and physicians. *The Accounting Review,* 57(4), 785-793.

Ryan, R. & Deci, E. (2000). Intrinsic and extrinsic motivations: Classic definitions and new directions. *Contemporary Educational Psychology*, 25(1), 54–67.

Soni, G., & Kodali, R. (2012). A critical review of empirical research methodology in supply chain management. *Journal of Manufacturing Technology Management,* 23(6), 753-779.

Sugahara, S. & Boland, G. (2006). Perceptions of the certified public accountants by accounting and non-accounting tertiary students in Japan. *Asian Review of Accounting,* 14(1/2), 149-167.

Sugahara, S., Hiramatsu, K. & Boland, G. (2009). The factors influencing accounting school students' career intention to become a Certified Public Accountant in Japan. *Asian Review of Accounting*, 17(1), 5–22.

Smyth, E., & Hannan, C. (2000). Spoilt for choice? School factors and third-level entry. Seminar paper. Dublin: The Economic and Social Research Institute.

Tan, L. M. & Laswad, F. (2006). Students' beliefs, attitudes and intentions to major in accounting. *Accounting Education*, 15(2), 167–187.

Ulmer, J. T., & Wilson, M. S. (2003). The potential contributions of quantitative research to symbolic interactionism. *Symbolic Interaction,* 26(4), 531-552.

Wilder, M. & Stocks, M. H. (2004). Are accounting graduates favourably recruited for entry-level management positions? *Management Accounting Quarterly,* 5(3), 21–28.

Zyl, C. V. & Villiers, C. D. (2011). Why some students choose to become chartered accountants (and others do not). *Meditari Accountancy Research,* 19(1/2), 56-74.

[Type here]

GOODS AND SERVICES TAX (GST): THE IMPORTANCE OF COMPREHENSION TOWARDS ACHIEVING THE DESIRED AWARENESS AMONG MALAYSIAN

Rani Diana Othman*, Nur Farahah Mohd Pauzi, Siti Anis Nadia Abu Bakar, Afidah Sapari
Mohd Syazwan Karim
Faculty of Accountancy, Universiti Teknologi MARA Melaka

ABSTRACT

In line with the implementation of value added tax (VAT) by 160 countries, Malaysia has taken its own path with the introduction of goods and services tax (GST) starting from 1 April 2015 to replace its previous tax system. This new tax system has been announced on 25 October 2013, which has given about 17 months for businesses and people in Malaysia to make a complete preparation for GST. GST is proven to be the fairer tax system to overcome the previous one which was not only limited in scope, but also came with a number of inherent weaknesses. Therefore, in order to achieve the desired awareness regarding GST, it is essential that all Malaysians have the required level of understanding in terms of its concepts, scopes, charges and mechanisms. This paper discusses the benefits of GST for the nation in the long run as well as the negative perceptions from the public. This should reveal the additional steps to be taken by the regulators in ensuring a better comprehension of this new tax system among the citizens as a support for the government's tax reform, which has been programmed for Malaysia to move towards a high-income nation.

Keywords*:* Goods and Services Tax, Sales and Services Tax, Tax Reform

INTRODUCTION

The implementation of Goods and Services Tax (GST) in Malaysia has been commented by many parties including from the Taxand Malaysia Sdn Bhd Chairman, Dr Veerinderjeet Singh, who said that, *"Malaysia is currently in a period of uncertainty and it was not the right time to reduce income tax, even with GST in place"* (Singh, 2013). Such comment was made since Malaysia has been facing fiscal deficit without a booming economic climate and slower future growth for a number of previous years. The following were 2012 Malaysia's fiscal positions, which have been highlighted as vulnerable indicators of Malaysian economic condition by Othman & Mirakhor (2012):

*Corresponding author's email: diana@bdrmelaka.uitm.edu.my

DOI: 10.1051/shsconf/20173600036

Table 1: Fiscal position of Malaysia on 2012

	Weak fiscal position	Malaysia's position
1.	Overall fiscal balance as a share of gross domestic product (GDP)	1. Persistent deficit for the past 15 years, where Malaysia's budget deficit in 2011 stood at 5.3% of GDP, while national debt was 53% of GDP. 2. Malaysia is still running fiscal deficits even when Malaysia's economy is growing by more than 5% in 2010. *(Debt and growth figures based on www.tradingeconomics.com)*
2.	Net financial debt as a share of GDP	1. National debt was reported to be 53% of GDP in 2011 and improved to 52% in 2012. 2. However, it was projected to climb to 54% in 2013. 3. The increase in issuance of government securities has led to an increase in interest payment from 7.6% in 2012 to 10.3% in 2013 which considered a 36% of debt service ratio (Bank Negara Report, 2011). *(External debt service ratio as percentage of export goods and service)*
3.	Sizeable uncovered contingent liabilities	The size of debts is estimated to be approximately 67% of GDP if contingent liabilities are considered. *(Based on contingent liability of RM117 billion reported in the Malaysia Insider, 20 June 2012)*
4.	Large share of non-discretionary spending and/or transfer	Incomparable of operating expenditure to development expenditure. In 2010, operating expenditure was RM152 billion compared to development expenditure of RM45 billion. The operating expenditure has increased to RM182 billion in 2012 compared to only RM48 billions of development expenditure of the same year. This shows that Malaysia is running deficits to fund spending which does not add value or productivity in the long run.

Due to weak fiscal positions mentioned above, suggestions have been made to improve the Malaysian economic conditions. A short term fiscal policy was proposed to be designed by taking into consideration the review of the sustainability indicators and the role strengthening of the long-term estimates (Abdullah, Muszafarshah & Dahalan, 2012). The objective of developing the fiscal policies by governments across the world is to boost the economic condition and the policy, which can either improve the usage of expenditure or revenue collection. Othman & Mirakhor (2012) proposed a framework focusing on a tax structure that can improve the revenue and the public sector financing which mobilises non-debt-creating sources to finance development expenditure. Another suggestion was also made by Professor Datuk Muhammed Ariff, a renowned Malaysian economist, in his statement as quoted by The Star, on January 7, 2012, *"To balance the books, the Government must rein in its operating expenditure and introduce tax reform"*. This is to confirm that a policy reform needs to be considered by looking at a simpler structure, where the levy of tax is based on the ability to pay.

One of the major tax reforms by the government of Malaysia was the introduction of GST, which has been implemented on April 1, 2015 with an intention to replace the Sales and Service Tax (SST). SST that has been used in the country for several decades was a form of indirect tax imposed on "taxable services" and "taxable goods" which were strictly prescribed by the act (Sales Tax Act 1972 and Service Tax Act 1975). Sales tax is operated in the same negative concept as GST, where all goods were subject to it unless specifically prescribed. Unlike the former, service tax operated in the opposite concept where only services specifically prescribed by the act were taxable. Both taxes were known as single stage tax, where sales tax became due and payable when there was a sale or disposal, and service tax was only due when payment was received. However, if payment was not received, the service tax was accounted for at the end of the 12-month period from the date of invoice issued.

*Corresponding author's email: diana@bdrmelaka.uitm.edu.my

DOI: 10.1051/shsconf/20173600036

GST is an indirect tax administered by The Royal Malaysian Customs Department (RMCD) and governed by the Goods and Services Tax Act 2014. GST is also commonly referred to as Value Added Tax (VAT) and collected at various stages of production. The collection of GST is based on output tax, which will be charged by the supplier on taxable goods as well as services, and input tax, which is incurred by the businesses on goods and services purchases. According to Tan Sri Ariffin Zakaria (2010), Chief Justice of Malaysia, the introduction of GST will be a major tax reform in the country, which is able to enhance the capability, effectiveness, and also transparency of the tax administration and management. This new tax system is also expected to prepare savings for the country should it experience inflation in the future (Palil & Ibrahim, 2011). Their research has also highlighted that the most salient feature of GST is its regressive nature with respect to income.

GST has been implemented in nearly 160 countries with different experiences of achievements and obstacles. Our neighbouring country, Singapore, has adopted GST on April 1994 and has succeeded based on advantages such as the fact that the tax is on consumption, which encourages savings and investments, rewards enterprise, and strengthens economic resilience. The system is also claimed to stimulate the competitiveness of a country (Low & Carol, 1994). As for India, after various trials in reforming the indirect tax structure, its government did not encounter many problems since there were no differences in term of levying the tax towards the GST as both are imposed on the same rate (Syed, Amit & Ravindra, 2013). Implementation of GST in New Zealand showed that the original rate in 1986 was 12.5 percent and increased to 15 percent in 2010. The advantage of GST in New Zealand is that the "zero-rate" is charged on businesses exporting goods and services, which permits those businesses to claim back the input tax as GST return because the non-New Zealand based consumers do not pay the tax.

Despite the initiative of introducing GST as an effort to reduce federal government deficit, it is aimed to diversify the nation's sources of income, provide a constant flow of earnings as to elevate public service, and also improve economic stability (McGowan & Billings, 1997). Overall, Malaysians are not fully convinced with the new tax scheme. Csontos, Kornai and Toth (1998) highlighted that such a perception was due to pessimistic view, which refers to citizens who have overestimated tax burden or inaccurately estimated public service burden, with the tax imposed to the public. This has further caused a misconception on tax requirements, in particular towards the recently introduced tax reform. Therefore, having sufficient understanding of the new system is essential for the taxpayers to support GST as a way to overcome the national fiscal deficit problem and economic downturn.

GST TO OVERCOME THE WEAKNESSES OF SST

Malaysian Indirect Tax System under the Previous Regime of SST

Indirect tax was considered as the most convenient tax for the purpose of revenue generation compared to the direct tax from the public (Barjoyai, 1993). Malaysian indirect tax structure consisted of the collection of various revenues such as excise duties on alcohol, tobacco and a few other products, import duties and stamp duties, export duty, levy, as well as others. However, the most broadly based indirect tax in Malaysia was the SST (Kasipillai & Sinnakkannu, 2008). Sales tax was governed by the Sales Tax Act of 1972 while Service Tax was governed by the Service Tax Act of 1975.

Throughout the years, SST became the important source of revenue to the country. The source of revenue estimates from indirect taxes of the federal government are presented in Table 1

*Corresponding author's email: diana@bdrmelaka.uitm.edu.my

below, which shows the source of revenue from SST comprising of 45 percent in 2013 and 46 percent in 2014 from the total indirect tax collected.

Table 2: Estimates of Federal Government's Revenue (2013 and 2014)

Source of revenue		**2013**		**2014**
	(RM million)	**(RM million)**	**(RM million)**	**(RM million)**
Direct Tax		127,020		133,148
Indirect Tax		36,486		38,822
- Export duty	2,099		2,105	
- Import duty	2,417		2,502	
- Excise duty	12,728		13,442	
- Sales tax	10,199		10,986	
- Service tax	6,161		6,810	
- Levy	150		150	
- Others	2,732		2,827	
Non tax revenue		54,428		50,627

Source: Estimates of Federal Government's Revenue Report (2014), Ministry of Finance

SST was a single stage ad valorem tax imposed based on the respective scope of charges that were accountable to the Royal Malaysian Customs Department (RMCD). The sales tax was imposed on taxable locally manufactured and/or imported goods. It was considered as a consumption tax, and therefore, under the system, the onus was on the manufacturers to levy charges and collect tax from their customers (Mansor, Tayib & Yusof, 2005). For imported goods, sales tax was collected from the importers upon the release of goods from customs control, which was computed, based on the value of taxable goods sold, used or disposed of, or imported. With effect from January 2003, the valuation method for locally manufactured goods was based on the transaction value as determined by the World Trade Organisation's (WTO) method of valuation (Mansor et al., 2005). Furthermore, starting from October 31, 2012, the sales tax rate varied from 0 percent, 5 percent, and 10 percent to 20 percent, depending on the categories of goods subject to sales tax. Certain goods such as food and fruits for general lifestyle were nil rates on sales tax, food and fruits for expensive lifestyle, as well as beer and wine were 5 percent on sales tax, whereas other goods were taxed standard rate of 10 percent (Choong, 2014). Specific sales tax rates were also imposed on certain classes of petroleum such as refined petroleum, diesel fuel, gas oil, and liquefied natural gas (Sapiei & Abdullah, 2015).

Section 13 of Sales Tax Act 1972 stated that every manufacturer of taxable goods shall apply for a sales tax license (JKED No. 1) (Choong, 2014). However the exemption from licensing applied to manufacturers whose respective turnovers were below RM100,000 and contractors whose respective turnovers were below RM20,000. Apart from exported goods, an exemption from sales tax also applied to goods listed in Schedule A to the Sales Tax (Exemption) Order 1980 (Kasipillai & Sinnakkannu, 2008), goods listed in Schedule B of the Sales Tax (Exemption From Licensing) Order 1997, Free Zone (Section 2A), licensed warehouse and licensed manufacturing warehouse (s2B), as well as goods manufactured in Labuan, Langkawi and Tioman (Choong, 2014).

The service tax had to be charged and levied on the taxable services provided by the taxable persons based on various rates for services such as restaurants, hotels, and health centres. It had been expanded to include other professional services such as lawyers, valuators, forwarding agencies, insurance companies, accountants, engineers, architects, and private hospitals. It also included the telecommunications services, security guard services, parking services, car rental and other taxable services, whose respective annual turnovers reached the threshold limit between RM150,000 and RM300,000, depending on the group of taxable

*Corresponding author's email: diana@bdrmelaka.uitm.edu.my

persons (Kasipillai & Sinnakkannu, 2008; Choong, 2014). The service tax applied at a single rate of 6 percent (with effect from 1 January 2011) throughout Malaysia except Langkawi, Labuan, Tioman, and Joint Development Area as defined in s2 Malaysia – Thailand Joint Authority Act 1990 (Choong, 2014).

Weaknesses of SST to be overcome by GST

The new regime of GST needs to be implemented in Malaysia due to the weaknesses of the previous SST system as highlighted by the former Deputy Minister of Finance, Datuk Chan Kong Choy (Mansor et al., 2005), who disclosed the fact that tax authorities were unable to collect a total of RM4.2 billion of taxes inclusive of the unpaid sales tax of RM75.7 million, service tax of RM46.8 million, custom duties of RM46.7 million, and excise duties of RM0.57 million as of April 2002. Furthermore, based on RMCD's statistics, since 2005 until 31 December 2008, RM335.95 million of taxes were not collected or failed to be paid by the taxpayers (Bidin, Shalihen, Shamsudin M & Othman, 2011). Kasipillai & Sinnakkannu (2008) stated that the sales tax was a complex tax on the manufacturing of goods and the importation of manufactured goods. The complexity included the exemption mechanism since it depended on the turnovers of licensed manufacturers and the types of manufacturing activity. Besides that, the service tax was also found to have an arbitrary effect on consumers' prices. The tax only applied to a set of defined activities classified as "taxable services" whose value had exceeded a certain annual threshold.

The weaknesses of the SST also included high incidence of tax avoidance, double taxation along the value-chain that pushed prices up, as well as extensive exemptions and this explained the low collection of revenue from SST (Zhou, Tam & Heng, 2013). From the total estimation of the federal government revenue in 2013, sales tax represented only 4.6 percent while service tax represented only 3.1 percent, whereby in total, SST contributed only 7.7 percent of revenue. Therefore a new form of tax was needed to both expand the tax base that would increase the revenue collection of the country, as well as gradually replace oil-related revenue and excise duties as it was already known that the major source of government revenue has come from petroleum and taxation. The increase in revenue for the country is important to finance the government expenditure as well as to help providing more socio-economic programmes (Alappatt & Shaikh, 2014).

Indirect tax under the new regime of GST

GST is a broad-based consumption tax on which a person pays when he or she spends money (Alappatt & Shaikh, 2014; Sapiei & Abdullah, 2015). More than 120 countries have already implemented GST (Alappatt & Shaikh, 2014). It is considered as a multi-staged value added tax, which means tax is charged on any taxable goods and services at all levels in the supply chain, namely production, manufacturing, wholesale, retail, and through to the final consumer. In contrast, the previous sales tax was imposed on the manufacturer before the service tax was imposed on the consumer at the point of purchase. Thus, double taxations in SST pushed up prices and reduced consumer welfare (Zhou et al., (2013). However, since GST is collected and paid at all stages of the supply chain, the registrants will charge and collect the tax on taxable goods and services they provide, which are the taxable supplies. The GST paid on their purchases, known as input tax credit, will be set off from the GST they charged and collected, that is the output tax. If the output tax exceeds the input tax, the difference is to be remitted to

*Corresponding author's email: diana@bdrmelaka.uitm.edu.my

the customs department. On the other hand, if the output tax is less than the input tax, a refund will be given by the Customs authorities (Alappatt & Shaikh, 2014).

A "taxable person" in GST means a person registered under Section 19, Goods and Services Tax Act 2014, "goods" means any kind of moveable and immovable properties but excludes money except a bank note or coin, whereas "supply" means all forms of supply, including supply of imported services [S.4(1)] (Sapiei & Abdullah, 2015, p 641). Thus GST can only be charged on the taxable supply made by a registered business. A supply must either be standard rated or zero-rated; exempt and out of scope supplies that are not taxable supplies (Department RMC, 2014).

The current standard GST rate is 6 percent and it is collected by the business and paid to the government, whereas zero-rated supply is applicable to any goods or services determined by the Minister [GST (Zero-Rated Supply) Order 2014] and to any goods if they are exported (Department RMC, 2014). For the zero-rated supply, a registered business cannot charge a tax to the customer, but it can claim credit for the GST paid on their business inputs. Mostly, the items under zero-rated supply comprise of basic foods such as rice, sugar, flour, fish, chicken, cooking oil, and egg. Meanwhile, exempted supply refers to non-taxable items that are not subject to GST [GST (Exempt Supply) Order 2014] such as public transportation, health services, public education, as well as funeral services. Out of scope supply consists of any items that are considered as out of scope and not subject to GST such as supply by state and federal government, supply by a business below the RM500,000 threshold, supply of goods outside Malaysia, supply of services outside Malaysia by a person not in Malaysia, and supply of goods and services made within tax-free areas in Malaysia, namely Labuan, Langkawi and Tioman.

According to Zhou et al., (2013, p.2), GST offers a single unified system where the tax burden is equally shared between the services and the manufacturing industries, whilst simultaneously broadening the tax base. This will help to minimise tax exemptions as well as the compounding effects of pyramiding tax, tax erosion, transfer pricing, and value shifting. It will also ensure a level of stability, as GST is less susceptible to the fluctuations inherent in the commodity markets. Besides that, GST is claimable by the businesses, thus the lesser incentive to evade tax.

WHY GST?

Improved Standard of Living towards Building the Nation

Taxation is a very important element and a major source of revenue for the economic development and growth. For most countries, the revenue is generated through taxation, that comprises of direct and indirect tax. The contribution to national fiscal revenue through the implementation of VAT by the developing countries, such as China, is much more than the developed countries (Xing & Whalley, 2014) and VAT has become the major source of tax revenue for China (Liu & Lu, 2015). In Malaysia, the implementation of GST is aimed to improve the revenue collection and reduce the nation's growing budget deficit, which ultimately could increase the standard of living among Malaysians as a whole.

From 1960s to 1970s, only a few countries adopted GST and most of them still relied on the direct taxes such as corporate tax and personal income tax. A very high tax rate was set in the early 1970s, such as United Kingdom with 75 percent, South Korea with 70 percent, and Taiwan with 60 percent (Low & Carol, 1994). As a consequence, high corporate taxes tend to

*Corresponding author's email: diana@bdrmelaka.uitm.edu.my

discourage initiatives of foreign investors to conduct international trade with those countries. According to Low & Carol (1994), one of the advantages for the implementation of GST in Singapore is its help in raising the national income. On the other hand, among the reasons for the implementation is to sustain the lower corporate tax rate in the country. Therefore, this shall inculcate the spirit of entrepreneurship among Singaporeans, which in turn will add values to its business environment and further attract foreign investors. Besides that, the implementation of GST could reduce the reliance on direct taxation as a main source of revenue. GST clearly could be a more stable source of revenue, enabling the government to channel back the benefits to the people. Therefore, the tax reform program adopted by the Malaysian government intends to earn a more stable income and improve standard of living among Malaysians. It is indeed a good economic development strategy in moving towards a high-income nation by the year 2020.

Enhanced Compliance

Since the Malaysian government introduced GST to replace SST, it has created an integrated system that would enhance the level of conformity among the tax payers. In the 2005 budget speech, the former Prime Minister, Tun Abdullah bin Haji Ahmad Badawi, mentioned that GST could improve tax compliance:

"… The government proposes to replace both of these taxes (Sales Tax and Service Tax) with a single consumption tax, based on the value-added concept. The new tax, known as the Goods and Services Tax (GST), will be more comprehensive, efficient, transparent and effective, thereby enhancing tax compliance…" (KPMG, 2013)

Previously, there were two separate acts in governing the sales and service tax, which were both handled by the RMCD. Therefore, it indicated a complex system of laws and regulations that caused a decrease of compliance among the tax payers. Initially, Malaysians showed a resistance towards GST implementation as the businesses needed to adapt with the changes in administration works that consequently incurred high cost. However, the government has taken significant steps to enhance the tax compliance by ensuring that all businesses between the thresholds have registered as GST registrants and accordingly update their accounting systems, providing the necessary trainings to those companies to assure a thorough understanding of GST key element that include the concepts, scopes, charges, and mechanisms, as well as offering advices and close monitoring on the treatment of transitional issues upon the implementation (malaysia-gst.my). Hence, it is a good effort of the Malaysian government to implement GST under the tax reform program as it enhances tax compliance and reduce tax avoidance nationwide.

Fair Pricing to Consumers

The supply chain network for business transactions would usually consist of manufacturers, suppliers, warehouses, distribution centres, dealers, and more (Zhang & Huang, 2014). According to Xing & Whalley (2014), the government will levy the VAT to the value added at each of the chain in order to avoid double taxation. Therefore, the net tax payable would represent the one on the value-added goods processed and distributed or services rendered (Carvalho & Lian, 2010). The implementation of GST in Malaysia is also designed to eliminate double taxation under the previous indirect tax charges of SST so that consumers will pay fairer prices for most goods and services.

*Corresponding author's email: diana@bdrmelaka.uitm.edu.my

After the implementation of GST, 22 products which were previously subject to SST have been exempted from the GST. While 273 products which were earlier imposed with SST of 10 percent are now subjected to the GST of 6 percent. Nevertheless, the main problem to be resolved is the traders' manipulation of the prices (The Star Online, May 2015). GST would no longer be an issue to the consumers once the prices of all consumers' products could be controlled and stabilised.

Greater Transparency

GST is designed to benefit consumers unlike the previous SST whereby they will exactly know whether the goods and services are subject to tax and the prices stated conform to the specified rates. Accordingly, it would reveal the hidden tax by the manufacturers (Whalley & Fretz, 1990), which ultimately could increase on the competitiveness among the businesses. According to Keen & Lockwood (2010), VAT has been introduced as part of a program that modernises tax administration. It develops self-assessment method, which would reduce the difficulty in administration and compliance. Therefore, businesses become the tax collectors of GST until the final remittance to the RMCD after setting off the correct amount of input tax credit. A strict adherence is required relating to timing and remittance as well as invoicing requirements (Beh & Ng, 2013). This is essential to reach the neccessary level of transparency in order to overcome the previous tax system that was vulnerable to evasion and avoidance.

In Malaysia, a notice is displayed at all supermarkets to help customers identify the goods with 6 percent GST, those with zero-rate, and items that have been exempted from GST. This will help customers to plan their spending, thus avoiding unnecessary payment on indirect tax. In addition, each of the transactions made by the customers will be accompanied by the simplified tax invoice as reference. Under the previous tax regime of SST, the final consumers would not know the amount of indirect tax they had to pay as the charges had been included in the final prices of goods and services, with no detail stated on the receipt of purchase (Malaysia Kini, 2015). The disclosure of indirect tax is essential for customers to realise that GST is a compulsory consumption tax and they have no choice but to pay as per what they consume (The Star Online, April 2015).

NEGATIVE PERCEPTION OF GST

Increase in Price for Goods and Services

Consumers are always concern that the implementation of GST or VAT will cause the price to hike up (Mohd Rashid & Shaik Asgar, 2014). However, the implementation of VAT in Australia showed no significant difference in the price within eight capital cities (Valadkhani & Layton, 2004). On the other hand, in Canada, the price only hiked up once, but not continuously after the implementation of VAT (Dombrovski & Hodžić, 2010). Even though there was an increment in price after the implementation of GST due to the unstable transitional stage from the previous tax system to the new tax system, Singapore has proven that they can maintain the inflationary pressure from rising up (Palil & Ibrahim, 2011).

According to Deputy Minister of Finance, Datuk Haji Ahmad Maslan, the implementation of GST will not increase prices, but further reduce or retain them. This has been proven with the effort by one of the biggest grocery and merchandise retailers in Malaysia, Mydin Mall, that managed to maintain the same prices as well as announcing the absorbtion of indirect tax for the products which have been charged with GST (Malaysia Kini, 2015). The decision of GST

*Corresponding author's email: diana@bdrmelaka.uitm.edu.my

absorption by Mydin Mall has been followed by many grocery and merchandise retailers as well as other businesses in Malaysia. As a result, there has been an increased competitiveness in the market that would continue in the long run.

Roles of Tax Authority

A few days after the implementation of GST in Malaysia, the news media reported drastically increased prices for consumers' products. This had happened due to the greed of businessmen, traders, and hawkers. Consequently, various complaints were recorded over seven days of implementation whereby the Domestic Trade, Cooperatives, and Consumerism Minister had received about 2,833 complaints on price increase in the market (Bernama, April 2014). This has aroused a curiosity among Malaysians on the role of tax authority as well as the government in handling these issues. They were indeed very intense at the beginning of GST implementation, subsequent to the political issues related to the new system.

In order to manage the problem, the Prime Minister, Dato' Seri Mohd Najib bin Tun Abdul Razak, has set up a special court of anti-profiteering in each capital state. The court handles special cases related to any offences by profiteers who exploit GST by unreasonably increasing the prices of goods and services (Bernama, June 2015). This is indeed an essential step to show the comprehensive efforts by the government to overcome the issue of profiteering at the beginning of GST implementation. Besides the courts, the fine and punishment charged on those convicted guilty indicated the severity of the crime. Any individuals that make unreasonable profits in selling or supplying goods are liable to a fine not exceeding RM100,000 or to imprisonment not exceeding three years, or both. Meanwhile, for any companies that are guilty of the same offences, they may be fined up to RM500,000 for the first offence and RM1,000,000 for the subsequent offence (The Star Online, March 2015). This is an indirect yet significant effort by the government to warn the public on the compliance issue.

Administrative Problem

The administrative problem lies within the complexity of the documents involved. GST is totally a new tax system which requires details of specific information on each part of the distribution channel. Therefore, a lot of documents are needed to cope with the procedures and guidelines in this new system. Nevertheless, according to Juddoo (2014), GST is a simple tax system that is essential for a better understanding by individuals and corporate taxpayers. A simple system also means fewer resources that are still able to effectively manage socially unproductive activities such as tax planning and litigation.

In particular, the administrative issues arise due to the unfamiliarity with the new system. This is a common problem faced by any organisations. Therefore, they may need time not only to familiarise themselves with the whole system, but also to reassess their approaches in managing the emerging issues. The understanding of how GST will affect their production processes will specifically help them in dealing with the compliance matters. Meanwhile, the government has undertaken the proactive measures by providing a free seminar to those taxable suppliers (www.customs.gov.my) as well as allowing them to claim for a special deduction under corporate taxation. All the training fees are claimable under Human Resources Development Fund's Scheme for attending a private training organised by a professional body in order to gain more knowledge and new adjustment for GST accounting treatment (Palil & Ibrahim, 2011).

*Corresponding author's email: diana@bdrmelaka.uitm.edu.my

CONCLUSION

GST is the new regime of indirect tax which differs significantly from the previous system. It brings a fundamental conversion to a multi-stage consumption tax, with vast procedural differences. This new approach covers from administrative matters including registration and invoicing, to penalty and interest rates. Such an overhaul will undoubtedly change the landscape of the commercial operation in Malaysia. In order to ensure the success of this modern consumption tax in the country, several recommendations have been made, such as, firstly, raising awareness on the exemption list and limit the expansion of the list; secondly, strengthen the cooperation with local consumer societies in order to fight profiteering activities; thirdly, commit to a five-year timeline and review GST with possibility of raising the rate by the year 2020; and finally, enhance the campaign on the importance of reducing budget deficit in order to raise public acceptance on arising issues (Zhou, Tam & Heng-Contaxis, 2013).

However, the most important factor to be considered is to solidify the comprehension among the public towards the scopes, charges, and mechanisms of the new GST system. Mohani (2003) and Tan & Chin-Fatt (2000) have asserted that the ways to increase public awareness is through tax knowledge that can be imparted by a general understanding on the tax regulations. Therefore, the government must put extra efforts in order to ensure that Malaysian taxpayers truly understand GST and gather their commitment of acceptance. In conclusion, a further research needs to be done in order to investigate the sufficiency of GST awareness campaign conducted by the government through various mass and social media. It is also claimed that the taxpayers' level of awareness has been associated with their tax knowledge (Palil & Ibrahim, 2011). Future research should also evaluate the efficacy of this tax reform in terms of its ability in enhancing the transparency of Malaysian tax administration and management. The most important aspect is that whether this new tax system is able to improve the nation's economic condition in preparing Malaysia to be a developed country by the year 2020.

REFERENCES

Abdullah, H., Muszafarshah, M. M. & Dahalan, J. (2012). An empirical study on fiscal sustainability in Malaysia. *International Journal Academic Research in Business and Social Sciences,* 2(1), 72-90.

Alappatt, M. & Shaikh J. M. (2014). Forthcoming procedure of goods and service tax (GST) in Malaysia. *Issues in Business Management and Economics*, 2(12), 210-213.

Bank Negara Malaysia (2011). *Bank Negara Malaysia Annual Report 2011.*

Beh, Y. & Ng, K. (2013). Implementation of goods and services tax in Malaysia. *Client Alert*, Wong and Partners, Member Firm of Baker and Mckenzie International.

Bernama (2014). 98 Offences recorded since implementation of GST, April 7. Retrieved from http://www.bernama.com.

Bernama (2015). Najib visits anti-profiteering operations centre at KPDNKK, June 29. Retrieved from http://www.bernama.com.

*Corresponding author's email: diana@bdrmelaka.uitm.edu.my

Bidin, Z., Shalihen, M. S., Shamsudin, F. M. & Othman, Z. (2011). Factors influencing intention to comply with local sales tax in Malaysia. *SSRN Electronic Journal.*

Carvalho, E. & Lian, Z. (2010). The impact of goods and services tax on the pattern of Canadian concumer spending and saving. *The Journal of Economic Asymmetries*, 17(1), 123-136.

Choong, K.F. (2014). *Malaysian taxation: Principles and practice.* 20th Edition, Malaysia: Infoworld.

Csontos, L., Kornai, J. & Toth, I. G. (1998). Tax awareness and reform of the welfare state: Hungarian survey result. *Economics of Transition*, 6(2), 287-312.

Department RMC (2014). "About GST" What is GST. Retrieved from http://gst.customs.gov.my/en/gst/Pages/gst_wi_aspx.

Dombrovski, R. & Hodžić, S. (2010). Impact of value added tax on tourism, *The International Business and Economics Research Journal*, 10, 131-138.

http://malaysia-gst.my/
http://www.customs.gov.my
Juddoo, K. (2014). The compliance cost of value added tax (VAT): The case of the republic of Mauritius. *eJournal of Tax Research*, 12(2), 499-521.

Kasipillai, J. & Sinnakkannu, J. (2008). Distributive effects of the introduction of GST in Malaysia, *International VAT Monitor*, 359-366.

Keen, M. & Lockwood, B. (2010). The value added tax: Its causes and consequences. *Journal of Development Economics*, 92, 138-151.

KPMG (2013. GST in Malaysia.
Liu, Q. & Lu, Y. (2015). Firm investment and exporting: Evidence from China's value-added tax reform. *Journal of International Economics*.

Low, S. P. & Carol, P. W. L. (1994). Implementation of the goods and services tax (GST) in the Singapore construction industry. *Journal of Property Finance,* 5(3), 41-58.

Malaysia Kini (2015). About Malaysia's GST. Retrieved from http://aboutgst.malaysia.com.

Mansor, M., Tayib, M. & Yusof, R. (2005). Malaysian indirect tax administration system: An analysis of efficiency and taxpayers' perceptions. *IJMS*, 12 (2), 19-40.

McGowan, J. R. & Billings, B. A. (1997). An analysis of the European community VAT: Implication for U.S tax policy. *Journal of Accounting, Auditing and Taxation*, 6(2), 131-148.

Mohani, A. (2003). Income tax non-compliance in Malaysia. Petaling Jaya: Prentice Hall.

Othman, A. & Mirakhor, A. (2012). Sustainable fiscal position Malaysia: A proposal for reform. 2nd ISRA Colloquium. "Islamic Finance in a Challenging Economy: Moving Forward".
Mohd Rashid, S. N. & Shaik Asgar, R. K. B. (2014). Study on the possible impact of GST towards Malaysia using selected economic indicators: Case of Singapore, Thailand and

*Corresponding author's email: diana@bdrmelaka.uitm.edu.my

Indonesia as model countries. *International Conference on Business and Economic Research Proceeding*. Sarawak, Malaysia.

Palil, M. R. & Ibrahim, M. A. (2011). The impacts of goods and services tax (GST) on middle income earners in Malaysia. *World Review of Business Research*, 1(3), 192-206.

Sapiei, M. A. & Abdullah, M. (2015). Veerinder on Malaysia tax theory and practice. 3th Edition, Malaysia: Wolters Kluwer.

Singh, V. (2013). Tax compliance and ethical decision making: A Malaysian perspective. Petaling Jaya: Longman.

Syed, M. A. T., Amit, K. S. & Ravindra, K. S. (2013). Challenges and opportunities of goods and service tax (GST) in India. *Indian Journal of Applied Research,* 3(5), 413-415.

Tan, L. M. & Chin-Fatt, C. (2000). The impact of tax knowledge on the perception of tax fairness and attitudes towards compliances. *Asian Review of Accounting*, 8, 44-58.

The Star Online (2015). Take stern action against profiteers, March 7.

The Star Online (2015). GST – Show me the money, show us our money, April 25.

The Star Online (2015). With GST, 295 products should be cheaper, May 12.

Valadkhani, A. & Layton, A. P. (2004). Quantifying the effect of the GST on inflation in Australia's capital cities: An intervention analysis. *Australian Economic Review*, 37(2), 125-138.

Whalley, J. & Fretz, D. (1990). The economics of the goods and services tax. *Canadian Tax Paper*, 88.

Xing, W. & Whalley, J. (2014). The golden tax project, value-added tax statistics and the analysis of internal trade in China. *China Economic Review*, 30, 448-468.

Zhang, Y. & Huang, J. H. (2014). Cost-based pricing model with value-added tax and corporate income tax for a supply chain network. *Applied Mathematical Modelling*, 38, 268-280.

Zhou, L. Z., Tam, J. & Heng-Contaxis, J. (2013). The Introduction of Goods and Services Tax in Malaysia: A Policy Analysis. *Centre for Public Policy Studies (CPPS) Policy Paper Series*, Asian Strategy & Leadership Institute.

*Corresponding author's email: diana@bdrmelaka.uitm.edu.my

HUMAN CAPITAL MANAGEMENT AND ACCOUNTABILITY OF SOCIAL ENTERPRISE

Siti Anis Nadia Abu Bakar*, Roshayani Arshad[1], Nur Farahah Mohd Pauzi, Siti Nurulhuda Mamat, Norliza Omar[2]
**Faculty of Accountancy, Universiti Teknologi MARA Melaka*
[1]Accounting Research Institute, Faculty of Accountancy, Universiti Teknologi MARA
[2]Faculty of Accountancy, Universiti Teknologi MARA Puncak Alam

ABSTRACT

The emergence of social business enterprises has been attributed to the increased demand for sustainability among the non-profit organisations (NPOs) due to diminishing funding from traditional sources and increased competition for the scarce resources. To date, there is growing interest on social businesses (SBs) as innovation in providing public services that contributes to the social wellbeing of community. This in turn can be linked directly or indirectly to economic development as the sustainability level of the country is measured specifically towards the development of human capital and solving the social problems rather than the economics alone. Indeed, having both economic and non-economic aims, SBs would be the potential solutions to address a range of societal issues and increase the efficiency and quality of public sector services that focus on equitable growth with ecological sustainability. Therefore, it is important to explore the relationship between capability of SBs and the social impact created. This paper explores on capability based on human capital management and social impact or value created through SBs. This is expected to provide some insights on the sustainability and credibility of SBs to support the government efforts in moving towards high nation income.

Keywords: Social Enterprise, Human Capital, Accountability

INTRODUCTION

There was no clear, consensual definition of social enterprise in the literature (Hackett, 2010) and there is no consistently applied definition of social enterprise. In fact, it appears that each discipline or field defines social enterprise in its own image or view. Indeed different authors have used the label to apply to a wide range of different organisational types which has rendered conceptualisation problematic (Simmons, 2008).

Social entrepreneurship emerges as a response to the complex society needs whereby the concept of social entrepreneurship has been rapidly emerging in the private, public and non-profit sectors over the last few years, and has embarked growing interest in social entrepreneurship as a whole (Johnson, 2000). Besides, it is a global phenomenon for social entrepreneurship to employ their innovative approaches to solve social problems which ultimately giving impact to the whole society (Robinson, Mair and Hockerts, 2009). Research on social enterprise has been gaining prominence in recent years and attracting attention in a number of different disciplines.

Sustainable development refers to the process whereby improvement is undertaken in such a way that interests of various stakeholders of nature and diverse social groups including future

*Corresponding author's email:ctanisnadia@melaka.uitm.edu.my

generations are protected and preserved. It deals with presenting our nature of the present country to our future generations and at the same time taking care of socio-economic interests of all sections of society. Sustainability is not a new concept as its roots from a growing economic of the country. However, the level of sustainability of the country is measured not only in terms of economics alone, but for developing countries the human capital aspect also plays an important role as the qualities of the generation become a major contributor to the country's development. Over the recent years, English governments have promoted social enterprise and social entrepreneurship as potential solutions to address a range of societal issues and to increase the efficiency and quality of public sector services and economy outcome (Hall, Miller and Millar, 2012).

Tapsell and Woods (2010) offer a working definition of social entrepreneurship as follows: "the construction and pursuit of opportunities for transformative social change through innovative activities occurring within or across economic and social communities in a historical and cultural context". Furthermore, social enterprise is a new concept of applying business expertise and market-based skills such as developing innovative approaches, selling products and offering services in trading business in order to generate income (Jiao, 2011). This is also supported by (Rajput and Chopra, 2014) stating that the economies all over the world are realising that if we are somehow able to revert back to our original commercial path by taking up activities involving manpower rather than machines, our sustainability efforts shall be paced up.

In the context of social enterprises or social organizations, efficiency might well imply that many such organisations exist to deliver social and other community services cheaper than the public sector. They allow for cost savings by having volunteers as they have lower overheads count. An alternative effectiveness perspective might contend that their purpose is to deliver these services better, perhaps higher in quality, faster, more flexibly than a bureaucratic public sector by embracing entrepreneurial people in smaller and leaner organisations (Thompson, 2011). Therefore, it is important to develop and apply social entrepreneurship to sustain social development and enhance human capital.

However, some scholars consider the mission when defining social entrepreneurship. For instance, Dees (2001) believed that social entrepreneurs play the role of change agents in the social sector, by adopting a mission to create and sustain social value (not just private value), recognizing and relentlessly pursuing new opportunities to serve that mission, engaging in a process of continuous innovation, adaptation and learning, acting boldly without being limited by resources currently in hand and exhibiting heightened accountability to the constituencies served and for the outcomes created. Therefore, social value is the most important factor in this process although social entrepreneurship is often viewed as business with a social purpose that earns income for the non-profit sector. Furthermore, Dees (2003) said he leaned toward another definition of social entrepreneurship, one that emphasized innovation and impact, not income, in dealing with social problems. Much of the early social enterprise literature has focused on defining social enterprises, and adapting theories to explain their recent emergence.

DRIVING TOWARDS SOCIAL ENTREPRENEURSHIP

The concept of entrepreneurship that has emerged earlier has been modernized to emphasis on social change. As a consequence, social entrepreneurship has emerged to become subject of interest to many researchers in the present days. It is driven to the fact that not only individuals are having entrepreneurship spirit but they are also aiming towards helping the society voluntarily instead of going solely for figures.

The emergence of social enterprises in East Asian countries is closely linked to the socio-economic changes of the late 1990s and to the early attempts of civil society organisations to cope with new social problems. As far as public authorities are concerned, they began to pay attention to those new initiatives as they had to consider new policy measures and programmes to meet the growing welfare needs (Defourny & Kim,2011).

Apart from that, Defourny (2001) in his writing mentioned three principles that define social sector; one related to human capital being priority on the redistribution of profit to people and workforce over capital. Moreover, in terms of control over the management, social economy environment refers to one where a democratic structure will be taking place. Members are equally involved in the decision making process and have a say in any accounting and administrative matter (Defourny, 2001). This evidenced the development in human skills which is one element of human capital.

In another study setting, Jiao (2011) constructed a social model where sustaining social development and improve human well-being as a key success factor developed from the application of social entrepreneurship activity. It has been highlighted in the paper that social entrepreneurs will be knowledgeable because they emphasize the benefit of end-beneficiary which is the community. They will have the skills in understanding the customers on their needs and to then satisfy them in innovative way. Social business also develops the capability of the human in resource utilization skills. Social entrepreneurship impact had attracted increased attention. People find social entrepreneurs job are compelling as this job offers these extraordinary people with creativity and innovative ideas in producing goods and services. The interest in social entrepreneurship signals a transformed change in the elements of human capital namely, knowledge, ability to integrate and utilize resource and development of skills.

There were many researches carried out on these voluntary organizations looking at different way. Defourny (2001) looked at certain behavioral features of enterprises to be called social. Taking 15 European Union (EU) countries as sample, four features were identified from the study; 1) continuous production of goods and services, 2) high degree of autonomy, 3) operation with economic risk via own funding and 4) workers are either volunteers or minimum paid workers. On the other hand, Overall et al. (2010) discussed the concept of cultural context which led to the emergence of social entrepreneurship. The study in one local Maori social enterprise brought the idea that reconnecting the Maori people to their communities in sustaining culture through innovations from the Young Maori by providing roadmaps in New Zealand created the spirit of social entrepreneurship.

There were really a wide range of diverse opinions about social entrepreneurship. Yu (2011) studied the motivation factors toward the success of social enterprises in China considering the tight state control ruled by the government. They pointed out that the government's newly effort to privatize public services, the initiative taken by the third sector in probing socio-economic problems, the spirit of private firms engaging in corporate social responsibility and foreign players' interest over social business fostering the kind of entrepreneurship activity. Meanwhile, Sodhi and Tang (2011) narrowed down the concept of social entrepreneurship towards creating business for the poor and help them via enabling good supply chain. The paper was drawn from the exploring of success social enterprises in Asia, South Africa, United States and many more. Apart from that, Hall et al. (2012) explored the motives of public sector to reform into social enterprise from health care perspective. Respondent of the study viewed social business as strategic improvement to deliver different kind of clinical and management services which could not happen in a bureaucratic public sector.

Cook and Inman (2012), on the other hand, discussed the success and loophole in some of the voluntary sectors for conservation in England. Without denying the social good of volunteering, they also highlighted issues on underfunding, task allocation, independence and potential undermining of labour market. However, Moreau and Mertens (2013) undertook different way of researching on social enterprise. They have developed a management competency model of social enterprise to ensure effective handling of this so-called social business. They were of the view that competency in social enterprise arise from competency in knowledge, skills and behaviour. Scholars have recognized the business move towards social entrepreneurship which makes social enterprise relationship very clear. Despite its increasing attention, has come concern over why exactly social entrepreneurs do

HOW DOES SOCIETY BENEFIT?

Social enterprise aims at solving social problems. The concept lies in this entrepreneurship differentiate their motives from public and private sector at large. Many studies highlighted concern raised in social business that any surplus generated from the so-called entrepreneurship activity must go back to the society, rather than keeping them solely within the owners and shareholders. Engelke, Mauksch, Darkow and Gracht (2015) conducted qualitative discussions involving experts in Germany about the development of social enterprise as problem solving solution in industrialized countries. They viewed social business as enterprise where profit stays in the business to support for social objectives. Jiao (2011) proposed that greater level of social entrepreneurship is positively related to society's social impact. Social business promotes social value where the application of the entrepreneurship is to be aligned with the need to solve social problem like the unemployment and the environment matters.

A study by Cheng (2015) asserted that social enterprises in Cambodia have emerged from critiques raised upon their non-governmental organizations (NGOs). The NGOs were seen as defeating their purpose where the poor still remains where they were, stressing over own financial survival and they have lowering the level of trust by the government. Furthermore, Daya (2014) also explored the social enterprise craft industry in the South Africa, focusing on how these entrepreneurs saved the poor and black women in particular and issues revolving them. Additionally, Sawmy (2015) explored the concept of inclusive business in extractive industry to sustain the business. The social business model implies the need to include the poor in the business chain to accommodate their corporate social responsibility by way of link them as local business partners, hire them as employees or create them as business producers.

Numerous papers stressed that helping the society are forces behind the emergence of social enterprise. Connolly and Kelly (2011) on his view about third sector and social sector explained that social sector defined their business where social objective are the central of their activities and reject any personal gain. A survey done with the service users of a social enterprise engaging in rural transportation highlighted the significant contribution that this service made to the residents. O'Shaughnessy, Casey and Enright (2011) found that elder residents mostly benefited from the social service; easy access to health care services, shops and post offices, removed the feeling of being isolated from other people, eliminate dependency on surrounding people to travel and many more. Researchers are on the idea that social entrepreneurs are more fascinated towards probing social problems. Social problems can be caused by social market failure too. Hackett (2010) discussed on the ability of social enterprise in eliminating unemployment, poverty, income gap of the society caused by failure in market that ordinary business failed to address. On its basic level, there is something appealing about social enterprise in driving social change which is to improve people's lives.

CONCLUSION

In recent years, we noticed a remarkable growth in the third sector. Social enterprise through its entrepreneurial activities is a mechanism in improving social value by reconciling the gaps in wealth which ultimately would influence the development of human capital and solve the social problem. In this paper, we reviewed literatures to specifically clarify the inconsistencies definition of social entrepreneurship that emerge from many thoughts and views. Then, we discussed the factors that initiated entrepreneurship that could benefit the society at large and also its relationship in developing the human capital. Finally, we discussed the consequences of social entrepreneurship to the society.

The government support for the entrepreneurship activities is really meaningful towards the effort of expanding social capital of social entrepreneur that can lead to a more harmonious society. The combination of different stakeholders in the social enterprise organizations that are more oriented could benefit the whole community (Defourny, 2001). This is vital importance to understand what social enterprise explicitly could bring in relation to the development of human capital who could be maintained as long term asset as the qualities of them become a major contributor to the organization and community at large.

Apart from that, given the increasing roles and profile, social enterprise is inextricably as the potential solutions to address a range of societal issues (Hall et al., 2012). Meanwhile, creating the opportunities for the poor is absolutely enabling good supply chain (Sodhi and Tang, 2011) that is essential to support sustainability efforts (Rajput and Chopra, 2014). The entrepreneurship by the social business requires proactive approaches whereby its application should be aligned in dealing with solving social problem like the unemployment issue in order to promote social value (Jiao, 2011) and as the main key to measure the social impact as a whole.

REFERENCES

Cheng, I. H. (2015), Re-modelling and Reconceptualising Skills Development in Cambodia: How are Social Enterprises Preparing Young People for Successful Transitions between Learning and Work? International Journal of Educational Development, Vol. 43, pp.134-141.

Connolly, C. and Kelly, M. (2011), Understanding Accountability in Social Enterprise Organisations: A Framework. Social Enterprise Journal, Vol.7 No.3, pp. 224-237.

Connolly, C. and Kelly, M. (2011), Understanding Accountability in Social Enterprise Organisations: A Framework, Social Enterprise Journal, Vol. 7 No. 3, pp. 224-237.

Cook, H. and Inman, A. (2012), The Voluntary Sector and Conservation for England: Achievements, Expanding Roles and Uncertain Future, Journal of Environmental Management, Vol. 112, pp. 170-177.

Daya, S. (2014), Saving the Other: Exploring the Social in Social Enterprise, Geoforum, Vol. 57, pp. 120-128.

Dees, G. J. (2001), The Meaning of Social Entrepreneurship, Available at: www.caseatduke.org/documents/dees_sedef.pdf.

Dees, G. J. (2003), New Definitions of Social Entrepreneurship: Free Eye Exams and Wheelchair Drivers, Knowledge@Wharton Newsletter, Vol. 12 No.10, pp. 3-16.

Defourny, J. (2001), From Third Sector to Social Enterprise, The Emergence of Social Enterprise, London and New York, Routledge, pp. 1-28.

Defourny.J and Kim.S.Y (2011), Emerging models of social enterprise in Eastern Asia: a cross-country analysis, Social Enterprise Journal Vol. 7 No. 1, 2011,pp. 86-111.

Engelke, H., Mauksch, S., Darkow, I. L. and Gracht, H. A. (2015), Opportunities for Social Enterprise in Germany - Evidence from an Expert Survey, Technological Forecasting and Social Change, Vol. 90, pp. 635-646.

Hackett, M. T. (2010), Challenging Social Enterprise Debates in Bangladesh. Social Enterprise Journal, Vol. 6 No.3, pp. 210-224.

Hackett, M. T. (2010), Challenging Social Enterprise Debates in Bangladesh, Social Enterprise Journal, Vol. 6 No.3, pp. 210-224.

Hall, K., Miller, R. and Millar, R. (2012), Jumped or Pushed: What Motivates NHS Staff to Set Up a Social Enterprise? Social Enterprise Journal, Vol. 8 No. 1, pp. 49-62.

Jiao, H. (2011), A Conceptual Model for Social Entrepreneurship Directed Toward Social Impact on Society, Social Enterprise Journal, Vol. 7 No.2, pp. 130-149.

Johnson, S. (2000), Literature Review on Social Entrepreneurship. Available at: www.bus.ualberta.ca/ccse/Publications/Publications/Lit. Review SE November 2000.rtf

Moreau, C. and Mertens, S. (2013), Managers' Competences in Social Enterprises: Which Specificities? Social Enterprise Journal, Vol. 9 No.2, pp. 164-183.

O'Shaughnessy, M., Casey, E. and Enright, P. (2011), Rural Transport in Peripheral Rural Areas: The Role of Social Enterprises in Meeting the Needs of Rural Citizens, Social Enterprise Journal, Vol.7 No.2, pp.183-190.

Overall, J., Tapsell, P. and Woods, C. (2010), Governance and Indigenous Social Entrepreneurship: When Context Counts. Social Enterprise Journal, Vol.6 No.2, pp. 146-161.

Rajput, N. and Chopra, P. (2014), Social Entrepreneurship and Social Sustainability: An Analytical Study. Global Journal of Finance and Management, Vol. 6 No.9, pp. 961-966.

Robinson, J. A., Mair, J. and Hockerts, K. (Eds) (2009), International Perspectives of Social Entrepreneurship, Palgrave, London.

Sawmy, M. V. (2015), Growing Inclusive Business Models in the Extractive Industries: Demonstrating a Smart Concept to Scale Up Positive Social Impacts, The Extractive Industries and Society, Article In Press.

Short, J., Moss, T. and Lumpkin, G. (2009), Research in Social Entrepreneurship: Past Contributions and Future Opportunities, Strategic Entrepreneurship Journal, Vol. 3 No.2, pp. 161-194.

Simmons, R. (2008), Harnessing Social Enterprise for Local Public Services: The Case of New Leisure Trusts in the UK, Public Policy and Administration, Vol. 23 No.3, pp. 177-202.

Sodhi, M. M. and Tang, C. S. (2011), Social Enterprises as Supply-Chain Enablers for the Poor, Socio-Economic Planning Sciences, Vol. 45 No4, pp. 146-153.

Thompson, J. (2011), Reflections on Social Enterprise and the Big Society, Social Enterprise Journal, Vol. 7 No.3, pp. 219-223.

Yu, X. M. (2011), Social Enterprise in China: Driving Forces, Development Patterns and Legal Framework, Social Enterprise Journal, Vol.7 No.1, pp. 9-32.

Defourny.J and Kim.S.Y (2011), Emerging models of social enterprise in Eastern Asia: a cross-country analysis, Social Enterprise Journal Vol. 7 No. 1, 2011, pp. 86-111.

REINFORCEMENT TOOL OF WHISTLEBLOWING TO ERADICATE FRAUD IN PUBLIC SECTOR

Siti Aisyah Basri*[1], Ahmad Daud Marsam, Rozaiha Ab Majid[1], Nor 'Asyiqin Abu[1], Nafsiah Mohamed[2]
Faculty of Accountancy, Universiti Teknologi MARA Melaka[1], Accounting Research Institute, Universiti Teknologi MARA Malaysia[2]

ABSTRACT

Whistleblowing is an internal control system that be a way to reduce or eradicate fraud. This paper focuses on investigating the reasons for whistleblowing is not a preferred method to eradicate fraud. This paper starts with defining the term "whistleblowing". In its simplest form, whistleblowing involves the act of reporting improper conduct within an organisation to internal or external parties. This paper explains the issues and effects concerning whistleblowing. One of the issues is the violation of public trust. Lastly, this paper will describe the reasons for resistance of people to blow the whistle. Among the reasons is the fear of reprisal. This is a concept paper and it aims to conduct interview session with the Head for every local authority in Melaka. The findings of this paper can be used by the Government to create a new model related to whistleblowing to eradicate fraud in public sector. The Government could use this paper as the support to enhance the achievement of National Key Result Areas (NKRAs) Against Corruption.

Keywords: Whistleblowing, Public Sector

INTRODUCTION

The increasing numbers of fraud cases is becoming a burden to the local authorities in Malaysia. There are many types of fraud such as corruption, bribery and misappropriation of assets. Eaton and Akers (2007) states that problems actually exist in the government and non-profit sectors and not only in the corporate sector. Recent alleged problems include kickbacks, payoffs, bribery, embezzlement and collusive bidding.

Rozaiha, Nafsiah, Azizah and Zanariah (2010) concludes that there is a risk of misappropriation of assets occurring in local government authorities in Malaysia. The misappropriation of assets could significantly affect the finance of the local authorities. Future misappropriation should be ceased to happen if strict internal control is to be implemented within the local authorities.

According Tan and Ong (2011), whistleblowing is one of the mechanism for internal control system. The study claims that whistleblowing should be adopted by both public and private sectors to achieve good corporate governance practices. University's graduates should be prepared to face challenges in the accounting profession and this could be done through the inculcation of the strong ethical sense and critical thinking skill (Nur Barizah, Suhaiza and Suaniza, 2008). Similarly, both academic and non-academic staffs should also be equipped

*Corresponding author's email: Aisyah936@melaka.uitm.edu.my

with these skills. This is essential to prepare them to be able to make judgement on the situation and time that requires whistleblowing.

Whistleblowing is an internal control system that can be seen as a way to reduce or may eliminate the misappropriation of assets. Organisations are encouraged to develop whistleblowing policies because they are seen as part of their internal control (Brennan and Kelly, 2007). Whistleblowing protects society from any corporate behaviour that could result from companies' irresponsible action (Hazlina, 2011). It is very unlikely that employees will blow the whistle if they do not have any knowledge on whistleblowing. Some of them might have a certain level of understanding on what whistleblowing is, however, they might be afraid that if they whistleblow, they are not well-protected.

The 2014 Malaysian Anti-Corruption Commission (MACC) has clearly indicated the connection between whistleblowing and National Key Result Area (NKRA). In order to obtain the status of a developed country, the economy of this country must be stable. The country needs to have the trust from foreign investors to achieve economic stability. Therefore, it is crucial to have a good image in the eyes of those investors. NKRA Against Corruption is tasked to look into the matter and to ensure clean and transparent business in the public and private sectors. Hence, eventually the whistleblowing is in need to achieve NKRA Against Corruption.

However, people are still afraid and refuse to whistleblow (Tan & Ong, 2011). The reasons of this resistance to whistleblow should be investigated because the resistance could prevent the success of NKRA Against Corruption. This research suggests to fight the corruption by reinforcing whistleblowing.

What is whistleblowing?

Near and Miceli (1985) defines whistleblowing as the disclosure by former or current organisation members of illegal, immoral or illegitimate practices under the control of their employers, to persons or organisations that may be able to affect action.

Whistleblower Protection Act (WPA) 2010 defines "whistleblower" as:

> "Any person who makes a disclosure of improper conduct to the enforcement agency under Section 6".

The WPA 2010 further defines "improper conduct" as any conduct which if proved, constitutes a disciplinary offence or a criminal offence.

Whistleblowing involves the act of reporting improper misconduct within an organization to internal or external parties.

According to Hazlina (2011), whistleblowing is an act of an informant who exposes wrongdoing within an organization with the aim to stop the wrongdoing. A whistleblower will report the wrongdoing to proper medium and authorities that have the power and are presumed willing to take corrective actions. The wrongdoings include any kind of violation of law, rules and regulation such as fraud and corruption.

*Corresponding author's email: Aisyah936@melaka.uitm.edu.my

DOI: 10.1051/shsconf/20173600038

If the whistleblowing policies and procedures are implemented successfully in an organisation, it would be a good early warning system to the organisation to eliminate improper conduct before the matter becomes worse and worst (Tan & Ong, 2011).

Francis, Armstrong and Foxley (2015) states that the purposes of whistleblowing include to expose wrongdoing, to stop wrongdoing, to prevent wrongdoing and to defend one's position. Eventually, whistleblowing aims to reduce corruption, to save money and to prevent problems associated with the wrongdoing.

The issues and effects concerning whistleblowing

Misconduct in an organisation is associated with declining organisational performance and it concerns with the issues that affect the society as a whole. These issues include issues related to public safety and/or purported violation of public trust. Francis *et al* (2015) states whistleblowing need to be conducted because wrongdoing may violate the personal and/or corporate financial gain, danger to the public, danger to someone and abuse of authority.

Therefore, whistleblowing is seen as a method to reduce the negative effect and issues of the misconduct. According to Near and Miceli (1985), the whistleblower may provide valuable information that is very helpful in improving organization effectiveness. This is supported by Lewis (2005) which concludes that there is an emerging concept of having whistleblowing as public interest reporting. Besides that, Hazlina (2011) believes that whistleblowing promotes good corporate governance and accountability in a corporation.

Malaysia is in the journey of becoming a developed country as a high-income nation by 2020, thus the Government has introduced the 6 NKRAs as a support. In order to achieve this vision, it is crucial to have stable economy with zero corruption to eventually have the foreign trust to invest in Malaysia. The foreign investment contributes significant effect on the economy of a country. Thus, obtaining this foreign trust is a must. Hence, reinforcing whistleblowing could be seen as one of the way to eliminate fraud in the public sector. Whistleblowing is a crucial method to support the NKRA Against Corruption and eventually to eradicate fraud in public sector.

The resistance to whistleblow

Whistleblowing is not as simple as it sounds. Despite the benefits of whistleblowing in detecting and preventing improper conduct, not many are willing to whistleblow on their employers' wrongdoing (Tan & Ong, 2011). Their unwillingness might be due to the perception that whistleblowing may result in loss of career, loss of family and social life. Whistleblowers might feel in danger and often whistleblowing requires exceptional courage.

O'Leary and Cotter (2000) finds that becoming a whistleblower is still perceived as unacceptable. It appears as if whistleblowing on the employers is still unacceptable but whistleblowing on a friend is almost inconceivable. Employees would not blow the whistle because they are afraid of reprisals.

Many potential whistleblowers are not confident that they will get statutory protection from employment tribunals and the courts under Part IV A Employment Rights Act 1996 (Lewis, 2005). Besides that, organisation may encourage the internal reporting of concerns about wrongdoing that do not have a public dimension to further its private interest.

*Corresponding author's email: Aisyah936@melaka.uitm.edu.my

DOI: 10.1051/shsconf/20173600038

A study by Brennan and Kelly (2007) finds that trainee auditors in organisations that have adequate formal structures for whistleblowing have greater confidence that by whistleblowing internally their careers would not be adversely affected. At the same time, trainee auditors showed a reluctance to whistleblow externally. The trainee auditors also expressed little confidence in the protection offered by legislation for whistleblowers.

A study done by Nur Barizah *et al* (2008) has shown that 60% of the respondents indicated that they will not whistleblow regardless realising the act is unethical. Fortunately, 37.6% of the respondents indicated that they will take immediate action by informing the relevant corporate authorities regarding the unethical act.

Nur Barizah *et al* (2008) claims that people are not willing to whistleblow because they are still worried about the lack of protection if they whistleblow. Other than that, the resistance might be due to the perception of others on the whistleblower. In most cases, the whistleblower may find themselves isolated from the others. This might be very difficult especially loyalty is a very strong trait in the Asian culture. Whistleblower might contemplate between acting for the society as a whole or conforming to being loyal to the management and to their corporate leaders. When there is a possibility that necessary action will be taken on the wrongdoers, the percentage of the respondents who were willing to whistleblow has increased to 44%. This shows that the existence of laws, rules and regulations may help build the confidence in the people who intend to whistleblow.

The same finding has been found by a study by Rafik (2008) that claims the respondents generally felt that whistleblowing was necessary in cases of fraud, although they were slightly less likely to do it themselves. This might be due to the high costs involved such as a retaliation and the difficulty to find future job in the same profession. Other than that, Francis *et al* (2015) found that people are not willing to whistleblow because they believe that no action will be taken, fear of retaliation and loss of the means of livelihood.

Soni, Maroun and Padia (2015) finds that higher level of distributive justice, interactional justice and procedural justice tend to increase the propensity to whistleblow. Distributive justice refers to the perceived fairness of outcomes. Interactional justice focusses on how the employee perceive his interactions with his superior. Procedural justice refers to the perceived fairness of the processes followed to achieve those outcomes. On the other hand, age, gender, seniority, performance evaluation and religiosity do not have a significant effect on decision to whistleblow. However, this is not in line with the previous study by Brennan and Kelly (2007) which finds that as the trainee auditors progress in their careers they become less reluctant to whistleblow externally.

In order to encourage Malaysian to whistleblow, the Parliament has passed WPA 2010. WPA 2010 intends to combat corruption and other wrongdoings by encouraging and facilitating disclosures of improper conduct in the public and private sector. The purpose of this Act is also to protect persons making those disclosures from unfavourable action, to provide for the matter disclosed to be investigated and dealt with and to provide for the remedies connected therewith.

*Corresponding author's email: Aisyah936@melaka.uitm.edu.my

PROSPECTIVE RESEARCH METHODOLOGY

The research is a qualitative research. It focuses on the local authorities in Melaka, namely Majlis Bandaraya Melaka Bersejarah (MBMB), Majlis Perbandaran Alor Gajah (MPAG), Majlis Perbandaran Jasin (MPJ) and Majlis Perbandaran Hang Tuah Jaya (MPHTJ).

An interview will be held with the Head for every local authority. The interview will focus on the examination of whistleblowing policies and procedure implemented by the Government to eradicate fraud in public sector in their local authorities.

The content of the interview will be organised and analysed using NVivo 10 for Windows.

CONCLUSION

Those at the managerial or executive level hold the key to the companies' success and ironically, most of corporate misconduct is done by those at these levels. Hence, it is very crucial to have ethical work environment by reinforcing whistleblowing.

The WPA 2010 came into force on 15th December 2010. However, there is still a bridging gap between whistleblower protection and good governance practices in Malaysia. As mentioned earlier, people have little confidence in the protection offered by legislation. Government has the vital role to create conducive condition for whistleblowers. If this does not happen, the Government might have problems to achieve successfully NKRA Against Corruption. The Government could use this research as the support to enhance the achievement of NKRA Against Corruption.

ACKNOWLEDGEMENT

The authors are grateful to the management of Universiti Teknologi MARA (UiTM), Accounting Research Institute (ARI) and Fundamental Research Grant Scheme (FRGS) by Ministry of Higher Education, Malaysia for their beliefs and continuous support.

REFERENCES

Brennan, N. and Kelly, J. (2007). A Study of Whistleblowing among Trainee Auditors. *The British Accounting Review,* Vol. 39, pp. 61 – 87

Eaton, T. V. and Akers, M. D. (2007). Whistleblowing and Good Governance. *The CPA Journal*, June 2007, Vol. 77, pp. 66 – 71

Francis, R.D., Armstrong, A. F. and Foxley, I. (2015). Whistleblowing: A Three Part View. *Journal of Financial Crime*, Vol. 22, Issue 2, pp. -

Hazlina, S. M. N. A. (2011). Whistleblowing and Corporate Social Responsibility. *SEGi Review*, Vol. 4, No. 2 pp 58 – 66

*Corresponding author's email: Aisyah936@melaka.uitm.edu.my

Lewis, D. (2015). Is a public interest test for workplace whistleblowing in society's interest? *International Journal of Law and Management*, Vol. 57, No. 2, pp. 144 – 158

Malaysian Anti-Corruption Commission (2014) website retrieved on 10th May 2015 from http://www.sprm.gov.my/index.php/en/

Near, J. P. and Miceli, M. P. (1985). Organisational Dissidence: The Case of Whistleblowing. *Journal of Business Ethics*, Vol. 4, No. 1, pp 1 - 16

Nur Barizah, A. B., Suhaiza, I. and Suaniza, M. (2008). Ethics of Future Accounting Professionals: Evidence from Malaysia. *Journal of Financial Reporting and Accounting*, Vol. 6, pp 21 – 23

O'Leary, C. and Cotter, D. (2000). The Ethics of Final Year Accountancy Students: An International Comparison.d *Management Auditing Journal,* 15(3), pp. 108 - 115

Rafik, E. (2008). Auditing Students' Professional Commitment and Anticipatory Socialization and their Relationship to Whistleblowing. *Managerial Auditing Journal*, Vol. 23, Issue 3, pp 283 – 294

Rozaiha, A. M., Nafsiah, M., Azizah, A. and Zanariah, M. (2010). An Exploratory Study on the Possibility of Misappropriation of Assets occurring in a Local Authority. *International Conference on Science and Social Research (CSSR) 2010*, Kuala Lumpur, Malaysia

Soni, F., Maroun, W. and Padia, N. (2015). Perceptions of Justice as a Catalyst for Whistle-blowing by Trainee Auditors in South Africa. *Meditari Accountancy Research*. Vol. 23, No. 1, pp 118 - 140

Tan, P. M. and Ong, S. F. (2011). Comparative Analysis of Whistleblower Protection Legislations in England, USA and Malaysia. *African Journal of Business Management*, Vol. 5, No. 26, pp 10446 – 10455

Whistleblower Protection Act 2010

*Corresponding author's email: Aisyah936@melaka.uitm.edu.my

INTENDED USE OF PROCEEDS AND THE PERFORMANCE OF INITIAL PUBLIC OFFERINGS (IPO).

Siti Azrina Adanan*, A'ieshah Abdullah Sani, Khair Syakira Bustamam , Amilia Saidin
Faculty of Accountancy, Universiti Teknologi MARA Melaka

ABSTRACT

Other than debt financing, Initial Public Offerings (IPO) become a common method of capital raising among firms for various purposes. Prospectus is issued on the pre-issuance of the IPO disclosing various mandatory and non-mandatory voluntary information to help the potential investors in their decision making process. Other than financial information on previous and forecasted future performance, the intended use of the IPO's proceeds information is one of the crucial information for the stakeholders especially the investors as they will be the financial contributor for the purposes. In one hand, the issuers may decide to issue highly priced IPO to fulfil their financial needs and other going-public motives. In the other hand, the investors willing to subscribe for the IPO to meet their investment target. The gap between the objectives of these two main stakeholders create uncertainties that may jeopardize the interest of the investors. Thus, the objective of this paper is to investigate the relationship between the intended uses of the IPO's proceed disclosed in the prospectus and their subsequent performance of the stocks. The result of the study can provide input for the investors and other stakeholders for their decision making process to meet their respective objectives.

Keywords: Initial Public Offerings, Proceeds, Market Performance

INTRODUCTION

Around the world, firms may employ various methods to raise funds to meet their short-term and long-term objectives to be sustainable and grow. Two common methods of financing are equity financing and debt financing. While the first does not put the firms into obligation to promise the provider of the funds with returns, but the cost of financing can be too costly to the issuing firms. The later in the other hand comes with some obligations to the firms, but it is less costly.

Even though the cost of issuance of the stocks or equity is expensive, it also comes with advantages. One of the advantages is it can allow outside investor to profit as well which is possible when the stock is under priced (Nelissen, 2016). Equity financing for a firm which issued shares for the first time is called Initial Public Offering (IPO). Most of the IPOs are under priced. Under priced happened when the offer or subscription price determined by the underwriters on behalf of the firms is lower than listing price. Listing price is the price on the first day the share traded in the stock exchange. Various studies conducted to support the under priced of the IPOs such as Rock (1986), Welch (1989), Habib & Ljungqvist, (1998), Ljungqvist and Wilhelm (2003), and (Leone, Rock, & Willenborg, 2007). Thus, the under pricing allows the investors to profit from the IPO issuance.

*Corresponding author's email: azrina@bdrmelaka.uitm.edu.my

DOI: 10.1051/shsconf/20173600039

Firms decide to go to public to raise external equity capital for a variety of reasons. One of them is when a firm has insufficient internal fund to make a value-added investment which will accelerate the firms' profit and stock value. Such investments come from various branches such as acquisition of the other firms either for expansion, cost reduction, and elimination of rival purposes. The equity market will provide avenue for the firms to raise fund to implement their plans. An alternative use of proceeds is to repay debt obligations. Since debt financing can sometime be more expensive, some firms may decide to payback its debt by using the proceeds from the IPO (Autore, Bray, & Peterson, 2009). In Malaysia, the other popular use of the proceeds is to finance their working capital to efficiently manage their cash flows in ensuring operation can be run smoothly.

The intended use of the IPO proceeds can be found in the prospectus which will be issued prior to the issuance of the IPO to give investors information to help them in their investment decision. In Malaysia, all of the firms which seek to be listed in Kuala Lumpur Stock Exchange (KLSE) have to follow the Security Commission Malaysia (SCM) requirement on what information that needs to be discloses in the prospectus. Most of the firms in Malaysia are seeking for funding their investments, payback their debts and to finance their working capital by using the proceeds generated from the IPO issuance. This information on the intended use of the proceeds provide insight about firms' motives and objectives (Autore et al., 2009). Thus, the information might have influence of the stock performance where it can give a picture on how the stocks will be priced in the secondary market (Leone et al., 2007).

The purpose of this study is to investigate the linkages between intended use of the IPO proceeds and the performance of the IPO when it is listed in the stock exchange. We would like to look into how the market react on the information. This can further explain the investors' perception on the decision made by the firms on how the firms intend to utilise the IPO proceeds.

The paper is organised as follows: Section 2 elaborates the literature review on the relevant IPO issues. Section 3 explains the sample selections, proceeds classification and methodology. Lastly, section 4 presents the conclusion, limitation of the study and recommendation for further research.

LITERATURE REVIEW

The concept of Initial Public Offering (IPO)

An initial public offering (IPO) occurs when a formerly unlisted company pursues to raise funds from the public capital market by issuing its own equity securities (Balatbat & Bertinshaw, 2008; Nelissen, 2016). It is the process when a private entity becomes a public company. Accordance to Securities and Exchange Commission (SEC) rules, a company intended to go public must disclose the uses to which it intends to pertain the initial public offering (IPO) proceed (Bray, 2010; Leone et al., 2007) in the prospectus issued. The disclosure is vital for the investors to make sound and informed decision making (Balatbat & Bertinshaw, 2008).

The intended use of IPO proceeds

There are different motives why firms raise equity capital. One of the basic doctrines of corporate financial theory is that firms should raise external capital to invest and grow when

*Corresponding author's email: azrina@bdrmelaka.uitm.edu.my

potentially value added investment projects arise with little or none internal funds available for the projects. The equity markets provide a means for firms to sell stock and use the proceeds for investment tenacities, such as acquisitions, capital expenditures, also for research and development. Fundamentally the use of proceeds filed SEC are investment, debt repayment, or general corporate purposes (Autore et., al 2009). The result is coherent with Walker & Yost (2008) that examined 438 Seasoned Equity Offering (SEO) companies and documented the firm's intended purpose for the proceeds, had identified three specific uses which are investment, debt repayment, and general corporate purposes. Walker and Yost further elaborate the definition of these three categories; 1) Investment; including any proceeds the firm intends to use for increasing the asset base of the firm, the acquisition of other companies and planned spending on research and development. 2) Debt repayment; includes proceeds the firm intends to use to repay any outstanding debt, including revolving credit borrowings. 3) General corporate purposes are reserved for those firms that do not detail up the specific use of funds.

According to (Andriansyah & Messinis, 2016) the intended use of proceeds is classified into five categories as adopted in (Leone et al., 2007); fixed assets investment; working capital financing; investment in shares of stock; debt repayments and secondary shares. Autore et al., (2009) further stated that some companies might not have a specific purpose for raising equity and might not wish to divulge the intended purpose to the public. However, for the purpose of this study, to focus on three intended proceed namely financing for investment, debt repayment and working capital financing as opposed by simplified version of (Andriansyah & Messinis, 2016).

The relationship between intended use of IPO proceeds with subsequent share price performance

Prior empirical studies have been conducted to measure the intended IPO proceeds with the post-issue operating performance (Andriansyah & Messinis, 2016; Autore et al., 2009; Walker & Yost, 2008, Busaba, Benveniste, & Guo (2001) and yet the information has very limited in exploiting the context of post-share price performance of IPO firms especially in Malaysia as stated in (Autore et al., 2009). Thus, future finding of this paper will add contribution towards the literatures as well for the investors and other stakeholders decision making.

A recent comprehensive empirical research has been conducted by (Andriansyah & Messinis, 2016) that investigates the link between intended use of proceeds and post-issue operating performance of IPO firms among Indonesian equity market over a period of 10 years (2000-2010). The finding indicates that investment in fixed assets and in stock market shares lead to better performance while other usages lead to insignificant performance. Fixed asset investment and investment in shares in stock, is more notable for average and high performing firms. The above finding is parallel with Autore et al., (2009) that discovered fixed assets investment desirably affect high performing firms. On the other hand, low performing firms might benefit from allocation to working capital financing, debt repayment and secondary shares. The unfavourable effect on low performing firms may relate to the lack of capacity by these firms to manage investment projects effectively.

Walker and Yost (2008) find that regardless of the stated intended use of proceeds, all three classifications (investment, debt repayment, and general corporate purposes) increase investment levels during the three year-period beginning the year prior to the issuance year (years -1, 0, and +1). Furthermore, that issuers stating specific investment plans experience a relatively favorable market reaction at the offer announcement and an improvement in industry-adjusted operating performance.

*Corresponding author's email: azrina@bdrmelaka.uitm.edu.my

DOI: 10.1051/shsconf/20173600039

Autore, Bray and Peterson (2009) suggest that a comprehensive and clear disclosure on intended use of IPO proceed especially in investment classification will attract investors' interest. Firms which provide solid and distinct disclosure on intended use of proceeds perform better than firms with less disclosure of their use of proceeds (Balatbat & Bertinshaw, 2008). However, Leone et al. (2007), find lower under pricing when there is greater specificity about the intended use of proceeds.

Meanwhile, Ljungqvist and Wilhelm (2003) reported, when the proceed is to fund operating expenses as opposed to capital expenditure is will lead to greater under pricing which benefited the investors more to meet their respective objectives.

Nevertheless, if the intended use is meant for debt repayment and general corporate purpose the market will not underperformed the benchmark because firm can arrange the timing of IPO issuance. This is consistent with Busaba, et al., (2001) when IPO intended use is for debt repayment, timing is more important rather than use the proceeds for investments.

SAMPLE SELECTION AND PROCEEDS CLASSIFICATION

Sample selections

The initial samples of firms which issued IPOs are derived from Bursa Malaysia website for year 2012 to 2015 contributed to the population of this study. Further consideration for the samples is all the samples must have daily closing price subsequent to the listing date which can be retrieved from DataStream database. The samples covered all types of industries. Total firms that can be included in the samples are xxx altogether for the four years both from the Main market and ACE market. The tabulation of the samples is presented in the following table:

Table 1: Sample distribution by year

Year	Main Market	ACE Market	Total
2012	14	3	17
2013	16	1	17
2014	12	3	15
2015	9	4	13

Table 1 presents the number of IPO issuing firms where in year 2012 and 2013 reported 17 numbers of firms each and in year 2014 and 2015 the numbers are 15 and 13 respectively.

Proceed classifications

We then manually collect the data on the intended use of the IPO proceeds for each firm from the prospectus issued and newspaper released prior to the IPO issuance. For categorisation purpose, we classify main three specific classifications for the use of the proceeds which represent the largest proportion of the IPO proceeds. They are financing for investment, debt repayment and working capital financing. This is a simplified version of Andriansyah & Messinis (2016) which includes five classification of intended use; fixed asset investment, working capital financing, investment in shares of stocks, debt repayment and secondary shares.

Issuers in the first classification are those that largest portion of the proceeds will be used for investment purpose (i.e., acquisition of non-current assets such as machineries, premises and

*Corresponding author's email: azrina@bdrmelaka.uitm.edu.my

subsidiaries). Issuers in the second classification meant for firms which the largest portion of the proceeds will be used for debt repayment purpose and issuers in the third classification are those firms that the largest portion of the proceeds will be used for working capital financing. Most of the firms in Malaysia issued IPO for these three main purposes.

CONCLUSION, LIMITATION AND RECOMMENDATION

Conclusion

As the IPO issuance become one of the most popular method to raise fund among public firms, this study may provide insight for the issuing firms, policy makers, underwriters and investors on the management of the IPO proceeds. The study may contribute to the shortage of studies on motives of IPO issuance in Malaysia particularly on their influence and ability to direct the reactions of the market. The results of this study provide at least a starting point for the conduct of further research to enhance the knowledge of IPO issuance motives influence on market reactions. Furthermore, this study may provide useful information to investors on how other market participants react on such information. In addition, the results of this study also may help directors and other top managements of a firm to decide how they should manage the proceeds from the IPO issuance to make sure that the benefit of shareholders is not jeopardised as maximizing shareholders wealth is the ultimate goal of a public firm. The results will give the firms a general overview of what investors want.

Limitations

There are several limitations that can be highlighted since they can be used as a guidance and opportunity for future research. First, limited availability of data and time constraint has made this limitation inevitable. Moreover, the data also may be contaminated by other contemporaneous events. Second, the study only classifies the firms into its classification based on their intended use of the IPO proceeds without any other factors and variables which may influence the investors' reaction. In addition to this, there is no analysis has been done to empirically test the samples selected above.

Recommendations

The stock performance is closely related to operating performance. Theories proposed that positive earning and operating performance leads to positive investors' reaction which will be the basis for the stock price to be higher and vice versa. Thus, it is worth to investigate further the after-issuance operating performance of the firms in relation to their intended use of initial public offering proceeds. Other than that, it is also worth to further classify the firms into their market capitalization to look into possible impacts on how investors react based on their perception and interpretation of the information released in the prospectus. Hence, further empirical study is required to solve the puzzling issues in IPO performance.

REFERENCES

Andriansyah, A., & Messinis, G. (2016). Paci fi c-Basin Finance Journal Intended use of IPO proceeds and fi rm performance : A quantile regression approach ☆. *Pacific-Basin*

*Corresponding author's email: azrina@bdrmelaka.uitm.edu.my

Finance Journal, *36*, 14–30. http://doi.org/10.1016/j.pacfin.2015.12.001

Autore, D. M., Bray, D. E., & Peterson, D. R. (2009). Intended use of proceeds and the long-run performance of seasoned equity issuers. *Journal of Corporate Finance*, *15*(3), 358–367. http://doi.org/10.1016/j.jcorpfin.2008.12.003

Balatbat, M. C., & Bertinshaw, S. . (2008). Use of proceeds disclosures in IPO prospectuses : do issuers come clean ? *The FINSIA Journal of Applied Finance*, (2), 17–22.

Bray, D. E. (2010). Two Essays on the Intended Use of Proceeds of Seasoned Equity Offerings.

Busaba,W.Y., Benveniste, L.M., Guo, R., 2001. The option to withdraw IPOs during the premarket: empirical analysis. Journal of Financial Economics 60, 73–102.

Habib, M. A., & Ljungqvist, A. P. (1998). Underpricing and IPO proceeds : a note, *61*, 381–383.

Leone, A. J., Rock, S., & Willenborg, M. (2007). Disclosure of Intended Use of Proceeds and Underpricing in Initial Public Offerings, *45*(1). http://doi.org/10.1111/j.1475-679X.2006.00229.x

Ljungqvist, A., Wilhelm,W., 2003. IPO pricing in the dot-com bubble. *Journal of Finance* 58, 723–752

M.D. Walker, K. Yost . Seasoned equity offerings: What firms say, do, and how the market reacts *Journal of Corporate Finance* 14 (2008) 376–386.

Nelissen, B. B. (2016). First-day overperformance of IPOs, (May).

Welch, I., (1989). Seasoned offerings, imitation costs, and the under pricing of initial public offerings. *Journal of Finance* 44, 421-450.

*Corresponding author's email: azrina@bdrmelaka.uitm.edu.my

INCORPORATING ROLE OF STAKEHOLDERS INTO CORPORATE CSR STRATEGY FOR SUSTAINABLE GROWTH: AN EXPLORATORY STUDY

Vani Tanggamani*, Siti Anis Nadia Abu Bakar, Rani Diana Othman
Faculty of Accountancy, Universiti Teknologi MARA Melaka

ABSTRACT

In today's modern day context, Corporate Social Responsibility (CSR) become the mantra for businesses as it can be seen as a strategic approach for firms to be succeed in their business endeavours. Hence, it remains the most widely used concept to refer to organizational-stakeholder relationships. An understanding about a stakeholder approach to CSR is an important means for a firm to enhance their commitment to operate in an economically, socially and environmentally sustainable manner. In support with above arguments, this article presents a theoretical proposition based on stakeholder theory for better CSR and firm performance. By drawing upon classic work in the field, the paper offers conceptual discussion and then systematically develops a means of stakeholder approach into corporate CSR strategy. The aim of this paper is to demonstrates the need for a firm to enhance further understanding about the role of stakeholders in the context of corporate CSR strategy, which is increasingly necessary in view of the fact that business entities are absolutely essential for economic development, but at the same time, their business activities pose a huge impact to society and the environment. Thus, firm's CSR disclosure is a pivotal tool to establish a relationship of working together with the stakeholders that ensures mutual benefit and continue to be a firm that is needed by society. This article contributes to the literature by providing a fundamental explanation of how a business should embrace responsibility for the impact of its activities on the stakeholders across various levels of the value chain. By doing so, firms are offered a means to take a much more proactive approach to CSR through the stakeholder approach which is precisely helpful in measuring the effectiveness of any CSR initiative on the society to foster business sustainability.

Keywords: Corporate Social Responsibility, Stakeholder Relationships, Stakeholder Theory

INTRODUCTION

The concept of Corporate Social Responsibility (CSR) remains open to various definitions and understanding (Lu, 2013). However, in line with the literatures definition of CSR, it can be concluded that CSR is a firm's obligation and accountability for incorporating social and environmental responsibilities into the business decision-making process as well as to use its resources in ways to contribute to the well being of society (Caroll, 1979; Gill, 2008; Caroll & Shabana, 2010).

Despite its controversial nature, CSR has become an avid interest of the millennial generation within the business world, mainly due to globalization. Until the late 1970s, CSR was derided as a joke, an oxymoron and a contradiction in terms of the investment and by the business community (Whalen, 2013; Lu, Chau, Wang & Pan, 2014). However, as for now, CSR become

*Corresponding author's email: vani@melaka.uitm.edu.my

the mantra for businesses to ensure that the company stays in business for a long time to come (Ditlev-Simonsen & Midttun, 2011).

Today's businesses are under increasing pressure to demonstrate and integrate their core financial operations with the engagement of social and environmental concerns for the benefit of its stakeholders (Amran, 2006; Pop,O.et al, 2011), in order to develop a sustainable and true presence (Chen, Feldmann, & Tang, 2015). Thus, the overall demand has driven many companies that are currently practicing corporate social responsibility. Hence, the business's purpose now is not only to earn profit but the welfare of the society as well (Zhu, 2011). Therefore, the question that firm ask now days is not "why should we get involved in CSR?" but rather "how should we do it?" and "how can we benefit the environment and contribute to the community?" (Ghotb, 2011). As a result, CSR is getting famous and business organisations are gradually more concerned with the way they operate their business and the impact of their action on society and environment (Lu, 2014).

During this era, CSR is an action tool of top management for long-term success, as described in earlier research by Gray, Owen and Adams (1996), since many corporate organisations believe that success with CSR strategies will lead to success in organisation performance too, as it constitutes a benefit not only to the society but to the business itself also (Ditlev-Simonsen & Midttun, 2011; DiSegni, Huly, & Akron, 2015). Hence, a larger number of companies are engaged in a serious effort to define and integrate CSR strategies into all aspects of their businesses due to the globalization (Chen et al., 2015).

CSR and Stakeholder Relationships

Businesses are an essential part of the communities in which they operate (Belkaoui, 2000; Dkhili & Ansi, 2012). Good management know that their long-term success is based on continued good relations with a wide range of stakeholders because business can't succeed in societies that are failing due to social or environmental challenges (Zhu, 2011). As a result, when operate in businesses in present environment, the most important area that firms must concern is businesses are under increasing pressure to demonstrably engage in activities which are described as CSR.

CSR is the firm's social responsibility that gives a picture of how a firm take charge of their business processes in order to produce an overall positive impact to the society (Neal & Cochran, 2008). As such, firm's responsibility extends beyond the shareholders as it does not operate in isolation of its surrounding, namely the society and the environment. In fact, business success is increasingly affected by stakeholder relationships, which covers a wide range of interests, the most pivotal of which being how the organisation works with social and environmental issues (Gray et al., 1996; Russo & Perini, 2010).

This implying that firm should concern with the society ethically and not only pays attention to their profit, and also should more care about their stakeholders' expectations and what they are desired for. This is due to the fact that most of the investors are invest in firms that have involve greatly in CSR activities (Raza, Ilyas, Rauf & Qamar, 2012). As a result, corporations have adopted the concept of CSR, in fact, the corporation's concern is seen to be shifting to broader components of CSR, which includes stakeholders as one of the most important focus (Carroll, 1979; Gray et al., 1996; Orlitzky, Schmidt & Rynes, 2003; Amran, Zain, Sulaiman, Sarker & Ooi, 2013).

*Corresponding author's email: vani@melaka.uitm.edu.my

In a nutshell, CSR disclosure is becoming as an important tool to maintain the positive relationship with stakeholders (Esa & Mohd Ghazali, 2012) in view of the fact that the satisfaction of all stakeholders enhances the image and reputation of the company, thus improving the financial performance (Freeman, 1984; Donaldson & Preston, 1995). Strategically speaking, business can only flourish when the CSR program is implemented effectively because CSR is essential resource for a firm to improve their stakeholders relationships.

Stakeholder Theory

Stakeholder theory, as introduced by Freeman (1984), has been analysed exclusively from the perspective of an organisation (Amran et al, 2013). This theory has been used as a tool to identify those stakeholders to whom an organisation should be accountable as to recognises their rights. Accordingly, Freeman (1984) defines a stakeholder as *"any group or individual who can affect or is affected by the achievement of the organisation's objectives"* (p. 46). Similarly, Murray and Vogel (1997), define stakeholder as any entity, typically outside the firm, that the organisation aims to influence and that has an impact on the organisation.

The aim of this theory is to develop a CSR framework using stakeholders due to the fact that stakeholders are recognised as the group of people interested in the company's activities (Freeman, 1984). Therefore, stakeholder theory suggests that the purpose of a business should be to create as much value as possible for stakeholders and ensure the expectations of stakeholders aligned and going in the same direction (O'Riordan & Fairbrass, 2008). In this regard, Table 1.1 illustrates the primary stakeholders in an organisation and their most important objectives or expectations. By knowing this, a firm would be possible to develop business operations in tune with the frame of reference in which the firm operates.

Table 1:Stakeholders in the organisation and their objectives or expectations.

Stakeholder	Objectives
Management	Governance, corporate culture, strategies, and overall performance
Shareholders	Profit, and the value of the stock in the stock market
Customers	Characteristics of the product and quality of services
Government	Security of society, sovereignty, respect for the laws, and paying taxes
Community	Environmental and social effects
Employees	Fairly compensated employment, working conditions, and job satisfaction
Suppliers	Price and volume of procurement, and sustainability
Banks	Payment systems and their flexibility
Investors	Information about the liquidity and viability of payment

Source: Adapted from Caroll, 1979; Clarkson, 1995; Cramer & Bergmans, 2003

A stakeholder approach to CSR suggests that the expectations of all primary stakeholders need to be met, and not just the shareholders (Elijido-Ten, 2007). This practically means an organisation must be able to combine the profit maximization with all stakeholder's benefits and expectations in order to draw a complete map of all these relationships as to have a clearer view of all the needs of its stakeholders (Freeman, 1984). Accordingly, stakeholder theory has been applied to explain the motivations of firms regarding CSR disclosure (Freeman, 1984; Donaldson & Preston, 1995; McWilliams & Siegel, 2001; Abd Mutalib, Muhammad Jamil &

*Corresponding author's email: vani@melaka.uitm.edu.my

Wan Hussin, 2014). Under the umbrella of CSR, the firm is expected to fulfil its responsibility to its primary stakeholders (Vormedal & Ruud, 2009; Hamid & Atan, 2011; Amran et al, 2013). As a result, firm's CSR disclosure has been used as a strategy to maintain the support of powerful stakeholders (Deegan & Blomquist, 2006). This implies that CSR represents an important element between the organisation and its stakeholders relationship (Murray & Vogel, 1997; Elijido-Ten, 2007).

In sum, stakeholder theory proposes that CSR performance results in improved firm performance (Orlitzky et al., 2003; Freeman, 2010). Therefore, organisations need to identify and prioritise their stakeholders' expectations through their business operation (Clifton & Azlan, 2010; Amran et al., 2013). In this respect, organisations must be able to secure their relationship with all stakeholders consistently, especially when they make CSR disclosure decision (Ahamed et al., 2014) since different stakeholders may have different impact on organisations.

How Do Firms Benefit?

CSR is concerned about dealing with the stakeholders of the firm ethically or in a socially responsible way. Stakeholders exist in a firm and outside of it (Hopkins, 1998). The stakeholder theory does take into account the wider consequences of the society (Freeman, 1984), thus, examine the nature of the CSR adoption and its link with the performance of the firm. Gyves and Higgins, (2008) have said that the relationship between CSR and firm performance really depends on how CSR is managed. Therefore, multiple stakeholder groups must be managed effectively for the well-being of the firm.

According to stakeholder theory, the satisfaction of various stakeholder groups is instrumental for a firm's performance (Donaldson & Preston, 1995). The rationale on why CSR affects firm performance is because a firm's social involvements signal the firm's genuine care and concern for the welfare of its stakeholders, including investors, employees, suppliers, customers, and the society at large. This public image helps build long-term relationships with those stakeholders, facilitating cooperation with and support from them, which in turn contributes to a higher level of firm performance. Indeed, the firm cannot continue to flourish and survive without the support of stakeholders (Freeman, 1984), therefore, this study utilises stakeholder theory as the governing theory to explain the relationship between CSR and firm's stakeholder, as depicted in Figure 1.1, since firms involved continuously in activities of CSR with stakeholders (Jones, 1995) on the basis of trust and cooperation.

*Corresponding author's email: vani@melaka.uitm.edu.my

 DOI: 10.1051/shsconf/20173600040

Figure 1

Focus of the Study: Scheme of the relationship between CSR and Firm Stakeholder according to Stakeholder Theory

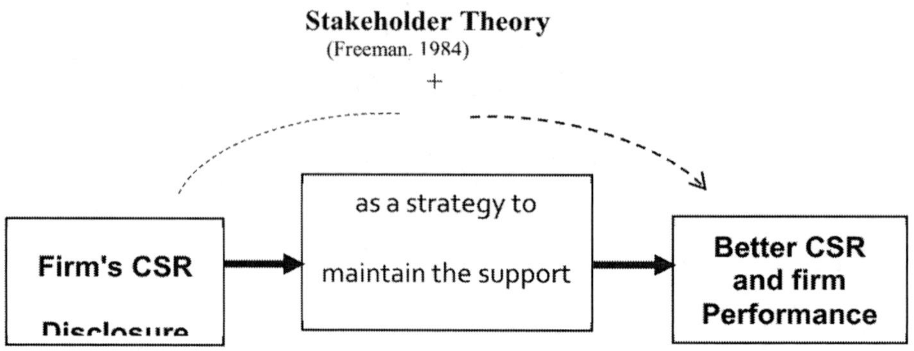

Source: Developed for this study

In other word, a firm's financial success is contingent on the ability to formulate and execute CSR strategy that manages effectively its relationships with stakeholders (Donaldson & Preston, 1995) as the stakeholders have their rights to seek for information about how the firm is impacting on them (Deegan & Unerman, 2006). Based on this perspective, Jones & Nisbet (2011) emphasized that business is under pressure from their stakeholders to report its CSR activities because these parties want to protect their long term interests in the firm. Thus, reporting firm's CSR activities is a major element that can be engaged by the firm to manage the stakeholder in order to gain their support and approval.

In pursuing social responsibility, firms need to acquire both tangible and intangible resources that allow them to greater spending on CSR activities in order to differentiate themselves and increase their competitiveness (Russo & Fouts, 1997; Schnietz & Epstein, 2005). In conclusion, this selection is grounded in the belief that firm stakeholder relationships are the essential assets that managers must manage (Post, Preston & Sachs, 2002) and the important role of stakeholders in addressing the issue of to whom a business should be accountable (Kakabase, Rozuel & Lee-Davies, 2005).

CONCLUSION

CSR is understood to be the way firms integrate social, environmental and economic concerns into their business processes, decision making, strategy and also operations in a transparent and accountable manner, thereby establish better practices within the firm, create wealth and improve society (Linh & Hieu, 2011). As such, the stakeholder theory is used as a basis to analyze those groups to whom the firm should be responsible and how far that scope of responsibility extends due to the fact that a firm's success is increasingly affected by stakeholders relationships (O'Riordan & Fairbrass, 2008). Thus, practically, firms still need to continually make commitment in implementing the CSR since there is a strong relationships between firm's CSR practices and its stakeholders (Waddock & Graves, 1997; Orlitzky et al., 2003; Ghelli, 2013; Abd-Mutalib et al., 2014).

*Corresponding author's email: vani@melaka.uitm.edu.my

REFERENCES

Abd-Mutalib, H., Jamil, C.Z.M., & Wan-Hussin, W.N. (2014). The Availability, Extent and Quality of Sustaiability Reporting by Malaysian Listed Firms: Subsequent to Manadatory Disclosure. Asian Journal of Finance & Accounting 6(2), 239-260.

Ahamed, W., Suhazeli, W., Almsafir, M. K., & AlSmadi, A. W. (2014). Does Corporate Social Responsibility Lead to Improve in Firm Financial Performance? Evidence from Malaysia.International Journal of Economics & Finance, 6(3)

Amran, A. (2006). Corporate Social Reporting in Malaysia: An Institutional Perspective. Unpublished PhD Thesis. University of Malaya.

Amran, A., Zain, M. M., Sulaiman, M., Sarker, T., & Ooi, S. K. (2013). Empowering Society for Better Corporate Social Responsibility (CSR): The case of Malaysia. Kajian Malaysia, 31(1), 57-78

Belkaoui, A.R. (2000). Accounting Theory, 4th edition, Academic Press, London

Caroll, A. B. (1979). A three-dimensional model of corporate social performance. *Academic of Management Review, 4*, 497-505.

Carroll, A., & Shabana, K. (2010). The business case for corporate social responsibility: A review of concepts, research and practice. International Journal of Management Reviews, 12(1), 85-105. http://dx.doi.org/10.1111/j.1468-2370.2009.00275.x

Clarkson, M. (1995). A stakeholder framework for analyzing and evaluating corporate social performance. Academy of management Review, 20(1), 92-117.

Clifton, D. and Azlan Amran. (2010). The stakeholder approach: A sustainability perspective. Journal of Business Ethics 98(1): 121–136.

Cramer, J., & Bergmans, F. (2003). Learning about corporate social responsibility : the Dutch experience. Amsterdam ; Washington, DC: IOS Press

Deegan, C. & Blomquist, C. (2006). Stakeholder influence on corporate reporting: An exploration of the interaction between WWF, Australia and the Australian minerals industry. Accounting, Organizations and Society, 31(4-5), 343-372.

Deegan, C. & Unerman, J. (2006). *Financial Accounting Theory*, McGraw Hill, Berkshire

Ditlev-Simonsen, C. D., & Midttun, A. (2011). What motivates managers to pursue corporate responsibility? a survey among key stakeholders. *Corporate Social Responsibility and Environmental Management, 18*, 25-38. doi: 10.1002/csr.237

D. M. DiSegni, M. Huly, and S. Akron, Corporate Social Responsibility, Environmental Leadership and Financial Performance, Social Responsibility Journal,11 (2015) 131-148.

Donaldson, T. and Preston, L.E. (1995), ``The stakeholder theory of the corporation: concepts, evidence, and implications", Academy of Management Review, Vol. 20, pp. 65-91.

Dkhili, H., & Ansi, H. (2012). The link between corporate social responsibility and financial

*Corresponding author's email: vani@melaka.uitm.edu.my

performance: The case of Tunisian companies. *Journal of Organizational Knowledge Management, 2*(4), 20–12.

Elijido – Ten, E (2007) Applying stakeholder theory to analyze corporate environmental performance: Evidence from Australian listed companies. Asian Review of Accounting,Vol. 15(2) PP 164-184

Esa, E., & Mohd Ghazali, (2012). Corporate social responsibility and corporate governance in Malaysian government-linked companies. Corporate Governance, 12(3), 292-305.

Freeman, R.E. (1984). *Strategic management: A stakeholder approach*. Boston: Pitman.

Freeman, I., & Hasnaoui, A. (2010). The meaning of corporate social responsibility: The vision of four nations. Journal of Business Ethics, 100, 419–443.

Gill, A. (2008). Corporate governance as social responsibility: A research agenda. Berkeley Journal of International Law, 26 (2), 452-478.

Ghelli, C. (2013). *Corporate social responsibility and social performance: An empirical evidence*. Copenhagen Business School

Ghotb, H. (2011). Corporate Social Responsibility (CSR) in catering industry: The perception of diners and their buying behavior. Master diss., University of Malaya, Kuala Lumpur, Malaysia

Gray R. H., Owen D. L. and Adams C. (1996). Accounting and accountability, changes and challenges in corporate social and environmental reporting. Prentice Hall, Hemel Hempstead

Gyves, S, & O.Higgins, E. (2008), .Corporate social responsibility: an avenue for sustainable benefit for society and the firm?., *Society and Business Review,* Vol. 3 No. 3, pp. 207-223.

Hamid, A F. Z., & Atan, R. (2011). Corporate Social Responsibility by the Malaysian Telecommunication Firms. International Journal of Business and Social Science, 2(5) [Special Issue], 198-208.

Hopkins M. (1998) .The Planetary Bargain: Corporate Social Responsibility Comes of Age., Macmillan, London.

Jones, T. M. (1995). Instrumental stakeholder theory: A synthesis of ethics and economics. *Academy of Management Review* 20: 404–437.

Jones, B. & Nisbet, P. (2011). Shareholder value versus stakeholder values: CSR and financialization in goal global food firms. *Socio-Economic Review*, 9, 287-314.

Kakabadse, K. N., Rozuel, C., & Davis, L. (2005). Corporate Social Responsibility and Stakeholder Approach: A Conceptual Review. International Journal of Business Governance and Ethics, 1(4), 277-302.

Linh, N. T., & Hieu, P. D. (2011). Impacts of Coprate Social Responsibility On The Financial Performance of The Firms. 1-38.

*Corresponding author's email: vani@melaka.uitm.edu.my

L.J. Chen, A. Feldmann, O. Tang. (2015). The relationship between disclosures of corporate social performance and financial performance: Evidences from GRI reports in manufacturing industry, International Journal of Production Economics 1-12.

Murray, K, & Vogel, C (1997), 'Using a Hierarchy-of-Effects Approach to Gauge the Effectiveness of Corporate Social Responsibility to Generate Goodwill Toward the Firm: Financial versus Nonfinancial Impacts', Journal Of Business Research, Vol. 38, Issue 2, pp. 141-159

McWilliams, A. and D. Siegel. (2001). Corporate social responsibility: A theory of the firm perspective. *Academy of Management Review* 25: 117–127.

Neal, R. & Cochran, P. L. (2008). Corporate social responsibility, corporate governance, and financial performance: Lessons from finance. Business Horizons, 51, 535-540

Orlitzky, M, Schmidt, FL and Rynes, SL (2003), Corporate Social Responsibility and Financial Performance: a meta analysis", *Organization Studies*, vol. 24, pp. 403-411.

O'Riordan, L. and Fairbrass, J. (2008) 'CSR - Theories, Models and Concepts in Stakeholder Dialogue – A Model for Decision-Makers in the Pharmaceutical Industry' Journal of Business Ethics 83(4), 754-758

Pop O., G. C. Dina, C. Martin (2011) Promoting the corporate social responsibility for a green economy and innovative jobs. Procedia Social and Behavioral Sciences 15 (2011) 1020–1023

Post, J. E., Preston, L. E. and Sachs, S. (2002) "Managing the Extended Enterprise: The New Stakeholder View", California Management Review, Vol. 45 No. 1, pp. 6-28.

Raza, A., Ilyas, M. I., Rauf, R., & Qamar, R. (2012). Relationship between Corporate Social Responsibility (CSR) and Corporate Financial Performance (CFP): Literature review approach. *Elixir Financial Management, 46*(9), 8404–8409.

Russo, M. and Fouts, P. (1997), .A resource-based perspective on corporate environmental performance and profitability., Academy of Management Journal, vol40, pp. 534-559.

R.M. Whalen, Corporate social responsibility and its relationship with financial performance, PhD thesis, Capella University, 2013

Schnietz, K., & Epstein, M. (2005). Exploring the financial value of a reputation for corporate social responsibility during a crisis. *Corporate Reputation Review, 7*(4), 327-345.

Vormedal I, and Ruud A.(2009). Sustainability reporting in Norway – an assessment of performance in the context of legal demands and sociopolitical drivers. Business Strategy and the Environment 18: 207–222.

Waddock, S.A., & Graves, S.B. (1997). The corporate social performance-financial performance link. *Strategic Management Journal, 18*, 303-319.

*Corresponding author's email: vani@melaka.uitm.edu.my

 DOI: 10.1051/shsconf/20173600040

W. S. Lu, K. W. Chau, H. D. Wang, and W. Pan, A Decade's Debate on The Nexus between Corporate Social and Corporate Financial Performance: A Critical Review of Empirical Studies 2002-2011, Journal of Cleaner Production, 79 (2014) 195-206.

Zhu, Y.L. (2011). An Empirical Study on Corporate Social Responsibility and Corporate Profitability. *Management Science and Industrial Engineering (MSIE),* 650-652.

*Corresponding author's email: vani@melaka.uitm.edu.my

ACCOUNTABILITY AND GOVERNANCE REPORTING BY NON-PROFIT ORGANIZATIONS

Nurfarahin Roslan*, Roshayani Arshad**, Nur Farahah Mohd Pauzi***

*Faculty of Accountancy, Universiti Teknologi MARA, Puncak Alam Selangor, Malaysia, **
*Accounting Research Institute, Universiti Teknologi MARA, Malaysia, ***
*Faculty of Accountancy, Universiti Teknologi MARA, Melaka, Malaysia. ****

ABSTRACT

Many non-profit organizations (NPOs) are facing funding cuts and reduction in voluntary donations during the current economic environment. In ensuring the sustainability of these organizations, the Malaysian government is currently promoting collaborative efforts between the government, social organisations and the private sector in providing social services. The potential advantages of such collaborations include, the risks of providing social services are shared between the government, social organizations and the private sector, build positive perceptions of government as enabler for social organizations, capitalise on social organizations' collective skills, enthusiasm and innovation in providing social services. However, the current reporting requirement for NPOs in Malaysia may give rise to information asymmetry that can hinder such efforts. Hence, this study aims to examine the current reporting practices of NPOs in Malaysia. Based on content analysis of annual reports of 205 NPOs registered with Registry of Societies for the financial year 2011, this study finds that the overall information reported by the NPOs is relatively weak. NPOs also disclosed relatively high non-financial information as compared to the financial information. Hence, there is a need for more comprehensive reporting guidelines for the NPOs that can facilitate the decision-making needs of the relevant stakeholders.

Keywords: Non-Profit Organizations, Accountability, Governance, Reporting

INTRODUCTION

Non-profit organizations (NPOs) fall under the third sector which is also known as charitable sector, voluntary sector or social sector that is comes after the public and private sector. There are 12 categories of NPOs in the system, namely, culture and recreation; education and research; health; social services; environment; development and housing; law advocacy and politics; philanthropic intermediaries and volunteerism promotion; international activities; religion; business; professional associations and unions; and lastly those that are not elsewhere classified (Salamon & Anheier, 1996). Sustainability of NPOs is a paramount issue in Malaysia and issues on transparency and accountability of NPOs have continuously been emphasized as the key requirement for good governance (Burger & Owens, 2010). Therefore, the annual report could be

*Corresponding author's email: nurfarahin1310@gmail.com

the best approach to demonstrate the extent of transparency and accountability through the information disclosed to the stakeholder as a whole (Flack & Ryan, 2002). As for Malaysia, there is no specific accounting standards dedicated mainly for NPOs, thus, Malaysian NPOs are subjected to a less stringent reporting requirements due to minimum regulatory requirement (Arshad, Abu Bakar, Sakri & Omar, 2012; Mohd Nasir, Othman, Said & Ghani, 2009 and Zainon, Hashim, Yahaya & Atan, 2013). On the other hand, the accounting bodies in the developed countries such as United Kingdom, United States, New Zealand, Australia and Canada have started to provide dedicated details guidelines in relation to reporting framework of non-profit organizations due to its remarkable growth. Currently, the Charity Commission in the UK and Republic of Ireland have come out with the Statement of Recommended Practices (SORP, 2013) to facilitate NPOs in preparing financial statements in accordance with the Financial Reporting Standard (FRS) whereas Canada released CICA Handbook and Financial Information Kit in 2010 to support the financial reporting of charities. Besides, Institute of Chartered Accountants in Australia has issued guidelines in producing financial reports to enhance NPOs annual reporting.

Traditionally, the main financial supports of NPOs have been derived through government funding and voluntary donations. According to Trussel & Parsons (2008), provision of financial support from the donors and grantors could be highly depending on the information disclosed by NPOs. There is no exception for NPOs in Malaysia having reduction in the traditional sources of funding which lead to increasing competition among NPOs for these limited resources during the current economic environment. Various initiatives have been taken by government agencies to ensure NPOs could obtain sufficient funding in order to deliver social services. To date, a Social Public-Private Partnership (SPPP) was introduced under the National Blue Ocean Strategy (NBOS) to address social disparities through delivering high value impact at low cost by aligning the government and the private sector resources in terms of funds and other in-kind contributions.

The loopholes in Malaysian reporting requirement have triggered problems such as low compliance level and inconsistency in preparing the financial statement of NPOs that can finally could be linked to the increment in the number of corporate scandals, mismanagement of fund and fraudulent activities involving NPOs (FATF, 2013 and Gordon, Khumawala, Kraut & Neely, 2010). Consequently, the information disclosed in the annual reports could subject to the discretion of the individual NPOs. Concerns over perceived lack of transparency and accountability in non-profit sector have triggered calls for the development of an accounting standards or recommended practices dedicated to the third sector which covers financial and non-financial disclosures in order to assist NPOs in producing comprehensive financial reports as a mechanism to discharge accountability and manage information asymmetry.

*Corresponding author's email: nurfarahin1310@gmail.com

OVERVIEW OF REPORTING REQUIREMENTS OF NON-PROFIT ORGANIZATIONS

Reporting requirements for non-profit organizations in Singapore

In Singapore, non-profit entity can be registered either as public company limited by guarantee, society or charitable trust. Public companies limited by guarantee are registered with Accounting and Corporate Regulatory Authority (ACRA) and are governed by the Singapore Companies Act 1967. For societies, they are required to be registered with the Registrar of Societies (ROS) and are governed under Singapore Societies Act 1966. Charitable Trusts are licensed by the Monetary Authority of Singapore (MAS) and are governed by the Singapore Trust Companies Act 2005. After the non-profit entity has obtained a legal status, they are allowed to apply for charity status from Commissioner of Charities and are governed under Singapore Charities Act 1994.

For disclosure and reporting requirements, non-profit entity with a charity status are required to submit annual reports (including audited statement of accounts) and governance evaluation checklist (compliance with the Code of Governance for Charities 2011) within six months from the end of the financial year to the Commissioner of Charities or Sector Administrators. The annual report should be prepared in accordance with the requirement set out in the Charities (Accounts and Annual Report) Regulations 2011 (http://statutes.agc.gov.sg).

Under the Charities Regulations 2011, the preparation of statement of accounts should comply with either Charities Accounting Standards (CAS) or Financial Reporting Standards (FRS). Statement of accounts should contain an income and expenditure account together with a balance sheet showing the financial position of the last financial year. Other statements such as notes explanation of the accounting policies, the details of transactions relating to charity various funds, notes on fund utilization, analyses of the information contained in the income and expenditure account and balance sheet are also required. However, for charity with gross income in any financial year does not exceed $50,000, the entity may elect to prepare a receipts and payments account and a statement of assets and liabilities, for that particular year. Besides submitting annual report to the Commissioner of Charities, non-profit entity registered either with ROS or ACRA are also required to submit annual return to whom the entity are registered.

Reporting requirements for non-profit organizations in Canada

In Canada, NPOs can be registered either as companies or societies with the Corporate Registry. For non-profit companies, they are governed under the Companies Act 1970, while societies are governed under Society Act 2000. In order for the entities to become registered charities, an application must be made to and approved by Canada Revenue Agency (CRA). Registered charities are required to comply with Income Tax Act and Provincial Legislation. For disclosure and reporting requirements, entities have reporting obligation under the Canada Not-for-Profit Corporation Act 2009 to submit an annual return within the 60 days following the financial year end together with financial statements and public accountants report not less than 21 days before

*Corresponding author's email: nurfarahin1310@gmail.com

each annual general meeting (https://www.ic.gc.ca). The financial statements must be prepared in accordance to the Canadian Generally Accepted Accounting Principles (GAAP) as set out in the Canadian Institute of Chartered Accountants (CICA) Handbook.

Reporting requirements for non-profit organizations in New Zealand

Charities in New Zealand need to be registered with the Charities Services which governed under the Charities Act 2005. Charities Services' roles include registering and monitoring charities as well as processing Annual Return (a yearly report which details the income, expenditure and activities). The charities register will list down all the registered charities in New Zealand which summarized charity's purposes, activities, sector and Annual Return that to be filed within six months after the financial year end date (www.charities.govt.nz/charities-in-new-zealand/the-charities-register). The registration and deregistration of charities are done by the independent three-person board in the Charities Registration Board (www.charities.govt.nz/about-charities-services).

Not-for-profit entities and registered charities will have significant changes in the financial reporting landscape with the first time of financial reporting obligations starting from 1 April 2015 (www.charities.govt.nz/new-reporting-standards/). In conjunction with that, PBE Standards for public entities (including both public sector entities and not-for-profit entities) based on the International Public-Sector Accounting Standards (IPSAS) issued by the International Public-Sector Accounting Standards Board (IPSASB) are as follows:

	Not-for-profit: Criteria of Tiers	Accounting Standard	Key Dates
TIER 1	• Publicly accountable • Large (i.e. annual expenses $>$ $30 million)	PBE Standards (with NFP enhancements)	**Effective date:** Periods beginning on or after 1 April 2015
TIER 2	• No public accountability • Expenses over $2 million up to $30 million	PBE Standards (RDR) (with NFP enhancements)	**Early adoption:** Permitted, however until NFP versions are issued, must use the public-sector version of the PBE Standards (i.e. without NFP enhancements)
TIER 3	• No public accountability • Expenses $\leq$ $2 million	PBE Simple Format (Accrual) (Not-for-profit)	
TIER 4	• No public accountability, and permitted by an Act (of Parliament) to report on cash basis • Operating payments $<$ $125,000	PBE Simple Format (Cash) (Not-for-profit)	

Source: KPMG (2014)

All of the transactions, balances and activities should be included in the financial statements of reporting entities which comprise of: (1) all transactions and activities undertaken using the reporting entity's identity, (2) transactions undertaken and balances held in the name of the reporting entity, (3) funds and associated transactions solicited in the name of and with the approval of the reporting entity by a third party fundraiser acting as an agent of the reporting entity and (4) account for the relationship under 'control', 'joint control', 'significant influence' or 'other relationship' with other entities. The financial reporting requirements should be included with the annual return within six months of the financial year end date (Deloitte, 2015).

*Corresponding author's email: nurfarahin1310@gmail.com

Reporting requirements for non-profit organizations in United States

Nonprofit organizations in US are regulated by local, state and federal laws (www.hg.org/nonprofit-organizations.html). The incorporation of nonprofit in US would be different accordingly from state to state in terms of law and regulation. However, a general overview for all the states would require, (1) filing of Articles of Incorporation (supplemental information such as Certificate of Disclosure, proof of corporate name and filing fees might be needed), (2) publishing the Articles of Incorporation a certain number of times in a local newspaper and the proof of publication might need to be filed with a state agency, and lastly, (3) filing for Federal Tax-Exempt Status (www.councilofnonprofits.org/tools-resources/how-start-nonprofit). There are all together 29 types of nonprofits which are tax exempt under Internal Revenue Code Section 501 (c) (www.charitynavigator.org).

For the federal filings, tax-exempt nonprofits are required to file an annual report Form 990 with the Internal Revenue Service (IRS) on the 15th day of the 5th month after the close of the nonprofit's fiscal year. Form 990 is opened to public inspection and most of the nonprofits will post the documents from a link on their website in order to demonstrate a commitment to transparency and make it easier for those seeking financial information to view the documents. Whereas, for the state filings, most states require nonprofit organizations to file one or more reports every year. In general, there are four types of filing requirements, (1) corporate filings, (2) financial reports, (3) fundraising registrations and (4) state tax-exemption filings (www.councilofnonprofits.org/tools-resources/public-disclosure-requirements-nonprofits).

Comparison with Reporting Requirements of Non-Profit Organizations in Malaysia

Non-profit organizations in Malaysia can be registered with two main regulatory bodies, namely Companies Commission of Malaysia (CCM) and Registry of Societies (ROS). Non-profit organizations that are registered under the CCM are incorporated under the Companies Act 1965, whereas, Societies Act 1966 regulates non-profit organizations registered under ROS. The financial statement of the non-profit organizations registered under CCM are required to comply with the Malaysian Financial Reporting Standards (MFRS) issued by Malaysian Accounting Standard Board (MASB) and must be audited which is equivalent with the rest of other private entities in Malaysia. However, those that registered under ROS are only encouraged to comply with the MFRS and their financial statements are not compulsory to be audited but encouraged to get the financial statements audited on voluntary basis. Additionally, according to Section 14(d) of the Societies Act 1966 (Act 335) and Regulations Act 1984, the organizations need to submit the annual return (Form 9) together with the Statement of Receipt and Payments and Balance Sheet of the last financial year within sixty days after the annual general meeting. However, cash flow statement, statement of changes in general fund, notes on significant accounting policies and other supplementary statements are not required to be submitted.

*Corresponding author's email: nurfarahin1310@gmail.com

METHODOLOGY

Sample and Data Collection

The sample for this study consists of 205 registered NPOs with ROS from various background including religious, welfare, social and recreation, women, culture, mutual benefit societies, trade associations, youth, disable, education, political, employment associations, general and others. The research approach involves the content analysis of annual reports which consist of 1 year data under review collected from year 2011 and also Form 9 provided by the ROS. For the purpose of measuring voluntary and mandatory disclosures in annual report, content analysis has been widely used in prior studies (O'Donovan, 2002; Clemens & Douglas, 2006). This study emphasis on the data disclosed in relation to financial and non-financial information in order to examine the reporting practices by NPOs in Malaysia.

Development of Total Disclosure Index

Total disclosure index (TDI) was developed for this study in order to measure the level of financial and non-financial information disclosed in the annual reports. The identification of items to be included in the index was based on the Society Act 1966 reporting requirement, recommended practices in United Kingdom, Canada and Australia, and some prior studies that relate to NPOs (Arshad et al., 2012, Froelich, 1999, Irvine & Ryan, 2013 and Zainon, Atan & Wah, 2014). The Society Act 1966 provides brief guidelines on the mandatory requirement for NPOs in Malaysia on the need to prepare and submit annual return which comprises of details of Balance Sheet (BS) and Statement of Income and Expenditure (IS) which resulted in inconsistency in the preparation and presentation of information disclosed.

This study has further referred to the Statement of Recommended Practices (SORP) in accordance to Financial Reporting Standards (FRS) published by the UK Charity Commission. SORP was adopted as it provides up-to-date information relates to charities annual report preparation and presentation. Furthermore, SORP report emphasis on both financial and non-financial information such as objectives and activities, achievements and performance, financial review, structure, governance and management, reference and administrative details in order to come out with comprehensive disclosure in annual reports. Besides, the identification of index was also guided by a review of the best practices in "Combating the Abuse of NPOs" issued by the Financial Action Task Force, June 2013. This report emphasized on financial transparency especially on the disclosures of fund bank account, program budget details including the identity of donors and recipients in relation to the anti-money laundering and terrorism financing.

Finally, TDI items are grouped into the following categories, (1) Balance Sheet (13 items), (2) Statement of Financial Activities (15 items) and (3) Non-financial Reporting (11 items) as summarizes in Table 1 below.

*Corresponding author's email: nurfarahin1310@gmail.com

DOI: 10.1051/shsconf/20173600041

Table 1: Main Categories of Data Disclosure

Category 1	**Disclosure Items (Balance Sheet)**
Information on Financial Reporting	Non-Current Assets (2 items)
	Current Assets (3 items)
	Non-Current Liabilities (2 items)
	Current Liabilities (2 items)
	Equity Funds (4 items)
	Disclosure Items (Statement of Financial Activities)
	Income generated from Operation (3 items)
	Income from Fundraising Activities (1 items)
	Income from Grants (3 items)
	Income from Investment Activities (3 items)
	Other Income (1 items)
	Cost of Operation (1 items)
	Cost of Fundraising Activities (1 items)
	Administrative Expenses (1 items)
	Other expenses (1 items)
Category 2	**Disclosure Items**
Information on Non-Financial Reporting	Organization activities (1 items)
	Programme Accomplishment (3 items)
	Financial Review (1 items)
	Corporate Info, Structure and Governance (4 items)
	Contribution (2 items)
Total number of items	**39 items**

Table 2: Summary of Item Disclosures under Category 2

1	**Objectives**
Program Accomplishment	
2	Achievement
3	Performance
4	Future plan
5	**Financial Review**
Corporate Information, structure and governance	
6	Corporate Information
7	Corporate Profile
8	Financial Expertise
9	Auditor
Contribution	
10	Volunteer Disclosures
11	Contribution to/from other countries

*Corresponding author's email: nurfarahin1310@gmail.com

Table 3: Summary of Item Disclosures under Category 1

(A) Disclosure Items under Balance Sheet	
Non-current Assets	
1	Property, Plant and Equipment
2	Long-Term Investment
Current Assets	
3	Trade and Other Receivables
4	Short-Term Investment
5	Cash and Bank Balances
Current Liabilities	
6	Trade and Other Payables
7	Short-Term Loan and Borrowings
Non-current Liabilities	
8	Taxation
9	Long-Term Loan and Borrowings
Equity	
10	Restricted Funds
11	Unrestricted Funds
12	Accumulated Reserves
13	Total Accumulated Funds
(B) Disclosure Item under Statement of Financial Activities (SOFA)	
Income from operation	
1	Donation and contribution
2	Fees
3	Commercial Activities
Income from Fundraising Activities	
4	Fundraising
Income from Grants	
5	Government
6	Private
7	Others
Income from Investment Activities	
8	Dividend
9	Interest
10	Rental
11	**Other Income**
12	**Cost of Operation**
13	**Cost of Fundraising Activities**
14	**Administrative Expenses**
15	**Other Expenses**

Table 4: Disclosure Score

Score	Indicators
0	No disclosure
1	Disclosed

*Corresponding author's email: nurfarahin1310@gmail.com

DOI: 10.1051/shsconf/20173600041

RESULTS AND FINDINGS

NPOs Reporting and its accountability

Reporting of comprehensive information is crucial for NPOs since they are accountable towards a large number of stakeholders including the public and resource providers. In order for NPOs to sustain and successfully deliver their social mission, they are highly dependent on the contribution from external parties. These parties require NPOs to be more reliable and transparent in reporting to help encounter their concern on the effectiveness and efficiency of the funds management. Trussel & Parson (2008) and Zainon et al (2014) have agreed that by disclosing details of their financial standing and social performance, NPOs can enhance their credibility, build community trust and demonstrate their responsiveness towards stakeholders. Failure to communicate sufficient information can lead to negative perceptions regarding the organization's effectiveness leading to withdrawals of support by the donors and other stakeholders. Besides, the reluctant behavior to disclose full organization information could also trigger the emergence of corporate scandals, misappropriation of fund and fraudulent activities (Gordon et al, 2010). Thus, the proposal of increasing the reporting requirement for Malaysian NPOs is important to improve the credibility and sustainability of NPOs in the long run.

Table 5: Descriptive Statistics for Level of Reporting Practices

	No of sample	Minimum	Maximum	Mean
Financial Reporting - Balance Sheets	205	.08	.85	.3698
Financial Reporting - SOFA	205	.13	.73	.3652
Non-Financial Reporting	205	.18	1.00	.4449
Total Disclosures	205	.13	.69	.3901

Total disclosures tested in this study consist of two major components including **Financial Reporting**; Balance Sheet (BS) and Statement of Financial Activities (SOFA), and **Non-Financial Reporting**. Through the review of 205 NPOs annual reports, it reveals that NPOs in Malaysia disclosed at least 8% of Balance Sheet items, 13% of SOFA items and 18% of non-financial items. The maximum scores indicate that NPOs might disclose up to 100% of the non-financial information, 85% of the balance sheet information and 73% of statement of financial activities information. From this analysis, it shows that NPOs in Malaysia is better in terms of disclosing the non-financial information as compared to financial information. NPOs are motivated to provide details disclosure of non-financial information due to its mission that are more related to community and social services.

However, the overall disclosure was found to be weak for all the components since the mean values were below 50%. On average, Malaysian NPOs only disclosed 37% of Balance Sheet and SOFA items, and 44% of Non-financial items. The result is already expected since NPOs are only subjected to a minimum reporting requirement under the Malaysian environment.

*Corresponding author's email: nurfarahin1310@gmail.com

 DOI: 10.1051/shsconf/20173600041

Moreover, NPOs in Malaysia are still reluctant to disclose information especially on the financial aspects as compared to non-financial aspects which could reflect that NPOs are still not taking full advantage of the comprehensive reporting as a communication device to gain donors and public trust based on five best practices, namely, completeness, accessibility, transparency, full disclosure and relevance (Gordon et al, 2010). As a consequence, limited disclosure by the NPOs due to the incomplete of record keeping, limited financial monitoring and unstructured volunteer management would hinder NPOs growth, sustainability and effectiveness (Minzner, Klerman, Markovitz and Fink, 2014).

Financial Reporting

Disclosures of Balance Sheet Items

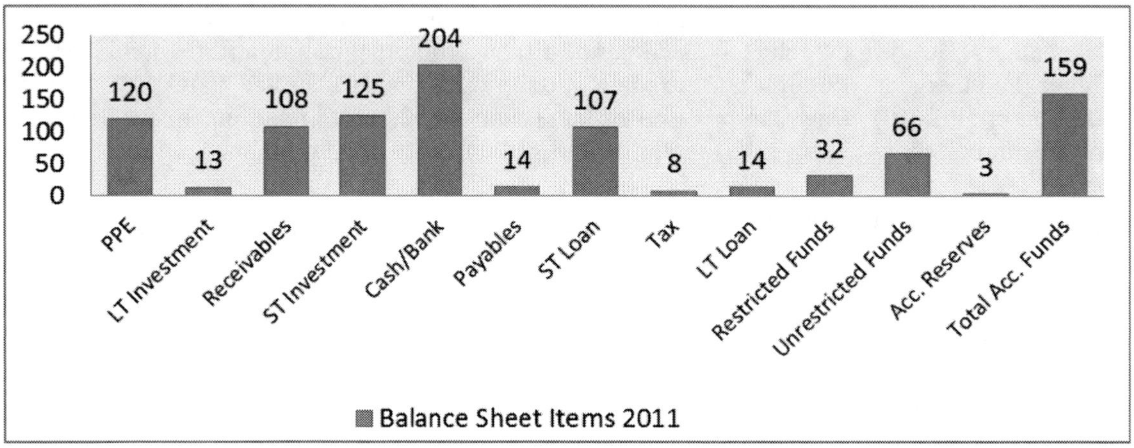

Figure 1: Disclosures of Balance Sheet Items

A detailed analysis was conducted on the disclosures of balance sheet items. Figure 1 shows the assets and liabilities disclosed by NPOs for the year 2011. Among the Balance Sheet items that are mostly disclosed are property, plant and equipment, current assets, short term loan and accumulated fund. Meanwhile, disclosures on long term investment, payables, taxation and equity funds is relatively low. In addition, results have showed less than half of Malaysian NPOs disclosed about their equity fund. This might provide perceptions either NPOs in Malaysia have a very limited fund since they are smaller in size as compared to other sectors or NPOs are reluctant to disclose information on the funds obtained. Disclosures on the equity fund are highly demanded since the stakeholders are the main contributor of NPOs income. According to Taniguchi and Marshall (2014), the inflow and outflow of funds need to be properly disclosed as individuals believe that their contribution will be put to good use based on their trust. Thus, providing useful and relevant information regarding NPOs financial position to the current and

*Corresponding author's email: nurfarahin1310@gmail.com

DOI: 10.1051/shsconf/20173600041

future resource providers is an important strategy to ensure sustainability of NPOs in the long run.

Disclosures of Statement of Financial Activities (SOFA) Items

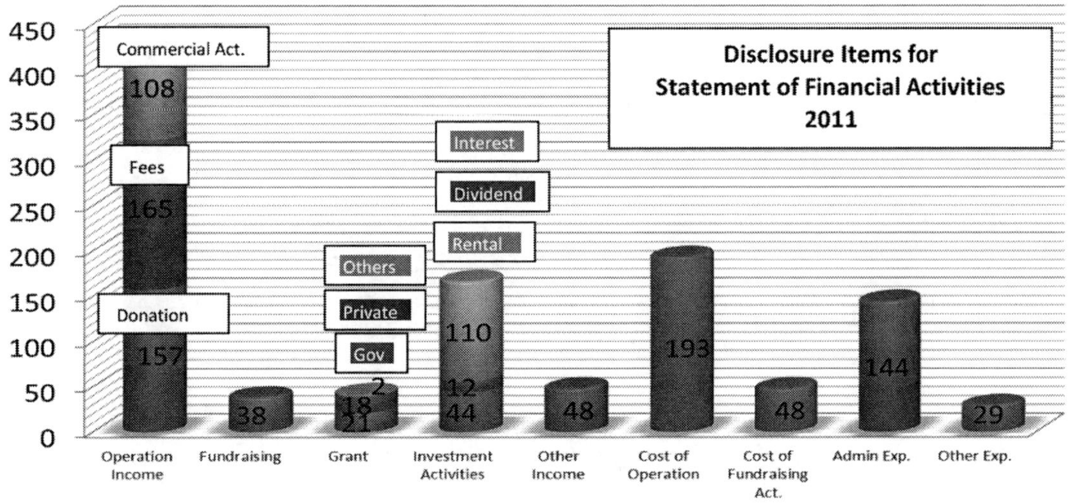

Figure 2: Disclosures of SOFA items

Figure 2 shows a detailed analysis on disclosures of Statement of Financial Activities by items. The results shows that, among the items that are highly disclosed in SOFA include the income from operation and investment activities, cost of operation and admin expenses. Besides, more than 50% of Malaysian NPOs disclosed their sources of income come from donation, membership fees and trading activities. This result provides a good signal that NPOs are trying to be transparent by disclosing the information related to the sources of income and expenditure. However, only a quarter of NPOs disclose about the cost of fundraising and the remaining disclosures are only limited to cost of operation and administrative. It is paramount for the NPOs to record and provide details information on the sources of revenue received such as donations, government grants, trading income and investment income which ultimately would be scrutinized by the public at large. Since membership fees is one of NPOs major income sources, therefore, NPOs are accountable towards the members, to disclose information on the management of the money, namely in terms of the details spending. Besides that, NPOs that involved in trading activities are expected to have proper and comprehensive reporting of their activities as they are accountable towards a larger crowd of stakeholders including the customers and suppliers as their additional resource provider.

There are still a lot of rooms for improvement need to be taken into consideration by NPOs to be more accountable in reporting of their activities since they are being pressured from the external resource providers to present the evidence on how the resources are spent and achievement acquired although they can voluntarily decide to measure on the social impact for their own purpose (Arvidson and Lyon, 2014). According to McGann and Johnstone (2006), transparency

*Corresponding author's email: nurfarahin1310@gmail.com

DOI: 10.1051/shsconf/20173600041

should emphasize on the three areas which include, transparency on how funds are spent, transparency in the method of intervention and to conduct regular monitoring and evaluation involving the stakeholders or making the outcomes of these known to the stakeholders. In addition, NPOs should aware that the extent of reporting made particularly on the spending of public fund would have a significant effect on donations (Marudas, 2004). Pfeffer and Salancik, (1978) have highlighted that NPOs would survive substantially from public funds if they can manage the flow of resources and reduces the uncertainty in their environment.

Non-Financial Reporting

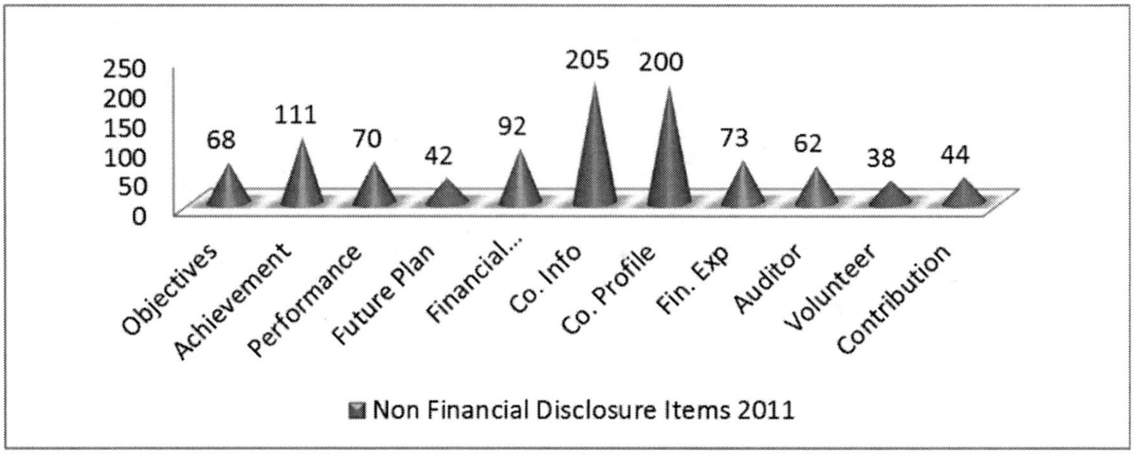

Figure 3: Disclosures of Non-financial items

In terms of non-financial reporting, results from Figure 3 indicate that among the most items disclosed by NPOs are corporate information and profile. However, more than half of the Malaysian NPOs did not disclosed information about their objectives, program achievement, performance, financial review, financial expertise and auditor. In addition, NPOs seems to neglect the information on the volunteers that gives significant contribution to the charities and disclosures of contribution received/paid from/to countries other than Malaysia.

Statement of Recommended Practices (SORP, 2013) highlighted that good reporting is when the charities provides a coherent explanation of the charity's objective, policies and strategies, also explaining the activities undertaken throughout the year. NPOs should disclose the non-financial information especially their objectives, policies and strategies in detailed sentences including how often it is being reviewed. Disclosures regarding the programs and activities conducted, the people involved and the achievements are important because it helps the resource provider or public to monitor the management of the fund given and the effectiveness of the programs conducted and beneficial bring to the society. Besides, any opinion on the improvement of the performance of NPOs is essential to be disclosed to the stakeholders in order to prove the ability of NPOs in providing social services in order to remain sustainable and to show their mission are not driven by the profit motive.

*Corresponding author's email: nurfarahin1310@gmail.com

CONCLUSION

This paper provides a descriptive analysis on the level of information disclosed in the annual reports of Malaysian NPOs, focusing on the financial and non-financial information. The quality of information disclosed in annual reports could enhance NPOs accountability towards multiple stakeholders. This is indeed an essential part to maintain NPOs credibility in their long run and to create the awareness of the importance of third sector to the community at large. This study found that the levels of financial and non-financial information disclosed among Malaysia NPOs registered with ROS are relatively weak. However, NPOs disclosed relatively high non-financial information as compared to the financial information.

These indicate several important insights relating to the non-profit sector. Firstly, since there is no proper mechanism introduced to fill in the reporting gap for NPOs in Malaysia, it has resulted to NPOs' reluctant behavior to disclose detailed information regarding their activities. In addition, as stated by the regulatory authority, NPOs are only subjected to voluntary compliance. Secondly, comprehensive disclosures integrating financial and non-financial information can potentially help to improve NPOs' sustainability through the showcase of accountability and the impact that NPOs can provide to the public as a whole. In other words, comprehensive disclosures may act as a tool for NPOs to communicate their activities, achievement and future plans which in turn allows NPOs to build community trust and demonstrate the responsiveness towards the resource providers. Lastly, the introduction of a comprehensive reporting guideline can be use by the Registry of Society as a monitoring device towards the organizations within the third sector.

Taking into accounts of the characteristics of NPOs in Malaysia, they still need some guidance with regards to the preparation and presentation of their annual reports. Reporting guidance or best practices promoted by regulatory authorities are expected to enhance the accountability of NPOs. Therefore, the development of comprehensive framework is needed to facilitate the NPOs in preparing and presenting their annual reports.

REFERENCES

Arshad, R., Abu Bakar, N., Sakri, F. H. & Omar, N. (2012). Organization characteristics and NPO reporting. *Asian Social Science,* 9(1), 1-9.

Arvidson, M. and Lyon, F. (2014) Social Impact Measurement and Non-profit Organisations: Compliance, Resistance and Promotion, International Society for Third-Sector Research, 25, 869-886.

Burger, R. & Owens, T. (2010). Promoting transparency in the NGO sector: Examining the availability and reliability of self-reported data. *World Development,* 38(9), 1263-1277.

*Corresponding author's email: nurfarahin1310@gmail.com

Clemens, B. & Douglas, T. J. (2006). Does coercion drive firms to adopt voluntary green initiatives? Relationship among coercion, superior firm resources and voluntary green initiatives. *Journal of Business Research,* 59, 483-541.

Deloitte (2015). Financial reporting by not-for-profit entities in New Zealand: Your questions answered. (May), 1-36.

FATF. (2013). Combating the abuse of non-profit organisations. (June), 1-17.

Flack, T. & Ryan, C. (2002). The role of the annual report in a charitable organization: A stakeholder perspective. Working Paper No. 2004-011, Queensland University of Technology.

Froelich, K. A. (1999). Diversification of revenue strategies: Evolving resource dependence in nonprofit organizations. *Nonprofit and Voluntary Sector Quarterly,* 28(3), 246-268.

Gordon, T. P., Khumawala, S. B., Kraut, M. & Neely, D. G. (2010). Five dimensions of effectiveness for nonprofit annual reports, *NonProfit Management and Leadership,* 21, 209-228.

Gonzalez, L. I. A., Vijande, M. L. S. and Casielles, R. V. (2002) The Market Orientation Concept in the Private Nonprofit Organization Domain, International Journal of Nonprofit and Voluntary Sector Marketing, 7(1), 55-67.

Irvine, H. & Ryan, C. (2013). Accounting regulation for charities: Internatinal responses to IFRS adoption. *Pacific Accounting Review,* 25(2), 124-144.

KPMG (2014). Accounting advisory services: New Zealand financial reporting framework. (June), 1-28. (kpmg.com/nz)

Marudas, N. (2004) Effects of Nonprofit Organizations' Wealth and Efficiency on Private Donations to Large Nonprofit Organizations, Research in Government and Nonprofit Accounting, 11, 71-91.

McDonald, R. E. (2007) An Investigation of Innovation in Nonprofit Organizations: The Role of Organizational Mission, Nonprofit and Voluntary Sector Quarterly, 36(2), 256-281.

McGann, J. and Johnstone, M. (2006) The Power Shift and NGO Credibility Crisis, International Journal of Not-for-Profit Law, 8, 65-77.

Mohd Nasir, N., Othman, R., Said, J. & Ghani, E. (2009). Financial reporting practices of charity organisations: A Malaysian evidence. *International Bulletin of Business Administration,* 6, 19-27.

*Corresponding author's email: nurfarahin1310@gmail.com

Minzner, A., Klerman, J. A., Markovitz, C. E. and Fink, B. (2014) The Impact of Capacity-Building Programs on Nonprofits: A Random Assignment Evaluation, Nonprofit and Voluntary Sector Quarterly, 43(3), 547-569.

O'Donovan, G. (2002). Environmental disclosures in the annual report: extending the applicability and predictive power of legitimacy theory. *Accounting, Auditing and Accountability Journal,* 15(3), 344-371.

Pfeffer, J. and Salancik, G. R. (1978) The External Control of Organizations: A Resource Dependence Perspective. Upper Saddle River: Pearson Education.

Reheul, A-M., Caneghem, T. V. and Verbruggen, S. (2014) Financial Reporting Lags in the Non-profit Sector: An Empirical Analysis, International Society for Third-Sector Research, 25, 352-377.

Salamon, L. M. & Anheier, H. K. (1996). The international classification of nonprofit organizations: ICNPO-Revision 1. Working Papers of the Johns Hopkins Comparative Nonprofit Sector Project, No. 19. Baltimore: The Johns Hopkins Institute for Policy Studies.

SORP, Statement of Recommended Practices. (2013). Charities SORP (FRSSE), (January) 1-183.

Taniguchi, H. and Thomas, L. D. (2014) The Effects of Social Trust and Institutional Trust on Formal Volunteering and Charitable Giving in Japan, International Society for Third-Sector Research, 25, 150-175.

Trussel, J. M. & Parsons, L. M. (2008). Financial reporting factors affecting donations to charitable organizations. *Advances in Accounting,* 23(7), 263-285.

Zaiton, S., Atan, R. & Wah, Y. B. (2014). An empirical study on the determinants of information disclosure of Malaysian non-profit organizations. *Asian Review of Accounting,* 22(1), 35-55.

Zainon, S., Hashim, M., Yahaya, N. & Atan, R. (2013). Annual reports of non-profit organizations (NPOs): An analysis. *Journal of Modern Accounting and Auditing,* 9(2), 183-192.

*Corresponding author's email: nurfarahin1310@gmail.com

TAXPAYERS' COMPETENCY LEVEL IN GOVERNING GOODS SERVICES TAX (GST) COMPLIANCE: MALAYSIA SCENARIO.

Siti Syaqilah Hambali* and Amrizah Kamaluddin **
*Faculty of Accountancy Universiti Teknologi MARA Puncak Alam**
*Accounting Research Institute Universiti Teknologi MARA Shah Alam***

ABSTRACT

GST implementation in Malaysia in April 2015 is part of the tax reformation with the goal of having a more efficient and effective taxation system. A successful tax reformation requires a high acceptance from taxpayers. In order to aid in the acceptance of GST implementation, taxpayers need to have a good competency level reflected through proper knowledge and awareness of the GST system, also known as GST literacy. Thus, the main objective of this study is to assess the level of GST literacy of taxpayers. Secondly, the current study aims to identify the relationship between GST literacy level with the GST acceptance of the taxpayers. Questionnaires were distributed to 150 secondary school teachers in Johor, one of the states in Malaysia. The finding indicates that the respondents' literacy rate towards the GST implementation in Malaysia is fair. The results imply that secondary school teachers have a moderate knowledge level and high awareness level on GST implementation in Malaysia. The study proves statistically that there is a positive significant relationship between GST literacy with GST acceptance.

Keywords: Competency, GST implementation, GST literacy, GST acceptance, GST knowledge and GST awareness

INTRODUCTION

Goods and Services Tax (GST) is a broad-based tax charged on goods and services in all sectors within the economy. Malaysia has implemented GST to replace the Sales and Services Tax (SST) with effect first April 2015. To this date, more than 160 countries have implemented GST. Out of the 160 countries, 7 countries represents ASEAN, 19 countries are from Asia, 53 countries are from Europe, 7 countries are from Oceania, 44 countries are from Africa, 11 countries are from South America, 19 countries are from Caribbean, Central and North America and the balance are countries that do not belong in United Nation (Royal Malaysia Customs Department).

GST is a non-claimable tax charged to consumers. In other words, the consumers cannot claim tax rebate on their purchase of the goods and services. Meanwhile, for the suppliers of goods and services, they carry the responsibility to collect GST on behalf of the government (Fatt, 2006).

> *"GST shall be levied and charged on the taxable supply of goods and services made in the course or furtherance of a business in Malaysia by a taxable person. GST is also charged on the importation of goods and services"*

*Corresponding author's email: syaqilah9972@salam.uitm.edu.my

(Royal Malaysia Customs Department)

The proposal to implement GST was made in the Budget Speech 2005 to replace SST which should have taken into effect on 1January 2007. However, the GST implementation was delayed due to many criticisms and non-readiness issues. It was then tabled again on 16December 2009 as the first reading in Dewan Rakyat and then again was delayed (Shamsuddin, Ruslan, Halim, Zahari, & Fazi, 2014). Subsequently, on 25October 2013, the SixthPrime Minister of Malaysia, Dato' Sri NajibTunRazak in the Budget presentation 2014 has announced that the GST will take effect on 1 April 2015.

PROBLEM STATEMENTS

The GST implementation in Malaysia is expected to have an impact on the price of the goods and services. It is also expected to affect the lower income consumers' purchasing power, an increase in the tax rate and an increase in the government responsibility in utilising GST money (Narayanan, 2014). Such impacts have been discussed and established in other countries and as such, it is expected that Malaysia may likely be facing the same experience. Therefore, the Malaysian government needs to take appropriate measures to alleviate these impacts once GST is implemented. Consideration needs to be made particularly on the link between the price of the goods and services and the income level of the consumers. It is expected that the increase in price of goods and services would reduce the lower income consumers' purchasing power. The significant increase in the price of goods and services such as transportations which are commonly used by the middle and lower income earners would impact their financial conditions (Narayanan, 2014).

Studies have shown that the public in general have limited knowledge and lack of awareness on the various kinds of tax in Malaysia. The public are also unaware of how the government utilises the tax money for true expenditure (Moomal and Zakaria (2014). Such lacking of GST knowledge might lead to a false calculation of GST charged by the businesses that wanted to deceit their customers. Of consequence, GST evasion cases would increase and the government would experience a low tax collection. In addition, the various GST charges may also confuse the public. The effectiveness of disseminating information on GST system to the public should be emphasized to enhance GST compliance and effectiveness. Sufficient knowledge is an essential requirement to gain the public acceptance and enhance compliance in achieving successful GST implementations. Therefore, knowledge in GST system is important to increase the public literacy, acceptance and confidence.

It is evident that to obtain the public attention and acceptance of GST, the country may find difficulty in achieving consensus among the public due to its large population. According to Mansor and Ilias (2013), changes can be difficult without proper preparation and resistance often occurs when changes take place. However, such resistance can be changed by addressing the mind-set of the public particularly the taxpayers. While, developing a new system is one of important preparation for GST and will involve in computerized accounting system for changes to take place.

Literally, the Malaysia public is at doubt and may not be easily convinced because of the different thinking and perspectives, and different political backgrounds either from the opposition parties or among the public. The pivot deal that always being discussed is how GST system benefits or harms them. Mansor and Ilias (2013) suggested that the acceptance of a new

*Corresponding author's email: syaqilah9972@salam.uitm.edu.my

tax reform can be alleviated through providing strong efforts in convincing the consumers by way of educating and explaining clearly how GST works and benefits them in a long run. Arguably, the importance of literacy on GST system among consumers is important.

The literacy level on GST among the public is evident because it leads to the acceptance of this tax reformation made by the government. There are several studies in the accounting literature that have reported that tax literacy among the Malaysians on average is still at medium level (Bhushan & Medury, 2013; Kamaluddin & Madi, 2005; Latiff, Noordin, Omar, & Harjito, 2005; Madi et al., 2010). It is worth to note that the responsibility in increasing the tax literacy among the people does not lie entirely in the government's hands. It is a shared responsibility in which the basic knowledge should be provided as early as possible since taxation is a universal knowledge that is applicable throughout the world.

Arguably, tax education should not only focus on students in the accounting and business related disciplines but also include students in other disciplines (Palil, Akir, & Ahmad, 2013). This is because these students will eventually become taxpayers in the future and therefore, are also in need of knowledge in taxation. In order to develop knowledgeable taxpayers, education from an early stage is necessary. Therefore, teachers' tax literacy should be developed in order to ensure that the secondary school students will received sufficient and useful information on taxation. In the future, taxpayers will be more opened up towards the taxation system as they have been educated earlier.

Palil et al. (2013) noted that literate taxpayers require sufficient tax knowledge and this knowledge should be provided starting from the early education in schools and not when they have enrolled in the tertiary level. GST knowledge is part of the tax education. Gaining GST knowledge should not only be the responsibility of the consumers but also is the responsibility of the companies. In order to increase the compliance rate among the companies, consumers and the public as a whole, they need to possess GST knowledge as early as possible. Once they have become the taxpayers or run their own businesses, they would be able to comply with at least the basic stage of taxation and of consequence, lead to higher compliance rate in Malaysia.

A group of studies have found that the awareness towards GST implementation in Malaysia is at a medium level (Latiff et al., 2005; Moomal & Zakaria, 2014; Palil & Ibrahim, 2011; Saira, Zariyawati, & Yoke-May, 2010; Shamsuddin et al., 2014). Among the reasons to such lack of awareness are lack of understanding and knowledge and this in turn has led them to oppose and reject GST implementation (Saira et al., 2010). The false concept of awareness in underestimating tax load and overestimating tax burden (Cullis & Jones, 1992) has provide further misconception on this new tax requirement (Saira et al., 2010).

LITERATURE REVIEW

GST in Malaysia works with the standard rate of 6% of the goods and services. Companies with annual turnover exceeding RM500, 000 are mandatory to register GST and collect the tax on behalf of the government. Voluntary registration of GST can be performed by the companies to charge GST on the goods and services provided that companies remain in the system for at least 2 years. However, to avoid over burdening the consumers, the government has taken the initiative to categorise the GST into three main categories. (refer table 1)

Table 1: Three Categories of GST

*Corresponding author's email: syaqilah9972@salam.uitm.edu.my

Categories	Examples
6% GST	For all items that not listed as GST exempted or zero-rated supply items.
0% GST	Foods items: *Rice, flour, sugar, cooking oils, spices, white bread and whole meal bread, coffee powder, tea dust, cocoa powder, yellow mee, kueyteow, seafood and other unprocessed foods.* Agriculture products and livestock's: *Paddy, fresh vegetables, fruits, cows, goats, poultry.* Utilities and export: *Water supply (domestic), electric supply for the first 300 units (domestic), goods and services exported.* The national essential medicines: *2900 medicines brands and medicines to treat 30 types of diseases such as heart failure, diabetes, high blood pressure, cancer and etc.* Reading materials: *Child's colouring book, textbooks, dictionary, religious book, newspaper and exercise and reference book.* Public Transport: *Rail services, bus services, taxi services, toll highways and bridges and water transportation services.*
GST exempted	Housing loans: *Sales and purchase / lease of residential property, agriculture land and land for general use* Other services: *Private healthcare and education services, selected financial services*

(Royal Malaysia Customs Department, 2015)

In recent years, Malaysian researchers have shown great interest in researching GST. To this date, Malaysia is the latest country to implement GST. Research on the effectiveness of GST is among the highest areas that received interest among researchers in the accounting field. Studies (Benge, Pallot, & Slack, 2013; Breen, Bergin-Seers, Roberts, & Sims, 2002; Hooper & Smith, 1997; James, 2000; James & Alley, 2008; Jenkins & Khadka, 1998; McGowan & Billings, 1997) in other countries have focused on the effectiveness and benefits of GST implementation to the consumers, companies and the country.

GST Literacy

According to the United Nations Education, Scientific and Cultural Organisations (UNECSO), literacy is defined 'as a person that can understand a content and writing a simple note from his reading' (Latiff et al., 2005, p. 2). On the other hand, illiterate person can be defined as the inability of a person to read and write a simple note for his own benefits. Following the definition, this study refers literacy as the ability of a person to read and understand GST system in Malaysia.

Latiff et al. (2005) defined tax literacy as 'one who own sufficient ability to read, write and understand tax matters'. Madi and Kamaluddin (2003) defined literacy as 'the ability to fill the tax form and calculate their tax liabilities independently'. In another study, Kamaluddin and Madi (2005, p. 6) extended their definition of a literate taxpayer as 'having an average understanding of basic tax knowledge but need assistance in determining their tax liability'. Another studies by Bhushan and Medury (2013, p. 76) defined tax literacy as 'knowledge possess by individual to manage the issues concerning their personal taxation effectively'.

Mahdavi (2008) concluded that improving tax literacy rate can increase the level of taxation because he found that a high level of education has a positive affect towards sales and excise taxes. On the other hand, there is positive relationship between the effective literacy rates on tax revenues (Book, 2003). The maverick taxpayers may use the advantage of difference in

*Corresponding author's email: syaqilah9972@salam.uitm.edu.my

taxes and languages in conducting tax evasion thus this will reduces tax revenues book (Book, 2003). High tax literacy can eliminate tax evasion, thus brings revenue to the country. A successful tax system can be maintained by developing tax literacy among the taxpayers to obtain information about domestic or international tax system (James, 2000).

In the context of taxation, many researchers (Bhushan & Medury, 2013; Book, 2003; James, 2000; Kamaluddin & Madi, 2005; Latiff et al., 2005; Mahdavi, 2008) as discussed above have defined tax literacy as related to the process of completing tax return from either corporate or individual tax liability since it is one of the major tax contributors to the government that affects the government financially. In brief, this study will examine the tax literacy in the perspective of GST.

GST Knowledge

Knowledge is defined as 'an information, understanding and skills that gain through learning and experience' (Oxford Student's Dictionary of English, 2007). Following this definition, this study refers knowledge as the information and understanding on GST implementation as part of the taxation system in Malaysia where the knowledge of taxation has become a major need for all taxpayers and potential taxpayers (Bardai, 1992). Tax knowledge such as charging GST on items, paying income tax and etc. is also required by businesses in their transactions. Tax knowledge that needs to be understood by businessman is because their daily business transactions include collecting in charging tax money to the customers or from the suppliers. Many companies in Malaysia realise that GST system is complex and they lack of knowledge in GST system at using the correct policies and rate in their business transactions (Palil, Ramli, Mustapha, & Hassan, 2013). The complexity may harm the successful of implementation process and may increase the rate of tax evasions.

The introduction of a new tax system such as self-assessment system beforehand requires the taxpayers to obtain sufficient tax knowledge (Loo, Evans, & McKerchar, 2010). Arguably, this can be applied to GST introduction. Saira et al. (2010) conducted a study to explore the GST awareness in Malaysia. They found that 95% of the respondents are aware that tax is one of the major incomes for the government. The respondents who are from the accounting and non-accounting background possessed a high percentage of basic taxation knowledge. Since the findings show that the knowledge on basic taxation is high, Saira et al. (2010) indicated that the respondents with an accounting background are aware of GST. They also found that having a basic tax knowledge would lead to a different level of understanding on GST among the respondents. They also found that the respondents have a balance understanding on GST since they are unable to define GST in general and are unable to identify the GST rate proposed by the government. Their findings indicated that having a basic knowledge of taxation is not sufficient to explain a person's level of understanding on the tax system.

Prior to the implementation of GST in Singapore, the government has taken an initiative to educate their taxpayers by providing various programs related to the rational and effect of GST to the public and businesses. The first program was getting the businesses ready (Jenkins & Khadka, 1998). Various programs were conducted such as dialogues and seminars between the Inland Revenue Authority of Singapore (IRAS) and the trade associations in Singapore. The purpose of the programs was to obtain input from the businesses on their actual business activities in order to design the GST system. The feedbacks gathered from the seminars and dialogues have assisted the government in tuning the GST system to meet the needs and conditions of the consumers and local businesses. A special GST division was established to examine the special aspects of specialised industries activities to avoid dispute before the GST

*Corresponding author's email: syaqilah9972@salam.uitm.edu.my

was implemented. Other programs were conducted to educate the public on GST (Jenkins & Khadka, 1998). IRAS also took the initiatives to educate the young college students and local officers so that they could explain and resolve disputes among family members who are also the consumers.

`GST Awareness

Moomal and Zakaria (2014) defined awareness as the action context of a person imparted from the level of understanding upon certain activities. The general definition of awareness is a consciousness or interest (Oxford Student's Dictionary of English, 2007). Following this definition, this study defines awareness as the interest of the taxpayers towards understanding GST implementation in Malaysia. Cullis and Jones (1992) identified two types of awareness among the public namely, optimistic and pessimistic. Optimistic refers to the underestimation of tax load by the public servers whilst pessimistic refers to the overestimation of tax burden (Moomal & Zakaria, 2014).

Saira et al. (2010) have also examined the awareness of GST between respondents with accounting background and respondents without accounting background. They found a different level of awareness where 95.74% of the respondents with accounting background are aware and understand GST. This is understandable since tax education is one of the subjects taught in accounting courses. The awareness is naturally developed based on the understanding and knowledge among the respondents with accounting background (Fatt, 2006). However, Moomal and Zakaria (2014) found that respondents with low education level are more aware of GST compared to the respondents with high education level.

Shamsuddin et al. (2014) examined the educators' awareness on GST implementation. Their results showed that 93.8% of the respondents are aware of GST proposed by the government. However, only 79.6% of the respondents understand why GST needs to be implemented in Malaysia. They concluded that acceptance of GST can only be achieved when the educators fully understand GST.

Al-Mamun et al. (2014) found that a significant negative relationship between the respondents' awareness and education level. In other words, the respondents' high level of education leads to a lesser awareness on GST. In contrast, Moomal and Zakaria (2014) found a no significant relationship between education level and awareness level although they found the acceptance level of GST varies among the respondents.

Moomal and Zakaria (2014) found that 76% their Malaysian respondents are aware of GST. The overall mean score of awareness level among the respondents showed a moderate to low level of awareness. The reasons to the low level of awareness are fear of price increases, occurrence of inflation, lack of information and unclear objectives (Moomal & Zakaria, 2014). However, their study was conducted before GST was implemented. The results may be different if the study had been conducted after the GST implementation. This is because awareness on GST may increase from time to time since certain GST issues may have been solved and some new issues may have also emerged. Of consequence, this may influence the public interest towards GST and affect their awareness.

*Corresponding author's email: syaqilah9972@salam.uitm.edu.my

GST Acceptance

GST is part of the tax system which is not recent and it shares the same concept as VAT. As in other tax systems, there would be people opposing to GST implementation. Benge et al. (2013) claimed that in gaining the public acceptance on tax system, the system must prove to be beneficial to the low-income families and beneficiaries, and a subsequent tax reformation process.

In Malaysia, GST implementation has faced some difficulties in terms of acceptance from the public (Choo, 2014; Fadzell, 2015; The Star Online, 2013). This calls for the government to strategise ways to inform the public on the benefits of GST in order to protect their revenue and enhance the tax system. It is clear that the government would not impose a policy that is not beneficial to the public and the country. The GST implementation in April 2015 may cause evasion and a burden to the average income earners (The Star Online, 2013). The consumers need to know which goods and services are not taxable and the exclusion list must be clear. Enforcement would ensure that the entire groups of profit-making organisations do not take advantage on this issue. Nurhayati, Halimatusadiah, and Diamonalisa (2015) added that 'a successful VAT/GST system depends upon proper planning and preparation, sound accounting practices, systematic organizational audit, fully acceptance of people, trained manpower, sound ethical business practices in business and ethical government representatives' (Nurhayati et al., 2015, p. 148).

Shamsuddin et al. (2014) found that the educators' acceptance towards GST implementation in Bandar Muadzam Shah Pahang is moderate. In relation to the mean score of the individual acceptance items, all items were above the average mean scores indicating that the educators are in between moderate to high acceptance level. However, Saira et al. (2010) provided a contrasting finding when investigating GST awareness in Malaysia. Their results showed 60.53% of the respondents oppose GST implementation in Malaysia and surprisingly, 66.67% out of the 60.53% of the respondents who oppose GST implementation are with an accounting background. Their results also showed that a high percentage of agreement on GST implementation was found from non-accounting background. They found that the respondents oppose to GST implementation was influenced by the increase in the price of goods and services. Such perception comes mainly from the low-income earners. Furthermore, Ahmad, Ahmad, and Alwie (2015) concluded that perceptions among Malaysian have a significant strong association with acceptance. However, interestingly, this is contrary to a study conducted by Olatunji (2013) where the perception of Nigerian on VAT has no significance effect on its acceptance.

Palil, Akir, et al. (2013) conducted a survey among enterprises in Malaysia on GST acceptance. They found that due to the low level of readiness, the acceptance level among the enterprises is still very low. The enterprises are not ready to accept GST due to lack of understanding GST as they perceived GST implementation would lead to high compliance cost. They concluded that only 17 % of their respondents have purchased the GST system software during the training sessions.

Low level of acceptance among the consumers was also found in Palil and Ibrahim (2011) and Moomal and Zakaria (2014). Palil and Ibrahim (2011) conducted their study using middle income earners located in the big cities in Malaysia. Only 12% of the respondents accept GST even though they were not aware of GST implementation and 39% of the respondents accept GST provided that they are aware of the GST implementation. Such low level of acceptance is

*Corresponding author's email: syaqilah9972@salam.uitm.edu.my

due to the perception that GST would increase their cost of living. Meanwhile, Moomal and Zakaria (2014) in their study on GST acceptance of Malaysian consumers found a moderate to low level of acceptance (53.5%). Such resistance could be attributed by the fear of price increased. The consumers also perceived that they received less information and were not clear on the purpose of GST implementations in Malaysia.

In New Zealand, a uniform rate approach of broad based GST system has influenced the acceptance level in the country. It becomes a moderating factor to the economy, help the poor and proven to reduce a number of special cases (Benge et al., 2013). The implementation of GST in New Zealand has received a quick acceptance among the public. Prior to GST implementation, 30 to 35% of public approval was obtained and the acceptance level increased to 65% two weeks after implementation (Benge et al., 2013; Breen et al., 2002). The high level of acceptance has largely come from a comprehensive publicity, education campaign and identified potential problems.

The government of New Zealand has provided duration of time for planning purpose before GST implementation took place. The GST Coordinating Office started to develop and coordinate the GST system for the country. The factors considered in implementing GST are overseeing the initiatives in the GST progress that focus on business needs, conducted a program for business and general public as a whole, and fairly view the misunderstanding on GST system and law in all aspects (Breen et al., 2002). These factors are seen to have provided a smooth transition in the introduction period despite the negative perception among the lower revenue companies due to lack of understanding GST. The lower revenue businesses felt that GST implementation is burdensome and unfair. Therefore, the burden and unfair treatment felt at all levels of income can influence the success of GST system either the public or companies and subsequently, low acceptance level.

Meanwhile, Singapore has its own strategy in gaining public acceptance towards GST implementation. GST was introduced in 1994 at 3% which is considered a low rate and this rate was not raised within the next 5 years. An increased in tax rate started in 2003 at 4% and then was further raised in 2007 to 7% (Narayanan, 2014). They started at a low level of tax rate in order to gain public acceptance due to their acknowledgment that public trust would influence the success of new tax reformation process in Singapore. The introduction of GST was supported by the reduction in income tax rate and property tax rate. GST was also introduced at a low rate because Singapore has a stable economy and thus, did not require additional funds to govern the country during that time. In addition, the increase in the GST rate provides an offset to the lower middle income levels such as GST Voucher Scheme and to the companies such as tourist refund scheme and special scheme related to export to minimise its impact at the early implementation stage (Jenkins & Khadka, 1998). The developed countries have made Singapore a benchmark for GST system. Therefore, Malaysia government can take the success initiative from New Zealand and Singapore in structuring their tax system in order to achieve high acceptance from the citizen.

GST Literacy and Relationship with GST Acceptance

The current study focuses on knowledge and awareness as proxies to literacy. Kamaluddin and Madi (2005) in their survey of tax literacy of knowledge possessed by individuals on personal tax have found that their respondents have a low level of knowledge towards tax compliance. On the other hand, Palil, Akir, et al. (2013) has examined the role of religiosity in relation to

*Corresponding author's email: syaqilah9972@salam.uitm.edu.my

the tax education and tax knowledge towards tax compliance. They found no significant relationship between tax education and tax compliance but a significant relationship between tax knowledge and tax compliance. This indicates that knowledge is a major determinant of tax compliance but not education level. Therefore, a negative perception on GST would probably come from the low level of knowledge which indirectly affects the acceptance level of GST implementation (Ahmad et al., 2015; Shamsuddin et al., 2014). Research finding by Bardai (1992) also points towards 30-50% of taxpayers in Malaysia are illeterate and it is associated with inadequate tax knowledge.

Using a questionnaire survey, Saira et al. (2010) found that the GST awareness among the respondents in Malaysia are inconsistent. They found that the respondents have a poor understanding of GST that led them to oppose GST implementation. The awareness among the public on GST needs to be raised in order to alleviate the misconception of GST attributed by the contradiction of optimistic and pessimistic awareness. Failure to rectify the misconception would influence the public acceptance towards GST. More recent study done by Ahmad et al. (2015) have confirmed that government effort in increasing the awareness towards GST among Malaysian has a strong association with consumers understanding on GST and their level of willingness in accepting GST. In addition, taxpayers' knowledge gains either in formal or non-formal education will have a positive impact on awareness level of taxpayers (Nurhayati et al., 2015).

RESEARCH METHOD

This study chose secondary school teachers who are currently working in BatuPahat, Johor Darul Takzim as the sample study. Secondary school teachers are chosen because previous study (Bahari & Ling, 2009) has only focused on educators in the higher institutions. No studies have examined the GST system from the perspective of the secondary school teachers. Since secondary school teachers are also GST taxpayers and contribute to the government's revenue through GST, this sample seems suitable.This study adopted the quantitative data analysis technique. This study also engaged in both descriptive and inference statistics. The primary data was collected through questionnaire survey. The questionnaires were distributed to secondary school teachers in BatuPahat. 91 valid data was then used for further research analysis. This study comprises of 4 variables namely GST acceptance, and GST literacy, where there are 2 dimensions of GST literacy namely GST knowledge and GST awareness. . In this study, 150 questionnaires were distributed to the respondents. Out of 150 questionnaires, 100 questionnaires were returned. From the returned questionnaires, 2 sets were rejected due to incomplete answers, resulting in 65.33% response rate.

RESULTS AND DISCUSSION

The reliability of indicators in the questionnaire survey in this study was measured by using the interim consistency reliability of Cronbach's alpha coefficient.

Table 2: Table of Reliability Statistics of Each Variable

Items	Cronbach's Alpha	N of Items

*Corresponding author's email: syaqilah9972@salam.uitm.edu.my

DOI: 10.1051/shsconf/20173600042

GST Knowledge	0.726	13
GST Awareness	0.699	9
GST Acceptance	0.925	13

Table 2 presents the reliability statistics of each variable. The result illustrates the reliability test of each variable. Each variable has a Cronbach's alpha coefficients of above 0.6. According to Sekaran and Bougie (2013), if the alpha value of a variable is within the range of 0.6 to 0.79, then the measurement is considered acceptable. Therefore, the results of GST knowledge and GST awareness indicators in this study are considered reliable for further analysis. The Cronbach alpha of GST acceptance presents a high coefficient level at 0.925. This indicates that this variable is good in its internal consistency and reliable for further analysis.

The mean analyses of GST Literacy and GST acceptance

Table 3 shows that the total mean score was then converted into percentage scale in order to measure the literacy level as defined and used by Kamaluddin and Madi (2005).

Table 3: Description of GST Literacy Levels

Mean Total Percentage	Literacy Level
0%-24%	Very illiterate
25%-49%	Illiterate
50%-74%	Literate
75%-100%	Very literate

Source: Kamaluddin and Madi (2005)

This study adapts the measurement of GST acceptance from Shamsuddin et al. (2014). The Likert-scale statements in the questionnaire were analysed in mean and using the total mean score shown in Table 3.3 to determine the level of acceptance.

Table 4: Description of GST Acceptance level

Total Mean Score	Acceptance Level
42-65	High
22-41	Moderate
1-21	Low

Source: Shamsuddin et al. (2014)

GST literacy is measured based on the mean scores of the GST knowledge and GST awareness. Meanwhile, GST acceptance is measured based on 13 indicators tested to determine the level of acceptance towards GST implementation in Malaysia.

Table5: Mean Score analyses of GST Literacy and GST Acceptance

No	Variables	Mean

*Corresponding author's email: syaqilah9972@salam.uitm.edu.my

1	GST Literacy mean score	3.37
2	GST Acceptance summated mean score	33.38

Based on Table 5, the result shows that the mean score for GST literacy is 3.37 over Likert Scale of 5. This study GST knowledge and GST awareness dimensions in measuring GST literacy. According to Kamaluddin and Madi (2005) GST literacy level in table 3, the result suggests that the secondary school teachers as the taxpayers are literate taxpayers (literacy level 3.37 / 5 =67.4%). Such result indicates that they have an average knowledge on GST but in need of more assistance to understand the GST system in Malaysia (Kamaluddin & Madi, 2005). They are also in need of assistance from the government to increase the public awareness on GST implementation in Malaysia so that their students will become literate in future. This study concludes that the secondary school teachers are literate towards GST implementation in Malaysia.

The Likert-scale responses provided by the respondents for the indicators in the questionnaire survey were then summated to obtain an overall mean score and subsequently divided into three levels. The results depict that the GST acceptance overall total mean score is 33.38 (refer table 5), a result that falls under moderate acceptance (refer table 4). Thus, such result indicates that the secondary school teachers have a moderate acceptance towards GST implementation in Malaysia. Similar result was found in Shamsuddin et al. (2014) and Moomal and Zakaria (2014)on the moderate acceptance of GST implementation among the Malaysian consumers. This study concludes that the secondary school teachers have a moderate acceptance level towards GST implementation in Malaysia.

Table 6 presents the summary of Pearson correlation results of GST knowledge, GST awareness, GST literacy and GST acceptance. The result indicates that GST awareness has a positive correlation with GST awareness of r=0.457 at a significant level of 0.01. As for GST literacy, the result shows a significant positive correlation with GST knowledge (r=0.852) and GST awareness (r=0.855) whilst for GST acceptance, the result shows a significant positive correlation with GST knowledge (r=0.371), GST awareness (0.375) and GST literacy (0.437) respectively at a significant level of 0.01.

Table 6: Pearson Correlation Test

	GST Knowledge	GST Awareness	GST Literacy	GST Acceptance
GST Knowledge	1			
GST Awareness	0.457**	1		
GST Literacy	0.852**	0.855**	1	
GST Acceptance	0.371**	0.375**	0.437**	1

Note: **Significant at 0.01 level

In sum, the correlation coefficients between all variables show positive significant correlations. Such result indicates that a high level of literacy reflect the level of GST acceptance. That is high level of literacy is associated with high level of GST knowledge and GST awareness.

Table 7:Multiple Regression Results

Variables	Coefficient	T-statictics	P-value

*Corresponding author's email: syaqilah9972@salam.uitm.edu.my

Intercept	-0.524	-0.753	0.453
GST knowledge	0.455***	2.344	0.021**
GST awareness	0.462***	2.406	0.018**
R	0.191		
Adj. R	0.173		
F-statistics	10.382		

Note: *** significant at 0.01 level

Table 7 shows that the R-square of GST knowledge and GST awareness explains 19.1% of the GST acceptance. In this study, the adjusted R-square 0.173 or 17.3% indicates that regression model explains as 17.3% of the variance in GST acceptance. The model is moderately fit and can predict GST acceptance. According to Muijs (2004) rule of thumb, the adjusted R-square range of 0.11-0.3 is considered as moderately fit model. Therefore, the model proposed for this study can be considered as fit to be used in examining the relationship between independent variables and dependent variable.

The result also shows the f-value of 10.382 of p-significant at 0.00. Since the p-significant value is less than 0.05, the model is considered as significant in predicting the GST knowledge and GST awareness on GST acceptance on GST implementation in Malaysia.

This study indicated that there is a significant relationship between the GST literacy and GST acceptance. The results signified the dimensions of GST literacy having positive significant relationships between the GST knowledge with GST acceptance, and GST awareness with GST acceptance.

CONCLUSION

In the current study, the competency level of the taxpayers is measured through GST literacy and GST awareness. GST literacy was measured based on two dimensions namely, GST knowledge and GST awareness. The results indicate that the school teachers are literate on the issues of GST implementation and GST system in Malaysia. A literate teacher is a teacher that that understands and has knowledge on GST system in Malaysia. The teachers understand the GST terminology, mandatory turnover limit which requires business to register under GST, types of GST, GST standard rate, pricing requirement with relevant to GST displayed and tax reformation process.

The results also reveal that teachers have high level of awareness on GST implementation in Malaysia. High awareness reflects that they are aware of receiving receipt upon purchase GST items, aware that the rate charged must be reflected on the receipts and aware of the responsibility to report to respective author of any deception by sellers. The respondents also aware that the GST knowledge and GST education are important for secondary school students, as they are the future taxpayers. A literate teacher is a person who able to read and understand GST system in Malaysia.

The acceptance level is divided into three categories namely, high, medium and low acceptance. The results show a moderate acceptance level of GST implementation in Malaysia among the secondary school teachers. This study suggests that the level of acceptance of the

*Corresponding author's email: syaqilah9972@salam.uitm.edu.my

secondary school teachers before and after GST implementation remains unchanged. This may due to the stage of GST implementation has not reached a period of one year when this study was conducted.

The taxpayers moderately accept GST implementation in Malaysia as they believe that GST contributes to the government's revenues. They believe that unclear objectives and lack of information provided by the government led them to have a low level of acceptance. They also moderately agree on the issues benefits and fairness that GST can offer to the consumers, businesses and country.

The results also found that the GST literacy level of the secondary school teachers affect their GST acceptance level. This study suggests that the level of literacy among the secondary school teachers on GST implementation has a positive effect their level of acceptance.

This study suggests that future study should examine the relationship of GST literacy levels and GST acceptance levels of teachers or educators in other states in Malaysia using a larger sample. More time should be allocated to collect and analyse the data. The future study could use a combination of instruments such as interviews, focused group interviews and documents analysis. In addition, the interaction between GST literacy level and GST compliance level in the perspective of gender, races, age, income levels, and other variables can be conducted.

REFERENCES

Ahmad, R., Ahmad, R., & Alwie, S. M. (2015). Instilling Malaysian Consumers' Mind-Set in Accepting GST: An Analysis on the Effectiveness of Communication used by the Government. *Innovation, Finance, and the Economy, 1*, 121-134. doi: 10.1007/978-3-319-15880-8_10

Al-Mamun, A., Entebang, H., Mansor, S. A., Yasser, Q. R., Nathan, T. M., & Rahman, M. A. (2014). The Impact of Demographic Factors on Tax Compliance Attitude and Behavior in Malaysia. *Journal of Finance, Accounting and Management, 5*(1), 109-124.

Alappatt, M., & Shaikh, J. M. (2014). Forthcoming Procedure of Goods and Services Tax (GST) in Malaysia. *Issues in Business Management and Economics, 2*(12), 210-213.

Bahari, A. B. M., & Ling, L. M. (2009). Introducing Tax Education in Non-Accounting Curriculum in Higher Education: Survey Evidence. *Journal of Financial Reporting & Accounting, 7*(1), 37-51.

Band, D. (1992). Corporate Governance: Why Agency Theory is not Enough. *European Management Journal, 10*(4), 453-459.

Bardai, B. (1992). Tax Illiteracy in Malaysia: Problems Solutions. *Journal of ACCAMADIA*, 7-31.

Benge, M., Pallot, M., & Slack, H. (2013). Possible Lessons For The United States From New Zealand's GST. *National Tax Journal, 66*(2), 479-498.

Bhushan, P., & Medury, Y. (2013). Determining Tax Literacy of Salaried Individuals - An Empirical Analysis. *IOSR Journal of Business and Management, 10*(6), 76-80.

*Corresponding author's email: syaqilah9972@salam.uitm.edu.my

Book, L. (2003). The Poor and Tax Compliance: One Size Does Not Fit All. *Kansas Law Review, 51*, 1145.

Breen, J., Bergin-Seers, S., Roberts, I., & Sims, R. (2002). The Impact of the Introduction of the GST on Small Business In Australia. *Asian Review of Accounting, 10*(1), 89-104.

Choo, T. S. (2014). The battle for GST acceptance, *The Sun Daily*. Retrieved from http://www.thesundaily.my/node/251014

Cullis, J., & Jones, P. (1992). *Public Finance and Public Choice: Analytical Perspectives*: McGraw-Hill International (UK) Limited.

Cvrlje, D. (2015). Tax Literacy as An Instrument of Combating and Overcoming Tax System Complexity, Low Tax Morale and Tax Non-Compliance. *The Mactotheme Review, 4*(3), 156-167.

Fadzell, A. (2015). By-election will show if GST accepted or not, *The Sun Daily*. Retrieved from http://www.thesundaily.my/news/1406721

Fatt, C. K. (2006). Towards goods and services tax in Malaysia : A Preliminary Study. *Business & Economics Society International Conference*, 1-17.

Field, A. (2013). *Discovering Statistics Using IBM SPSS Statistics* (4th ed.). London: SAGE Publications Inc.

Fishbein, M., & Ajzen, I. (1975). *Belief, Attitude, Intention, and Behavior: An Introduction to Theory and Research* Reading, MA: Addison-Wesley.

Hair, J. F., Black, W. C., Babin, B. J., Anderson, R. E., & Tatham, R. (2006). *Multivariate Data Analysis* (6th ed.): Upper Saddle River, NJ: Prentice-Hall.

Hooper, P., & Smith, K. A. (1997). A value-added tax in the U.S.: An argument in favor. *Business Horizons, 40*(3), 78-83. doi: http://dx.doi.org/10.1016/S0007-6813(97)90056-2

James, S. (2000). VAT/GST: The UK Experience Revisited. *Revenue Law Journal, 10*(1), 72-87.

James, S., & Alley, C. (2008). Successful Tax Reform: The Experience Of Value Added Tax In The United Kingdom And Goods And Services Tax In New Zealand. *Journal of Finance and Management in Public Services, 8*(1), 35-47.

Jenkins, G. P., & Khadka, R. (1998). Value Added Tax Policy And Implementation In Singapore. *International Bureau of Fiscal Documentation, 9*(2), 35-47.

Kamaluddin, A., & Madi, N. (2005). Tax Literacy and Tax Awareness of Salaried Individuals in Sabah and Sarawak. *Journal of Financial Reporting and Accounting, 3*(1), 71-89.

Kumar, N. (2014). Goods And Services Tax in India: A Way Forward. *Global Journal of Multidisciplinary Studies, 3*(6), 216-225.

*Corresponding author's email: syaqilah9972@salam.uitm.edu.my

Latiff, A. R. A., Noordin, B. A. A., Omar, M. R. C., & Harjito, D. A. (2005). Tax Literacy Rate Among Taxpayers: Evidence From Malaysia. *Jurnal Akuntansi & Auditing Indonesia, 9*(1), 1-11.

Lindholm, R. W. (1970). The Value Added Tax: A Short Review Of The Literature. *Journal of Economic Literature, 8*(4), 1178-1189.

Loo, E. C., Evans, C., & McKerchar, M. (2010). Challenges in Understanding Compliance Behavior of Taxpayers in Malaysia. *Asian Journal of Business and Accounting, 3*(1), 145-161.

Madi, N., & Kamaluddin, A. (2003). Tax literacy among salaried individuals in Sarawak. *Bureau of Research and Consultancy MARA University of Technology Shah Alam Selamgor.*

Madi, N., Kamaluddin, A., Janggu, T., Ibrahim, M. B. A., Samah, A. B. A., & Jusoff, K. (2010). Tax Literacy among Employees: Sabah and Sarawak's Perspective. *International Journal of Economics and Finance, 2*(1), 218-223.

Mahdavi, S. (2008). The level and composition of tax revenue in developing countries: Evidence from unbalanced panel data. *International Review of Economics & Finance, 17*(4), 607-617.

Mansor, N. H. A., & Ilias, A. (2013). Goods and Services Tax (GST): A New Tax Reform in Malaysia. *International Journal of Economics Business and Management Studies, 2*(1), 12-19.

McGowan, J. R., & Billings, B. A. (1997). An Analysis of the European Community VAT: Implications for U.S. Tax Policy. *journal of International Accounting Auditing & Taxation, 6*(2), 131-148.

Moomal, A., & Zakaria, Z. (2014). Awareness, Acceptance and Expectations of Malaysia Consumers Towards Good and Service Tax (GST). *Journal of Contemporary Management Sciences, 4*(2), 15-27.

Muijs, D. (2004). *Doing Quantitative Research in Education with SPSS*: SAGE Publications Ltd.

Narayanan, S. (2014). THE IMPACT OF THE GOODS AND SERVICES TAX (GST) IN MALAYSIA: LESSONS FROM EXPERIENCES ELSEWHERE (A NOTE). *The Singapore Economic Review, 59*(1), 1-15.

Nathan, T. M., & Rahman, M. A. (2014). The Impact of Demographics Factors on Tax Compliance Attitude and Behavior in Malaysia. *Journal of Finance, Accounting and Management, 5*(1), 109-124.

Nurhayati, N., Halimatusadiah, E., & Diamonalisa. (2015). Influence of tax officer service quality and knowledge of tax on individual taxpayer compliance in tax office (kpp) Bojonagara Bandung. *International Journal of Applied Research, 1*(8), 805-809.

*Corresponding author's email: syaqilah9972@salam.uitm.edu.my

Olatunji, O. C. (2013). Value Added Tax (VAT) and Inflation in Nigeria (1990 to 2003). *Asian Journal of Humanities and Social Sciences (AJHSS), 1*(1), 123-135.

Oxford Student's Dictionary of English. (Ed.) (2007). Malaysia: Oxford Fajar Sdn. Bhd.

Palil, M. R. (2005). Taxpayers Knowledge: A Descriptive Evidence On Demographics Factors In Malaysia. *Jurnal Akuntansi & Keuangan, 7*(1), 11-21.

Palil, M. R., Akir, M. R. M., & Ahmad, W. F. W. (2013). The Perception of Tax Payers on Tax Knowledge and Tax Education with Level of Tax Compliance: A Study the Influences of Religiosity. *ASEAN Journal of Economics, Management and Accounting, 1*(1), 118-129.

Palil, M. R., & Ibrahim, M. A. (2011). The Impacts of Goods and Services Tax (GST) on Middle Income Earners in Malaysia. *World Review of Business Research, 1*(2), 192-206.

Palil, M. R., Ramli, R., Mustapha, A. F., & Hassan, N. S. A. (2013). Elements of Compliance Costs: Lesson from Malaysia Companies towards Goods and Services Tax (GST). *Asian Social Science, 9*(11), 135-147.

Pallant, J. (2011). *SPSS SURVIVAL MANUAL A step by step guide to data analysis using SPSS* (4th Ed.): Allen & Unwin.

Rashid, E. R. B. A., Mohammed, N. F., & Saheh, N. M. (2005). *EXPLORING LEVEL OF AWARENESS OF SMALL AND MEDIUM SIZED ENTERPRISES TOWARDS GOODS AND SERVICES TAX IMPLEMENTATION IN MALAYSIA* Research Universiti Teknology MARA (UiTM).

Rauf, J. b. A., & Yusuf, M. Z. b. M. (2014). *Guide on Pre-Tertiary Education*. (03-88822111).

Royal Malaysia Customs Department. (15th February 2013). Understanding GST. Retrieved 3rd September, 2015, from http://gst.customs.gov.my/en/gst/Pages/gst_un.aspx

Royal Malaysia Customs Department. (2015). Goods and Services Tax (GST). In R. M. C. Department (Ed.).

Saira, K., Zariyawati, M. A., & Yoke-May, L. (2010, 13-15 July 2010). *An Exploratory Study of Goods and Services Tax Awareness in Malaysia.* Paper presented at the Proceedings Seminar on Nasional Resilience (SNAR 2010) "Political Managements and Policies in Malaysia", Bayview Hotel Langkawi.

Sekaran, U., & Bougie, R. (2013). *Research Methods for Business: A Skill-Building Approach* (6th ed.): John Wiley & Sons Ltd.

Shamsuddin, A., Ruslan, M. I. M., Halim, A. A., Zahari, N. F., & Fazi, N. F. M. (2014). Educators' Awareness and Acceptance Towards Goods and Services Tax (GST) Implementation in Malaysia: A Study in Bandar Muadzam Shah, Pahang. *International Journal of Business, Economics, and Law, 4*(1), 1-10.

*Corresponding author's email: syaqilah9972@salam.uitm.edu.my

Sumartaya, D., & Hafidiah, A. (2014). The influence of taxpayers's awareness and tax morale toward tax evasion. *International Journal of Business, Economics, and Law, 5*(1), 60-68.

Tan, L. M., & Chin-Fatt, C. (2000). The Impact of Tax Knowledge on the Perceptions of Tax Fairness and Attitudes Towards Compliance. *Asian Review of Accounting, 8*(1), 44-58.

The Star. (2015, 23rd August 2015). GST will help country when oil prices fall, says Najib, *The Star*. Retrieved from http://www.thestar.com.my/News/Nation/2015/08/23/Najib-Umno-GST-good-for-revenue/

The Star Online. (2013). Getting GST acceptance will be taugh, *The Star Online*. Retrieved from http://www.thestar.com.my/Business/Business-News/2013/05/25/Getting-GST-acceptance-will-be-tough-Minister-hints-at-commitment-to-reduce-corporate-and-personal/?style=biz

Wolk, H. I., Dodd, J. L., & Tearney, M. G. (2004). *Accounting Theory: Conceptual Issues in a Political and Economic Environment* (6 ed.): Thomson South Western.

*Corresponding author's email: syaqilah9972@salam.uitm.edu.my

AUTHOR INDEX